WITHDRAWN
UTSA LIBRARIES

RENEWALS 458-4574

DATE DUE

PROGRESS IN BRAIN RESEARCH

VOLUME 158

FUNCTIONAL GENOMICS AND PROTEOMICS
IN THE CLINICAL NEUROSCIENCES

Other volumes in PROGRESS IN BRAIN RESEARCH

Volume 122: The Biological Basis for Mind Body Interactions, by E.A. Mayer and C.B. Saper (Eds.) – 1999, ISBN 0-444-50049-9.
Volume 123: Peripheral and Spinal Mechanisms in the Neural Control of Movement, by M.D. Binder (Ed.) – 1999, ISBN 0-444-50288-2.
Volume 124: Cerebellar Modules: Molecules, Morphology and Function, by N.M. Gerrits, T.J.H. Ruigrok and C.E. De Zeeuw (Eds.) – 2000, ISBN 0-444-50108-8.
Volume 125: Transmission Revisited, by L.F. Agnati, K. Fuxe, C. Nicholson and E. Syková (Eds.) – 2000, ISBN 0-444-50314-5.
Volume 126: Cognition, Emotion and Autonomic Responses: the Integrative Role of the Prefrontal Cortex and Limbic Structures, by H.B.M. Uylings, C.G. Van Eden, J.P.C. De Bruin, M.G.P. Feenstra and C.M.A. Pennartz (Eds.) – 2000, ISBN 0-444-50332-3.
Volume 127: Neural Transplantation II. Novel Cell Therapies for CNS Disorders, by S.B. Dunnett and A. Björklund (Eds.) – 2000, ISBN 0-444-50109-6.
Volume 128: Neural Plasticity and Regeneration, by F.J. Seil (Ed.) – 2000, ISBN 0-444-50209-2.
Volume 129: Nervous System Plasticity and Chronic Pain, by J. Sandkühler, B. Bromm and G.F. Gebhart (Eds.) – 2000, ISBN 0-444-50509-1.
Volume 130: Advances in Neural Population Coding, by M.A.L. Nicolelis (Ed.) – 2001, ISBN 0-444-50110-X.
Volume 131: Concepts and Challenges in Retinal Biology, by H. Kolb, H. Ripps and S. Wu (Eds.) – 2001, ISBN 0-444-506772.
Volume 132: Glial Cell Function, by B. Castellano López and M. Nieto-Sampedro (Eds.) – 2001, ISBN 0-444-50508-3.
Volume 133: The Maternal Brain. Neurobiological and Neuroendocrine Adaptation and Disorders in Pregnancy and Post Partum, by J.A. Russell, A.J. Douglas, R.J. Windle and C.D. Ingram (Eds.) – 2001, ISBN 0-444-50548-2.
Volume 134: Vision: From Neurons to Cognition, by C. Casanova and M. Ptito (Eds.) – 2001, ISBN 0-444-50586-5.
Volume 135: Do Seizures Damage the Brain, by A. Pitkänen and T. Sutula (Eds.) – 2002, ISBN 0-444-50814-7.
Volume 136: Changing Views of Cajal's Neuron, by E.C. Azmitia, J. DeFelipe, E.G. Jones, P. Rakic and C.E. Ribak (Eds.) – 2002, ISBN 0-444-50815-5.
Volume 137: Spinal Cord Trauma: Regeneration, Neural Repair and Functional Recovery, by L. McKerracher, G. Doucet and S. Rossignol (Eds.) – 2002, ISBN 0-444-50817-1.
Volume 138: Plasticity in the Adult Brain: From Genes to Neurotherapy, by M.A. Hofman, G.J. Boer, A.J.G.D. Holtmaat, E.J.W. Van Someren, J. Verhaagen and D.F. Swaab (Eds.) – 2002, ISBN 0-444-50981-X.
Volume 139: Vasopressin and Oxytocin: From Genes to Clinical Applications, by D. Poulain, S. Oliet and D. Theodosis (Eds.) – 2002, ISBN 0-444-50982-8.
Volume 140: The Brain's Eye, by J. Hyönä, D.P. Munoz, W. Heide and R. Radach (Eds.) – 2002, ISBN 0-444-51097-4.
Volume 141: Gonadotropin-Releasing Hormone: Molecules and Receptors, by I.S. Parhar (Ed.) – 2002, ISBN 0-444-50979-8.
Volume 142: Neural Control of Space Coding, and Action Production, by C. Prablanc, D. Pélisson and Y. Rossetti (Eds.) – 2003, ISBN 0-444-509771.
Volume 143: Brain Mechanisms for the Integration of Posture and Movement, by S. Mori, D.G. Stuart and M. Wiesendanger (Eds.) – 2004, ISBN 0-444-513892.
Volume 144: The Roots of Visual Awareness, by C.A. Heywood, A.D. Milner and C. Blakemore (Eds.) – 2004, ISBN 0-444-50978-X.
Volume 145: Acetylcholine in the Cerebral Cortex, by L. Descarries, K. Krnjević and M. Steriade (Eds.) – 2004, ISBN 0-444-511253.
Volume 146: NGF and Related Molecules in Health and Disease, by L. Aloe and L. Calzà (Eds.) – 2004, ISBN 0-444-51472-4.
Volume 147: Development, Dynamics and Pathology of Neuronal Networks: From Molecules to Functional Circuits, by J. Van Pelt, M. Kamermans, C.N. Levelt, A. Van Ooyen, G.J.A. Ramakers and P.R. Roelfsema (Eds.) – 2005, ISBN 0-444-51663-8.
Volume 148: Creating Coordination in the Cerebellum, by C.I. De Zeeuw and F. Cicirata (Eds.) – 2005, ISBN 0-444-51754-5.
Volume 149: Cortical Function: A View from the Thalamus, by V.A. Casagrande, R.W. Guillery and S.M. Sherman (Eds.) – 2005, ISBN 0-444-51679-4.
Volume 150: The Boundaries of Consciousness: Neurobiology and Neuropathology, by Steven Laureys (Ed.) – 2005, ISBN 0-444-51851-7.
Volume 151: Neuroanatomy of the Oculomotor System, by J.A. Büttner-Ennever (Ed.) – 2006, ISBN 0-444-516964.
Volume 152: Autonomic Dysfunction after Spinal Cord Injury, by L.C. Weaver and C. Polosa (Eds.) – 2006, ISBN 0-444-51925-4.
Volume 153: Hypothalamic Integration of Energy Metabolism, by A. Kalsbeek, E. Fliers, M.A. Hofman, D.F. Swaab, E.J.W. Van Someren and R. M. Buijs (Eds.) – 2006, ISBN 978-0-444-52261-0.
Volume 154: Visual Perception, Part I, Fundamentals of Vision: Low and Mid-Level Processes in Perception, by S. Martinez-Conde, S.L. Macknik, L.M. Martinez, J.M. Alonso and P.U. Tse (Eds.) – 2006, ISBN 0-444-52966-7.
Volume 155: Visual Perception, Part II, Fundamentals of Awareness, Multi-Sensory Integration and High-Order Perception, by S. Martinez-Conde, S.L. Macknik, L.M. Martinez, J.M. Alonso and P.U. Tse (Eds.) – 2006, ISBN 0-444-519-270.
Volume 156: Understanding Emotions, by S. Anders, G. Ende, M. Junghofer, J. Kissler and D. Wildgruber (Eds.) – 2006, ISBN 0-444-52182-8.
Volume 157: Reprogramming of the Brain, by A.R. Møller (Ed.) – 2006, ISBN 0-444-51602-6.

PROGRESS IN BRAIN RESEARCH

VOLUME 158

FUNCTIONAL GENOMICS AND PROTEOMICS IN THE CLINICAL NEUROSCIENCES

EDITED BY

S.E. HEMBY

Department of Physiology and Pharmacology, Medical Center Boulevard, Wake Forest University School of Medicine, Winston-Salem, NC 27157, USA

S. BAHN

Cambridge Centre for Neuropsychiatric Research, Institute of Biotechnology, University of Cambridge, Tennis Court Road, Cambridge CB2 1QT, UK

AMSTERDAM – BOSTON – HEIDELBERG – LONDON – NEW YORK – OXFORD
PARIS – SAN DIEGO – SAN FRANCISCO – SINGAPORE – SYDNEY – TOKYO

Elsevier
Radarweg 29, PO Box 211, 1000 AE Amsterdam, The Netherlands
The Boulevard, Langford Lane, Kidlington, Oxford OX5 1GB, UK

First edition 2006

Copyright © 2006 Elsevier B.V. All rights reserved

No part of this publication may be reproduced, stored in a retrieval system
or transmitted in any form or by any means electronic, mechanical, photocopying,
recording or otherwise without the prior written permission of the publisher

Permissions may be sought directly from Elsevier's Science & Technology Rights
Department in Oxford, UK: phone (+44) (0) 1865 843830; fax (+44) (0) 1865 853333;
e-mail: permissions@elsevier.com. Alternatively you can submit your request online by
visiting the Elsevier web site at http://elsevier.com/locate/permissions, and selecting
Obtaining permission to use Elsevier material

Notice
No responsibility is assumed by the publisher for any injury and/or damage to persons
or property as a matter of products liability, negligence or otherwise, or from any use
or operation of any methods, products, instructions or ideas contained in the material
herein. Because of rapid advances in the medical sciences, in particular, independent
verification of diagnoses and drug dosages should be made

Library of Congress Cataloging-in-Publication Data
A catalog record for this book is available from the Library of Congress

British Library Cataloguing in Publication Data

Functional genomics and proteomics in the clinical neurosciences. - (Progress in brain research; v. 158)
1. Neurogenetics - Methodology 2. Proteomics - Methodology
3. Nervous system - Diseases - Genetic aspects
I. Hemby, S. E. II.Bahn, S.
612.8
ISBN-13: 9780444518538
ISBN-10: 0444518533

ISBN-13: 978-0-444-51853-8 (this volume)
ISBN-10: 0-444-51853-3 (this volume)
ISBN-13: 978-0-444-80104-3 (Series)
ISBN-10-0-444-80104-9 (Series)
ISSN: 0079-6123 (Series)

For information on all Elsevier publications
visit our website at books.elsevier.com

Printed and bound in The Netherlands

06 07 08 09 10   10 9 8 7 6 5 4 3 2 1

Working together to grow
libraries in developing countries

www.elsevier.com | www.bookaid.org | www.sabre.org

ELSEVIER   BOOK AID International   Sabre Foundation

Erratum to Progress in Brain Research Vol. 158

Functional Genomics and Proteomics in the Clinical Neurosciences
edited by Scott E. Hemby and Sabine Bahn

In the above publication, in Chapter 12, entitled "Functional genomics of sex hormone-dependent neuroendocrine systems: specific and generalized actions in the CNS" by A.W. Lee, N. Devidze, D.W. Pfaff and J. Zhou, on page 250, the three equations are missing the "$\delta A =$" in front of them. This was inadvertently left out in the final proofs.

The equations should read as shown below:

$$\delta A = F_1(As_1) + F_2(As_2) \cdots + \cdots F_n(As_n) + F_g(A_g) \tag{1}$$

$$\delta A = F_1(As_1) \bullet F_2(As_2) \cdots \bullet \cdots F_n(As_n) \bullet F_g(A_g) \tag{2}$$

$$\delta A = F_1(As_1) \bullet e^{F_g(A_g)} \tag{3}$$

Readers are requested to insert/substitute the new/correct equations in that chapter and to ignore the current p. 250. Please note that for convenience p. 249 has been reprinted unaltered on the verso of p. 250 for those readers who physically replace these pages.

The Publisher apologises for this error and any inconvenience that this may have caused.

Theory of orchestrated genomic responses to hormones by functional modules (GAPPS)

From this series of individual gene inductions by estrogens acting in the basal forebrain, and the recounting (above) of downstream genes and their physiological routes of action, there emerges a theoretical molecular 'formula', as quoted here from Mong and Pfaff (2004) which appears to account for some of the causal relations between sex hormones and female sex behaviors (Fig. 3). First, there is a hormone-dependent *growth* response, which permits hormone-facilitated, behavior-directing hypothalamic neurons a greater range of input/output connections and, thus, physiological power. Second, progesterone can *amplify* the estrogen effect, in part through the downstream genes listed above. Then, through indirect behavioral means — the reduction of anxiety and a partial analgesia — the female as an organism is *prepared* for engaging in reproductive behavior sequences. Here the genes for oxytocin (and its receptor) as well as the genes for the opioid peptide enkephalin (and its receptors) are important. Next, neurotransmitter receptor induction by estradiol *permits* the neural circuit for lordosis behavior to be activated. The noradrenergic α-1 receptor and the muscarinic acetylcholine receptors are key here, in the ventromedial nucleus of the hypothalamus. Finally, induction of the decapeptide which triggers ovulation, GnRH as well as its cognate receptor acts to *Synchronise* mating behavior with ovulation in a biologically adaptive fashion.

This theoretical formulation is intended to tie together disparate results from several transcriptional systems into one set of modules. Even so, the genomic mechanisms uncovered so far probably represent only a subset of the full range of neurochemical steps underlying sex behaviors.

From lordosis to sexual arousal to generalized CNS arousal

Success in analyzing mechanisms underlying a simple sex behavior, lordosis, emboldened us to look ahead to attack the deeper concepts of sexual motivation, sexual arousal, and generalized CNS arousal. The ability of sex hormones to turn on sexual behaviors when all other aspects of the experiment are held constant gives proof-positive of the ability of sex hormones to raise sexual motivation (argued in Bodnar et al., 2002). In turn, motivational forces are divided into two types: specific motivational states (such as sex, hunger, thirst, fear, etc.) and generalized states that account for the activation of behavior (generalized arousal). With this train of reasoning, it was exciting to approach mechanisms for generalized arousal because it underlies the activation of all behaviors.

Generalized arousal can be assayed in the laboratory in its natural form (Frohlich et al., 2001), and has a permanence across the life span of the animal or human being. Because it occurs first in any chain of behavioral responses, its alterations can be causal to later behavioral alterations, while the reverse is not true. Assisted by a precise and complete operational definition (below), we have found generalized arousal easy to study in a genetically tractable organism like the mouse. It appears to be universal among higher animals in that it is triggered by giant medullary reticular neurons like Mauthner cells (Faber et al., 1991; Lee and Eaton, 1991; Lee et al., 1993; Zottoli and

---

MODULAR SYSTEMS DOWNSTREAM FROM HORMONE-FACILITATED TRANSCRIPTION RESPONSIBLE FOR A MAMMALIAN SOCIAL BEHAVIOR: "GAPPS".

- *Growth* (rRNA, cell body, synapses).
- *Amplify* (pgst/PR → → downstream genes).
- *Prepare* (indirect behavioral means; analgesia (ENK gene) and anxiolysis (OT gene).
- *Permit* (NE alpha-1b; muscarinic receptors).
- *Synchronize* (GnRH gene, GnRH Rcptr gene ------synchronizes with ovulation).

Fig. 3. Modular systems downstream from hormone-facilitated transcription responsible for a mammalian social behavior: 'GAPPS'. The 'GAPPS' model emphasizes the ability of sex hormones to make nerve cells *Grow*, to set up their own *Amplification*, to initiate *Preparative* behavioral steps, to have *Permissive* actions on the rest of the lordosis circuit, and to *Synchronize* sex behavior with ovulation. From Mong and Pfaff, (2004)

Faber, 2000; Canfield, 2003), and can even be tied to stimulus salience and dopamine neurotransmission in Drosophila (van Swinderen and Greenspan, 2003; Kume et al., 2005). Generalized arousal is important in that its disorders in humans are disastrous, ranging from comatose vegetative and fatigue states, through attention deficit disorders, through problems with vigilance and mood (Pfaff, 2005). CNS arousal decline probably contributes to mental problems during aging. Although we usually think of problems related to arousal states that are too low, the opposite disorders can also cause trouble. Sleep disorders are frequent among modern adults, and difficulties with surgical anesthesia deserve attention.

Given that CNS arousal is crucial, its conceptualization has sometimes been vague. Therefore, we need to achieve an operational definition which is precise, complete and which yields quantitative, physical measures. The following operational definition has been proposed (Pfaff, 2005): a more aroused animal or human being (i) is more alert to sensory stimuli in all sensory modalities; (ii) emits more voluntary motor activity; and (iii) is more reactive emotionally. In addition to reviewing hormonal, neural and genetic mechanisms underlying generalized arousal (below), we conceptualize it in mathematical terms as follows.

During the last several years we have been seeking to formulate a mathematical description of arousal-related processes in the mammalian CNS. First, a meta-analysis of experimental data from five studies with mice, using principal components analysis, yielded the estimate that among arousal-related measures there is a generalized arousal component that accounts for about one-third of the variance (Garey et al., 2003). In that same paper, we presented the simplest form of an equation portraying the state of arousal in the mammalian brain as an increasing function not only of generalized arousal (accounting for about one-third of the data), but also of several specific forms of arousal (sexual, hunger, thirst, salt hunger, fear, pain, etc., accounting for the rest of the data related to arousal). That formulation was incomplete. At least three equations may be needed to describe three different levels of the molecular and biophysical forces leading to the activation of behavior. Working with Professor Martin Braun (Department of Mathematics at Queens College, CUNY) we can think of three families of equations which may serve to 'state the set of problems' ripe for molecular neurobiological discovery. Eqs. (1), (4), and (7) are likely to be the most important, for a reason stated below. The mathematics of arousal is open to investigation.

First, inputs. Small changes in the state of arousal (A) of the mammalian CNS can be described as a compound increasing function (F) of a generalized arousal force supplemented by many specific forms of arousal (sexual, hunger, thirst, salt hunger, fear, pain, temperature, etc.). The manner in which these various forces augment each other is not known. Below are equations that hypothetically picture their relations to each other as additive (Eq. (1)), multiplicative (Eq. (2)) or exponential (Eq. (3)).

$$\delta A = F_1(As_1) + F_2(As_2) \cdots + \cdots F_n(As_n) + F_g(A_g) \quad (1)$$

$$\delta A = F_1(As_1) \bullet F_2(As_2) \cdots \bullet \cdots F_n(As_n) \bullet F_g(A_g) \quad (2)$$

$$\delta A = F_1(As_1) \bullet e^{F_g(A_g)} \quad (3)$$

Where A is the state of arousal of the nervous system at any moment, the As designation refers to various specific forms of arousal and Ag refers to generalized CNS arousal. These simplified formulations are intended to show the type of problem we have to solve by obtaining the appropriate experimental data.

Second. What about the internal operations and mechanisms within arousal systems in the brain itself? We can think of three equations that conceptualize small changes in a collective arousal mechanism (M) as an increasing function (G) of upward, ascending arousal pathways (U) in the CNS and descending (D) arousal pathways, both of them operating in a set of mechanisms that have a particular initial condition. Relations between ascending and descending pathways could be additive (Eq. (4)), multiplicative (Eq. (5)) or exponential (Eq. (6)).

List of Contributors

A. Adach, The Nencki Institute of Experimental Biology, Polish Academy of Sciences, Pasteur 3, 02-093 Warsaw, Poland

S. Alesci, Clinical Neuroendocrinology Branch, National Institute of Mental Health, NIH, Bethesda, MD 20892, USA

S. Bahn, Cambridge Centre for Neuropsychiatric Research, Institute of Biotechnology, University of Cambridge, Tennis Court Road, Cambridge CB2 1QT, UK

F.M. Benes, Program in Structural and Molecular Neuroscience, McLean Hospital, Mailman Research Center, 115 Mill Street, Belmont, MA 02178, USA

E.B. Binder, Department of Psychiatry and Behavioral Sciences and, Department of Human Genetics, Emory University School of Medicine, 101 Woodruff Circle, Suite 4000, Atlanta, GA 30322, USA

S. Che, Center for Dementia Research, Nathan Kline Institute and Department of Psychiatry, New York University School of Medicine, 140 Old Orangeburg Road, Orangeburg, NY 10962, USA

S.E. Counts, Department of Neurological Sciences, Rush University Medical Center, Chicago, IL 60612, USA

H. Creely, Max-Planck Institute for Evolutionary Anthropology, Deutscher Platz, D-04103 Leipzig, Germany

M. Dabrowski, The Nencki Institute of Experimental Biology, Polish Academy of Sciences, Pasteur 3, 02-093 Warsaw, Poland

N. Devidze, Laboratory of Neurobiology and Behavior, Box 275, The Rockefeller University, New York, NY 10021, USA

S.D. Ginsberg, Center for Dementia Research, Nathan Kline Institute, Departments of Psychiatry, and Physiology and Neuroscience, New York University School of Medicine, 140 Old Orangeburg Road, Orangeburg, NY 10962, USA

S.E. Hemby, Department of Physiology and Pharmacology, Wake Forest University, School of Medicine, Medical Center Boulevard, Winston-Salem, NC 27157, USA

I. Ilias, Department of Pharmacology, Faculty of Medicine, University of Patras, Rion-Patras, Greece

P. Khaitovich, Max-Planck Institute for Evolutionary Anthropology, Deutscher Platz, D-04103 Leipzig, Germany

S.H. Koslow, Office of Neuroinformatics, 6001 Executive Boulevard, Room 6161, MSC 9613, Bethesda, MD 20892-9613, USA

A.W. Lee, Laboratory of Neurobiology and Behavior, Box 275, Rockefeller University, New York, NY 10021, USA

D.A. Lewis, Departments of Psychiatry and Neuroscience, University of Pittsburgh, 3811 O'Hara Street, W1652 BST, Pittsburgh, PA 15213-2593, USA

K. Lukasiuk, The Nencki Institute of Experimental Biology, Polish Academy of Sciences, Pasteur 3, 02-093 Warsaw, Poland

H.K. Manji, Laboratory of Molecular Pathophysiology, National Institute of Mental Health, NIH, Building 35–Room 1C917, 35 Convent Drive, MSC 3711, Bethesda, MD 20892-3711, USA

K. Mirnics, Departments of Psychiatry and Neurobiology, School of Medicine, University of Pittsburgh, W1655 Biomedical Science, Pittsburgh, PA 15261, USA

E.J. Mufson, Department of Neurological Sciences, Rush University Medical Center, 1735 W. Harrison Street, Suite 300, Chicago, IL 60612, USA

C.B. Nemeroff, Department of Psychiatry and Behavioral Sciences, Emory University School of Medicine, 101 Woodruff Circle, Suite 4000, Atlanta, GA 30322, USA

D.W. Pfaff, Laboratory of Neurobiology and Behavior, Box 275, The Rockefeller University, New York, NY 10021, USA

J.H. Phan, The Wallace H. Coulter Department of Biomedical Engineering, Georgia Institute of Technology and Emory University, 313 Ferst Drive, UA Whitaker Building – Suite 4106, Atlanta, GA 30332-0535, USA

A. Pitkänen, A.I. Virtanen Institute for Molecular Sciences, University of Kuopio, PO Box 1627, FIN-70 211 Kuopio, Finland

C.-F. Quo, The Wallace H. Coulter Department of Biomedical Engineering, Georgia Institute of Technology and Emory University, 313 Ferst Drive, UA Whitaker Building – Suite 4106, Atlanta, GA 30332-0535, USA

M. Rodak, NIH, Clinical Neuroendocrinology Branch, National Institute of Mental Health, NIH, Bethesda, MD 20892, USA

N.S. Tannu, Department of Physiology and Pharmacology, Wake Forest University, School of Medicine, Medical Center Boulevard, Winston-Salem, NC 27157, USA

M.D. Wang, The Wallace H. Coulter Department of Biomedical Engineering, Georgia Institute of Technology and Emory University, 313 Ferst Drive, UA Whitaker Building – Suite 4106, Atlanta, GA 30322, USA

M.T. Wayland, Cambridge Centre for Neuropsychiatric Research, Institute of Biotechnology, University of Cambridge, Tennis Court Road, Cambridge CB2 1QT, UK

M.J. Webster, Department of Psychiatry, Stanley Laboratory of Brain Research, Uniformed Services University of the Health Sciences, 4301 Jones Bridge Road, Bethesda, MD 20814-4799, USA

J. Zhou, Laboratory of Neurobiology and Behavior, Box 275, The Rockefeller University, New York, NY 10021, USA

R. Zhou, Department of Psychiatry, Uniformed Services, University of the Health Sciences, Bethesda, MD, USA

Foreword

Stephen H. Koslow

Office of Neuroinformatics Department of Health and Human Services, NIH, National Institute of Mental Health, Bethesda, MD 20892-9613, USA

This new monograph critically addresses important issues in genomics and proteomics as applied to the study of the brain and brain disorders. This publication is appearing 50 plus years after the 1953 breakthrough discovery of DNA by Watson and Crick. *Functional Genomics and Proteomics in the Clinical Neurosciences* is extremely timely as we are now starting to harvest the results of the Watson and Crick discovery and will ultimately unravel the genetic mechanisms underlying both diseases and normal development of all species, and this current publication should help refocus and enhance some of the research in this area. The true impact of the uncovering of DNA was not fully realized until the mapping of the human genome was completed in 2004. While the assembly of the constituent organization of the human genome is seminal, the technical developments which made this mapping possible are now hastening the understanding of human disorders.

Mapping the genome required the development of a variety of high-throughput technologies, computer databases, and the creation of the field of informatics. With these three capabilities in place, it became possible, in a short period of time, to determine the sequence of the tens of thousands of genes on chromosomes of all life forms. With the full panoply of new technologies developed, it was now achievable to store these data for future further analysis and annotation and to analyze these genes for functional significance using the innovative informatics tools. The genomic and proteomic approaches presented in this book should lead to a system for the understanding of brain function and development. Through this ability to understand interactions between the genome and proteome, it will be possible to define specific biological markers which, by monitoring their changes over periods of time, will prove useful in understanding and predicting the progression of both developmental and disease processes. These biological markers will be useful in the diagnosis and prognosis of disease onset as well as therapeutic response, relapse, or cure.

Genomic and proteomic methods are being applied to understand genetic networks which regulate cellular processes signaling the production of protein molecules which mediate the physiological process. While this approach is now used in all fields of biomedical research, the most challenging application is to understand those mechanisms involved in human brain disorders and illnesses. Identification of the genes responsible for human genetic diseases requires large clinical studies using the latest technological advances and paradigms available for mapping, sequencing, and interpreting human and other vertebrate genomes. Currently, the most important method for large-scale analysis of gene expression patterns is microarray analysis. *Functional Genomics and Proteomics in the Clinical Neuroscience* brings together outstanding research teams who provide detailed monographs on laboratory and informatics methods of analysis in specific applications to clinical disorders of the nervous system. Consideration is also given to the future implications of these findings on the fields of Neurology and Psychiatry.

Functional Genomics and Proteomics in the Clinical Neuroscience is conceptually divided into three sections. The first section, "Methodologies," presents in-depth expert opinions on tissue preparation for genomics and proteomics as well as the present approaches to data mining and informatics. The second section focuses on microarray results in "Applications of Genomics and Proteomic Technologies to Clinical Neuroscience". It covers a diverse group of illnesses, including mood disorders, schizophrenia, developmental disorders, substance abuse, neurodegenerative disorders, epilepsy, and disorders of the neuroimmune and steroid hormone systems. The last section, "Future Directions", appropriately focuses on the future impact of these studies on the research direction of genomics and proteomics, the clinical practice of Psychiatry, Neurology and Neuroimmunology, as well as an overview of what will be learnt from comparative neurogenomics and proteomics.

Those individuals interested in genomic and proteomic approaches to the understanding of the nervous system and its disorders should benefit greatly from the outstanding material which is drawn together and well presented by the experts in this challenging field of research.

Contents

List of Contributors v

Foreword by S.H. Koslow vii

Section I. Methodologies

1. Tissue preparation and banking
   M.J. Webster (Bethesda, MD, USA). 3

2. Functional genomic methodologies
   S.D. Ginsberg and K. Mirnics (Orangeburg, NY and Pittsburgh, PA, USA) 15

3. Methods for proteomics in neuroscience
   N.S. Tannu and S.E. Hemby (Winston-Salem, NC, USA) 41

4. Functional genomics and proteomics in the clinical neurosciences: data mining and bioinformatics
   J.H. Phan, C.-F. Quo and M.D. Wang (Atlanta, GA, USA). 83

5. Reproducibility of microarray studies: concordance of current analysis methods
   M.T. Wayland and S. Bahn (Cambridge, UK). 109

Section II. Applications of Genomics and Proteomic Technologies to Clinical Neuroscience

6. The genomics of mood disorders
   S. Alesci, M. Rodak, I. Ilias, R. Zhou, H.K. Manji (Bethesda, MD, USA and Rion-Patras, Greece). 129

7. Transcriptome alterations in schizophrenia: disturbing the functional architecture of the dorsolateral prefrontal cortex
   D.A. Lewis and K. Mirnics (Pittsburgh, PA, USA) 141

8. Strategies for improving sensitivity of gene expression profiling: regulation of apoptosis in the limbic lobe of schizophrenics and bipolars
   F.M. Benes (Belmont and Boston MA, USA) 153

9.	Assessment of genome and proteome profiles in cocaine abuse S.E. Hemby (Winston-Salem, NC, USA)	173
10.	Neuronal gene expression profiling: uncovering the molecular biology of neurodegenerative disease E.J. Mufson, S.E. Counts, S. Che and S.D. Ginsberg (Chicago, IL, and Orangeburg, NY, USA)	197
11.	Epileptogenesis-related genes revisited K. Lukasiuk, M. Dabrowski, A. Adach and A. Pitkänen (Warsaw, Poland and Kuopio, Finland)	223
12.	Functional genomics of sex hormone-dependent neuroendocrine systems: specific and generalized actions in the CNS A.W. Lee, N. Devidze, D.W. Pfaff and J. Zhou (New York, NY, USA)	243

Section III. Future Directions

13.	Implications for the practice of psychiatry E.B. Binder and C.B. Nemeroff (Atlanta, GA, USA)	275
14.	Human brain evolution H. Creely and P. Khaitovich (Leipzig, Germany)	295

Subject Index 311

SECTION I

Methodologies

CHAPTER 1

Tissue preparation and banking

Maree J. Webster*

Stanley Laboratory of Brain Research, Department of Psychiatry, USUHS, 4301 Jones Bridge Rd., Bethesda, MD 20814, USA

Abstract: With the increasing application of genomic and proteomic technologies to the research of neurological and psychiatric disorders it has become imperative that the postmortem tissue utilized be of the highest quality possible. Every step of the research design, from identifying donors, acquiring sufficient information for accurate diagnosis, to assessing tissue quality has to be carefully considered. In order to obtain high-quality RNA and protein from the postmortem brain tissue a standardized system of brain collection, dissection, and storage must be employed and key ante- and postmortem factors must be considered. Reliable RNA expression and protein data can be obtained from postmortem brains with relatively long postmortem intervals (PMIs) if the agonal factors and acidosis are not severe. While pH values are correlated with RNA integrity number (RIN), a higher pH does not guarantee intact RNA. Consequently RNA integrity must be assessed for every case before it is included in a study. An analysis of anti- and postmortem factors in a large brain collection has revealed that several diagnostic groups have significantly lower pH values than other groups, however, they do not have significantly lower RIN values. Moreover, the lower pH of these groups is not entirely due to agonal factors and/or smoking, indicating that these subjects may have additional metabolic abnormalities that contribute to the lower pH values.

Introduction

The development and increasing application of genomic and proteomic technologies to the research of neurological and psychiatric disorders has made it imperative that the postmortem tissue utilized be of the highest quality possible. Every step of the research design, from identifying donors, acquiring sufficient information for accurate diagnosis, to assessing tissue quality has to be carefully considered. In this review, we will briefly outline the steps necessary to acquire useful brain tissue. Many of these steps have been reviewed in great detail in other publications (Kleinman et al., 1995; Vonsattel et al., 1995; Green et al., 1999; Kittel et al., 1999; Shankar and Mahadven, 1999; Lewis, 2002; Bunney et al., 2003; Hynd et al., 2003;

Katsel et al., 2005) and so will only briefly be discussed here. Demographic factors that affect the integrity of the tissue, and the RNA in particular, will be discussed in more detail.

Identifying subjects

Recruitment

Brain donations are generally obtained by direct donations from individuals affiliated with patient advocacy groups, from hospices, from state psychiatric hospitals or veteran's administration (VA) hospitals or from medical examiner's offices. Each source has advantages and disadvantages. For example, donations through patient advocacy groups allow accurate diagnosis of living subjects but the tissue is obtained from multiple sources, and often harvested with nonstandardized

---

*Corresponding author. Tel.: +1-301-295-0794; Fax: +1-301-295-6276; E-mail: websterm@stanleyresearch.org

techniques. Brain tissue obtained from the hospice setting has very short postmortem intervals (PMI) but tends to have prolonged agonal states that may adversely affect the quality of the tissue. Donations from the state psychiatric hospitals or VA hospitals allow for very accurate diagnosis of the living patients; however, the patients are often of very advanced age. The medical examiner's office is a source of relatively young subjects, many suicide victims and mentally ill, with short PMI, however, obtaining all the medical records to make an accurate diagnosis is often problematic.

Prospective/retrospective assessment

Prospective studies allow an accurate diagnosis to be made in the living patient. However, as mentioned above the studies are frequently confounded by age. The advanced age may be associated with superimposed neurodegenerative disorders and neuropathology (Healy et al., 1996). However, retrospective studies, establishing diagnosis after death can be difficult. Many cases cannot be confirmed by existing records or by follow-up interviews with family members or physicians. Regardless of the type of assessment, many diagnoses may be confounded by alcoholism or substance abuse.

Control subjects

Brain tissue from unaffected controls with no neurological or psychiatric symptoms should be included in all studies. A structured telephone interview with a first-degree family member should be carried out in all cases. Information should be obtained about birth and development, education, jobs, family history of neurological or mental illness, drug and alcohol use, smoking, and medical problems and medications. Open-ended questions seeking information about personality traits should also be included (Torrey et al., 2000). An additional control group can also be included when it is necessary to account for a particular variable. For example, because almost all patients with schizophrenia have taken neuroleptic medications an additional neuroleptic-treated control group (bipolar (BP) disorder, psychotic depression, Huntington's disease) may be included.

Ethical issues

Informed consent must be obtained from the legally designated next of kin for any tissue donations. Regulations for consent vary by state, but in most states they may be obtained over the telephone if telephonically witnessed by a third party or recorded on tape. The permit must be detailed and include permission to take the brain, cranial contents, blood and any other specimens required by the protocol. While donor cards are often available in prospective studies they are often considered to be insufficient evidence for the pathologist, particularly in a case of a mentally ill patient, and therefore should be followed up with a phone call to the next of kin. Researchers are obligated to inform the next of kin of the project for which the tissues will be used and are responsible for ensuring each donor's anonymity and privacy of all medical information.

Collection and harvesting tissue

Handling tissue

It is important to have all tissue for any one study collected in the same manner, e.g. freezing, dissection, and the preparation of samples. All personnel involved in the removal of the brain and the processing of it should be trained centrally by the same team to ensure that each specimen is handled in exactly the same manner.

Freezing/fixing

A commonly used protocol is to have the cerebrum hemisected, one hemisphere fixed in formalin and the other cut into 1 cm thick coronal slices and frozen in an isopentane/dry ice slurry, then stored at −80 °C. The right and left hemispheres may be randomly alternated for formalin fixing or freezing. However, every study will determine the protocol that ensures the most efficient use of the whole brain for their particular needs.

Neuropathological screening

A standard neuropathological examination should be conducted and a macroscopic and microscopic description provided for every case. Cases should be screened for the presence of cerebrovascular disease, hemorrhage, trauma, tumors, or other pathology and confirmed by examination of appropriate sections from suspect areas. Cases should be screened for Alzheimer's disease, Parkinson's disease, ethanol-induced changes, and anoxic/hypoxic-related alterations.

Toxicology

Toxicology screening may be especially important in studies of substance abuse and other psychiatric disorders where it is necessary to have an indication of the substances consumed around the time of death. Urine and blood toxicology screens can be performed by the medical examiner's office but the protocols often differ between offices so that a standardized protocol may need to be established. Additional toxicological analysis can be performed on the postmortem brain tissue by a forensic toxicologist to screen for ethanol or psychoactive substances. A more-detailed analysis of substance abuse history can be obtained from segmental hair analysis but this is not a routine practice.

Dissection and storage

Anatomical areas of interest can be dissected from the frozen slabs, however, it is important that the tissue not thawed because ice crystals will form in the tissue, and protein denaturation may occur. RNA integrity is particularly vulnerable to freeze/thaw cycles. Small pieces of cortex can be removed with a sterile razor blade or chisel. Larger subcortical structures can be dissected carefully and precisely with a 'hack-saw' or fine dental drill. Close attention must be paid to the exact anatomical location of each dissection. Large variation in gene expression has been detected between different frontal cortical areas (Ryan et al., 2006). Moreover, samples with different white/gray matter ratios of tissue may generate data that is incorrectly interpreted as differential expression of glial-related genes (Mirnics et al., 2001).

Whether an area is going to be dissected and sectioned onto slides or whether it will be homogenized for RNA or protein extraction will depend on the study design and how many researchers are expected to be able to use the samples. Providing slide-based sections of areas that are relatively small, but in high demand, to multiple researchers makes the most efficient use of the tissue. Fresh frozen sections can be used for in situ hybridization and immunohistochemistry for both qualitative and quantitative analysis. The sections preserve the normal anatomy, provide landmarks and allow for the precise localization of the RNA or protein to a particular cell type. The sections can also be used successfully to extract RNA from cells that have been microdissected with laser capture technology (Altar et al., 2005). The RNA can then be used for microarray studies and PCR. Specific regions of interest e.g. the hippocampus or amygdala can be carefully dissected and then frozen sections can be cut on a cryostat and stored at $-80\,°C$. One section for every 0.5 mm should be stained to allow for the construction of an anatomical atlas through each structure for each case so that the exact anatomical location within each structure can be matched across cases. An additional way to ensure the most-efficient use of the tissue is to extract RNA and protein from a particular structure and make it available to multiple researchers. This is a more-efficient use of the tissue than sending tissue pieces to individual researchers and it eliminates variability between studies associated with different extraction techniques. It also eliminates variability due to the anatomical heterogeneous nature of a structure since the entire structure can be homogenized not just a portion of it.

The formalin-fixed hemisphere should have the various regions of interest dissected and embedded in paraffin, ideally within 1 month of harvesting. The paraffin sections are popular for immunohistochemistry but longer fixation times affect the quality of the staining. Therefore, the mean fixation times for groups in a study need to be matched. Alternatively fixed specimens can be processed through a graded series of sucrose solutions and stored in a 30% sucrose buffer solution at 4 °C. The sucrose is a cryoprotectant and sections can then be cut frozen on a cryostat. Storage times can be extended without losing immunoreactivity to the extent that formalin fixation effects the staining.

Documenting

Demographic database

Each case is assigned a number and entered with all relevant variables into a demographic database. Examples of variables include age, race, sex, date of birth, date of death, cause of death, rate of death, refrigeration interval, postmortem interval, pH of brain, RNA quality (e.g. RNA integrity number (RIN) number), left brain (frozen or fixed), brain weight, weight of deceased, and height of deceased. Other variables deemed necessary to include would depend on the nature of the diagnostic groups included in the collection. Typical variables included for a collection of mentally ill subjects would include diagnosis (e.g. primary axis 1 DSMIV), secondary diagnosis (if any), tertiary diagnosis (if any), marital status, education (years), number of children (if any), suicide status, any known significant problems during birth or pregnancy, any known significant problems during development in childhood, number of siblings, age of onset of disorder, duration of illness, total hospitalization time (years), exacerbation of illness at time of death, global severity of illness, lifetime alcohol use, lifetime drug use (it is useful to use some kind of ratings scale to quantify alcohol and drug use), smoking at time of death, family history (of relevant disorder), awareness of illness/insight, whether the illness included psychotic features, estimated lifetime intake of antipsychotics (in fluphenazine mg equivalents), on or off antipsychotics or mood stabilizers at time of death, name of the antipsychotic taken at time of death, mood stabilizer taken at time of death, and significant medical illnesses.

Designating cohorts

Once the diagnosis is verified and the tissue quality is deemed suitable for research the brain can then be placed into a cohort of subjects for study. In some instances it is necessary to use different pieces from one brain in multiple cohorts, however it is generally easier and more efficient to assign the whole brain to a designated cohort. Data from multiple areas and multiple studies can be collected from the cohort and then analyzed as a meta-analysis. This approach allows for the examination of inter-assay variability on the reproducibility of postmortem findings as well as the extensive characterization of a specific cohort of patients (Knable et al., 2001a, b, 2002a, b, 2004; Torrey et al., 2005).

Matching groups in a cohort

Groups of subjects between which gene expression is being compared, must be matched for those variables that are known to affect gene expression. The issue of pH is a major one. Not only does pH affect RNA integrity but it also affects individual gene expression (Johnston et al., 1997; Miller et al., 2004; Webster et al., 2002a; Preece and Cairns, 2003; Li et al., 2004; Prabakaran et al., 2004; Altar et al., 2005). Age is another major factor that affects gene expression (Nichols et al., 1993; Harrison et al., 1995; Castensson et al., 2000; Romanczyk et al., 2002; Webster et al., 2002b; Law et al., 2003; Preece and Cairns, 2003; Erraji-Benchekroun et al., 2005) and PMI has also been shown to affect the expression level of some genes (Burke et al., 1991; Harrison et al., 1995; Castensson et al., 2000; Trotter et al., 2002; Inoue et al., 2002). Gender influences gene expression in certain brain areas (Castensson et al., 2000; Fernandez-Guasti et al., 2000; Kruijver et al., 2001; Preece and Cairns, 2003). Ethnicity may also influence expression of certain genes as may laterality (Halpern et al., 2005). The period of time that specimens are stored in the freezer before RNA is extracted may also have some effect on RNA quality and gene expression (Burke et al., 1991; Leonard et al., 1993; Eastwood et al., 1995).

Thus when designing studies to measure gene expression, equal numbers of cases and controls must be matched for the individual variables. Alternatively, a match-pairs design can be employed where each case has a gender-matched control that also matches as closely as possible for age, race, PMI, pH, and hemisphere. The confounding variables can also be corrected for in the statistical analysis by including the variables as covariates in an ANCOVA analysis.

Tissue inventories

Regardless of whether the tissue collected is going to be used by the same research group or distributed to multiple groups, it is important to keep an inventory of which structures on each brain have been dissected and distributed. Records should be kept on how the structure was processed, the date it was distributed and to whom. The date the data is returned should also be included if that is a requirement of the protocol. All returned data should be inventoried in such a way that future meta-analysis of the data can be conducted.

RNA integrity

Factors affecting RNA integrity

The emergence of genomic technologies requires that intact RNA be recovered from postmortem specimens. As each brain enters the collection, RNA is extracted and evaluated. If the RNA is degraded it may not be an efficient use of resources to further process the tissue and retrieve and analyze the medical records. The Stanley Medical Research Institute (SMRI) Laboratory has assessed total RNA quality on over 450 cases using the RIN determined by electrophoresis using the Agilent 2100 Bioanalyzer (Agilent Technologies, Palo Alto, CA, USA). We then used the RIN numbers to determine if demographic variables, including diagnosis, PMI, pH, agonal state at death, cause of death, or substance abuse would reliably predict overall RNA quality.

The RIN is determined from the entire electrophoretic trace of the RNA sample, including the ratio of the ribosomal bands and the presence or absence of degradation products. RIN is a valuable tool to screen cases to determine which will be suitable for future studies and which should be excluded because of poor quality RNA. A RIN of one indicates completely degraded RNA whereas a RIN of 10 indicates perfectly intact RNA. While PMI was initially considered a major factor in predicting RNA integrity (Bauer et al., 2003), numerous publications have shown that it is not a highly correlated factor (Johnson et al., 1986; Perrett et al., 1988; Harrison et al., 1995; Yasojima et al., 2001). SMRI studies have shown that PMI is weakly correlated with RIN ($r = -0.1943$; $p < 0.001$, Fig. 1A).

Many studies have now shown that the pH of a postmortem brain specimen is a better predictor of RNA quality than PMI (Harrison et al., 1995; Kingsbury et al., 1995; Johnston et al., 1997). SMRI has found that pH and RIN are more significantly correlated ($r = 0.26165$; $p < 0.0001$) than PMI and RIN but there is no specific pH value below which the RNA will definitely be degraded (Fig. 1B).

During the course of collecting the specimens it has become apparent that cases within certain diagnostic groups have more degraded RNA. When dividing all SMRI cases by diagnosis, ANOVA has shown a significant effect of diagnosis on both pH ($df = 5, 455$; $F = 3.84$; $p = 0.002$) and RIN ($df = 5, 457$; $F = 3.65$; $p = 0.003$). The pH was significantly lower in the BP group as compared to the depressed, unaffected controls and psychiatric disorder NOS groups ($p = 0.03, 0.02$, and 0.04 respectively; Fig. 2A). The pH was also significantly lower in the schizophrenia group as compared to the depressed, unaffected controls and psychiatric NOS groups ($p = 0.002, 0.0008$, and 0.008 respectively, Fig. 2A). In contrast, the RIN was significantly lower in the depressed group as compared to the unaffected controls and the schizophrenia group ($p = 0.003$ and 0.05 respectively, Fig. 2B). RIN was also significantly lower in the psychiatric not otherwise specified (NOS) group as compared to the unaffected controls and the schizophrenia group ($p = 0.003$ and 0.03 respectively, Fig. 2B). Surprisingly, the groups that have the lowest pH are not the groups with the lower RIN numbers.

Several studies had shown that prolonged agonal states reduce both tissue pH (Yates et al., 1990; Hardy et al., 1985) and RNA integrity (Harrison et al., 1991; Barton et al., 1993; Harrison et al.,

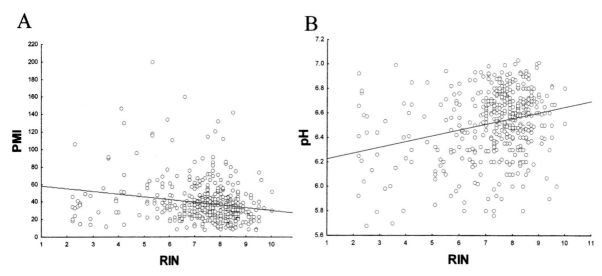

Fig. 1. (A) PMI is negatively correlated with RIN, ($N = 449$, $r = -0.1943$, $p < 0.001$). (B) pH is positively correlated with RIN, ($N = 451$, $r = 0.2617$, $p < 0.0001$).

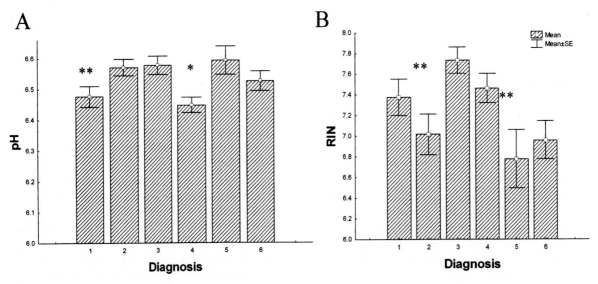

Fig. 2. Effect of diagnosis on brain pH and RIN. (A) Brain pH was significantly lower in groups 1 (BP, $N = 81$) and 4 (schizophrenia, $N = 116$) as compared to group 3 (unaffected controls, $N = 87$). (B) RIN was significantly lower in groups 2 (depressed, $N = 76$) and 5 (psychiatric NOS, $N = 31$) as compared to the group 3 (unaffected controls). Group 6 (other, $N = 62$). *$p < 0.05$, **$p < 0.01$.

1995; Kingsbury et al; 1995; Tomita et al., 2004). Therefore, SMRI actively discouraged collecting from any individuals who had undergone a prolonged agonal process or who had been on a mechanical ventilator prior to death. Nevertheless, ANOVA showed a significant effect of a subjective measure of 'rate of death' on pH ($df = 3$, 455; $F = 3.78$; $p = 0.01$; Fig. 3A). Those individuals who suffered possible anoxia at death (e.g. carbon monoxide (CO) poisoning or pneumonia) had significantly lower pH than those dying suddenly ($p = 0.01$). Those dying more slowly (e.g. from cancer) also had a significantly lower pH than those dying suddenly ($p = 0.04$). This was not

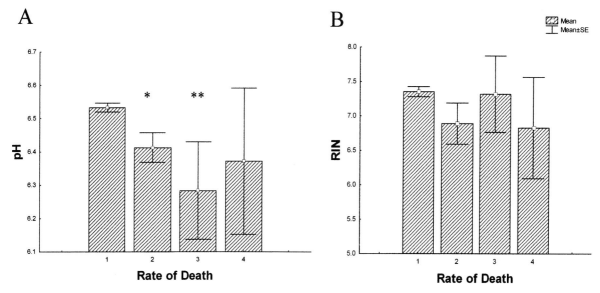

Fig. 3. Effect of 'rate of death' on brain pH and RIN. (A) Brain pH was significantly lower in groups 2 (possible anoxia at death e.g. pneumonia, CO poisoning, $N = 35$) and 3 (slower death e.g. cancer, $N = 5$) as compared to group 1 (sudden, $N = 405$). (B) RIN did not differ significantly between groups. Group 4 (mechanical ventilator prior to death, $N = 4$). $*p < 0.05$, $**p < 0.01$.

reflected in the RIN numbers, however, where no significant effect of 'rate of death' on RIN was seen ($df = 3, 448$; $F = 1.13$; $p = 0.34$; Fig. 3B).

Subjects were also categorized into groups based specifically on 'type of death'. ANOVA revealed a significant effect of 'type of death' on both pH ($df = 7, 443$; $F = 6.29$; $p < 0.00001$; Fig. 4A) and RIN ($df = 7, 445$; $F = 2.256$; $p = 0.03$; Fig. 4B). Those individuals dying from pneumonia and chronic obstructive pulmonary disease (COPD), chronic conditions such as cancer and those dying in hospitals, all had significantly lower pH values than the group dying suddenly ($p = 0.0003, 0.01, 0.00001$, respectively). In contrast, the groups that died from a drug overdose (OD) or from carbon monoxide poisoning (CO) had significantly lower RIN than the group that died suddenly ($p = 0.01$ and 0.003, respectively). Again, the groups with the lowest pH values are not necessarily the groups with the lower RIN values.

'Type of death' appears to reflect the differences in pH and RIN values that were found between the various diagnostic groups, e.g. the BP and schizophrenia groups, that have the lowest pH values, also have the highest percentage of cases dying from COPD or pneumonia, chronic conditions

and while hospitalized (10% BP, 21% schizophrenia, 6% depressed, 6% unaffected controls, 6% NOS, and 19% other). However, the 'other' group should also have significantly lower pH values given the high percentage of cases that died from these conditions. Similarly, the groups with the lowest RIN values, depression and NOS, have the highest percentage of cases dying from CO poisoning (17% depression, 10% NOS as compared to 7% BP, 0% unaffected controls, 0% schizophrenia, and 6% for other). However, when combining those who died from OD and CO poisoning there is less difference between the groups (54% BP, 66% depression, 0% unaffected controls, 24% schizophrenia, 58% NOS, and 55% other). The control and schizophrenia groups both have the highest RIN values and the lowest percentage of OD and CO deaths. However, one would expect the RIN values to be low in the BP and the 'other' group based on the percentage of OD and CO deaths in each group. Thus, while pH and agonal factors do not definitively predict RNA integrity, it is clear that they have an important effect.

Given the fact that those individuals dying from COPD and pneumonia have lower pH values it was not surprising to find that those individuals

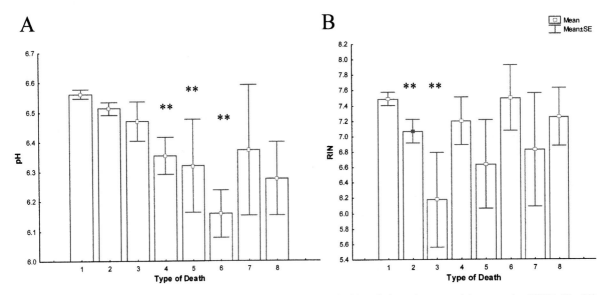

Fig. 4. Effect of 'type of death' on brain pH and RIN. (A) Brain pH was significantly lower in groups 4 (pneumonia, COPD, $N = 23$), 5 (chronic condition, $N = 8$) and 6 (hospital death without ventilator, $N = 9$) as compared to group 1 (sudden, $N = 246$). (B) RIN was significantly lower in groups 2 (drug overdose, $N = 114$) and 3 (CO poisoning, $N = 12$) as compared to group 1 (sudden). Group 7 (mechanical ventilator, $N = 4$), group 8 (unknown, $N = 5$). **$p < 0.01$.

who were smoking at the time of death had significantly lower pH values than those not smoking ($p < 0.03$, Fig. 5A). In contrast, there was no difference in RIN between the smokers and non-smokers ($p = 0.6$; Fig. 5B). The low pH values found in BP and schizophrenia may reflect the higher percentage of individuals who smoke in those groups. SMRI does not have smoking history on every case, but of those who were definite smokers there were 54% in the BP group, 47% in depression, 31% in unaffected controls, 72% in schizophrenia, 58% in NOS and 36% in 'other'. The fact the BP and schizophrenia groups have more individuals dying from COPD and pneumonia and more are smokers, indicates that they may be experiencing a mild hypoxia before and during the death process that leads to lower pH values (Hardy et al., 1985; Harrison et al., 1991, 1995; Li et al., 2004). However, the psychotic NOS group also has a high percentage of smokers but their pH value is not lower than controls. Thus the BP and schizophrenia group may have additional metabolic abnormalities that contribute to the lower pH values (Hamakawa et al., 2004; Karry et al., 2004; Konradi et al., 2004; Prabakaran et al., 2004; Altar et al., 2005; Kato, 2005).

Dividing the cases into five groups based on the history of alcohol use showed no effect of alcohol on pH values or RIN numbers. Similarly history of drug use had no effect on pH or RIN (data not shown).

Thus, while high-quality RNA is more likely to come from individuals who have died suddenly, who have no agonal factors or evidence of hypoxia and who have a short PMI, this does not guarantee good RNA. The RNA from every case must be assessed individually before the case is assigned to a study cohort.

Protein integrity

Although a considerable number of publications have appeared on the proteomic analysis of human postmortem brain samples (Edgar et al., 1999a, b; Greber et al., 1999; Edgar et al., 2000; Johnston-Wilson et al., 2000; Schonberger et al., 2001; Butterfield and Castegna, 2003; Lubec et al., 2003; Butterfield, 2004; Lewohl et al., 2004; Prabakaran et al., 2004; Swatton et al., 2004; Newcombe et al., 2005) surprisingly little is known about quantitative postmortem changes in the brain protein

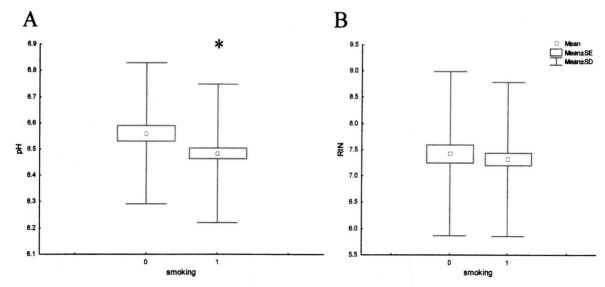

Fig. 5. Effects of smoking on brain pH and RIN. (A) Brain pH is significantly lower in group 1 (smokers, $N = 160$) as compared to group 0 (nonsmokers, $N = 87$), *$p = 0.03$). (B) RIN did not differ significantly between groups.

profile. It is generally believed that proteins are more stable than RNA (Harrison et al., 1991b, 1995; Hynd et al., 2003; Voshol et al., 2003; Kim et al., 2004). However, proteins can be degraded, truncated or modified by postmortem effects (Fountoulakis et al., 2001). Most changes are detected after a PMI of 24 h or more. Proteolytic products are detected after 72 h in certain classes of proteins. Whether other factors such as pH or agonal state are important is not known. Johnston-Wilson et al. (2000) reported that extended storage time and mode of death had an effect on the expression of some proteins but PMI was not examined in this study. A recent initiative by the Human Proteome Organization (HUPO) to conduct proteome analysis of human brain from biopsies and autopsies will add valuable data to assess protein stability in postmortem tissue (Hamacher et al., 2004).

Conclusions

High-quality RNA and protein can be extracted from postmortem brain tissue provided a standardized system of brain collection, dissection, and storage is employed and key ante- and postmortem factors are considered. Reliable RNA expression and protein data can be obtained from postmortem brains with relatively long PMIs if the agonal factors and acidosis are not severe. While pH values are correlated with RIN values, a high pH does not guarantee intact RNA, consequently RNA integrity should be assessed for every cases before it is included in a study. Schizophrenia and BP disorder groups have significantly lower pH values than the other groups examined here, but do not have significantly lower RIN values. The lower pH of these groups is not entirely due to the agonal factors and/or smoking, indicating that subjects with schizophrenia and bipolar disorder may have additional metabolic abnormalities that contribute to the lower pH values.

References

Altar, C.A., Jurata, L.W., Charles, V., Lemire, A., Liu, P., Bukham, Y., Young, T.A., Bullard, J., Yokoe, H., Webster, M.J., Knable, M.B. and Brockman, J.A. (2005) Deficient expression of proteasome, ubiquitin, and mitochondrial genes in hippocampal neurons of multiple schizophrenia patient groups. Biol. Psychiatry, 58: 85–96.

Barton, A.J.L., Pearson, R.C.A., Najlerahin, A. and Harrison, P.J. (1993) Pre- and postmortem influences on brain RNA. J. Neurochem., 61: 1–11.

Bauer, M., Gramlich, I., Polzin and Patzelt, D. (2003) Quantification of mRNA degradation as possible indicator of postmortem interval — a pilot study. Legal Med., 5: 220–227.

Bunney, W.E., Bunney, B.G., Vawter, M.P., Tomita, H., Li, J., Evans, S.J., Choudary, P.V., myers, R.M., Jones, E.G., Watson, S.J. and Akil, H. (2003) Microarray technology: a review of new strategies to discover candidate vulnerability genes in psychiatric disorders. Am. J. Psychiatry, 160: 657–666.

Burke, W.J., O'Malley, K.L., Chung, H.D., Harmon, S.K., Miller, J.P. and Berg, L. (1991) Effect of pre- and postmortem variables on specific mRNA levels in human brain. Mol. Brain Res., 11: 37–41.

Butterfield, D.A. (2004) Proteomics: a new approach to investigate oxidative stress in Alzheimer's disease brain. Brain Res., 1000: 1–7.

Butterfield, D.A. and Castegna, A. (2003) Proteomics for the identification of specifically oxidized proteins in brain: technology and application to the study of neurodegenerative disorders. Amino Acids, 25: 419–425.

Castensson, A., Emilsson, L., Preece, P. and Jazin, E. (2000) High-resolution quantification of specific mRNA levels in human brain autopsies and biopsies. Genome Res., 10: 1219–1229.

Eastwood, S.L., McDonald, B., Burnet, P.W.J., Beckwith, J.P., Kerwin, R.W. and Harrison, P.J. (1995) Decreased expression of mRNAs encoding non-NMDA glutamate receptors GluR1 and GluR2 in medial temporal lobe neurons in schizophrenia. Mol. Brain Res., 29: 211–223.

Edgar, P.F., Douglas, J.E., Cooper, G.J.S., Dean, B., Kydd, R. and Faull, R.L.M. (2000) Comparative proteome analysis of the hippocampus implicates chromosome 6q in schizophrenia. Mol. Psychiatry, 5: 85–90.

Edgar, P.F., Douglas, J.E., Knight, C., Cooper, G.J.S., Faull, R.L.M. and Kydd, R. (1999a) proteome map of the human hippocampus. Hippocampus, 9: 644–650.

Edgar, P.F., Schonberger, S.J., Dean, B., Faull, R.L.M., Kydd, R. and Cooper, G.J.S. (1999b) A comparative proteome analysis of hippocampus from schizophrenic and Alzheimer's disease individuals. Mol. Psychiatry, 4: 173–178.

Erraji-Benchekroun, L., Underwood, M.D., Arango, V., Galfalvy, H., Pavlidis, P., Smyrniotopoulos, P., Mann, J.J. and Sibille, E. (2005) Molecular aging in human prefrontal cortex is selective and continuous throughout adult life. Biol. Psychiatry, 57: 549–558.

Fernandez-Guasti, A., Kruijver, F.P., Fodor, M. and Swaab, D.F. (2000) Sex differences in the distribution of androgen receptors in the human hypothalamus. J. Comp. Neurol., 425: 422–435.

Fountoulakis, M., Hardmeier, R., Hoger, H. and Lubec, G. (2001) Postmortem changes in the level of brain proteins. Exp. Neurol., 167: 86–94.

Greber, S., Lubec, G., Cairns, N. and Fountoulakis, M. (1999) Decreased levels of synaptosomal associated protein 25 in the brain of patients with Down Syndrome and Alzheimer's disease. Electrophoresis, 20: 928–934.

Green, R., Clark, A., Hickey, W., Hutsler, J. and Gazzaniga, M. (1999) Braincutting for psychiatrists: the time is ripe. J. Neuropsychiatry Clin. Neurosci., 11: 301–306.

Hardy, J.A., Wester, B., Winlad, B., Gezelius, C., Bring, G. and Eriksson, A. (1985) The patients dying after long terminal phase have acidotic brains: implications for biochemical measurements on autopsy tissue. J. Neural Transm., 61: 253–264.

Halpern, M.E., Gunturkun, O., Hopkins, W.D. and Rogers, L.J. (2005) Lateraliization of the vertebrate brain: taking the side of model systems. J. Neurosci., 25: 10351–10357.

Hamacher, M., Klose, J., Rossier, J., Marcus, K. and Meyer, H.E. (2004) Does understanding the brain need proteomics and does understanding proteomics need brain?" — Second HUPO HBPP workshop hosted in Paris. Proteomics, 4: 1932–1934.

Hamakawa, H., Murashita, J., Yamada, N., Inubushi, T., Kato, N. and Kato, T. (2004) Reduced intracellular pH in the basal ganglia and whole brain measured by ^{31}P-MRS in bipolar disorder. Psychiatry Clin. Neurosci., 58: 82–88.

Harrison, P.J., Procter, A.W., Barton, A.J.L., Lowes, S.L., Najlerahim, A., Bertolucci, P.H., Bowen, D.M. and Pearson, R.C.A. (1991) Terminal coma affects messenger RNA detection in post mortem human temporal cortex. Mol. Brain Res., 9: 161–164.

Harrison, P.J., Heath, P.R., Eastwood, S.L., Burnet, P.W.J., McDonald, B. and Pearson, R.C.A. (1995) The relative importance of premortem acidosis and postmortem interval for human brain gene expression studies: selective mRNA vulnerability and comparison with their encoded proteins. Neurosci. Lett., 200: 151–154.

Healy, D.J., Sima, A.A.F., Tapp, A., Watson, S.J. and Meador-Woodruff, J.H. (1996). J. Psychiat. Res., 30: 45–49.

Hynd, M.R., Lewohl, J.M., Scott, H.L. and Dodd, P.R. (2003) Biochemical and molecular studies using human autopsy brain tissue. J. Neurochem., 85: 543–562.

Inoue, H., Kimura, A. and Tuji, T. (2002) Degradation profile of mRNA in a dead rat body: basic semi-quantification study. Forensic Sci. Internat., 130: 127–132.

Johnson, S.A., Morgan, D.G. and Finch, C.E. (1986) Extensive postmortem stability of RNA from rat and human brain. J. Neurosci. Res., 16: 267–280.

Johnston, N.L., Cervenak, J., Torrey, E.F. and Yolken, R. (1997) Multivariate analysis of RNA levels from postmortem human brains as measured by three different methods of RT-PCR. J. Neurosci. Methods, 77: 83–92.

Johnston-Wilson, N.L., Sims, C.D., Hofmann, J-P., Anderson, L., Shore, A.D., Torrey, E.F. and Yolken, R.H. (2000) Disease-specific alterations in frontal cortex brain proteins in schizophrenia, bipolar disorder, and major depressive disorder. Mol. Psychiatry, 5: 142–149.

Karry, R., Klein, E. and Ben Shachar, D. (2004) Mitochondrial complex 1 subunits expression is altered in schizophrenia: a postmortem study. Biol. Psychiatry, 55: 676–684.

Kato, T. (2005) Mitochondrial dysfunction in bipolar disorder: from ^{31}P-magnetic resonance spectroscopic findings to their molecular mechanisms. International Rev. Neurobiol., 63: 21–40.

Katsel, P.L., Davis, K.L. and Haroutunian, V. (2005) Large scale microarray studies of gene expression in multiple regions of the brain in schizophrenia. Int. Rev. Neurobio., 63: 41–82.

Kim, S.I., Voshol, H., van Oostrum, J., Hastings, T.G., Cascio, M. and Glucksman, M.J. (2004) Neuroproteomics: expression profiling of the brain's proteomes in health and disease. Neurochem. Res., 29: 1317–1331.

Kingsbury, A.E., Foster, O.J.F., Nisbet, A.P., Cairns, N., Bray, L., Eve, D.J., Lees, A.J. and Marsden, C.D. (1995) Tissue pH as an indicator of mRNA preservation in human post-mortem brain. Mol. Brain Res., 28: 311–318.

Kittel, D.A., Hyde, T.M., Herman, M.M. and Kleinman, J.E. (1999) The collection of tissue at autopsy: practical and ethical issues. In: Dean, B., Kleinman, J.E. and Hyde, T.M. (Eds.), Using CNS Tissue in Psychiatric Research: A Practical Guide. Harwood Academic Publishers, Amsterdam, The Netherlands, pp. 1–18.

Kleinman, J.E., Hyde, T.M. and Herman, M.M. (1995) Methodological issues in the neuropathology of mental illness. In: Bloom, F.E. and Kupfer, D.J. (Eds.), Psycopharmacology: The Fourth Generation of Progress. Raven Press, Ltd., New York, pp. 859–864.

Knable, M.B., Barci, B.M., Bartko, J.J., Webster, M.J. and Torrey, E.F. (2002a) Molecular predictors of major psychiatric illness: classification and regression tree (CART) analysis of post-mortem prefrontal markers. Mol. Psychiatry, 7: 392–404.

Knable, M.B., Barci, B.M., Bartko, J.J., Webster, M.J. and Torrey, E.F. (2002b) Abnormalities of the cingulate gyrus in bipolar disorder and other severe psychiatric illness: post-mortem findings from the Stanley Foundation Neuropathology Consortium and literature review. Clin. Neurosci. Res., 2: 171–181.

Knable, M.B., Barci, B.M., Webster, M.J. and Torrey, E.F. (2001a) Summary of prefrontal molecular abnormalities in the Stanley Foundation Neuropathology Consortium. In: Agam, G., Everall, I. and Belmaker, R.H. (Eds.), The Post-Mortem Brain in Psychiatric Research. Kluwer, Boston, pp. 105–137.

Knable, M.B., Barci, B.M., Webster, M.J. and Torrey, E.F. (2004) Molecular abnormalities of the hippocampus in severe psychiatric illness: Post-mortem findings from the Stanley Neuropathology Consortium. Mol. Psychiatry, 9: 609–620.

Knable, M.B., Torrey, E.F., Webster, M.J. and Bartko, J. (2001b) Multivariate analysis of prefrontal cortical data from the Stanley Foundation Neuropathology Consortium. Brain Res. Bull., 55: 651–660.

Konradi, C., Eaton, M., MacDonald, M.L., Walsh, J., Benes, F.M. and Heckers, S. (2004). Molecular evidence for mitochondrial dysfunction in bipolar disorder, 61: 300–308.

Kruijver, F.P., Fernandez-Guasti, A., Fodor, M., Kraan, E.M. and Swaab, D.F. (2001) Sex differences in androgen receptors of the human mamillary bodies are related to endocrine status rather than to sexual orientation or transsexuality. J. Clin. Endocrinol. Metab., 86: 818–827.

Law, A., Shannon-Weickert, C., Webster, M.J., Herman, M.M., Kleinman, J.E. and Harrison, P.J. (2003) Expression of NMDA receptor NR1, NR2A and NR2B subunit mRNAs during development of the human hippocampal formation. Eur. J. Neurosci., 18: 1197–1205.

Leonard, S., Logel, J., Luthman, D., Casanova, M., Kirch, D. and Freedman, R. (1993) Biological stability of mRNA isolated from human postmortem brain collections. Biol. Psychiatry, 33: 456–466.

Lewis, D.A. (2002) The human brain revisited: Opportunities and challenges in postmortem studies of psychiatric disorders. Neuropsychopharm, 26: 143–154.

Lewohl, J.M., van Dyk, D.D., Craft, G.E., Innes, D.J., Mayfield, R.D., Cobon, G., Harris, R.A. and Dodd, P.R. (2004) The application of proteomics to the human alcoholic brain. Ann. N.Y. Acad. Sci., 1025: 14–26.

Li, J.Z., Vawter, M.P., Walsh, D.M., Tomita, H., Evans, S.J., Choudary, P.V., Lopez, J.F., Avelar, A., Shokoohi, V., Chung, T., Mesarwi, O., Jones, E.G., Watson, S.J., Akil, H., Bunney, W.E. and Meyers, R.M. (2004) Systematic changes in gene expression in postmortem human brains associated with tissue pH and terminal medical conditions. Hum. Mol. Genet., 13: 609–616.

Lubec, G., Krapfenbauer, K. and Fountoulakis, M. (2003) Proteomics in brain research: potentials and limitations. Prog. Neurobiol., 69: 193–211.

Miller, C.L., Diglisic, S., Leister, F., Webster, M.J. and Yolken, R.H. (2004) Evaluating RNA status for RT/PCR in extracts of postmortem human brain tissue. BioTechniques, 36: 628–633.

Mirnics, K., Middleton, F.A., Lewis, D.A. and Levitt, P. (2001) Analysis of complex brain disorders with gene expression microarrays: schizophrenia as a disease of the synapse. Trends Neurosci., 24: 479–486.

Newcombe, J., Eriksson, B., Ottervald, J., Yang, Y. and Franzen, B. (2005) Extraction and proteomic analysis of protein from normal and multiple sclerosis postmortem brain. J. Chromatog. B, 815: 191–201.

Nichols, N.R., Day, J.R., Laping, N.J., Johnson, S.A. and Finch, C.E. (1993) GFAP mRNA increases with age in rat and human brain. Neurobiol. Aging, 14: 421–429.

Perret, C.W., Marchbanks, R.M. and Whatley, S.A. (1988) Characterisation of messenger RNA extracted post-mortem from the brains of schizophrenic, depressed and control subjects. J. Neurol. Neurosurg. Psych., 51: 325–331.

Prabakaran, S., Swatton, J.E., Ryan, M.M., Huffaker, S.J., Huang, J.T.-J., Griffin, J.L., Wayland, M., Freeman, T., Durbridge, F., Lilley, K.S., Karp, N.A., Hester, S., Trachev, D., Mimmack, M.M., Yolken, R.H., Webster, M.J., Torrey, E.F. and Bahn, S. (2004) Mitochondrial dysfunction in schizophrenia: evidence for compromised brain metabolism and oxidative stress. Mol. Psychiatry, 9: 684–697.

Preece, P. and Cairns, N.J. (2003) Quantifying mRNA in postmortem human brain: influence of gender, age at death, postmortem interval, brain pH, agonal state and inter-lobe mRNA variance. Mol. Brain Res., 118: 60–71.

Romanczyk, T.B., Shannon Weickert, C., Webster, M.J., Akil, M., Herman, M.M. and Kleinman, J.E. (2002) Full-length trkB mRNA increases during young adulthood and decreases during aging in the human prefrontal cortex. Eur. J. Neurosci., 15: 269–280.

Ryan, M.M., Lockstone, H.E., Wayland, M.T., Huffacker, S.J., Webster, M.J. and Bahn, S. (2006) Gene expression analysis of bipolar disorder reveals down-regulation of the ubiquitin cycle and alterations in synaptic genes. Molecular Psychiatry (In Press).

Schonberger, S.J., Edgar, P.F., Kydd, R., Faull, R.L.M. and Cooper, G.J.S. (2001) Proteomic analysis of the brain in Alzheimer's disease: molecular phenotype of a complex disease process. Proteomics, 1: 1519–1528.

Shankar, S.K. and Mahadevan, A. (1999) Relevance of human brain banking in neuroscience — a national facility. Ann. Ind. Acad. Neurol., 2: 59–70.

Swatton, J.E., Prabakaran, S., Karp, N.A., Lilley, K.S. and Bahn, S. (2004) Protein profiling of human postmortem brain using 2-dimensional fluorescence difference gel electrophoresis (2-D DIGE). Mol. Psychiatry, 9: 1–16.

Tomita, H., Vawter, M.P., walsh, D.M., Evans, S.J., Choudary, P.V., Li, J., Overman, K.M., Atz, m.E., Myers, R.M., jones, E.G., Watson, S.J., Akil, H. and Bunney, W.E. (2004) Effects of agonal and postmortem factors on gene expression profile: quality control in microarray analysis of postmortem human brain. Biol. Psychiatry, 55: 346–352.

Torrey, E.F., Webster, M., Knable, M., Johnston, N. and Yolken, R.H. (2000) The Stanley Foundation brain collection and neuropathology consortium. Schizophr. Res., 44: 151–155.

Torrey, E.F., Barci, B.M., Webster, M.J., Bartko, J.J., Meader-Woodruff, J.H. and Knable, M.B. (2005) Neurochemical markers for schizophrenia, bipolar disorder, and major depression in postmortem brains. Biol. Psychiatry, 57: 252–260.

Trotter, S.A., Brill, L.B. and Bennett, J.P. (2002) Stability of gene expression in postmortem brain revealed by cDNA gene array analysis. Brain Res., 942: 120–123.

Vonsattel, J.P.G., Aizawa, H., Ge, P., DiFiglia, M., McKee, A.C., MacDonald, M., Gusella, J.F., Landwehrmeyer, G.B., Bird, E.D., Richardseon, E.P. and Hedley-Whyte, E.T. (1995) An improved approach to prepare human brains for research. J. Neuropath. Exp. Neurol., 54: 42–56.

Voshol, H., Glucksman, M.J. and Oostrum, J. (2003) Proteomics in the discovery of new therapeutic targets for psychiatric disease. Curr. Mol. Med., 3: 447–458.

Webster, M.J., Knable, M.B., O'Grady, J., Orthmann, J. and Shannon Weickert, C. (2002a) Regional specificity of brain glucocorticoid receptor mRNA alterations in subjects with schizophrenia and mood disorders. Mol. Psychiatry, 7: 985–994.

Webster, M.J., Shannon Weickert, C., Herman, M.M. and Kleinman, J.E. (2002b) BDNF mRNA expression during postnatal development of the human prefrontal cortex. Dev. Brain Res., 139: 139–150.

Yasojima, K., McGeer, E.G. and McGeer, P.L. (2001) High stability of mRNAs postmortem and protocols for their assessment by RT-PCR. Brain Res. Protoc., 8: 212–218.

Yates, C., Butterworth, J., Tennant, M.C. and Gordon, A. (1990) Enzyme activities in relation to pH and lactate in postmortem brain in Alzheimer-type and other dementias. J. Neurochem., 55: 1624–1630.

CHAPTER 2

Functional genomic methodologies

Stephen D. Ginsberg[1,2,]* and Károly Mirnics[3,4]

[1]*Center for Dementia Research, Nathan Kline Institute, New York University School of Medicine, 140 Old Orangeburg Road, Orangeburg, NY 10962, USA*
[2]*Department of Psychiatry, and Department of Physiology and Neuroscience, New York University School of Medicine, 140 Old Orangeburg Road, Orangeburg, NY 10962, USA*
[3]*Department of Psychiatry, School of Medicine, University of Pittsburgh, Pittsburgh, PA 15261, USA*
[4]*Department of Neurobiology, School of Medicine, University of Pittsburgh, Pittsburgh, PA 15261, USA*

Abstract: The ability to form tenable hypotheses regarding the neurobiological basis of normative functions as well as mechanisms underlying neurodegenerative and neuropsychiatric disorders is often limited by the highly complex brain circuitry and the cellular and molecular mosaics therein. The brain is an intricate structure with heterogeneous neuronal and nonneuronal cell populations dispersed throughout the central nervous system. Varied and diverse brain functions are mediated through gene expression, and ultimately protein expression, within these cell types and interconnected circuits. Large-scale high-throughput analysis of gene expression in brain regions and individual cell populations using modern functional genomics technologies has enabled the simultaneous quantitative assessment of dozens to hundreds to thousands of genes. Technical and experimental advances in the accession of tissues, RNA amplification technologies, and the refinement of downstream genetic methodologies including microarray analysis and real-time quantitative PCR have generated a wellspring of informative studies pertinent to understanding brain structure and function. In this review, we outline the advantages as well as some of the potential challenges of applying high throughput functional genomics technologies toward a better understanding of brain tissues and diseases using animal models as well as human postmortem tissues.

Keywords: microarray; RNA amplification; gene expression; molecular fingerprint; QPCR; transcriptome; brain

Introduction

Conventional molecular biology methods have enabled the research community to assess gene expression levels over a multitude of paradigms including normative conditions, development, experimental manipulations, and pathological states. These methods include Southern analysis (DNA detection), Northern analysis (RNA detection), polymerase-chain reaction (PCR; DNA detection), reverse transcriptase-PCR (RT-PCR; RNA detection), ribonuclease (RNase) protection assay (RNA detection), and in situ hybridization (RNA detection) among others. These well-accepted methods typically quantitate the abundance of individual transcripts or molecules one at a time (or a few at a time). Developments in high throughput functional genomic technologies enable the assessment of dozens to hundreds to thousands of genes simultaneously in a coordinated and potentially stochastic fashion (Schena et al., 1995; Lockhart et al., 1996; Brown and Botstein, 1999; Eisen and Brown, 1999). Transcriptome profiling exper-

*Corresponding author. Tel.: +1-845-398-2170; Fax: +1-845-398-5422; E-mail: ginsberg@nki.rfmh.org

iments using microarrays are data-driven approaches with the potential power to uncover unanticipated relationships between gene expression alterations and experimental manipulations or neuropsychiatric/neuropathological conditions. One of the difficulties of applying microarray and related genomic studies to neuroscience-based disciplines is that many classes of transcripts as well as individual genes tend to be expressed at relatively low levels (with corresponding low copy number for specific genes). In addition, gene expression patterns, or mosaics, often are cell-type specific and reflect the intricate connectivity of the brain, making global studies of gene expression patterns difficult to interpret and the relevance of performing these types of studies comes into question (Galvin, 2004; Ginsberg et al., 2004).

High-throughput expression profiling technologies have effectively altered the experimental design of gene expression assessments (Schena et al., 1995; Velculescu et al., 1995; Lockhart et al., 1996; Brenner et al., 2000a; Mirnics and Pevsner, 2004; Ginsberg et al., 2006c). Microarray platforms are at the center of this paradigm shift, as they provide an affordable, simultaneous assessment of expression levels of a sample preparation in a single experiment. For example, microarray analysis has been utilized in the hippocampus and basal ganglia for a variety of downstream genetic analyses in animal models, including studies of alcohol consumption (Saito et al., 2002), aging (Blalock et al., 2003), amyloid overexpression (Dickey et al., 2003; Reddy et al., 2004), antidepressant administration (Drigues et al., 2003), caloric restriction (Weindruch et al., 2001; Park and Prolla, 2005; Prolla, 2005), development (Elliott et al., 2003), electroconvulsive seizures (Altar et al., 2004), epilepsy (Becker et al., 2003; Newton et al., 2003), exercise (Molteni et al., 2002), experimental parkinsonism (Napolitano et al., 2002; Greene et al., 2005), fear conditioning (Mei et al., 2005), hippocampal cytoarchitecture (Zhao et al., 2001; Lein et al., 2004), hypergravity (Del Signore et al., 2004), hypoxia/ischemia (Gilbert et al., 2003; Qiu et al., 2004), inbred mouse strains (Fernandes et al., 2004), learning and memory (Cavallaro et al., 2002a, b), nerve agent exposure (Blanton et al., 2004), perforant path transection (Ginsberg, 2005a), and traumatic brain injury (Marciano et al., 2002; Matzilevich et al., 2002). Regional analyses (including dorsolateral prefrontal cortex, hippocampus, and substantia nigra) of gene expression has also been performed on postmortem human brain tissues including studies assessing Alzheimer's disease (AD) (Loring et al., 2001; Colangelo et al., 2002; Blalock et al., 2004; Lukiw, 2004; Katsel et al., 2005), epilepsy (Becker et al., 2002, 2003; Lukasiuk and Pitkänen, 2004), Parkinson's disease (PD) (Grunblatt et al., 2004; Hauser et al., 2005; Lu et al., 2005; Miller et al., 2006), and schizophrenia (Mirnics et al., 2000; Hakak et al., 2001; Middleton et al., 2002; Mimmack et al., 2002; Pongrac et al., 2002). An advantage of global and regional gene expression is that additional RNA amplification protocols are not necessary to generate significant hybridization signal intensity for array platforms. However, an obvious disadvantage is the lack of single cell resolution, as multiple subpopulations of neurons, nonneuronal cells, vascular elements, and epithelial cells will contribute to the total input RNA source (Ginsberg et al., 1999b, 2004; Galvin and Ginsberg, 2004; Kawasaki, 2004).

Input sources of RNA

The potential to understand gene changes under normal and pathophysiological conditions has expanded exponentially with the implementation of relatively new high-throughput expression analyses. Microarray analysis has emerged as an important and useful tool to evaluate gene transcript levels in a myriad of systems and paradigms. One drawback to these technologies is that significant amounts of high-quality input sources of RNA are required in order to achieve results with the desired precision and reproducibility. Whole organism and regional studies can generate significant input amounts of RNAs without any amplification procedures (Shaulsky and Loomis, 2002; McMurray and Gottschling, 2004; Fabrizio et al., 2005), however the quantity of RNA harvested from a single cell, estimated to be approximately 0.1–1.0 pg, is insufficient to be harvested accurately from standard RNA extraction procedures (Phillips and Eberwine,

1996; Sambrook and Russell, 2001). Thus, methods have been developed to increase the amount of input starting material for downstream genetic analyses including exponential PCR-based analyses and linear RNA amplification procedures.

Input sources of RNA can originate from a variety of in vivo and in vitro sources. Fresh, frozen, and fixed tissues are useful to varying degrees depending on the paradigm and tissue quality. For example, many laboratories isolate genomic DNA and total RNA from paraffin-embedded fixed tissues as well as fresh and frozen tissues (Fend et al., 1999; Tanji et al., 2001; Kabbarah et al., 2003; Su et al., 2004; Kinnecom and Pachter, 2005). In addition, genetic material preserved by fixation is a great resource because fresh and/or frozen tissues are not frequently available, whereas archived fixed human tissues exist in many brain banks and individual laboratories. With mRNA as a starting material, it cannot be emphasized enough the importance of preservation of RNA integrity in tissues and cells for downstream genetic analyses. RNA species are particularly sensitive to RNase degradation. RNases are found in virtually every cell type and are quite stable over a broad pH range (Farrell Jr, 1998). Therefore, it is prudent to employ RNase-free precautions during all RNA extraction and potential amplification procedures. In addition, extensive discussion centers on devising optimal methods to prepare brain tissues for downstream genetic analyses. A consensus on which protocol should be utilized has not yet been achieved. RNAs have been mined from tissue samples using cross-linking fixatives including 10% neutral buffered formalin and 4% paraformaldehyde as well as precipitating fixatives such as 70% ethanol buffered with 150 mM sodium chloride (Goldsworthy et al., 1999; Tanji et al., 2001; Ginsberg and Che, 2002; Su et al., 2004). A useful method for assessing RNA quality in tissue sections prior to performing expression-profiling studies is acridine orange (AO) histofluorescence. AO is a fluorescent dye that intercalates selectively into nucleic acids and has been used to detect RNA and DNA in brain tissues (Ginsberg et al., 1997; Mai et al., 1984; Mufson et al., 2002; Vincent et al., 2002). AO can be employed in combination with immunocytochemistry to identify cytoplasmic RNAs and specific antigens of interest, and is compatible with confocal microscopy (Tatton and Kish, 1997; Ginsberg et al., 1998). In brain tissue sections, AO-positive neurons are in contrast to the pale background of white matter tracts that lack abundant nucleic acids. Nonneuronal cells tend to have less AO histofluorescence as compared to neurons and brain tumor cells (Sarnat et al., 1987). Importantly, individual RNA species (e.g., mRNA, rRNA, and tRNA) cannot be delineated by AO histofluorescence. Rather, this method provides a diagnostic test that is employed on adjacent tissue sections to ensure the likelihood that an individual case has abundant RNA prior to performing expensive expression-profiling studies. A more thorough examination of RNA quality can be obtained via bioanalysis (e.g., 2100 Bioanalyzer, Agilent Technologies), which utilizes capillary gel electrophoresis to detect RNA quality and quantitate relative abundance (Che and Ginsberg, 2004, 2006; Ginsberg and Che, 2004). Bioanalysis displays the analytical assessment of RNAs in an electropherogram and/or digital gel format, with relatively high sensitivity. Bioanalysis platforms can also evaluate DNA and protein quality and abundance (Freeman and Hemby, 2004; Ginsberg et al., 2006c).

Gene expression profiling: toward an informed choice

Most high-throughput transcriptome-profiling studies are expensive, labor intensive, and time consuming. In order to generate meaningful datasets, an appropriate experimental design must be in place prior to starting the experimental series. A sound experimental design necessitates an informed choice of the experimental method, thus enabling reasonable assumptions about the investigated experimental system. To ensure generation of interpretable data, method selection and the necessary assumptions must be based on critical evaluation of all available information, and they will vary inherently from experiment to experiment.

It is important to reiterate that regional and cellular sources of mRNAs and proteins are different, and it is not unreasonable to demonstrate that transcriptome and proteome analyses of the same region can yield divergent data (Pongrac et al., 2002;

Mirnics and Pevsner, 2004). Axons projecting into distant sites traffic synthesized, cell-specific proteins. Thus, in a proteomic analysis, altered protein levels may be a result of disturbances in a distant region. In contrast, mRNA species originate principally from the somatodendritic cellular domains that reside in the harvested region. This divergence can be illustrated by two hypothetical examples. In the first example, dopamine transporter (DAT) levels are decreased in the proteomic analysis of the prefrontal cortex (PFC) yet, transcriptome analyses will barely detect the presence of DAT in the same tissue block of PFC, as somata residing in the PFC do not express DAT mRNA at high enough levels to be detected conclusively in regional microarray experiments. In the second example, transcriptome studies reveal a strong down regulation of synapsin I (SYN1) message in the PFC, but proteomic studies do not uncover such a change. Although one interpretation is that transcript changes do not necessarily translate into protein changes, an alternative explanation must also be considered that the intrinsic protein production of the brain region has decreased, but it is masked by the massive, unchanged levels of SYN1 from extrinsic sources (e.g., synaptic terminals of projection neurons residing in distant brain regions). In summary, as transcriptome and proteome studies measure different cellular compartments and different cellular events, they should be viewed as complementary approaches. When a principal investigator (PI) considers using transcriptome-profiling approaches to compare mRNA composition of control and experimental conditions, several pertinent questions arise that are addressed in the sections below.

When a high-throughput genomic method is appropriate for an experiment

Exploratory experiments, where an initial hypothesis is too generic or incomplete, are well suited for high-throughput transcriptome analyses. In this approach, the investigation focuses on scanning significant portions of the genome, and the experimental outcome relies on data-driven analyses. These types of datasets can be very informative, but generating a high-quality dataset can be costly, time consuming, and requires considerable expertise in functional genomics methodologies. Furthermore, regardless of the statistical significance, critical portions of the datasets will have to be verified by independent methods. Finally, the outcome of transcriptome studies can be platform-dependent, and negative findings must me interpreted with caution. Importantly, classical analytical methods of mRNA analysis (e.g., Northern hybridization, RNase protection assays, and in situ hybridization, among others) are not replaced with high-throughput transcription-profiling tools (Mirnics et al., 2001). Rather, these low-throughput methods often have important advantages over the high-throughput assays, especially when a specific hypothesis is being tested. For example, Northern analysis can uncover splice variants of specific transcripts, and in situ hybridization provide information regarding cellular distribution of the transcripts in question, potentially uncovering valuable information that is not gleaned routinely via high-throughput expression profiling using microarray platforms.

Sample size and transcript representation considerations

Once it becomes clear that the sample size and/or the number of investigated genes mandate a high-throughput transcriptome analysis approach, the strength and weaknesses of each available method should be considered. For example, techniques such as serial analysis of gene expression (SAGE) (Velculescu et al., 1995) or massively parallel signature sequencing (MPSS) (Oudes et al., 2005), which generate exhaustive and precise datasets, may be best suited for an in depth profiling a few precious samples originating from a rare human brain disease with unknown etiology/pathophysiology (Mirnics et al., 2001). In contrast, analyzing the expression levels of a 30-member transcript family across 400 samples from 10 different human brain diseases can be most efficiently achieved by qPCR (Bustin, 2002; Mimmack et al., 2004). Moreover, microarray analysis may be better suited for analyzing gene expression of 40 brain samples for 20,000 transcripts of interest across experimental and control conditions.

Level of sensitivity to detect the molecules of interest

Getting familiar with the advantages and limitation of expression profiling methods is essential for making an informed choice. Bulk profiling of brain tissue is inherently insensitive (Ginsberg et al., 2004, 2006c). Anatomical information and cell-type specific distribution of transcripts is lost during the material harvest, and the isolated RNA contains an admixture of glial, interneuronal, and projection neuron transcripts. Furthermore, mRNAs specific to rare cell types are diluted by higher levels of mRNA species originating from more abundant cell types. As a result, in a complex sample many of the rare transcripts can be diluted beyond detection (Mirnics et al., 2001; Pongrac et al., 2002; Mirnics and Pevsner, 2004). For example, what will be the final dilution of an mRNA species that is expressed at 10 copies/cell in a rare neuronal subtype with an abundance of one out of 2000 harvested cells? First, the desired mRNA species will be diluted by highly expressed transcripts from the cell of interest (e.g., actin, housekeeping genes, among other abundant transcripts), making initial dilutions in the range of 1:10,000 to 1:30,000 for the transcript of interest. This is followed by a second dilution, as only one out of 2000 cells express the rare transcript in the tissue, suggesting that at the whole tissue mRNA level the transcript of interest is present only in approximately one copy out of 20,000,000 mRNA molecules. Unfortunately, reliably measuring the abundance of such lowly expressed transcripts is beyond the sensitivity of the current high-throughput expression profiling methods. To remedy this, the cell type of interest must be enriched in the sample prior to expression profiling. Regional and cellular enrichment is often performed by targeted harvesting via microaspiration (Ginsberg et al., 2000; Ginsberg, 2005b; Hemby et al., 2002, 2003; Mufson et al., 2002), laser capture microdissection (Vincent et al., 2002; Kamme et al., 2003; Ginsberg and Che, 2004; Ginsberg et al., 2006c), or fluorescent automated cell-sorting (FACS) techniques (Galbraith et al., 2004; Bouchard et al., 2005).

Ability to detect an expression difference between two conditions

This is a difficult issue that demands in-depth assessment. For example, analysis of bulk tissue does not ensure that different cell types are equally affected by the disease process (Mirnics et al., 2001, 2006; Pongrac et al., 2002). For example, if 25% of the dopaminergic cells within the midbrain lose 50% of the tyrosine hydroxylase transcript, actual levels in bulk tissue isolates will be reduced by only 12.5% (e.g., 25% of 50%), and most of the high-throughput genomic methods will not be able to reliably uncover such changes. Moreover, transcriptome variability in the brains of normal control subjects is considerable (Lu et al., 2004; Tomita et al., 2004; Vawter et al., 2004), and appear to depend on both genetic (e.g., mutations, single-nucleotide polymorphisms (SNPs) and epigenetic factors (e.g., lifestyle, hormonal status, environmental influences, among others). Furthermore, human brain disease transcriptome data gathered to date underscore the observation that not all diseased individuals display identical transcriptome profile mosaics. Rather, human brain disorders under a single diagnosis (e.g., schizophrenia) encompass a continuum of clinical, genetic, and molecular phenotypes (Mirnics and Lewis, 2001; Mirnics et al., 2001, 2006). As individual variability in expression levels is considerable even in normal controls, and not all diseased subjects may show the same expression-level deficits, performing power calculations based upon reasonable assumptions become critical. Namely, assuming variance X, and a disease effect of 20% on the gene expression change, how many subjects do I need to analyze in order to detect at a 80–90% chance level the suspected gene expression changes? Such calculations are routinely performed for microarray studies, and determining the required sample size with the assistance of a statistician versed in microarray study design should be an integral part any experimental design.

In addition, even miniscule cross-hybridization between gene probes may prevent the discovery of bona fide expression changes in microarray experiments (Pongrac et al., 2002; Hollingshead et al., 2005; Mirnics et al., 2006). For example, if a dopamine 4 receptor (Unigene-NCBI annotation DRD4) probe is binding 0.5% of a glyceraldehyde-3-phosphate dehydrogenase (GAPDH) transcript, and if the GAPDH transcript is 500 times more abundant in the bulk sample than the DRD4 mRNA, the majority of the DRD4 probe signal will be derived from the cross-hybridizing GAPDH signal. In this case, although the DRD4 probe had 99.5% specificity for the DRD4 transcript, any DRD4 receptor expression change will be masked by the GAPDH signal.

Magnitude of expression-level changes in the brain

Wide variability exists in the ability to detect significant expression-level differences across organ systems and pathophysiological conditions. For example, in cancer research alterations of 3–100-fold are routinely observed via microarray analysis (Rickman et al., 2001; Brazma and Vilo, 2000; Jarvinen et al., 2004), whereas microarray analysis in brain typically yields expression-level changes of range 1.2–2.0-fold, with rare and/or occasional transcripts displaying larger fold differences in the 4–8-fold range (Colantuoni et al., 2000; Pongrac et al., 2002; Hollingshead et al., 2005; Mirnics et al., 2001, 2006). Thus, there is typically little room for measurement error in brain research paradigms. The experimental design will ideally incorporate a number of precautions that will reduce technical variability and enable valuable, predictive statistical analyses. These measures may include (but are not limited to) employing microarrays synthesized from the same production lot, processing tissue samples through a single operator, utilizing one hybridization station, using only one batch of buffers and fluorescent labels per experiment, and parallel processing pairs of experimental and control samples.

Analysis of transcripts in human postmortem brain tissue

Postmortem tissue is well suited for transcriptome profiling as evidenced by microarray analysis on a wide variety of neurodegenerative and neuropsychiatric disorders using both regional and single-cell tissue accession methods (Mirnics et al., 2000; Mirnics and Lewis, 2001; Hemby et al., 2002; Mufson et al., 2002; Ginsberg et al., 2006a, b, c). mRNAs in the brain are preserved at least up to 24–36 hours after death, and postmortem intervals of up to 72 h have been used successfully in functional genomics-based studies Leonard et al., 1993; (Harrison et al., 1995; Van Deerlin et al., 2002a; Tomita et al., 2004). However, RNA can be degraded due to a myriad of factors, and fragmented RNA does not generate reliable data. Important variables including agonal state, antemortem characteristics, clinical diagnostic criteria, duration of fixation, length of storage, and postmortem interval merit consideration when employing postmortem human tissues. Agonal state refers to the time between the terminal phase of onset of an illness and death. The agonal state can have profound effects on several parameters including protein degradation, RNA stability, and tissue pH (Leonard et al., 1993; Van Deerlin et al., 2000, 2002b; Bahn et al., 2001). Agonal sequelae including coma, hypoxia, and pneumonia are known to alter RNA and protein levels (Hynd et al., 2003; Li et al., 2004; Tomita et al., 2004). Even with intact RNA as an input source, prolonged agonal state can strongly bias the brain transcriptome (Tomita et al., 2004), and these samples should be avoided if possible.

Minimum starting material for functional genomic analysis

Modern molecular and cellular neuroscience approaches have enabled expression-profiling studies to take place at regional, cellular, and potentially subcellular levels (Eberwine et al., 2002; Ginsberg et al., 2006c). Therefore, expression-profiling experiments can be attempted successfully (see below) regardless of the amount of input RNA (Eberwine et al., 2001, 2002; Ginsberg and Che, 2002; Ginsberg et al., 2004). Starting material can be amplified linearly to the degree that a PI can investigate the mRNA makeup of a dendrite, neurite, or growth cone (Crino and Eberwine, 1996; Crino et al., 1998; Eberwine et al., 2002). However, sample amplification from minute amounts of starting material requires a more advanced molecular biology expertise and may result in moderately decreased follow-up assay sensitivity.

Use of biological replicates or technical replicates in an experiment

Technical replicates, performed on two independent RNA isolates from the same tissue block, will help to separate true data from false discovery. In contrast, analyzing many biological replicates will increase the statistical power of the study and reduce potential cohort bias. Simply stated, performing both technical replicates and biological replicates is optimal, providing that PI can afford to do both types of replicates in a given study, especially in qPCR and microarray experiments.

Performing a preliminary evaluation versus a full-blown microarray study

To perform a comprehensive investigation of transcript networks and molecular cascades, a PI should perform a "full" transcriptome-profiling study. This includes analyzing the correct number of samples and use of appropriate statistical analyses that incorporate false-discovery assessment (Mirnics and Pevsner, 2004; Mirnics et al., 2006). However, transcriptome studies can be also performed as pilot experiments. For example, in an exploratory microarray experiment, comparing one control and one experimental sample may provide interesting leads for future discovery. The observed differential expression is based on fold change (i.e., not on establishing probabilistic significance) and gives rise to many false-positive observations. If properly followed up, a PI can generate interesting leads and targets for future in-depth investigation. Pilot experiments "replace" statistical power with verification strategy, therefore these data must be considered false until validated by independent methods using a distinct, independently derived cohort of appropriate size.

Verification of expression-profiling analysis

Data analysis tools are evolving in parallel with transcriptome-profiling methods. Over the last several years, a large number of statistical analysis tools have been developed to support the accumulating genomics datasets (Pavlidis and Noble, 2001; Kotlyar et al., 2002; Nadon and Shoemaker, 2002; Aittokallio et al., 2003; Reiner et al., 2003). However, all analytical tools make certain assumptions and decisions, which greatly influence the final outcome of the analyses. Although none of them can achieve a perfect separation of true findings from experimental noise and their utility may differ across datasets, most of the widely used programs perform well in identifying true gene-expression changes (see discussion below). However, obtained data should be considered preliminary until a critical portion of the dataset has been verified by an independent method. We believe that in situ hybridization is, while labor intensive, an extremely informative data validation method (Singer and Ward, 1982; Mirnics and Lewis, 2001; Ausubel et al., 2005). In situ hybridization retains the cytoarchitecture of the brain, and outcomes are not affected by harvesting bias (e.g., amount of white matter versus gray matter), and the method provides information regarding cellular localization of gene expression changes (e.g., in situ hybridization results can differentiate between 100% of the cells losing 50% of the transcript of interest from 50% of the cells losing 100% of the respective transcript; Pongrac et al., 2002).

Potential for comparison of expression-profiling datasets

Sequencing-based methods (e.g., SAGE and MPSS) generate absolute data. Theoretically, datasets from these types of experiments are directly comparable, regardless of data origination (Velculescu et al., 1995; Brenner et al., 1999, 2000a). In contrast, cDNA and oligonucleotide microarray platforms generate relative expression data that depend significantly upon the preferences of the PI who conceived the experiment, generated experimental findings, and analyzed the datasets. Since results are influenced by tissue-harvesting methods, RNA isolation procedures, labeling protocols, microarray platforms, hardware, normalization procedures, and statistical analyses, the lists of differentially expressed genes based on a microarray experiment represent the *expert opinion* of the PI (Jarvinen et al., 2004; Qin and Kerr, 2004; Shippy et al., 2004; Hollingshead et al., 2005; Irizarry et al., 2005; Larkin et al., 2005; Mirnics et al., 2006). Thus, directly comparing microarray datasets, even when the same platform was used, is usually not possible. Rather, one can compare *outcomes* of experimental series, where the PI independently evaluates various datasets for gene expression-level changes. This discovery approach is supported by the NCBI data repositories, including the Gene Expression Omnibus (Ball et al., 2004a, b, c; Barrett et al., 2005; Boyle, 2005).

Interpretation of negative data

Within the majority of transcriptome-profiling approaches, especially microarray analysis, failure to find an expression profile difference of individual mRNAs should be interpreted cautiously. Many reasons exist where *bona fide* alterations in expression levels may not be revealed in any given high-throughput transcriptome-profiling experiment. In addition to the already discussed probe cross-hybridization and RNA dilution factors, poor probe performance and suboptimal RNA quality, as well as selecting inappropriate or too stringent statistical analyses may also arise to type II errors (see below). Nevertheless, negative data *does not constitute conclusive evidence* that two sample groups express comparable levels of mRNA (Hollingshead et al., 2005; Mirnics et al., 2006).

Conventional methods of analyzing gene expression: Northern hybridization

Three most commonly used classical methods for RNA quantification include Northern hybridization, RNase protection assay, and in situ hybridization histochemistry. Northern hybridization is a well-established method for assessing gene expression that begins with electrophoresis of total RNA or mRNA under denaturing conditions using an agarose-formaldehyde gel (Alwine et al., 1977; Ausubel et al., 2005). Once RNA species are separated by size, separated RNAs in the gel are transferred to a nylon or nitrocellulose membrane by upward capillary transfer. This is followed by hybridization analysis of the RNA sequences of interest using a labeled DNA or RNA probe. Detection of hybridization signal intensity is label dependent, and may include radioactivity assessment, chemiluminescence, or other methods. Dozens of samples can be analyzed simultaneously, but it is usually not recommended to assess the expression of multiple genes in the same reaction. Advantages of this method include the potential to assess the abundance of splice variants of different sizes, simplicity of procedures employed, fast turnaround of data, basic equipment needed, low cost, and re-usability. Drawbacks include low throughput, moderate sensitivity, challenging quantification (especially if different amounts of RNA species are loaded), and use of a single gene for normalization and relative expression-level quantitation.

Conventional methods of analyzing gene expression: RNase protection assay

RNase protection assay is an extremely sensitive technique for quantification of specific transcripts in solution (Melton et al., 1984; Calzone et al., 1987; Ausubel et al., 2005). A complementary, radioactive labeled in vitro transcript probe is hybridized in solution to the specific target RNA, creating a double-stranded structure. Once hybridization occurs, RNase is added to the solution. The probe–target duplex is stable, and will withstand RNase degradation, whereas single-stranded RNA species will be digested. The "probe-protected" duplex is analyzed on a high resolution denaturing polyacrylamide gel, and the hybridization signal intensity is directly proportional to the abundance of the RNA species. Advantages of this assay include very high sensitivity, measurement precision, high specificity, utilization of basic laboratory equipment, rapid turnaround, and low cost. Drawbacks include low throughput, inability to detect splice variants, relatively high-input RNA requirement per assay, and sensitivity to SNPs and mutations.

Conventional methods of analyzing gene expression: in situ hybridization

In situ hybridization is a method where a fragment of known complementary DNA, RNA or oligonucleotide is labeled with a reporter tag applied to tissue sections (e.g., 10–40 μm thick) (Brahic and Haase, 1978; Singer and Ward, 1982; Ausubel et al., 2005). The complementary probe will bind strongly to its target mRNA inside cells expressing the transcript of interest. Probe binding is visualized by exposure to autoradiographic film, dipping in photographic emulsion, peroxidase reaction, or fluorescence, and the distribution of the labeled cells can be analyzed microscopically. The main advantage of in situ hybridization lies in the power to provide information about the cellular localization of the transcript of interest. Drawbacks include the requirement of anatomically preserved tissue, need for more advanced expertise and equipment (e.g., cryostat and microscope) and inability to distinguish between multiple splice variants in the same reaction.

qPCR

Amplification of genetic signals can be performed at both DNA and RNA levels, and final amplified products are either DNA or RNA. A common method for DNA amplification is PCR (Mullis, 1990). Starting material for PCR reactions can originate from genomic DNA or cDNA reverse transcribed from RNA (e.g., RT-PCR). PCR is an effective method to amplify a DNA template. However, PCR is an exponential, nonlinear amplification, and variation can occur within individual mRNA species of different molecular mass and basepair (bp) composition. PCR-based methods tend to amplify abundant genes over rare genes and may distort quantitative relationships among gene populations (Phillips and Eberwine, 1996). Furthermore, amplified PCR products may not be proportional to the abundance of the starting material, potentially skewing relative gene expression-level comparisons. Real-time qPCR can quantitate PCR product formation during each cycle of amplification and has lessened concerns that are associated with conventional PCR methods. Other advantages of real-time qPCR include high-throughput capabilities, the ability to simultaneously multiplex reactions, enhanced sensitivity, reduced inter assay variation, and lack of post PCR manipulations (Bustin, 2002). A variety of dye chemistries are employed in qPCR assays including hydrolysis probes, molecular beacons, and double-stranded DNA-binding dyes (Kricka, 2002). A well-utilized example of a hydrolysis probe is the TaqMan assay. In this method, *Taq* polymerase enzyme cleaves a specific probe during the PCR extension phase. The PCR probe is dual-labeled with a reporter dye and a quenching dye at two separate ends. When the probe is intact (e.g., in its free form), fluorescence emission of the reporter dye is absorbed by the quenching dye by fluorescence resonance energy transfer (Giulietti et al., 2001; Shintani-Ishida et al., 2005). An increase of reporter fluorescence emission occurs when separation of the reporter and quencher dyes takes place during nuclease degradation in the PCR reaction (Cardullo et al., 1988). This process occurs during each PCR cycle, and does not interfere with the exponential accumulation of amplified product. Molecular beacons are probes that form a stem–loop structure from a single-stranded DNA molecule (Bonnet et al., 1999; Tan et al., 2005). Molecular beacons are particularly useful for identifying point mutations, as these probes can detect targets that differ by only a single nucleotide. DNA-binding dyes such as SYBR green (Invitrogen, Carlsbad, CA) incorporate selectively into double-stranded DNA. SYBR green emits undetectable fluorescence levels when it is in its free form. Once bound to double-stranded DNA, an intense fluorescent signal is emitted (Kricka, 2002). qPCR has emerged as an excellent companion technology to microarray analysis, as differential regulation of individual transcripts observed via array platforms can be tested with relative ease and quickness via qPCR-based assays (Mimmack et al., 2004).

Serial analysis of gene expression (SAGE)

SAGE is a sequencing-based analysis of overall gene expression patterns that does not require pre-existing clones or probes, and can identify and quantify expression of known and novel genes (Velculescu et al., 1995, 2000). SAGE rests on two basic principles. First, a 10 bp long oligonucleotide sequence defined by a specific restriction endonuclease at a fixed distance from the polyadenylated (poly(A)) tail of the mRNA, can uniquely identify any transcript. A 10 bp sequence theoretically gives rise to 1,048,576 different sequence combinations, and is sufficient to discriminate transcripts that encode the human genome. Second, end-to-end ligation (concatenation) of these unique oligonucleotides termed "tags" allows identification of many transcripts in a single-standard sequencing reaction. SAGE initiates with binding of the poly(A) tail of mRNA species to solid-phase oligo d(T) magnetic beads. This step is followed by cDNA synthesis directly onto the oligo d(T) beads. In the next step, cDNA is digested with the enzyme *Nla*III to reveal the closest restriction site to the oligo d(T) bead. *Nla*III digestion is predicted to occur at every 256 bp and is highly probable to be present on the vast majority of mRNA species. Next, the sample is divided into half and ligated into two different linker sequences, both containing the recognition site for *Bsm*FI, a type IIS restriction enzyme that cuts 10 bp 3′ from the *Nla*III enzyme recognition site. Thus, *Bsm*FI generates a unique oligonucleotide known as the SAGE tag (Velculescu et al., 1995, 2000). These tags are released from the beads, separated, blunted, and ligated to each other to form ditags. The ditags are amplified in a standard PCR reaction, released from the linkers, gel purified, serially ligated into concatomers, cloned, and sequenced using an automated sequencer (Velculescu et al., 1995, 2000). Transcript abundance is determined by tag count. The more often the same 10 bp tag (which corresponds to a single RNA species) is encountered, the more abundant the transcript is in the sample. Thus, tag counts are compared across different samples. The sensitivity of SAGE depends directly on the total number of uncovered tags, and this mandates sequencing of tens of thousands of concatomer clones. Recent modifications of the original SAGE technology enables the use of smaller amounts of starting material and improved library construction (e.g., SAGE-Lite, Peters et al., 1999; Vilain and Vassart, 2004 and Micro-SAGE Datson et al., 1999), as well as generation of 17 bp SAGE tags (Saha et al., 2002; Gowda et al., 2004; Heidenblut et al., 2004). Advantages of SAGE include expression assessment of the whole transcriptome including unknown transcripts and generation of immortal, sequence-based data (Brenner et al., 1999). The most-notable drawbacks of SAGE are very high cost per analyzed sample, low throughput, and slow data turnaround.

Massive parallel signature sequencing (MPSS)

The sequencing methodologies MPSS and SAGE share a number of similarities. Both rely on restriction digestion and sequence specificity of short tags from individual transcripts. Moreover, MPSS and SAGE generate immortal data, and the outcome of both methods is expressed as number of transcript of interest/million transcripts in a sample (Brenner et al., 2000a, b; Reinartz et al., 2002; Silva et al., 2004; Oudes et al., 2005). However, a typical SAGE experiment has approximately 50,000 tags sequenced, whereas MPSS generates data for around 500,000–1,000,000 sequence tags. MPSS is initiated with reverse transcribing the mRNA using a biotin-coupled oligo (d)T primer. The resultant cDNA is digested by cleaving at the GATC sequence closest to the poly(A) tail using *Dpn*II restriction digestion. This procedure is followed by ligation of an adapter of known sequence to the cut end. The mixture of adaptors includes every possible overhang that is annealed to a target sequence in order to enable only a perfectly complementary overhang to be ligated. The resulting sequences are captured onto streptavidin-coated microbeads and subjected to *Mme*I restriction digestion. This results in 20-nucleotide long fragments from each transcript. These tags are cloned into a specific vector generating a library. The cloned tags are captured onto microbeads and the sequences are amplified. The microbeads are arrayed in a flow cell, and each of the bead-bound DNA fragments are sequenced simultaneously for 20 residues using a proprietary procedure without requiring fragment separation and separate sequencing reactions (Brenner et al., 2000a; Reinartz et al., 2002). Sequence-dependent fluorescent signals from the microbeads are recorded by a CCD camera after each cycle for transcript identification. Advantages of MPSS include high sensitivity, specificity, and generation of immortal data. The main drawback is high cost of analysis per sample.

Total analysis of gene expression (TOGA)

TOGA enables identification of unique transcripts via eight nucleotide long sequences consisting of a four nucleotide restriction endonuclease cleavage site, an adjacent four nucleotide parsing sequence, and their respective distances from the 3' ends of mRNA (Sutcliffe et al., 2000; Thomas et al., 2001, 2003). The parsing sequences are used as parts of primer-binding sites in 256 PCR-based assays. Briefly, double-stranded cDNA is synthesized using a pool of *Not*I-containing biotinylated oligo (d)T primers degenerate in the 3' ultimate three nucleotide positions. After cleavage with *Msp*I, the biotinylated fragment is captured on streptavidin-coated magnetic beads and released from the beads by digestion with *Not*I. The *Msp*I–*Not*I generated fragments are cloned into an RNA expression vector. This is followed by linearization of the insert-containing plasmids and in vitro transcription with T3 RNA polymerase. A new reverse transcription follows using a primer that anneals to vector sequences. In a subsequent PCR step, the primer extends across the nonreconstituted *Msp*I/*Cla*I site by one of the four possible nucleotides and a universal 3' primer subdivides the cDNA species into four pools. A subsequent PCR in the presence of a fluorescent 3' primer and each (in separate reactions) of the 256 possible 5' primers that extends four nucleotides into the inserts subdivides the input species into 256 subpools for electrophoretic resolution. The resulting molecules are analyzed based on the length of electrophoretic bands. Thus, the final PCR products are identified based on an eight nucleotide sequence (e.g., $CCGGN_1N_2N_3N_4$) and the distance from the 3' end of the mRNA, as well as based upon the known length of vector-derived sequences (Sutcliffe et al., 2000). Advantages of TOGA include generation of sequence-based data, high sensitivity and specificity, as well as capability to identify unknown transcripts. The main drawbacks are low sample throughput, high assay cost, and requirement of technologically advanced and expensive equipment.

Sequencing by hybridization (SBH)

SBH is based on the fact that short oligonucleotide probes of known sequence preferentially hybridize with entirely complementary nucleic acid targets. Therefore, DNA sequence information can be obtained based on complementary binding rather than by actual sequencing, (Drmanac et al., 1989, 1990; Broude et al., 1994), and can be assayed subsequently by various methods. Multiple approaches of SBH have been used successfully in the past, and novel approaches are rapidly being developed (Drmanac et al., 2002; Shendure et al., 2004). In one implementation of SBH that is useful for gene expression studies, SBH is combined with a ligase reaction. For example, two sets of 6-mer nucleotides (4096 sequences for each), when end-ligated into a 12-mer, can theoretically resolve >16,000,000 unique DNA sequences. In this approach, one set of short probes is bound to a solid support, and the other set is labeled and floats in the solution, where the sample of interest is added. Once complementary hybridization occurs between the two probes and targets sample, DNA ligase is added to a reaction. If the immobilized and the floating probe are bound to adjacent regions of the sample transcript, their ends are adjacent, and the DNA ligase will ligate the two six-nucleotide long probes into a continuous 12-nucleotide long structure. This new nucleotide inherits the properties of the two six-nucleotide long probes, being anchored to a solid support, and is labeled by a fluorescent label. Fluorescence can be measured by creating a high-resolution optical image, and the signal intensity is proportional to the abundance of a transcript. By knowing the spatial arrangement of the anchored probes and labeled floating probes, the 12-nucleotide long sequence identity is matched to sequence databases, thus identifying the corresponding gene transcripts. Advantages of SBH include great method flexibility, identification of unknown transcripts, SNP and splice variants, and good resolution. Drawbacks include the high cost of instrumentation, technical expertise, requirement of advanced robotics, and low sample throughput.

Microarray platforms

Technical advances have fostered the development of high-density microarrays that allow for high-throughput analysis of hundreds to thousands of genes simultaneously. Microarrays represent miniaturized, high-density dot-blots that take advantage of complementary hybridization between nucleic acids (Schena et al., 1995; Lockhart et al., 1996; Brown and Botstein, 1999). Synthesis of cDNA microarrays entails adhering cDNAs or expressed sequence-tagged cDNAs (ESTs) to solid supports such as glass slides, plastic slides, or nylon membranes. A parallel technology uses photolithography to adhere oligonucleotides to array media (Lockhart et al., 1996). The anchored cDNA/EST/oligonucleotide is commonly called the microarray target. Target length varies from short oligonucleotides to chromosomal fragments, depending on the platform design. The sample RNA is used to generate labeled probes. When starting material is not amplified, probe generation is performed via reverse transcription, and a detection label is incorporated directly into the cDNA strand (Soverchia et al., 2005). When the amount of starting material is limited, such as homogeneous populations of cells or single cells, the RNA is amplified through an RNA amplification methodology (see below), and the amplified material is labeled and used to probe the microarrays (Ginsberg and Che, 2002; Ginsberg, 2005b). The target-labeled probe complex emits a measurable signal, and this signal is proportional to the abundance of the labeled probe in the sample. Arrays are washed to remove nonspecific background hybridization, and imaged using a laser scanner for biotinylated/fluorescently labeled probes and a phosphor imager for radioactively labeled probes. The specific signal intensity (minus background) of total (or amplified) RNA bound to each probe set (e.g., oligonucleotides or cDNAs/ESTs) is expressed as a ratio of the total hybridization signal intensity of the array, thereby minimizing variations across the array platform due to differences in the specific activity of the probe and the absolute quantity of probe present. Hybridization signal intensity is measured at high resolution over the whole microarray and stored as an optical image. In a process called image segmentation, using specific spatial coordinates, microarray images are divided into miniscule areas that correspond to each of the microarray gene targets.

Unfortunately, substantial variation exists between array platforms. There are significant differences in microarray production (e.g., deposition versus in situ synthesis), microarray target length (e.g., oligonucleotides versus cDNAs/ESTs), probe-labeling procedures (e.g., radioactive labeling, direct or indirect fluorochrome incorporation, and gold particle deposition), amplification procedures (e.g., PCR-based amplification, T7-based RNA amplification, and isothermal amplification), and a myriad of hybridization procedures. Each of these technical variables may have considerable effects on the experimental outcome (Quackenbush, 2002; Mirnics and Pevsner, 2004; Hollingshead et al., 2005).

Analyzing massive datasets

The main challenge of analyzing datasets with hundreds to thousands of data points lies in separating biologically true expression changes from experimental noise. Unfortunately, from two observations that show a similar statistical significance one may be a true biological change, while the other may represent false discovery. While perfect separation of "true" and "false" data cannot be achieved by any statistical method (Broberg, 2005), there are several approaches that can aid the PI in deciding upon the choice of statistical criteria that are most appropriate for any given dataset. Regardless of the exact statistical method used, increasing statistical stringency will decrease type I errors (false-positive findings), with a concomitant price of increasing type II errors (false-negative findings) (Mirnics et al., 2001, 2006; Unger et al., 2005a, b). Statistical analyses are absolutely critical for the proper interpretation of microarray findings, and we dedicate our discussion on analysis of expression gene array data.

Implementing a dual criteria of statistical significance (e.g., $p < 0.05$) and fold change (e.g., 1.2-fold, average log2 ratio > 0.263) will result in the elimination of large number of type I errors, but will have little impact on type II errors (Mirnics and Pevsner, 2004; Unger et al., 2005b). Expression changes that are minute can often have an artificially inflated significance due to a very low variance within repeated observations. The value of such observations is limited as they rarely reproduce on a new cohort of subjects, are challenging to measure and verify, and relevance and/or biological significance is challenged when a miniscule gene expression change that may not even exceed 5% is observed.

Estimating the false-discovery rate (FDR) is required (Aubert et al., 2004; Meuwissen and Goddard, 2004; Pounds and Cheng, 2004; Lin, 2005), and no surrogate measurements should be accepted. FDR is dataset specific, and it is erroneous to assume that FDR in a new dataset is similar to that found in a previously performed study. An optimal method to assess experimental variability is by using technical replicates, performed on material that is independently isolated from the same tissue block (Lazarov et al., 2005; Mirnics et al., 2005). This is best achieved by treating the technical replicates as independent experiments, performing the same analytical procedures on them, and assessing overlap between outcomes of independent analyses of the technical replicates at a range of different statistical stringencies. Thus, the PI can establish a list of differentially expressed genes based on an analysis that yields an acceptable combination of type I and type II error rates. If technical replicates are not available, permutation analysis of the microarrays can also help generate a reasonable FDR estimate (Lazarov et al., 2005; Mirnics et al., 2005; Unger et al., 2005a, b). In this approach, data analysis is performed between control and experimental samples at various statistical stringencies. This is followed by random assignment of microarrays into two symmetrical groups containing the same number of experimental and control arrays, and performing the same statistical analyses on them as on the initial experimental data. Random assignment of microarrays into two groups and subsequent analysis are continued until all permutations are exhausted, and true experimental data is compared to that obtained in the permutation. Finally, the PI can choose a statistical stringency at which the mean FDR obtained in the permutation analysis is the smallest compared to the experimental data. Several microarray analysis software packages have a built-in permutation FDR analysis (e.g., DChip and Jacknife) (Schadt et al., 2001; Lyons-Weiler et al., 2003; Zhong et al., 2003). In addition to replication and permutation approaches, there are many other biostatistical methods that provide good FDR estimates (Aubert et al., 2004; Meuwissen and Goddard, 2004; Pounds and Cheng, 2004; Broberg, 2005; Lin, 2005;). Although analyzing the same microarray dataset with different analytical tools will give rise to somewhat divergent data, the most robust and verifiable findings usually survive regardless of the exact normalization approach or data mining procedure. Thus, focusing on the overlap of data obtained by different analytical procedures may significantly reduce false positive findings.

Regional and single cell assessment

With continued technological advances furthering the microarray discipline, it is likely that the field will shift from bulk and/or global hybridization of brain tissues to regional and cellular analyses (Ginsberg et al., 2004; Mirnics and Pevsner, 2004). Specifically, expression profiling of bulk tissues tends to yield a molecular fingerprint of the most abundant cell type(s), whereas less abundant cell types within the tissue sample are underrepresented (Szaniszlo et al., 2004; Ginsberg and Che, 2005). Targeted isolation of relevant cell types within the central nervous system will improve the discriminative power of array analysis, potentially identifying important changes in gene expression that would otherwise be undetected. The brain is a complex structure with heterogeneous neuronal populations (e.g., pyramidal neurons and interneurons) and nonneuronal cell populations (e.g., glial cells, epithelial cells, and vascular cells). Ongoing advances in molecular neuroscience provide the tools necessary to assay gene expression from specific homogeneous cell populations within defined brain regions, laminae, and nuclei without contamination from intermingled neuronal subtypes and nonneuronal cells (Eberwine et al., 2002; Vincent et al., 2002; Galvin, 2004; Kawasaki, 2004;

Ginsberg et al., 2006c). However, gene expression profiling of homogeneous populations of cells is a difficult task that demands a multidisciplinary approach. With the advent of high-throughput microdissection technologies, it is now possible to isolate and evaluate DNA, RNA, and proteins from discrete regions of brain tissues and homogeneous cellular populations. Single cells or populations of cells, either identified physiologically in living preparations (Eberwine et al., 1992; Brooks-Kayal et al., 1998; Levsky et al., 2002), or by immunocytochemical or histochemical procedures in fixed cells in vitro or in vivo (Hemby et al., 2003; Kamme et al., 2003; Ginsberg and Che, 2005; Ginsberg et al., 2006a; Luo et al., 1999) can be employed for downstream genetic analyses. RNA amplification using PCR-based analyses have been employed for microarray studies (Brail et al., 1999; Mikulowska-Mennis et al., 2002; Van Deerlin et al., 2002a), notwithstanding the caveats associated with exponential amplification procedures. In addition to PCR-based amplification technologies, several linear RNA amplification procedures have been developed that maintain relative representation of each transcript present in the original input sample mRNA. Linear RNA amplification strategies including amplified antisense RNA (aRNA) (Eberwine et al., 1992, 2002; Ginsberg et al., 1999a, 2000), isothermal amplification (NuGEN; San Carlos, CA) (Dafforn et al., 2004; Kurn et al., 2005), and terminal continuation (TC) RNA amplification (Che and Ginsberg, 2004, 2006; Ginsberg, 2005b; Ginsberg et al., 2006c) have been used in combination with population cell and single cell microdissection procedures to enable the use of microarray analysis.

RNA amplification strategies: aRNA amplification

RNA amplification is a series of intricate stepwise molecular-based methods used to amplify genomic signals in a linear fashion from minute quantities of starting materials for microarray analysis and other downstream genetic applications (Fig. 1). aRNA amplification, developed by Eberwine and colleagues (Eberwine et al., 1992, 2001; Kacharmina et al., 1999) utilizes a T7 RNA polymerase-based amplification procedure that allows for quantitation of the relative abundance of gene expression levels. aRNA products maintain a proportional representation of the size and complexity of input mRNAs (VanGelder et al., 1990; Eberwine et al., 1992). The aRNA amplification procedure includes hybridization of a 66 bp oligonucleotide primer consisting of 24 thymidine triphosphates (TTPs) and a T7 RNA polymerase promoter sequence termed oligo d(T)T7 to mRNAs and conversion to an mRNA–cDNA hybrid using reverse transcriptase (Tecott et al., 1988; VanGelder et al., 1990). Upon conversion of the mRNA–cDNA hybrid to double-stranded cDNA, a functional T7 RNA polymerase promoter is formed. aRNA synthesis occurs with the addition of T7 RNA polymerase and nucleotide triphosphates (NTPs). Expression levels are quantitated via cDNA or oligonucleotide microarrays by probing with fluorescent and/or biotin-tagged NTPs incorporated into aRNA products. In addition, radiolabeled NTPs can be incorporated into aRNA products and used for hybridization to membrane-based arrays. Each round of aRNA results in an approximate 1000-fold amplification from the original amount of each poly(A) mRNA in the sample (Eberwine et al., 1992, 2001). Two rounds are typically necessary to generate sufficient quantities of aRNA for subsequent downstream genetic analyses. aRNA products are biased toward the 3′ end of the transcript because of the initial priming at the poly(A) RNA tail (Phillips and Eberwine, 1996; Kacharmina et al., 1999). aRNA products tend not to be of full length, but relative levels of gene expression can be compared across conditions and fidelity to the proportional representation of genes from the input sample is maintained (Phillips and Eberwine, 1996; Poirier et al., 1997; Eberwine et al., 2001; Che and Ginsberg, 2006; Ginsberg et al., 2006c). Thus, it is important to select clones with 3′ transcript representation for use on a microarray platform, or to assess multiple ESTs or oligonucleotide probes linked to the same gene (Ginsberg et al., 2000, 2006b; Hemby et al., 2002). Although aRNA is a complicated series of procedures, successful results have been generated with microaspirated animal model and postmortem human brain tissue samples utilizing a wide variety of microarray platforms (Ghasemzadeh et al., 1996; Brooks-Kayal

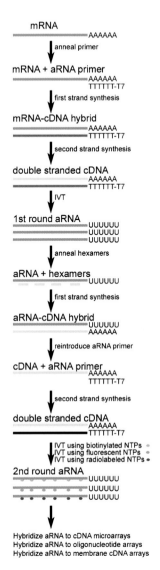

Fig. 1. Schematic representation of the aRNA amplification. An oligo d(T)T7 primer is hybridized directly to poly(A) mRNAs, and a double-stranded mRNA–cDNA hybrid is formed (first-strand synthesis) by reverse transcribing the primed mRNAs with dNTPs and reverse transcriptase. Double-stranded mRNA–cDNA hybrid is converted into double-stranded cDNA forming a functional T7 RNA polymerase promoter. A first round of aRNA synthesis occurs via in vitro transcription (IVT) using T7 RNA polymerase and NTPs. A second round of aRNA amplification occurs by annealing random hexamers to the newly formed aRNA, and performing first strand synthesis. The aRNA primer is then reintroduced which binds to the poly(A) sequence on the newly synthesized cDNA strand. A double-stranded cDNA template is formed by second-strand synthesis. A second round of aRNA products is produced by IVT using biotinylated, fluorescent, or radiolabeled NTPs. Adapted from Ginsberg (2005b) with permission from the publisher.

et al., 1998; Chow et al., 1998; Ginsberg et al., 2000; Eberwine et al., 2001; Ginsberg and Che, 2002; Hemby et al., 2002, 2003; Ginsberg, 2005a). Modifications of the original aRNA procedure have been performed (Wang et al., 2000; Luzzi et al., 2001; Iscove et al., 2002; Xiang et al., 2003; Moll et al., 2004), and several kits that use aRNA technology to amplify small amounts of RNA are commercially available.

RNA amplification strategies: isothermal amplification

A novel technology for RNA-based single-primer isothermal amplification (Ribo-SPIA) yielding high-fidelity RNA amplification for gene expression analysis has been developed and marketed by NuGEN and is of interest to neuroscientists performing microarray evaluations. Ribo-SPIA technology is different from methods that use in vitro transcription with a bacteriophage transcription promoter such as T7 polymerase. Ribo-SPIA is a linear RNA amplification procedure that entails the formation of a double-stranded cDNA as a substrate for subsequent SPIA amplification, effectively generating multiple copies of single-stranded DNA products that are complementary to the initial input mRNA source (Dafforn et al., 2004; Kurn et al., 2005) (Fig. 2). The technology employs a chimeric RNA/DNA primer consisting of a known RNA sequence tag on the 5′ end of an oligo (d)T cDNA primer. Once the chimeric DNA/RNA primer is incorporated, the resulting double-stranded cDNA contains a short RNA/DNA hybrid segment of known sequence at the 5′ end of the first strand. The SPIA reaction is initiated by incubating with RNase H, a second chimeric DNA/RNA SPIA primer, and DNA polymerase (Dafforn et al., 2004; Kurn et al., 2005). RNase H only digests the short RNA segment leaving the corresponding cDNA portion exposed. In this manner, the SPIA primer is free to hybridize to its complementary sequence and prime the DNA polymerase to initiate synthesis of another cDNA strand, effectively displacing the original cDNA strand. Cycling of this process enables high fidelity linear amplification from each cDNA template. A

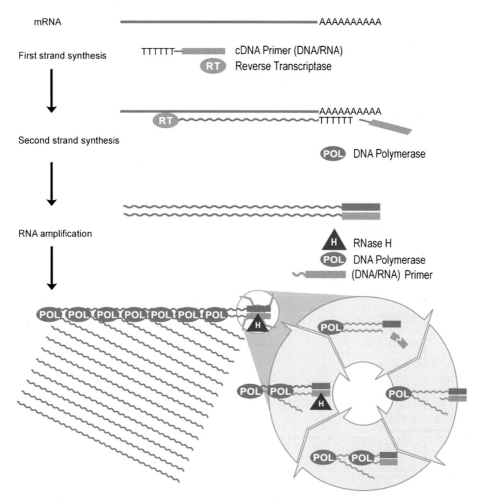

Fig. 2. Schematic representation of the isothermal RNA amplification Ribo-SPIA procedure developed by NuGEN. The novel linear and isothermal amplification method consists of first-strand cDNA synthesis employing a DNA/RNA chimeric primer, subsequent second-strand cDNA synthesis step, followed by SPIA-based cDNA amplification to generate single-stranded cDNA amplification products for downstream genetic analyses including microarray and qPCR studies. Adapted from Kurn et al. (2005), with permission.

single Ribo-SPIA amplification can generate amplified RNA for multiple hybridization experiments from less than about 5 ng of total RNA (Dafforn et al., 2004; Kurn et al., 2005). NuGEN has developed kits that use the Ribo-SPIA technology in concert with aminoallyl or biotin labeling for microarray applications, and also offers a kit that preamplifies mRNA for use in downstream qPCR studies and other functional genomics studies of gene expression from minute amounts of input RNAs.

RNA amplification strategies: terminal continuation (TC) RNA amplification

A variety of strategies have been developed by independent laboratories to improve RNA amplification efficiency (Brail et al., 1999; Matz et al., 1999; Wang et al., 2000; Zhumabayeva et al., 2001; Iscove et al., 2002). One of the principal obstacles is the problematic second-strand cDNA synthesis. This is an issue that plagues most of the current RNA amplification methods. Important facets to improving RNA amplification strategies include increasing the efficiency of second-strand cDNA synthesis and allowing for flexibility in the placement of bacteriophage transcriptional promoter sequences.

A new RNA amplification procedure has been developed that utilizes a method of TC (Che and Ginsberg, 2004, 2006; Ginsberg, 2005b). TC RNA amplification of genetic signals includes synthesizing first-strand cDNA complementary to the mRNA template, subsequently generating second-strand cDNA complementary to the first-strand cDNA, and finally in vitro transcription using the double-stranded cDNA as template (Che and Ginsberg, 2004, 2006; Ginsberg, 2005b). First-strand cDNA synthesis complementary to the template mRNA entails the use of two oligonucleotide primers, a poly d(T) primer and a TC primer. The poly d(T) primer is similar to conventional primers that exploit the poly(A) sequence present on most mRNAs. The TC primer contains a span of three cytidine triphosphates (CTPs) or guanosine triphosphates (GTPs) at the 3′ terminus (Che and Ginsberg, 2004). Adenosine triphosphates (ATPs) or TTPs do not perform well as constituents of the TC primer (Ginsberg and Che, 2004). In this configuration, second-strand cDNA synthesis can be initiated by annealing a second oligonucleotide primer complementary to the attached oligonucleotide (Che and Ginsberg, 2004), and can be performed with robust DNA polymerases, such as *Taq* polymerase. One round of amplification is sufficient for downstream genetic analyses (Che and Ginsberg, 2004; Ginsberg, 2005b). Additionally, TC RNA transcription can be performed using a promoter sequence (e.g., T7, T3, or SP6) attached to either the 3′ or 5′ oligonucleotide primers. Therefore, transcript orientation can be in an antisense orientation (similar to conventional aRNA methods) when the bacteriophage promoter sequence is placed on the poly d(T) primer or in a sense orientation when the promoter sequence is attached to the TC primer, depending upon the design of the experimental paradigm (Fig. 3) (Che and Ginsberg, 2006; Ginsberg et al., 2006c). Regional and single-cell gene expression studies within the brains of animal models and human postmortem brain tissues have been performed via microarray analysis coupled with TC RNA amplification (Ginsberg and Che, 2002, 2004, 2005; Mufson et al., 2002; Counts et al., 2006; Ginsberg et al., 2006a, b, c).

Additional considerations

The application of microarray technology in the neurosciences is still in its infancy. As such, precision and reproducibility of expression-profiling studies is a critical parameter that is improving as these methodologies become more refined and universally accepted, and research scientists become more adept at exploiting the significant potential of these high-throughput genomic techniques. Advances at the level of tissue dissection, RNA amplification, microarray platforms, and statistical applications will ultimately lead to greater utility and flexibility of these resources. For example, improvements include the utilization of pooled populations of individual cell types to reduce variability in expression levels yet maintain an expression profile for a single cell type (Ginsberg, 2005b; Ginsberg et al., 2006c). Validation of array results is critical, and several independent alternative techniques are quite useful to reproduce changes seen on an array platform such as qPCR, SAGE, MPSS, TOGA, and in situ hybridization, among others.

The identification of novel disease-related genes will eventually lead to new cell culture models and genetically engineered rodent models that may recapitulate specific aspects of disease pathogenesis that has not been evaluated previously. Furthermore, microarray analyses of these new models will be important. In this manner, the validity of models can be studied by microarray analysis. Specifically, if captured regions, populations of cells, and/or single cells obtained from cultures or animal models display expression profiles that differ significantly from the human condition it is designed to model, applicability and relevance are questioned. Ultimately, a cell culture application or animal model manipulation (e.g., genetic, environmental) may be deemed to be inappropriate for further study within the context of single cell or homogeneous population cell analysis

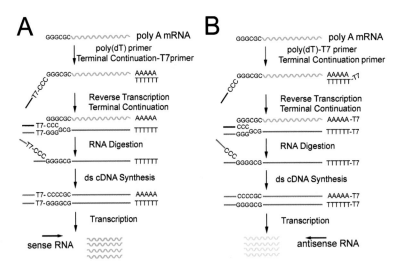

Fig. 3. Overview of the TC RNA amplification method. (A) A TC primer (containing a bacteriophage promoter sequence for sense orientation) and a poly d(T) primer are added to the mRNA population to be amplified (green rippled line). First-strand synthesis (blue line) occurs as an mRNA–cDNA hybrid is formed following reverse transcription and TC of the oligonucleotide primers. After an RNase H digestion step to remove the original mRNA template strand, second-strand synthesis (red line) is performed using *Taq* polymerase. The resultant double-stranded product is utilized as template for in vitro transcription, yielding high fidelity, linear RNA amplification of sense orientation (green rippled lines). (B) Schematic representation similar to (A) illustrating the TC RNA amplification procedure amplifying RNA in the antisense orientation (yellow rippled lines). Adapted from Che and Ginsberg (2004), with permisson.

based upon disparities in genomic and proteomic profiles from a human condition that they were designed to mimic.

Conclusions

When deciding whether or not to use functional genomic methodologies, the most crucial aspect to consider is the question the PI is interested in answering, and determining the method(s) that would be best suited to perform the experiment. Once a PI has decided that a microarray experiment is appropriate, much consideration needs to go into sample size and preparation, tissue and/or cell quality, and importantly, input amount of RNA that will likely be generated. Sample preparation, RNA amplification, array hybridization, and array analysis usually requires a long-term commitment, as many investigators have found out much to their chagrin. A qPCR, RNase protection assay, or in situ hybridization histochemistry experiment would be more useful, for instance, if a researcher is assessing the regulation of a single-gene product (or splice variants/isoforms of an individual gene family) in a particular study. Further, there is a strong impetus to supplement functional genomic data with measurements of protein expression by immunoblotting, immunocytochemical, ELISA, proteomic, or metabolomic techniques. In summary, a combination of multidisciplinary approaches is ideal for verification and broad application of gene expression level alterations, with the explicit knowledge that the sum of the evaluations may be more informative and reflect the actual biology of the system than an individual method.

Acknowledgements

We thank Dr. Nurith Kurn for helpful discussions with the isothermal amplification procedure. We thank Dr. Shaoli Che, and Ms. Irina Elarova for expert technical assistance. Support comes from the NIA (SDG, AG10668, AG14449, AG17617, and AG09466), NIMH (KM, MH067234, MH45156, and MH070786), and NINDS (SDG, NS43939, and NS48447). We also express our appreciation to the families of the patients studied here who made this research possible.

References

Aittokallio, T., Kurki, M., Nevalainen, O., Nikula, T., West, A. and Lahesmaa, R. (2003) Computational strategies for analyzing data in gene expression microarray experiments. J. Bioinform. Comput. Biol., 1: 541–586.

Altar, C.A., Laeng, P., Jurata, L.W., Brockman, J.A., Lemire, A., Bullard, J., Bukhman, Y.V., Young, T.A., Charles, V. and Palfreyman, M.G. (2004) Electroconvulsive seizures regulate gene expression of distinct neurotrophic signaling pathways. J. Neurosci., 24: 2667–2677.

Alwine, J.C., Kemp, D.J. and Stark, G.R. (1977) Method for detection of specific RNAs in agarose gels by transfer to diazobenzyloxymethyl-paper and hybridization with DNA probes. Proc. Natl. Acad. Sci. USA, 74: 5350–5354.

Aubert, J., Bar-Hen, A., Daudin, J.J. and Robin, S. (2004) Determination of the differentially expressed genes in microarray experiments using local FDR. BMC Bioinformatics, 5: 125.

Ausubel, F.M., Brent, R., Kingston, R.E., Moore, D.D., Seidman, J.G., Smith, J.A. and Struhl, K. (2005) Current Protocols in Molecular Biology. Wiley, New York.

Bahn, S., Augood, S.J., Ryan, M., Standaert, D.G., Starkey, M. and Emson, P.C. (2001) Gene expression profiling in the post-mortem human brain — no cause for dismay. J. Chem. Neuroanat., 22: 79–94.

Ball, C., Brazma, A., Causton, H., Chervitz, S., Edgar, R., Hingamp, P., Matese, J.C., Icahn, C., Parkinson, H., Quackenbush, J., Ringwald, M., Sansone, S.A., Sherlock, G., Spellman, P., Stoeckert, C., Tateno, Y., Taylor, R., White, J. and Winegarden, N. (2004a) An open letter on microarray data from the MGED Society. Microbiology, 150: 3522–3524.

Ball, C., Brazma, A., Causton, H., Chervitz, S., Edgar, R., Hingamp, P., Matese, J.C., Parkinson, H., Quackenbush, J., Ringwald, M., Sansone, S.A., Sherlock, G., Spellman, P., Stoeckert, C., Tateno, Y., Taylor, R., White, J. and Winegarden, N. (2004b) Standards for microarray data: an open letter. Environ. Health Perspect., 112: A666–A667.

Ball, C.A., Brazma, A., Causton, H., Chervitz, S., Edgar, R., Hingamp, P., Matese, J.C., Parkinson, H., Quackenbush, J., Ringwald, M., Sansone, S.A., Sherlock, G., Spellman, P., Stoeckert, C., Tateno, Y., Taylor, R., White, J. and Winegarden, N. (2004c) Submission of microarray data to public repositories. PLoS Biol., 2: E317.

Barrett, T., Suzek, T.O., Troup, D.B., Wilhite, S.E., Ngau, W.C., Ledoux, P., Rudnev, D., Lash, A.E., Fujibuchi, W. and Edgar, R. (2005) NCBI GEO: mining millions of expression profiles — database and tools. Nucleic Acids Res., 33: D562–D566.

Becker, A.J., Chen, J., Zien, A., Sochivko, D., Normann, S., Schramm, J., Elger, C.E., Wiestler, O.D. and Blumcke, I. (2003) Correlated stage- and subfield-associated hippocampal gene expression patterns in experimental and human temporal lobe epilepsy. Eur. J. Neurosci., 18: 2792–2802.

Becker, A.J., Wiestler, O.D. and Blumcke, I. (2002) Functional genomics in experimental and human temporal lobe epilepsy: powerful new tools to identify molecular disease mechanisms of hippocampal damage. Prog. Brain Res., 135: 161–173.

Blalock, E.M., Chen, K.C., Sharrow, K., Herman, J.P., Porter, N.M., Foster, T.C. and Landfield, P.W. (2003) Gene microarrays in hippocampal aging: statistical profiling identifies novel processes correlated with cognitive impairment. J. Neurosci., 23: 3807–3819.

Blalock, E.M., Geddes, J.W., Chen, K.C., Porter, N.M., Markesbery, W.R. and Landfield, P.W. (2004) Incipient Alzheimer's disease: microarray correlation analyses reveal major transcriptional and tumor suppressor responses. Proc. Natl. Acad. Sci. USA, 101: 2173–2178.

Blanton, J.L., D'Ambrozio, J.A., Sistrunk, J.E. and Midboe, E.G. (2004) Global changes in the expression patterns of RNA isolated from the hippocampus and cortex of VX exposed mice. J. Biochem. Mol. Toxicol., 18: 115–123.

Bonnet, G., Tyagi, S., Libchaber, A. and Kramer, F.R. (1999) Thermodynamic basis of the enhanced specificity of structured DNA probes. Proc. Natl. Acad. Sci. USA, 96: 6171–6176.

Bouchard, M., Grote, D., Craven, S.E., Sun, Q., Steinlein, P. and Busslinger, M. (2005) Identification of Pax2-regulated genes by expression profiling of the mid-hindbrain organizer region. Development, 132: 2633–2643.

Boyle, J. (2005) Gene-expression Omnibus integration and clustering tools in SeqExpress. Bioinformatics, 21: 2550–2551.

Brahic, M. and Haase, A.T. (1978) Detection of viral sequences of low reiteration frequency by in situ hybridization. Proc. Natl. Acad. Sci. USA, 75: 6125–6129.

Brail, L.H., Jang, A., Billia, F., Iscove, N.N., Klamut, H.J. and Hill, R.P. (1999) Gene expression in individual cells: analysis using global single cell reverse transcription polymerase chain reaction (GSC RT-PCR). Mutat. Res., 406: 45–54.

Brazma, A. and Vilo, J. (2000) Gene expression data analysis. FEBS Lett., 480: 17–24.

Brenner, S., Johnson, M., Bridgham, J., Golda, G., Lloyd, D.H., Johnson, D., Luo, S., McCurdy, S., Foy, M., Ewan, M., Roth, R., George, D., Eletr, S., Albrecht, G., Vermaas, E., Williams, S.R., Moon, K., Burcham, T., Pallas, M., DuBridge, R.B., Kirchner, J., Fearon, K., Mao, J. and Corcoran, K. (2000a) Gene expression analysis by massively parallel signature sequencing (MPSS) on microbead arrays. Nat. Biotechnol., 18: 630–634.

Brenner, S., Williams, S.R., Vermaas, E.H., Storck, T., Moon, K., McCollum, C., Mao, J.I., Luo, S., Kirchner, J.J., Eletr, S., DuBridge, R.B., Burcham, T. and Albrecht, G. (2000b) In vitro cloning of complex mixtures of DNA on microbeads: physical separation of differentially expressed cDNAs. Proc. Natl. Acad. Sci. USA, 97: 1665–1670.

Brenner, S.E., Barken, D. and Levitt, M. (1999) The PRESAGE database for structural genomics. Nucleic Acids Res., 27: 251–253.

Broberg, P. (2005) A comparative review of estimates of the proportion unchanged genes and the false discovery rate. BMC Bioinform., 6: 199.

Brooks-Kayal, A.R., Jin, H., Price, M. and Dichter, M.A. (1998) Developmental expression of GABA(A) receptor subunit mRNAs in individual hippocampal neurons in vitro and in vivo. J. Neurochem., 70: 1017–1028.

Broude, N.E., Sano, T., Smith, C.L. and Cantor, C.R. (1994) Enhanced DNA sequencing by hybridization. Proc. Natl. Acad. Sci. USA, 91: 3072–3076.

Brown, P.O. and Botstein, D. (1999) Exploring the new world of the genome with DNA microarrays. Nat. Genet., 21: 33–37.

Bustin, S.A. (2002) Quantification of mRNA using real-time reverse transcription PCR (RT-PCR): trends and problems. J. Mol. Endocrinol., 29: 23–39.

Calzone, F.J., Britten, R.J. and Davidson, E.H. (1987) Mapping of gene transcripts by nuclease protection assays and cDNA primer extension. Methods Enzymol., 152: 611–632.

Cardullo, R.A., Agrawal, S., Flores, C., Zamecnik, P.C. and Wolf, D.E. (1988) Detection of nucleic acid hybridization by nonradiative fluorescence resonance energy transfer. Proc. Natl. Acad. Sci. USA, 85: 8790–8794.

Cavallaro, S., D'Agata, V. and Alkon, D.L. (2002a) Programs of gene expression during the laying down of memory formation as revealed by DNA microarrays. Neurochem Res., 27: 1201–1207.

Cavallaro, S., D'Agata, V., Manickam, P., Dufour, F. and Alkon, D.L. (2002b) Memory-specific temporal profiles of gene expression in the hippocampus. Proc. Natl. Acad. Sci. USA, 99: 16279–16284.

Che, S. and Ginsberg, S.D. (2004) Amplification of transcripts using terminal continuation. Lab. Invest., 84: 131–137.

Che, S. and Ginsberg, S.D. (2006) RNA amplification methodologies. In: McNamara, P.A. (Ed.) Trends in RNA Research. Nova Science Publishing, Hauppauge, pp. 277–301.

Chow, N., Cox, C., Callahan, L.M., Weimer, J.M., Guo, L. and Coleman, P.D. (1998) Expression profiles of multiple genes in single neurons of Alzheimer's disease. Proc. Natl. Acad. Sci. USA, 95: 9620–9625.

Colangelo, V., Schurr, J., Ball, M.J., Pelaez, R.P., Bazan, N.G. and Lukiw, W.J. (2002) Gene expression profiling of 12633 genes in Alzheimer hippocampal CA1: transcription and neurotrophic factor down-regulation and up-regulation of apoptotic and pro-inflammatory signaling. J. Neurosci. Res., 70: 462–473.

Colantuoni, C., Purcell, A.E., Bouton, C.M. and Pevsner, J. (2000) High throughput analysis of gene expression in the human brain. J. Neurosci. Res., 59: 1–10.

Counts, S.E., Chen, E.Y., Che, S., Ikonomovic, M.D., Wuu, J., Ginsberg, S.D., Dekosky, S.T. and Mufson, E.J. (2006) Galanin fiber hypertrophy within the cholinergic nucleus basalis during the progression of Alzheimer's disease. Dement. Geriatr. Cogn. Disord., 21: 205–214.

Crino, P.B. and Eberwine, J. (1996) Molecular characterization of the dendritic growth cone: regulated mRNA transport and local protein synthesis. Neuron, 17: 1173–1187.

Crino, P.B., Khodakhah, K., Becker, K., Ginsberg, S.D., Hemby, S. and Eberwine, J.H. (1998) Presence and phosphorylation of transcription factors in dendrites. Proc. Natl. Acad. Sci. USA, 95: 2313–2318.

Dafforn, A., Chen, P., Deng, G., Herrler, M., Iglehart, D., Koritala, S., Lato, S., Pillarisetty, S., Purohit, R., Wang, M., Wang, S. and Kurn, N. (2004) Linear mRNA amplification from as little as 5 ng total RNA for global gene expression analysis. BioTechniques, 37: 854–857.

Datson, N.A., van der Perk-de Jong, J., van den Berg, M.P., de Kloet, E.R. and Vreugdenhil, E. (1999) MicroSAGE: a modified procedure for serial analysis of gene expression in limited amounts of tissue. Nucleic Acids Res., 27: 1300–1307.

Del Signore, A., Mandillo, S., Rizzo, A., Di Mauro, E., Mele, A., Negri, R., Oliverio, A. and Paggi, P. (2004) Hippocampal gene expression is modulated by hypergravity. Eur. J. Neurosci., 19: 667–677.

Dickey, C.A., Loring, J.F., Montgomery, J., Gordon, M.N., Eastman, P.S. and Morgan, D. (2003) Selectively reduced expression of synaptic plasticity-related genes in amyloid precursor protein + presenilin-1 transgenic mice. J. Neurosci., 23: 5219–5226.

Drigues, N., Poltyrev, T., Bejar, C., Weinstock, M. and Youdim, M.B. (2003) cDNA gene expression profile of rat hippocampus after chronic treatment with antidepressant drugs. J. Neural. Transm., 110: 1413–1436.

Drmanac, R., Drmanac, S., Chui, G., Diaz, R., Hou, A., Jin, H., Jin, P., Kwon, S., Lacy, S., Moeur, B., Shafto, J., Swanson, D., Ukrainczyk, T., Xu, C. and Little, D. (2002) Sequencing by hybridization (SBH): advantages, achievements, and opportunities. Adv. Biochem. Eng. Biotechnol., 77: 75–101.

Drmanac, R., Labat, I., Brukner, I. and Crkvenjakov, R. (1989) Sequencing of megabase plus DNA by hybridization: theory of the method. Genomics, 4: 114–128.

Drmanac, R., Strezoska, Z., Labat, I., Drmanac, S. and Crkvenjakov, R. (1990) Reliable hybridization of oligonucleotides as short as six nucleotides. DNA Cell Biol., 9: 527–534.

Eberwine, J., Belt, B., Kacharmina, J.E. and Miyashiro, K. (2002) Analysis of subcellularly localized mRNAs using in situ hybridization, mRNA amplification, and expression profiling. Neurochem Res., 27: 1065–1077.

Eberwine, J., Kacharmina, J.E., Andrews, C., Miyashiro, K., McIntosh, T., Becker, K., Barrett, T., Hinkle, D., Dent, G. and Marciano, P. (2001) mRNA expression analysis of tissue sections and single cells. J. Neurosci., 21: 8310–8314.

Eberwine, J., Yeh, H., Miyashiro, K., Cao, Y., Nair, S., Finnell, R., Zettel, M. and Coleman, P. (1992) Analysis of gene expression in single live neurons. Proc. Natl. Acad. Sci. USA, 89: 3010–3014.

Eisen, M.B. and Brown, P.O. (1999) DNA arrays for analysis of gene expression. Methods Enzymol., 303: 179–205.

Elliott, R.C., Miles, M.F. and Lowenstein, D.H. (2003) Overlapping microarray profiles of dentate gyrus gene expression during development- and epilepsy-associated neurogenesis and axon outgrowth. J. Neurosci., 23: 2218–2227.

Fabrizio, P., Li, L. and Longo, V.D. (2005) Analysis of gene expression profile in yeast aging chronologically. Mech. Ageing Dev., 126: 11–16.

Farrell Jr., R.E. (1998) RNA Methodologies (2nd edition). Academic Press, San Diego.

Fend, F., Emmert-Buck, M.R., Chuaqui, R., Cole, K., Lee, J., Liotta, L.A. and Raffeld, M. (1999) Immuno-LCM: laser capture microdissection of immunostained frozen sections for mRNA analysis. Am. J. Pathol., 154: 61–66.

Fernandes, C., Paya-Cano, J.L., Sluyter, F., D'Souza, U., Plomin, R. and Schalkwyk, L.C. (2004) Hippocampal gene expression profiling across eight mouse inbred strains: towards understanding the molecular basis for behaviour. Eur. J. Neurosci., 19: 2576–2582.

Freeman, W.M. and Hemby, S.E. (2004) Proteomics for protein expression profiling in neuroscience. Neurochem Res., 29: 1065–1081.

Galbraith, D.W., Elumalai, R. and Gong, F.C. (2004) Integrative flow cytometric and microarray approaches for use in transcriptional profiling. Methods Mol. Biol., 263: 259–280.

Galvin, J.E. (2004) Neurodegenerative diseases: pathology and the advantage of single-cell profiling. Neurochem Res., 29: 1041–1051.

Galvin, J.E. and Ginsberg, S.D. (2004) Expression profiling and pharmacotherapeutic development in the central nervous system. Alzheimer Dis. Assoc. Disord., 18: 264–269.

Ghasemzadeh, M.B., Sharma, S., Surmeier, D.J., Eberwine, J.H. and Chesselet, M.F. (1996) Multiplicity of glutamate receptor subunits in single striatal neurons: an RNA amplification study. Mol. Pharmacol., 49: 852–859.

Gilbert, R.W., Costain, W.J., Blanchard, M.E., Mullen, K.L., Currie, R.W. and Robertson, H.A. (2003) DNA microarray analysis of hippocampal gene expression measured twelve hours after hypoxia-ischemia in the mouse. J. Cereb. Blood. Flow. Metab., 23: 1195–1211.

Ginsberg, S.D. (2005a) Glutamatergic neurotransmission expression profiling in the mouse hippocampus after perforant-path transection. Am. J. Geriatr. Psychiatry, 13: 1052–1061.

Ginsberg, S.D. (2005b) RNA amplification strategies for small sample populations. Methods, 37: 229–237.

Ginsberg, S.D. and Che, S. (2002) RNA amplification in brain tissues. Neurochem Res., 27: 981–992.

Ginsberg, S.D. and Che, S. (2004) Combined histochemical staining, RNA amplification, regional, and single cell analysis within the hippocampus. Lab. Invest., 84: 952–962.

Ginsberg, S.D. and Che, S. (2005) Expression profile analysis within the human hippocampus: comparison of CA1 and CA3 pyramidal neurons. J. Comp. Neurol., 487: 107–118.

Ginsberg, S.D., Che, S., Counts, S.E. and Mufson, E.J. (2006a) Shift in the ratio of 3-repeat tau and 4-repeat tau mRNAs in individual cholinergic basal forebrain neurons in mild cognitive impairment and Alzheimer's disease. J. Neurochem., 96: 1401–1408.

Ginsberg, S.D., Che, S., Wuu, J., Counts, S.E. and Mufson, E.J. (2006b) Down regulation of trk but not p75NTR gene expression in single cholinergic basal forebrain neurons mark the progression of Alzheimer's disease. J. Neurochem, 97: 475–487.

Ginsberg, S.D., Crino, P.B., Hemby, S.E., Weingarten, J.A., Lee, V.M.-Y., Eberwine, J.H. and Trojanowski, J.Q. (1999a) Predominance of neuronal mRNAs in individual Alzheimer's disease senile plaques. Ann. Neurol., 45: 174–181.

Ginsberg, S.D., Crino, P.B., Lee, V.M.-Y., Eberwine, J.H. and Trojanowski, J.Q. (1997) Sequestration of RNA in Alzheimer's disease neurofibrillary tangles and senile plaques. Ann Neurol, 41: 200–209.

Ginsberg, S.D., Elarova, I., Ruben, M., Tan, F., Counts, S.E., Eberwine, J.H., Trojanowski, J.Q., Hemby, S.E., Mufson, E.J. and Che, S. (2004) Single cell gene expression analysis: implications for neurodegenerative and neuropsychiatric disorders. Neurochem Res., 29: 1054–1065.

Ginsberg, S.D., Galvin, J.E., Chiu, T.-S., Lee, V.M.-Y., Masliah, E. and Trojanowski, J.Q. (1998) RNA sequestration to pathological lesions of neurodegenerative disorders. Acta Neuropathol., 96: 487–494.

Ginsberg, S.D., Hemby, S.E., Lee, V.M.-Y., Eberwine, J.H. and Trojanowski, J.Q. (2000) Expression profile of transcripts in Alzheimer's disease tangle-bearing CA1 neurons. Ann. Neurol., 48: 77–87.

Ginsberg, S.D., Hemby, S.E., Mufson, E.J. and Martin, L.J. (2006c) Cell and tissue microdissection in combination with genomic and proteomic applications. In: Zaborszky, L., Wouterlood, F.G. and Lanciego, J.L. (Eds.), Neuroanatomical Tract Tracing 3: Molecules, Neurons, and Systems. Springer, New York, pp. 109–141.

Ginsberg, S.D., Schmidt, M.L., Crino, P.B., Eberwine, J.H., Lee, V.M.-Y. and Trojanowski, J.Q. (1999b) Molecular pathology of Alzheimer's disease and related disorders. In: Peters, A. and Morrison, J.H. (Eds.) Cerebral Cortex, Vol. 14. Neurodegenerative and Age-related Changes in Structure and Function of Cerebral Cortex. Kluwer, New York, pp. 603–653.

Giulietti, A., Overbergh, L., Valckx, D., Decallonne, B., Bouillon, R. and Mathieu, C. (2001) An overview of real-time quantitative PCR: applications to quantify cytokine gene expression. Methods, 25: 386–401.

Goldsworthy, S.M., Stockton, P.S., Trempus, C.S., Foley, J.F. and Maronpot, R.R. (1999) Effects of fixation on RNA extraction and amplification from laser capture microdissected tissue. Mol. Carcinog., 25: 86–91.

Gowda, M., Jantasuriyarat, C., Dean, R.A. and Wang, G.L. (2004) Robust-LongSAGE (RL-SAGE): a substantially improved LongSAGE method for gene discovery and transcriptome analysis. Plant Physiol., 134: 890–897.

Greene, J.G., Dingledine, R. and Greenamyre, J.T. (2005) Gene expression profiling of rat midbrain dopamine neurons: implications for selective vulnerability in parkinsonism. Neurobiol Dis., 18: 19–31.

Grunblatt, E., Mandel, S., Jacob-Hirsch, J., Zeligson, S., Amariglo, N., Rechavi, G., Li, J., Ravid, R., Roggendorf, W.,

Riederer, P. and Youdim, M.B. (2004) Gene expression profiling of parkinsonian substantia nigra pars compacta; alterations in ubiquitin-proteasome, heat shock protein, iron and oxidative stress regulated proteins, cell adhesion/cellular matrix and vesicle trafficking genes. J. Neural. Transm., 111: 1543–1573.

Hakak, Y., Walker, J.R., Li, C., Wong, W.H., Davis, K.L., Buxbaum, J.D., Haroutunian, V. and Fienberg, A.A. (2001) Genome-wide expression analysis reveals dysregulation of myelination-related genes in chronic schizophrenia. Proc. Natl. Acad. Sci. USA, 98: 4746–4751.

Harrison, P.J., Heath, P.R., Eastwood, S.L., Burnet, P.W., McDonald, B. and Pearson, R.C. (1995) The relative importance of premortem acidosis and postmortem interval for human brain gene expression studies: selective mRNA vulnerability and comparison with their encoded proteins. Neurosci Lett., 200: 151–154.

Hauser, M.A., Li, Y.J., Xu, H., Noureddine, M.A., Shao, Y.S., Gullans, S.R., Scherzer, C.R., Jensen, R.V., McLaurin, A.C., Gibson, J.R., Scott, B.L., Jewett, R.M., Stenger, J.E., Schmechel, D.E., Hulette, C.M. and Vance, J.M. (2005) Expression profiling of substantia nigra in Parkinson disease, progressive supranuclear palsy, and frontotemporal dementia with parkinsonism. Arch. Neurol., 62: 917–921.

Heidenblut, A.M., Luttges, J., Buchholz, M., Heinitz, C., Emmersen, J., Nielsen, K.L., Schreiter, P., Souquet, M., Nowacki, S., Herbrand, U., Kloppel, G., Schmiegel, W., Gress, T. and Hahn, S.A. (2004) aRNA-longSAGE: a new approach to generate SAGE libraries from microdissected cells. Nucleic Acids Res., 32: e131.

Hemby, S.E., Ginsberg, S.D., Brunk, B., Arnold, S.E., Trojanowski, J.Q. and Eberwine, J.H. (2002) Gene expression profile for schizophrenia: discrete neuron transcription patterns in the entorhinal cortex. Arch. Gen. Psychiat., 59: 631–640.

Hemby, S.E., Trojanowski, J.Q. and Ginsberg, S.D. (2003) Neuron-specific age-related decreases in dopamine receptor subtype mRNAs. J Comp Neurol, 456: 176–183.

Hollingshead, D., Lewis, D.A. and Mirnics, K. (2005) Platform influence on DNA microarray data in postmortem brain research. Neurobiol. Dis., 18: 649–655.

Hynd, M.R., Lewohl, J.M., Scott, H.L. and Dodd, P.R. (2003) Biochemical and molecular studies using human autopsy brain tissue. J. Neurochem., 85: 543–562.

Irizarry, R.A., Warren, D., Spencer, F., Kim, I.F., Biswal, S., Frank, B.C., Gabrielson, E., Garcia, J.G., Geoghegan, J., Germino, G., Griffin, C., Hilmer, S.C., Hoffman, E., Jedlicka, A.E., Kawasaki, E., Martinez-Murillo, F., Morsberger, L., Lee, H., Petersen, D., Quackenbush, J., Scott, A., Wilson, M., Yang, Y., Ye, S.Q. and Yu, W. (2005) Multiple-laboratory comparison of microarray platforms. Nat. Methods, 2: 345–350.

Iscove, N.N., Barbara, M., Gu, M., Gibson, M., Modi, C. and Winegarden, N. (2002) Representation is faithfully preserved in global cDNA amplified exponentially from sub-picogram quantities of mRNA. Nat. Biotechnol., 20: 940–943.

Jarvinen, A.K., Hautaniemi, S., Edgren, H., Auvinen, P., Saarela, J., Kallioniemi, O.P. and Monni, O. (2004) Are data from different gene expression microarray platforms comparable? Genomics, 83: 1164–1168.

Kabbarah, O., Pinto, K., Mutch, D.G. and Goodfellow, P.J. (2003) Expression profiling of mouse endometrial cancers microdissected from ethanol-fixed, paraffin-embedded tissues. Am. J. Pathol., 162: 755–762.

Kacharmina, J.E., Crino, P.B. and Eberwine, J. (1999) Preparation of cDNA from single cells and subcellular regions. Methods Enzymol., 303: 3–18.

Kamme, F., Salunga, R., Yu, J., Tran, D.T., Zhu, J., Luo, L., Bittner, A., Guo, H.Q., Miller, N., Wan, J. and Erlander, M. (2003) Single-cell microarray analysis in hippocampus CA1: demonstration and validation of cellular heterogeneity. J. Neurosci., 23: 3607–3615.

Katsel, P.L., Davis, K.L. and Haroutunian, V. (2005) Large-scale microarray studies of gene expression in multiple regions of the brain in schizophrenia and Alzheimer's disease. Int. Rev. Neurobiol., 63: 41–82.

Kawasaki, E.S. (2004) Microarrays and the gene expression profile of a single cell. Ann. N Y Acad. Sci., 1020: 92–100.

Kinnecom, K. and Pachter, J.S. (2005) Selective capture of endothelial and perivascular cells from brain microvessels using laser capture microdissection. Brain. Res. Protoc., 16: 1–9.

Kotlyar, M., Fuhrman, S., Ableson, A. and Somogyi, R. (2002) Spearman correlation identifies statistically significant gene expression clusters in spinal cord development and injury. Neurochem. Res., 27: 1133–1140.

Kricka, L.J. (2002) Stains, labels and detection strategies for nucleic acids assays. Ann. Clin. Biochem., 39: 114–129.

Kurn, N., Chen, P., Heath, J.D., Kopf-Sill, A., Stephens, K.M. and Wang, S. (2005) Novel isothermal, linear nucleic acid amplification systems for highly multiplexed applications. Clin. Chem., 51: 1973–1981.

Larkin, J.E., Frank, B.C., Gavras, H., Sultana, R. and Quackenbush, J. (2005) Independence and reproducibility across microarray platforms. Nat. Methods, 2: 337–344.

Lazarov, O., Robinson, J., Tang, Y.P., Hairston, I.S., Korade-Mirnics, Z., Lee, V.M., Hersh, L.B., Sapolsky, R.M., Mirnics, K. and Sisodia, S.S. (2005) Environmental enrichment reduces Abeta levels and amyloid deposition in transgenic mice. Cell, 120: 701–713.

Lein, E.S., Zhao, X. and Gage, F.H. (2004) Defining a molecular atlas of the hippocampus using DNA microarrays and high-throughput in situ hybridization. J. Neurosci., 24: 3879–3889.

Leonard, S., Logel, J., Luthman, D., Casanova, M., Kirch, D. and Freedman, R. (1993) Biological stability of mRNA isolated from human postmortem brain collections. Biol. Psychiatry, 33: 456–466.

Levsky, J.M., Shenoy, S.M., Pezo, R.C. and Singer, R.H. (2002) Single-cell gene expression profiling. Science, 297: 836–840.

Li, J.Z., Vawter, M.P., Walsh, D.M., Tomita, H., Evans, S.J., Choudary, P.V., Lopez, J.F., Avelar, A., Shokoohi, V., Chung, T., Mesarwi, O., Jones, E.G., Watson, S.J., Akil, H., Bunney Jr., W.E. and Myers, R.M. (2004) Systematic changes in gene expression in postmortem human brains associated with tissue pH and terminal medical conditions. Hum. Mol. Genet., 13: 609–616.

Lin, D.Y. (2005) An efficient Monte Carlo approach to assessing statistical significance in genomic studies. Bioinformatics, 21: 781–787.

Lockhart, D.J., Dong, H., Byrne, M.C., Follettie, M.T., Gallo, M.V., Chee, M.S., Mittmann, M., Wang, C., Kobayashi, M., Horton, H. and Brown, E.L. (1996) Expression monitoring by hybridization to high density oligonucleotide arrays. Nat. Biotechnol., 14: 1675–1680.

Loring, J.F., Wen, X., Lee, J.M., Seilhamer, J. and Somogyi, R. (2001) A gene expression profile of Alzheimer's disease. DNA Cell Biol., 20: 683–695.

Lu, L., Neff, F., Alvarez-Fischer, D., Henze, C., Xie, Y., Oertel, W.H., Schlegel, J. and Hartmann, A. (2005) Gene expression profiling of Lewy body-bearing neurons in Parkinson's disease. Exp. Neurol., 195: 27–39.

Lu, T., Pan, Y., Kao, S.Y., Li, C., Kohane, I., Chan, J. and Yankner, B.A. (2004) Gene regulation and DNA damage in the ageing human brain. Nature, 429: 883–891.

Lukasiuk, K. and Pitkänen, A. (2004) Large-scale analysis of gene expression in epilepsy research: is synthesis already possible? Neurochem. Res., 29: 1164–1173.

Lukiw, W.J. (2004) Gene expression profiling in fetal, aged, and Alzheimer hippocampus: a continuum of stress-related signaling. Neurochem. Res., 29: 1287–1297.

Luo, L., Salunga, R.C., Guo, H., Bittner, A., Joy, K.C., Galindo, J.E., Xiao, H., Rogers, K.E., Wan, J.S., Jackson, M.R. and Erlander, M.G. (1999) Gene expression profiles of laser-captured adjacent neuronal subtypes. Nat. Med., 5: 117–122.

Luzzi, V., Holtschlag, V. and Watson, M.A. (2001) Expression profiling of ductal carcinoma in situ by laser capture microdissection and high density oligonucleotide arrays. Am. J. Pathol., 158: 2005–2010.

Lyons-Weiler, J., Patel, S. and Bhattacharya, S. (2003) A classification-based machine learning approach for the analysis of genome-wide expression data. Genome Res., 13: 503–512.

Mai, J.K., Schmidt-Kastner, R. and Tefett, H.-B. (1984) Use of acridine orange for histologic analysis of the central nervous system. J. Histochem. Cytochem., 32: 97–104.

Marciano, P.G., Eberwine, J.H., Ragupathi, R., Saatman, K.E., Meaney, D.F. and McIntosh, T.K. (2002) Expression profiling following traumatic brain injury: a review. Neurochem. Res., 27: 1147–1155.

Matz, M., Shagin, D., Bogdanova, E., Britanova, O., Lukyanov, S., Diatchenko, L. and Chenchik, A. (1999) Amplification of cDNA ends based on template-switching effect and step-out PCR. Nucleic Acids Res., 27: 1558–1560.

Matzilevich, D.A., Rall, J.M., Moore, A.N., Grill, R.J. and Dash, P.K. (2002) High density microarray analysis of hippocampal gene expression following experimental brain injury. J. Neurosci. Res., 67: 646–663.

McMurray, M.A. and Gottschling, D.E. (2004) Aging and genetic instability in yeast. Curr. Opin. Microbiol., 7: 673–679.

Mei, B., Li, C., Dong, S., Jiang, C.H., Wang, H. and Hu, Y. (2005) Distinct gene expression profiles in hippocampus and amygdala after fear conditioning. Brain Res. Bull., 67: 1–12.

Melton, D.A., Krieg, P.A., Rebagliati, M.R., Maniatis, T., Zinn, K. and Green, M.R. (1984) Efficient in vitro synthesis of biologically active RNA and RNA hybridization probes from plasmids containing a bacteriophage SP6 promoter. Nucleic Acids Res., 12: 7035–7056.

Meuwissen, T.H. and Goddard, M.E. (2004) Bootstrapping of gene-expression data improves and controls the false discovery rate of differentially expressed genes. Genet. Sel. Evol., 36: 191–205.

Middleton, F.A., Mirnics, K., Pierri, J.N., Lewis, D.A. and Levitt, P. (2002) Gene expression profiling reveals alterations of specific metabolic pathways in schizophrenia. J. Neurosci., 22: 2718–2729.

Mikulowska-Mennis, A., Taylor, T.B., Vishnu, P., Michie, S.A., Raja, R., Horner, N. and Kunitake, S.T. (2002) High-quality RNA from cells isolated by laser capture microdissection. BioTechniques, 33: 176–179.

Miller, R.M., Kiser, G.L., Kaysser-Kranich, T.M., Lockner, R.J., Palaniappan, C. and Federoff, H.J. (2006) Robust dysregulation of gene expression in substantia nigra and striatum in Parkinson's disease. Neurobiol. Dis., 21: 305–313.

Mimmack, M.L., Brooking, J. and Bahn, S. (2004) Quantitative polymerase chain reaction: validation of microarray results from postmortem brain studies. Biol. Psychiatry, 55: 337–345.

Mimmack, M.L., Ryan, M., Baba, H., Navarro-Ruiz, J., Iritani, S., Faull, R.L., McKenna, P.J., Jones, P.B., Arai, H., Starkey, M., Emson, P.C. and Bahn, S. (2002) Gene expression analysis in schizophrenia: reproducible up-regulation of several members of the apolipoprotein L family located in a high-susceptibility locus for schizophrenia on chromosome 22. Proc. Natl. Acad. Sci. USA, 99: 4680–4685.

Mirnics, K., Korade, Z., Arion, D., Lazarov, O., Unger, T., Macioce, M., Sabatini, M., Terrano, D., Douglass, K.C., Schor, N.F. and Sisodia, S.S. (2005) Presenilin-1-dependent transcriptome changes. J. Neurosci., 25: 1571–1578.

Mirnics K. Levitt P. and Lewis, D.A. (2006) Critical appraisal of DNA microarrays in psychiatric genomics. Biol. Psychiatry (in press).

Mirnics, K. and Lewis, D.A. (2001) Genes and subtypes of schizophrenia. Trends Mol. Med., 7: 281–283.

Mirnics, K., Middleton, F.A., Lewis, D.A. and Levitt, P. (2001) Analysis of complex brain disorders with gene expression microarrays: schizophrenia as a disease of the synapse. Trends Neurosci., 24: 479–486.

Mirnics, K., Middleton, F.A., Marquez, A., Lewis, D.A. and Levitt, P. (2000) Molecular characterization of schizophrenia viewed by microarray analysis of gene expression in prefrontal cortex. Neuron, 28: 53–67.

Mirnics, K. and Pevsner, J. (2004) Progress in the use of microarray technology to study the neurobiology of disease. Nat. Neurosci., 7: 434–439.

Moll, P.R., Duschl, J. and Richter, K. (2004) Optimized RNA amplification using T7-RNA-polymerase based in vitro transcription. Anal. Biochem., 334: 164–174.

Molteni, R., Ying, Z. and Gomez-Pinilla, F. (2002) Differential effects of acute and chronic exercise on plasticity-related genes in the rat hippocampus revealed by microarray. Eur. J. Neurosci., 16: 1107–1116.

Mufson, E.J., Counts, S.E. and Ginsberg, S.D. (2002) Single cell gene expression profiles of nucleus basalis cholinergic neurons in Alzheimer's disease. Neurochem. Res., 27: 1035–1048.

Mullis, K.B. (1990) The unusual origin of the polymerase chain reaction. Sci. Am., 262: 56–65.

Nadon, R. and Shoemaker, J. (2002) Statistical issues with microarrays: processing and analysis. Trends Genet., 18: 265–271.

Napolitano, M., Centonze, D., Calce, A., Picconi, B., Spiezia, S., Gulino, A., Bernardi, G. and Calabresi, P. (2002) Experimental parkinsonism modulates multiple genes involved in the transduction of dopaminergic signals in the striatum. Neurobiol. Dis., 10: 387–395.

Newton, S.S., Collier, E.F., Hunsberger, J., Adams, D., Terwilliger, R., Selvanayagam, E. and Duman, R.S. (2003) Gene profile of electroconvulsive seizures: induction of neurotrophic and angiogenic factors. J. Neurosci., 23: 10841–10851.

Oudes, A.J., Roach, J.C., Walashek, L.S., Eichner, L.J., True, L.D., Vessella, R.L. and Liu, A.Y. (2005) Application of affymetrix array and massively parallel signature sequencing for identification of genes involved in prostate cancer progression. BMC Cancer, 5: 86.

Park, S.K. and Prolla, T.A. (2005) Lessons learned from gene expression profile studies of aging and caloric restriction. Ageing Res. Rev., 4: 55–65.

Pavlidis, P. and Noble, W.S. (2001) Analysis of strain and regional variation in gene expression in mouse brain. Genome Biol., 2: 0042.

Peters, D.G., Kassam, A.B., Yonas, H., O'Hare, E.H., Ferrell, R.E. and Brufsky, A.M. (1999) Comprehensive transcript analysis in small quantities of mRNA by SAGE-lite. Nucleic Acids Res., 27: e39.

Phillips, J. and Eberwine, J.H. (1996) Antisense RNA amplification: a linear amplification method for analyzing the mRNA population from single living cells. Methods Enzymol. Suppl., 10: 283–288.

Poirier, G.M., Pyati, J., Wan, J.S. and Erlander, M.G. (1997) Screening differentially expressed cDNA clones obtained by differential display using amplified RNA. Nucleic Acids Res., 25: 913–914.

Pongrac, J., Middleton, F.A., Lewis, D.A., Levitt, P. and Mirnics, K. (2002) Gene expression profiling with DNA microarrays: advancing our understanding of psychiatric disorders. Neurochem. Res., 27: 1049–1063.

Pounds, S. and Cheng, C. (2004) Improving false discovery rate estimation. Bioinformatics, 20: 1737–1745.

Prolla, T.A. (2005) Multiple roads to the aging phenotype: insights from the molecular dissection of progerias through DNA microarray analysis. Mech. Ageing Dev., 126: 461–465.

Qin, L.X. and Kerr, K.F. (2004) Empirical evaluation of data transformations and ranking statistics for microarray analysis. Nucleic Acids Res., 32: 5471–5479.

Qiu, J., Hu, X., Nesic, O., Grafe, M.R., Rassin, D.K., Wood, T.G. and Perez-Polo, J.R. (2004) Effects of NF-kappaB oligonucleotide "decoys" on gene expression in P7 rat hippocampus after hypoxia/ischemia. J. Neurosci. Res., 77: 108–118.

Quackenbush, J. (2002) Microarray data normalization and transformation. Nat. Genet., 32: 496–501.

Reddy, P.H., McWeeney, S., Park, B.S., Manczak, M., Gutala, R.V., Partovi, D., Jung, Y., Yau, V., Searles, R., Mori, M. and Quinn, J. (2004) Gene expression profiles of transcripts in amyloid precursor protein transgenic mice: up-regulation of mitochondrial metabolism and apoptotic genes is an early cellular change in Alzheimer's disease. Hum. Mol. Genet., 13: 1225–1240.

Reinartz, J., Bruyns, E., Lin, J.Z., Burcham, T., Brenner, S., Bowen, B., Kramer, M. and Woychik, R. (2002) Massively parallel signature sequencing (MPSS) as a tool for in-depth quantitative gene expression profiling in all organisms. Brief. Funct. Genomic. Proteomic., 1: 95–104.

Reiner, A., Yekutieli, D. and Benjamini, Y. (2003) Identifying differentially expressed genes using false discovery rate controlling procedures. Bioinformatics, 19: 368–375.

Rickman, D.S., Bobek, M.P., Misek, D.E., Kuick, R., Blaivas, M., Kurnit, D.M., Taylor, J. and Hanash, S.M. (2001) Distinctive molecular profiles of high-grade and low-grade gliomas based on oligonucleotide microarray analysis. Cancer Res., 61: 6885–6891.

Saha, S., Sparks, A.B., Rago, C., Akmaev, V., Wang, C.J., Vogelstein, B., Kinzler, K.W. and Velculescu, V.E. (2002) Using the transcriptome to annotate the genome. Nat. Biotechnol., 20: 508–512.

Saito, M., Smiley, J., Toth, R. and Vadasz, C. (2002) Microarray analysis of gene expression in rat hippocampus after chronic ethanol treatment. Neurochem. Res., 27: 1221–1229.

Sambrook, J. and Russell, D.W. (2001) Molecular Cloning: A Laboratory Manual (3rd edition). Cold Spring Harbor Laboratory Press, Cold Spring Harbor.

Sarnat, H.B., Curry, B., Rewcastle, N.B. and Trevenen, C.L. (1987) Gliosis and glioma distinguished by acridine orange. Can. J. Neurol. Sci., 14: 31–35.

Schadt, E.E., Li, C., Ellis, B. and Wong, W.H. (2001) Feature extraction and normalization algorithms for high-density oligonucleotide gene expression array data. J. Cell. Biochem. Suppl., 37: 120–125.

Schena, M., Shalon, D., Davis, R.W. and Brown, P.O. (1995) Quantitative monitoring of gene expression patterns with a complementary DNA microarray. Science, 270: 467–470.

Shaulsky, G. and Loomis, W.F. (2002) Gene expression patterns in Dictyostelium using microarrays. Protist, 153: 93–98.

Shendure, J., Mitra, R.D., Varma, C. and Church, G.M. (2004) Advanced sequencing technologies: methods and goals. Nat. Rev. Genet., 5: 335–344.

Shintani-Ishida, K., Zhu, B.L. and Maeda, H. (2005) TaqMan fluorogenic detection system to analyze gene transcription in autopsy material. Methods Mol. Biol., 291: 415–421.

Shippy, R., Sendera, T.J., Lockner, R., Palaniappan, C., Kaysser-Kranich, T., Watts, G. and Alsobrook, J. (2004)

Performance evaluation of commercial short-oligonucleotide microarrays and the impact of noise in making cross-platform correlations. BMC Genom., 5: 61.

Silva, A.P., De Souza, J.E., Galante, P.A., Riggins, G.J., De Souza, S.J. and Camargo, A.A. (2004) The impact of SNPs on the interpretation of SAGE and MPSS experimental data. Nucleic Acids Res., 32: 6104–6110.

Singer, R.H. and Ward, D.C. (1982) Actin gene expression visualized in chicken muscle tissue culture by using in situ hybridization with a biotinated nucleotide analog. Proc. Natl. Acad. Sci. USA, 79: 7331–7335.

Soverchia, L., Ubaldi, M., Leonardi-Essmann, F., Ciccocioppo, R. and Hardiman, G. (2005) Microarrays — the challenge of preparing brain tissue samples. Addict. Biol., 10: 5–13.

Su, J.M., Perlaky, L., Li, X.N., Leung, H.C., Antalffy, B., Armstrong, D. and Lau, C.C. (2004) Comparison of ethanol versus formalin fixation on preservation of histology and RNA in laser capture microdissected brain tissues. Brain Pathol., 14: 175–182.

Sutcliffe, J.G., Foye, P.E., Erlander, M.G., Hilbush, B.S., Bodzin, L.J., Durham, J.T. and Hasel, K.W. (2000) TOGA: an automated parsing technology for analyzing expression of nearly all genes. Proc. Natl. Acad. Sci. USA, 97: 1976–1981.

Szaniszlo, P., Wang, N., Sinha, M., Reece, L.M., Van Hook, J.W., Luxon, B.A. and Leary, J.F. (2004) Getting the right cells to the array: gene expression microarray analysis of cell mixtures and sorted cells. Cytometry A, 59: 191–202.

Tan, L., Li, Y., Drake, T.J., Moroz, L., Wang, K., Li, J., Munteanu, A., Chaoyong, J.Y., Martinez, K. and Tan, W. (2005) Molecular beacons for bioanalytical applications. Analyst, 130: 1002–1005.

Tanji, N., Ross, M.D., Cara, A., Markowitz, G.S., Klotman, P.E. and D'Agati, V.D. (2001) Effect of tissue processing on the ability to recover nucleic acid from specific renal tissue compartments by laser capture microdissection. Exp. Nephrol., 9: 229–234.

Tatton, N.A. and Kish, S.J. (1997) In situ detection of apoptotic nuclei in the substantia nigra compacta of 1-methyl-4-phenyl-1,2,3,6-tetrahydropyridine-treated mice using terminal deoxynucleotidyl transferase labelling and acridine orange staining. Neuroscience, 77: 1037–1048.

Tecott, L.H., Barchas, J.D. and Eberwine, J.H. (1988) In situ transcription: specific synthesis of complementary DNA in fixed tissue sections. Science, 240: 1661–1664.

Thomas, E.A., Danielson, P.E., Nelson, P.A., Pribyl, T.M., Hilbush, B.S., Hasel, K.W. and Sutcliffe, J.G. (2001) Clozapine increases apolipoprotein D expression in rodent brain: towards a mechanism for neuroleptic pharmacotherapy. J. Neurochem., 76: 789–796.

Thomas, E.A., George, R.C., Danielson, P.E., Nelson, P.A., Warren, A.J., Lo, D. and Sutcliffe, J.G. (2003) Antipsychotic drug treatment alters expression of mRNAs encoding lipid metabolism-related proteins. Mol. Psychiatry, 8: 983–993.

Tomita, H., Vawter, M.P., Walsh, D.M., Evans, S.J., Choudary, P.V., Li, J., Overman, K.M., Atz, M.E., Myers, R.M., Jones, E.G., Watson, S.J., Akil, H. and Bunney Jr., W.E. (2004) Effect of agonal and postmortem factors on gene expression profile: quality control in microarray analyses of postmortem human brain. Biol. Psychiatry, 55: 346–352.

Unger, T., Korade, Z., Lazarov, O., Terrano, D., Schor, N.F., Sisodia, S.S. and Mirnics, K. (2005a) Transcriptome differences between the frontal cortex and hippocampus of wild-type and humanized presenilin-1 transgenic mice. Am. J. Geriatr. Psychiatry, 13: 1041–1051.

Unger, T., Korade, Z., Lazarov, O., Terrano, D., Sisodia, S.S. and Mirnics, K. (2005b) True and false discovery in DNA microarray experiments: transcriptome changes in the hippocampus of presenilin 1 mutant mice. Methods, 37: 261–273.

Van Deerlin, V.M., Gill, L.H. and Nelson, P.T. (2002a) Optimizing gene expression analysis in archival brain tissue. Neurochem. Res., 27: 993–1003.

Van Deerlin, V.M.D., Ginsberg, S.D., Lee, V.M.-Y. and Trojanowski, J.Q. (2000) Fixed post mortem brain tissue for mRNA expression analysis in neurodegenerative diseases. In: Geschwind, D.H. (Ed.), DNA Microarrays: The New Frontier in Gene Discovery and Gene Expression Analysis. Society for Neuroscience, Washington, DC, pp. 118–128.

Van Deerlin, V.M.D., Ginsberg, S.D., Lee, V.M.-Y. and Trojanowski, J.Q. (2002b) The use of fixed human post mortem brain tissue to study mRNA expression in neurodegenerative diseases: applications of microdissection and mRNA amplification. In: Geschwind, D.H. and Gregg, J.P. (Eds.), Microarrays for the Neurosciences: An Essential Guide. MIT Press, Boston, pp. 201–235.

VanGelder, R., von Zastrow, M., Yool, A., Dement, W., Barchas, J. and Eberwine, J. (1990) Amplified RNA (aRNA) synthesized from limited quantities of heterogeneous cDNA. Proc. Natl. Acad. Sci. USA, 87: 1663–1667.

Vawter, M.P., Evans, S., Choudary, P., Tomita, H., Meador-Woodruff, J., Molnar, M., Li, J., Lopez, J.F., Myers, R., Cox, D., Watson, S.J., Akil, H., Jones, E.G. and Bunney, W.E. (2004) Gender-specific gene expression in post-mortem human brain: localization to sex chromosomes. Neuropsychopharmacology, 29: 373–384.

Velculescu, V.E., Vogelstein, B. and Kinzler, K.W. (2000) Analysing uncharted transcriptomes with SAGE. Trends Genet., 16: 423–425.

Velculescu, V.E., Zhang, L., Vogelstein, B. and Kinzler, K.W. (1995) Serial analysis of gene expression. Science, 270: 484–487.

Vilain, C. and Vassart, G. (2004) Small amplified RNA-SAGE. Methods Mol. Biol., 258: 135–152.

Vincent, V.A., DeVoss, J.J., Ryan, H.S. and Murphy Jr., G.M. (2002) Analysis of neuronal gene expression with laser capture microdissection. J. Neurosci. Res., 69: 578–586.

Wang, E., Miller, L.D., Ohnmacht, G.A., Liu, E.T. and Marincola, F.M. (2000) High-fidelity mRNA amplification for gene profiling. Nat. Biotechnol., 18: 457–459.

Weindruch, R., Kayo, T., Lee, C.K. and Prolla, T.A. (2001) Microarray profiling of gene expression in aging and its alteration by caloric restriction in mice. J. Nutr., 131: 918S–923S.

Xiang, C.C., Chen, M., Ma, L., Phan, Q.N., Inman, J.M., Kozhich, O.A. and Brownstein, M.J. (2003) A new strategy to amplify degraded RNA from small tissue samples for microarray studies. Nucleic Acids Res., 31: E53.

Zhao, X., Lein, E.S., He, A., Smith, S.C., Aston, C. and Gage, F.H. (2001) Transcriptional profiling reveals strict boundaries between hippocampal subregions. J. Comp. Neurol., 441: 187–196.

Zhong, S., Li, C. and Wong, W.H. (2003) ChipInfo: software for extracting gene annotation and gene ontology information for microarray analysis. Nucleic Acids Res., 31: 3483–3486.

Zhumabayeva, B., Diatchenko, L., Chenchik, A. and Siebert, P.D. (2001) Use of SMART-generated cDNA for gene expression studies in multiple human tumors. BioTechniques, 30: 158–163.

CHAPTER 3

Methods for proteomics in neuroscience

Nilesh S. Tannu* and Scott E. Hemby

Department of Physiology and Pharmacology, Wake Forest University School of Medicine, Winston-Salem, NC 27157, USA

Abstract: Proteomics reveals complex protein expression, function, interactions and localization in different phenotypes of neuron. As proteomics, regarded as a highly complex screening technology, moves from a theoretical approach to practical reality, neuroscientists have to determine the most-appropriate applications for this technology. Even though proteomics compliments genomics, it is in sheer contrast to the basically constant genome due to its dynamic nature. Neuroscientists have to surmount difficulties particular to the research in neuroscience; such as limited sample amounts, heterogeneous cellular compositions in samples and the fact that many proteins of interest are hydrophobic proteins. The necessity of exclusive technology, sophisticated software and skilled manpower tops the challenge. This review examines subcellular organelle isolation, protein fractionation and separation using two-dimensional gel electrophoresis (2-DGE) as well as multi-dimensional liquid chromatography (LC) followed by mass spectrometry (MS). The methods for quantifying relative gene product expression between samples (e.g., two-dimensional difference in gel electrophoresis (2D-DIGE), isotope-coded affinity tag (ICAT) and iTRAQ) are elaborated. An overview of the techniques used currently to assign post-translational modification status on a proteomics scale is also evaluated. The feasible coverage of the proteome, ability to detect unique cell components such as post-synaptic densities and membrane proteins, resource requirements and quantitative as well as qualitative reliability of different approaches is also discussed. While there are many challenges in neuroproteomics, this field promises many returns in the future.

Keywords: proteomics; brain; 2-DGE; 2D-DIGE; MuDPIT; ICAT; iTRAQ; mass spectrometry

Introduction

The comprehensive sequencing of human and other important genomes has immensely enhanced our insight of the cellular machinery of higher organisms. This has been largely accomplished by the innovations in large-scale analysis of mRNA expression, viz. microarrays, serial-analysis of gene expression (SAGE) and differential display into gene expression (Venter et al., 2001). It is desirable to complement the global gene expression analyses with studies examining the corresponding proteomes. The hypotheses-based approach has always been foremost to an unbiased approach to determine organized changes encompassing the expression of entire proteome. The early proteomic research, ~6000 scientific publications, have been directed toward cataloging proteins and constructing protein databases (Anderson et al., 2001). This stepping-up has made us understand that higher organisms have many differences in the controlling mechanisms of cellular function. The most obvious insight has been the understanding that a plethora of proteins are produced by a single gene in higher organisms. These rapid advancements have improved our understanding of the cellular machinery within the brain and its

---

*Corresponding author. Tel.: +1-336-716-8589; Fax: +1-336-716-8501; E-mail: ntannu@wfubmc.edu

DOI: 10.1016/S0079-6123(06)58003-3

role in health and disease. The dynamic nature of proteome; changing nature of protein-expression profiles during cell cycle and with the intra and extracellular stimuli, alternative splicing and the post-translational modifications (PTM) has made a paradigm shift in the neuroscience field to analyze cellular proteomes by various proteomics methodologies. To realize the elaborate neuroadaptive mechanism in health and disease, it has become elementary to determine the global alternations in proteins by simultaneous assessment of all proteins in a cell (Freeman and Hemby, 2004).

Proteins carry out the greater part of biological events in the cell, even though certain mRNAs can act as effector molecules. mRNA expression may not directly associate with protein expression as mRNA is not the working endpoint of gene expression (Anderson and Seilhamer, 1997; Gygi et al., 1999b). Also, there has been sufficient evidence recently that over 50% of all genes are subject to transcriptional variation by RNA splicing and editing accountable for production of specific isoforms in various cell/tissue types. Sufficient evidence also points toward significant transcriptional and PTM of proteins controlling cellular functions (Roberts and Smith, 2002). Nearly 30–50% of mammalian genes are expressed in the central nervous system (CNS). Proteomic analysis of brain regions may be useful to study the differential patterns of gene expression in order to investigate the complexity of CNS disorders (Fountoulakis, 2004). It is necessary to emphasize the fundamental difference in the study of proteomics as compared to genomics. Proteins have no base pairing and consequently there is no technique of protein amplification like the polymerase chain reaction (PCR) (Mullis, 1990) or antisense RNA (aRNA) (Van Gelder et al., 1990). Nucleic acid hybridization relies on base pairing and the construction of probes that will recognize a specific nucleic acid sequence of interest. However, for proteomic research the recognition of specific proteins is more difficult. The average protein concentration is 10^2–10^8 copies/cell; on the other hand, the rapid turnover of mRNAs is responsible for their average concentration to range from 10^{-4} to 10^2 copies/cell. Thus, proteomics technology has an inherent advantage over genomics for investigating small number of cells (Holland, 2002; Godovac-Zimmermann et al., 2005).

The expansion of proteomic technologies can be ascribed to the rapid development of mass spectrometry (MS), bioinformatics and the current accessibility of vast information from genome sequencing of many organisms. Rather than traditional approaches which examined one or a few proteins at a time, in a few samples, proteomics attempts to concurrently examine large numbers (thousands) of proteins in mutliple samples. Proteomics, a technology-driven science, involves the study of each and every protein in a biological system with respect to structure, change in expression level, protein–protein interactions, PTM as well as the study of multi-protein complexes, coined as structural-, functional- and expression-proteomics, respectively. Most of the initial effort in proteomics was directed toward protein identification and determination of relative abundances. The most-widely developed field, possible at large-scale in automatic mode, is of expression proteomics and is by default the preferred approach of most studies involving proteomics. However, changes in protein abundance exclusively do not define protein function as many of their vital activities are brought about by PTM and do not have the ability to resolve regulatory mechanisms that affect protein abundance and function such as protein–protein interactions and subcellular distribution. Therefore, it is unrealistic that a single strategy will suffice to unravel all the protein complexities in a tissue or cell type.

In addition to the classical approach of expression proteomics to study the various aspects of global proteomics, it is necessary to use variety of differing strategies The development of innovative strategies has been ongoing in neuroproteomics in particular for studying the PTM, mapping of proteins from multi-protein complexes and mapping of organelle proteomes (Dreger, 2003b). Significant innovations in MS continue to have a major impact as well as further the field of neuroproteomics. Routine unequivocal and high-throughput protein identification has been made possible by the nano-electrospray combined with the hybrid quadrupole time-of-light mass spectrometer tandem mass analyzer (ESI Q-TOF MSMS) as well as by matrix-assisted laser desorption-ionization (MALDI) Q-TOF MSMS tandem MS and MALDI-TOF-TOF tandem MS. To know the identities of thousands of different proteins in neurons along with their expression levels, their PTMs as well as protein–protein interaction maps would revolutionize neurobiology and medicine by detecting novel drug targets and diagnostic biomarkers (Fountoulakis, 2004). The present review will analyze applications to familiarize neuroscientists with the available tools for proteome research.

Subcellular fractionation

Biological samples subjected to proteomic analysis in neuroproteomics encompass cell populations, tissues and CSF. These samples are extremely complex as the protein constituents vary in charge, molecular mass, hydrophobicity, PTM and occurrence in complex and subcellular location. The coding genes for CNS oscillate between 25,000 and 30,000 (Southan, 2004). This added complexity of neuroproteome will be overwhelming if we hypothesize that each protein on an average has 10 splice variants, cleavage products and PTM leaving us to analyze almost 250,000–300,000 protein forms. Currently, there are no proteomic methods, which have the capacity to segregate and identify this many proteins at a time. Therefore, fractionation of the entire proteome into distinct fractions to be analyzed separately for content and phenotypic differences is of paramount importance. Each neuron has proteins, which are compartmentalized, providing distinct environments for biological processing such as protein synthesis, degradation, energy production, DNA replication, etc. Therefore, protein localization is normally linked to its function and subcellular fractionation reduces the complexity of the neuroproteome to be analyzed by segregating proteins based on their cellular locations. Tannu et al. (2004a) and many other groups have documented that the subcellular fractionation enables the potential enrichment of lower-abundance proteins (such as signaling molecules) by allowing higher starting amount (at least 3–8 times) than whole-cell proteome analysis.

Most of the organelles in the neuron have been initially characterized by subcellular fractionation and microscopy; however, a complete registry of the proteins in each organelle is yet to be made. Even in physiological states some proteins are translocated between different compartments, such as shuttling between nucleoplasm and cytoplasm. In many diseases the change in gene expression is preceded by translocation events which do not alter the overall abundance of proteins in the entire neuron. A very opportune example for a neuroscientist will be the shuttling of neurotransmitter receptors between the membrane and the cytoplasmic pools in a synapse, which have been associated with synaptic plasticity (Malinow and Malenka, 2002). From a neuroscience point of view, studying differing amounts of protein in different compartments has functional significance as compared to the study of total amount in a neuron. In studies involving subcellular proteomics, fractionation strategies are of prime importance with respect to the accuracy of the proteomics data which assigns new gene products to a particular subcellular location.

The most important caveats for subcellular fractionation are (1) the varying extent of enrichment, (2) differential isolation of cytoskeleton components with organelles and (3) current fractionation techniques which enrich one particular subcellular structure. An ideal proteome study involves monitoring multiple subcellular structures in parallel (Yates et al., 2005). Post-analyses methods, high-throughput prediction tools accessible though Internet are routinely used to validate the subcellular location. These are based on the in-silico analyses of the primary structure of unknown gene products giving cues to credible protein functions and subcellular locations. The programs routinely used and available publicly are: MitoProt (http://ihg.gsf.de/ihg/mitoprot.html) for mitochondrial, SignalP (http://www.cbs.dtu.dk/services/SignalP/) for signal sequences and TargetP (http://www.cbs.dtu.dk/services/TargetP/) and/or PSORT (http://psort.ims.u-tokyo.ac.jp/) for general subcellular location prediction (Dreger, 2003b). Notwithstanding the caveat, the study of neuroproteomics by subcellular fractionation-proteome analysis will help in mining low abundance and membrane proteins and at the same time assign most of the specific proteins to specific subcellular structures and provide valuable insight into the physiological mechanisms and the molecular basis of pathological events. Most of the fractionation strategies are similar for tissues as well as the cell populations. However, the CSF proteins are in a different environment, as compared to the tissues and cell populations, and their segregation strategies will be discussed separately (Dreger, 2003a; Swatton et al., 2004; Righetti et al., 2005; Yates et al., 2005; Wang and Hanash, 2005). It must be noted at this point that the discussion will include the studies involving non-neuronal tissues for instances where no data exists for neuronal tissues. One should also consider that detailed fractionation conditions are valid for particular tissues and may/may not work on different tissue. However, the outlined strategies should serve as an important starting point to standardize the protocol for tissue under consideration. The goal of neuroproteomics development is to gain the ability to simultaneously study nuclear envelope (NE), nuclear pore complex (NPC), nucleolus, golgi apparatus, mitochondria, perioxisomes, cytoplasm, membrane, synapse and post-synaptic density (PSD) fractions of different phenotypical neurons (Schirmer and Gerace, 2002; Schirmer and Gerace, 2005). The fractionation strategies which remain unique for brain tissue are for the synaptosomes, PSD and the CSF. These will be discussed separately to stress their importance in neuroscience.

Cerebrospinal fluid

Even though CSF, secreted by the choroid plexus in the lateral ventricles, is mostly found in the four ventricles it is an important determinant of the extracellular fluid (ECF) surrounding neurons and glia in the CNS. The changes in the brain which affect the proteins involved in biochemical pathways may be reflected in CSF (e.g., change in the CSF levels of total-tau protein and the light subtype of the neurofilament proteins (NF-L) after acute ischemic stroke (Hesse et al., 2001) and the elevation of the tau and phosphorylated tau in the CSF of Alzheimer's disease (AD) patients (Tapiola et al., 1997)). CSF is also found in the subarachnoid space flowing down the spinal canal as well as upwards over the brain convexities. The CSF composition is in steady state with the brain ECF and, therefore, has an important function of maintaining a constant external environment for neurons and glia. It also functions to remove harmful brain metabolites, provide mechanical cushion, and serve as a lymphatic system and as a conduit for peptide hormones secreted by hypothalamus. CSF contains proteins, peptides, enzymes, small molecules and salts which play an important role in many physiological processes e.g., CSF pH is an important regulator of pulmonary ventilation and cerebral blood flow. Even though most of the proteins in CSF are derived either directly from neuronal cells or actively transported by pinocytosis across the blood–brain barrier (BBB) some proteins are synthesized intrathecally e.g., prostaglandin D2 synthase and cystatin C. CSF is not simply an ultrafiltrate of serum but a highly specific repository of cellular byproducts, metabolites, neurotransmitters and proteolytic fragments making analysis of its proteome crucial for understanding neurobiology and neuropathology (Wildenauer et al., 1991; Zheng et al., 2003).

While the proteomic studies of neuronal tissue has the challenges of dealing with deterioration of the BBB in postmortem tissues (PMT) or perform invasive biopsies from antemortem tissues, CSF proteomics is amenable for serial analysis by minimal invasive lumbar puncture either ante or postmortem. The serial evaluation of CSF proteome in context of disease progression and treatment efficiency is a potential clinical application of proteomics that can be used from bench to bedside. A change in protein expression may yield important insight into various CNS diseases by improving our understanding of the molecular basis of disease as well as providing early-stage biomarkers. The CSF has a low protein concentration ($\sim$150–450 μg/ml) and a high salt concentration ($>$150 mmol/l). Despite the low total protein concentration, the concentration of albumin ($\sim$60% of the total CSF protein) (Hammack et al., 2003) and immunoglobulin is extremely high in the CSF. First, the salt concentration should be adjusted in the CSF in a way so that the final concentration of salts should be < 10 mmol/l in rehydration buffer. Second, the high-abundance proteins, albumin and immunoglobulin, should be depleted so that the lower-abundance proteins can be resolved. We briefly describe the strategies that have been tested to circumvent the above problems.

Desalting

Until recently, studies using desalting techniques have reported varying degrees of success with respect to the recovery of the proteins from CSF. A brief summary of the techniques used till date follow.

Protein precipitation: Protein precipitation has been undertaken using ice-cold ethanol, trichloroacetic acid (TCA) in acetone and acetone. Hansson et al., (2004) showed 100% recovery of CSF proteins using > 70% ice-cold ethanol for 2 h at −20°C. Ice cold 80% acetone precipitation for 2 h at −20°C, on the other hand showed mixed reproducibility of 40–50% (Sickmann et al., 2000) and 94% (Yuan et al., 2002 recovery. However, precipitation with TCA in acetone (4:10%:1, v/w/v) for 45 min at −20°C as well as chloroform/methanol (4:8:3, v/v/v) at room temperature for 2 h showed very low recovery of 23% CSF proteins (Yuan et al., 2002; Hansson et al., 2004).

Bio-Spin column (Bio-Rad): Bio-Spin polyacrylamide micro-column with a M_W cutoff of 6 kDa gave less recovery as compared with the ethanol precipitation for the Hansson group (Hansson et al., 2004). On the other hand, the study conducted by (Yuan 2002; Yuan et al., 2002), and Terry (Terry and Desiderio, 2003) showed a recovery of 91% and 99%, respectively.

Ultra filtration: The Ultrafree MC with a cutoff of 5 kDa and Centricon with a cutoff of 3 kDa (Millipore, Bedford, MA, USA) was used by Sickmann et al. (2000) and Hammack et al. (2003), respectively, for desalting of the CSF followed by protein concentration. This approach tested by them showed a recovery of more than 70% CSF proteins by both the groups.

Dialysis: This method was compared directly by Hammack et al. (2003) with the ultra filtration. There was a non-specific loss of 40–60% proteins using MWCO dialysis tubing (Spectrum, Rancho Dominguez, CA, USA) for 12 h. From a range of methods applied for desalting of CSF, 70% ice-cold ethanol shows the most promise. The protein recovery and the separation of proteins on a 2D-gel are crucial factors that decide the efficiency of the above methods. The ultrafiltration, which losses most of the protein due to their adsorption to the filter has a better protein recovery than dialysis. On the other hand, even though the acetone precipitation had a higher recovery of proteins, the vertical and horizontal streaking on the 2D-gels marred the image analysis quality. The Bio-Spin column has an overall better recovery of proteins with the image quality of 2D-gels well preserved.

Protein depletion

To have a realistic opportunity of analyzing the low-abundance CSF proteins, it becomes crucial to deplete the CSF sample from albumin and immunoglobulin which constitute 50–60% of the CSF protein concentration.

Affinity removal: There is a loss of albumin-binding proteins during albumin depletion. This was shown during depletion study of albumin and immunoglobulin using a Cibacron Blue F3G-A (Blue Sepharose 6 Fast Flow) and protein G (Prosep-G), respectively by Raymackers et al. (2000) and Hammack et al. (2003). The loss of proteins bound to albumin can be minimized by segregating them by a separate experiment. Another problem encountered was the low binding of lipoproteins and enzymes to Cibacron Blue F3G-A. As the above kits had been specifically designed for serum, preconcentration of the CSF is recommended. Recently highly specific immobilized anti-albumin and anti-immunoglobulin antibodies were developed by Pierce (ProteoSeek™) and Sigma (ProteoPrep®), which claim the depletion of approximately 90–95% of albumin and immunoglobulin. More recently ProteoPrep® 20 has been developed by Sigma to deplete the 20-most abundant proteins which constitute approximately 97% of the total protein amount.

Liquid-phase isoelectric focusing: Prefractionation method in which CSF proteins are segregated into different fractions based on charge. Each of these fractions is then run separately on a 2-D gel for further segregation. This technique has shown to facilitate detection of less-abundant protein components, reduce sample complexity, increase the protein load and the protein amount in each gel spot for MALDI-MS analysis as compared to unfractionated CSF by Davidsson et al. (2002).

Solid-phase extraction: This method utilizes the differential hydrophobic nature of proteins to separate CSF proteins into three different fractions using a solid-phase extraction cartridge (Yuan and Desiderio, 2005). This technique showed the enrichment of low-abundance CSF proteins at the same time resulting in the ability to resolve them well from high abundance proteins (Schirmer and Gerace, 2005).

Synaptosomes and post-synaptic density

Synapses can be purified in vitro and are called as synaptosomes. Synaptosomes constitute the entire pre-synaptic terminal (including mitochondria and synaptic vesicles) and portions of the post-synaptic terminal (including post-synaptic membrane and PSD) and are considered as highly specific intercellular junctions responsible for transmission of signals within CNS. The study of the proteomes of synapse as well as the PSD is an important starting point in neuroscience to understand complex brain functions.

Synaptosomes are subcellular membranous structures formed during mild disruption of brain tissue. The shearing forces cause the nerve endings to break off and subsequent resealing of the membranes form the synaptosomes. Synaptosomes have a complex structure equipped with components of signal transduction, metabolic pathways and organelles, as well as structural components required for vesicular transport. Schrimpf et al. (2005) characterized 1131 proteins from synaptosomal fractionation belonging to the following categories: proteins involved in exo- and endocytosis; guanine nucleotide-binding proteins and their regulators; synaptic adhesion molecules and ligands; PSD proteins; cytoskeletal proteins; enzymes involved in transmitter synthesis and degradation; transporter proteins; receptor proteins. Synaptosomes can be isolated from brain homogenate by differential and density-gradient centrifugation (Schrimpf et al., 2005). Briefly, brain tissue is homogenized in 5 mM HEPES and 320 mM sucrose (pH 7.4) using a Potter–Elvehjem homogenizer (800 rpm, 12 passes). The cell debris and nuclei are removed by centrifugation of the homogenate twice for 5 min at $1000 \times g$, and the combined supernatant is centrifuged for 20 min at $12000 \times g$. The resulting pellet consists of a colored layer comprising the mitochondria and a white pellet composed of the synaptosomes. On whirl mixing the synaptosomal pellet gets resuspended whereas the mitochondrial pellet remains. This is repeated twice and the suspended synaptosomal pellet is layered on a Ficoll gradient comprising of 4.8 ml of 12% Ficoll overlaid with 4.8 ml of 7.5% Ficoll. Centrifugation at $6,8999 \times g$ for 1 h, the synaptosomal fraction is enriched at the junction of 7.5%/12% Ficoll as a cream-colored layer. The synaptosomes are then recovered by aspiration, resuspended in Krebs' solution. The suspended synaptosomes are centrifuged for 20 min at $12,000 \times g$ to be recovered as a pellet.

The PSD is a disk-like structure with a thickness of ~30–40 nm and width of ~100–200 nm. The most important structures associated with the PSD are the cytoskeletal proteins, regulatory enzymes, and neurotransmitter receptors and associated proteins. There has been more than one fractionation method used for the segregation of PSD in proteomic studies. Recently, two groups successfully characterized 244 and 374 proteins, respectively from PSD fractions. Trinidad et al. (2005) as well as Peng et al. (2004) found proteins know to be in the PSD, NMDA receptor subunits NR1A and NR2A as well as the associated PSD-95, to be highly enriched whereas the pre-synaptic protein synaptophysin to be undetectable. The proteins identified in the PSD fractions belonged to the following groups: scaffold and adaptor proteins; signaling proteins; cytoskeletal and interacting proteins; phosphoproteins; proteins involved in trafficking; proteins involved in energy production and transfer; ubiquitination system; the receptors, ion channels and adhesion proteins; kinases, phosphotases and regulators. These studies were successful in characterization of many proteins that had not been previously associated with PSD, opening up possibilities of their involvement in synaptic morphology and signal transduction. A major concern of these studies was the contamination from subcellular structures unrelated to PSD such as pre-synaptic proteins, housekeeping proteins, mitochondrial proteins, glial cytoskeleton and myelin sheaths. It is imperative to follow up such studies with more focused experiments to confirm whether these are bona fide PSD proteins. As the methods are similar, for brevity we will discuss the method used by Peng et al. (2004). Briefly, the isolated adult rat forebrain was homogenized using Teflon homogenizer (12 passes) in buffer A comprising of 5 mM HEPES (pH 7.4), 1 mM $MgCl_2$, 0.5 mM $CaCl_2$, 1 mM NaF, 1 mM β-glycerophosphate, 0.1 mM PMSF, 1 µg/ml aprotonin, 1 µg/ml leupeptin, 1 mM benzamidine, 0.1 mM pepstatin and the phosphatase inhibitor mixture I (Sigma). All the purification steps were performed at 4°C. This homogenate was centrifuged for 10 min at $1400 \times g$ to obtain supernatant (S1) and pellet (P1). The pellet P1 was homogenized again with the Teflon homogenizer (5 strokes). After centrifugation at $700 \times g$, the supernatant (S1″) was pooled with S1 to be centrifuged for 10 min at $13800 \times g$ to collect pellet P2. A P2 suspension was created in buffer B (0.32 M sucrose, 6 mM Tris (pH 8.0), 1 mM NaF, 1 mM β-glycerophosphate, 0.1 mM PMSF, 1 µg/ml aprotonin, 1 µg/ml leupeptin, 1 mM benzamidine, 0.1 mM pepstatin) using 5 strokes of Teflon homogenizer. A discontinuous sucrose gradient comprising of 0.85 M/1 M/1.15 M sucrose solution in 6 mM Tris (pH 8.0) was loaded with P2 suspension and centrifuged at $82500 \times g$ for 2 h using SW-41 rotor. The synaptosomal fraction at the junction of 1 and 1.15 M sucrose was collected using a syringe and needle and was made up to 4 ml using the buffer B. An equal volume of buffer constituting of 6 mM Tris (pH 8.1) and 1% Triton X-100 was added to the above suspension mixed for 15 min and centrifuged for 20 min at $32,800 \times g$ using Ti70.1 rotor. The pellet was brought up in buffer made of 6 mM Tris (pH 8.1) and 0.5% Triton X-100, and centrifuged for 1 h at $201,800 \times g$. The resulting pellet is the PSD fraction used for further proteomic analysis.

Nuclei, mitochondria, cytoplasm and membrane

Several recent proteomics studies have employed fractionation methods that allow collection of multiple cellular components from one tissue source (Fountoulakis, 2004; Tannu et al., 2004a). The major benefit of these studies has been the ability to compare by enriching the low-abundance proteins, which were not detectable by analysis of the whole-cell proteome. This has enabled the analysis of important signaling molecules. The coverage of the proteome analyzed is also increased as the proteins spots from different fractions have an additive effect toward the whole proteome. As the fractions are from the same cellular subset they minimize the experimental variability. The crucial drawback has been the overlap of the proteins between fractions. It is important to differentiate between the proteins that are cross-contaminants. A general approach to segregate the above-mentioned fractions in a single experiment is schematized in Fig. 1 and described in detail previously (Tang et al., 2003; Tannu et al., 2004a; Fountoulakis, 2004). Briefly, tissue samples are dounce homogenized in 10 mM HEPES, 10 mM NaCl, 1 mM KH_2PO_4, 5 mM $NaHCO_3$, 1 mM $CaCl_2$, 0.5 mM $MgCl_2$, 5 mM EDTA and the following protease inhibitors (PI): 1 mM PMSF, 10 mM benzamidine, 10 µg/ml aprotinin, 10 µg/ml leupeptin, and 1 µg/ml pepstatin and centrifuged at $9645 \times g$ for 5 min. Supernatant (cytosol and membrane) is removed and the pellet (nuclei and debris) resuspended in 20 mM Tris HCl, 1 mM EDTA (pH = 8.0) with PIs and centrifuged at $9645 \times g$ for 5 min. This procedure is repeated twice and the pellet is resuspended in the solution and stored at −20°C (nuclear fraction). The supernatant is then centrifuged at $107\,170 \times g$ for

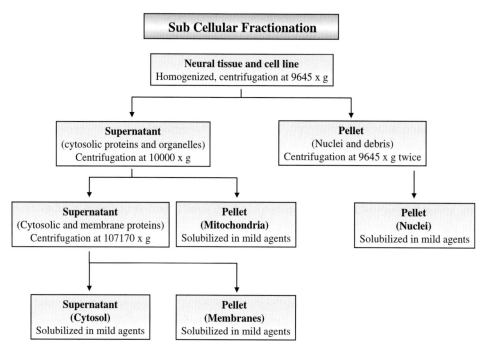

Fig. 1. Schematic representation of subcellular fractionation technique commonly used prior to neuroproteomic analysis (elaborate description in text). Adapted with kind permission from Wiley (Fountoulakis, 2004).

30 min at 4°C. Following, the supernatant containing the cytosolic fragment is removed and stored at −20°C (cytosolic fraction). The pellet is resuspended in 10 mM Tris (pH = 7.5), 300 mM sucrose, 1 mM EDTA (pH = 8.0), 0.1% NP40 and PIs and centrifuged at $4287 \times g$ for 5 min at 4°C. The supernatant is discarded and the pellet resuspended in the buffer and washed three times before resuspension in the buffer and protease inhibitors and storing the samples at −20°C (membrane fraction).

Mitochondria
The mitochondrion is a complex structure involved in fundamental processes, such as the TCA cycle, β-oxidation of fatty acids, urea cycle, electron transport, oxidative phosphorylation, apoptosis and heme synthesis, dysfunction of which is observed in many neurological degenerative diseases (Alzheimer, Parkinson and Huntington Disease). For example, mitochondria in frontal and temporal cortex of AD are decreased 25% of normals. Neuroproteomic analysis of the mitochondria has focused on the abundance of the mitochondrial proteins in different brain regions (Lovell et al., 2005; Zhang et al., 2005). Cataloging mitochondrial proteomes from different species and tissues have documented 400–700 mitochondrial-associated proteins (Mootha et al., 2003; Sickmann et al., 2003; Taylor et al., 2003). These datasets will enable scientists to compare protein orthologues between species to better understand the mitochondrial machinery. Mootha et al. (2003) identified 591 mitochondrial proteins of which approximately 170 proteins had not been previously linked to mitochondria. This important study enabled the identification of a single candidate gene (LRPPRC) for the French-Canadian-type of Leigh syndrome by integration of the proteomics and the RNA-expression data with the genotype data. We briefly present a method used by Mark A Lovell to fractionate mitochondria from primary rat cortical neuron cultures. A FractionPREP cell fractionation kit (BioVision, Mountain View, CA, USA) was used to isolate the mitochondria. The cell pellet lysed in 500 μl lysis buffer (on ice for 15 min) was centrifuged twice at $2000 \times g$ for 5 min at 4°C to pellet nuclei

and cell membranes. The supernatant was further centrifuged for 30 min at 11,000 × g (4°C). The mitochondria were then rinsed in ice-cold PBS (3–5 times) and resuspended in 200 μl distilled/deionized water (DW) and homogenized using micro-Dounce homogenizer on ice.

Membranes

Based on the recent sequencing information from several genomes, membrane proteins constitute 20–30% of all the cellular proteins. Plasma membrane proteins are involved in various important cellular processes in the brain including signal transduction, cell adhesion, exocytosis and metabolite and ion transport (Stevens et al., 2000). Even though the lipids form an important constituent of the membrane, the membrane-spanning proteins confer unique functions to the membranes forming the means of communication between separated structures. As the membrane proteins are amphipathic, their hydrophobic nature makes them difficult to study and necessitates different strategies for analysis as compared to the rest of the cellular proteins. Therefore, while great strides have been made toward the analysis of soluble cellular proteins, membrane proteins reported in majority of the proteomic analyses have been under-represented (Wilkins et al., 1998). The importance of characterization of membrane proteins cannot be over emphasized; they account for approximately 70% of all the drug targets and present an enormous challenge for the modern proteomics initiative (Wu et al., 2003). Even though some reports showed improved solubilization using different buffers (zwitterionic detergent MEGA 10 (decanoyl-N-methylglucamide), zwitterionic lipid LPC (1-lauroyl lysophosphatidylcholine)) and detergent (1, 2-diheptanoyl-*sn*-glycero-3-phosphatdiyl choline (DHPC)), the traditional proteomic approach of two-dimensional gel electrophoresis (2–DGE) has many limitations for analyzing membrane proteins (Churchward et al., 2005). The most important limitations are that most hydrophobic proteins are insoluble in non-detergent isoelectric focusing (IEF) sample buffer, the ones which are insoluble are precipitated at their isoelectric point (pI), they are in low abundance compared to hydrophilic proteins; the pIs of hydrophobic proteins are generally alkaline and even the use of extended pH gradients are difficult to resolve well at the basic ends. The carrier ampholytes inhibit the interaction between the hydrophobic proteins and the immobiline. This prevents the precipitation and the subsequent streaking of the basic end of the gels. Irrespectively, the liquid chromatography (LC)-MS methods have an ascendancy in resolving the issues of the hydrophobic membrane protein separation. To overcome the problems of 2-DGE associated with resolving membrane proteins, several studies have used SDS-PAGE followed by LC-MS to catalog membrane proteins (Ferro et al., 2002; Galeva and Altermann, 2002). The principle issue faced by this approach is the inaccessibility of cleavage sites of the membrane-spanning domains to trypsin and the limited sequence coverage, confidence, with which proteins are characterized.

An attractive alternative is shotgun/multi-dimensional protein identification technology (MuDPIT) proteomics where proteins are first digested by proteases and the complex peptide *mixture* is analyzed by LC-MS-based methods, although significant computing recourses are required for analysis. This strategy has greatly enhanced sensitivity for detecting mass changes due to covalent modifications making it amenable for detection of PTMs as well as giving insights into protein topology. The methods recently employed to address the issue of solubility of membrane proteins have used detergents (Navarre et al., 2002), organic solvents (Blonder et al., 2002) and organic acids (Washburn et al., 2001).

Organic acid: This method utilized 90% formic acid in the presence of cyanogen bromide to solubilize yeast membrane-enriched fraction. The concentrated formic acid provides the solubilization agent. The cyanogen bromide on the other hand is active in this acidic ambiance, when it is able to cleave the embedded membrane proteins. The fragments were further digested serially by proteinase LysC and trypsin followed by peptide characterization by MuDPIT. MuDPIT allows peptide separation by 2-DLC(separation based on charge and hydrophobicity) and identification by in- or off-line tandem MS. This qualitative method was successful in identifying 131 integral membrane proteins (Washburn et al., 2001).

Detergents: The detergent is used to solubilize the membrane proteins. Microsomal membrane-enriched fraction was boiled using 0.5% SDS. Labeling with isotope-coded affinity tag (ICAT) was followed by dilution of SDS, so that the SDS concentration was compatible with trypsin digestion. The peptide mixture was separated by sequential cation exchange chromatography, avidin affinity chromatography and reverse-phase microcapillary HPLC for MS analysis resulting in identification of 491 proteins. This method offers the ability to quantitatively analyze the sample while at the same time improving the recovery of low-abundance integral-membrane proteins.

Organic solvents: The organic solvents can also be used to solubilize the membrane proteins. Blonder et al. (2002) thermally denatured and sonicated membrane-enriched fraction of *Deinococcus radiodurans* in 60% methanol in presence of trypsin. This method was also used for quantitation as well as detection of low-abundance proteins by coupling with ICAT (Goshe et al., 2003).

High pH: The high pH method prevents the resealing of the membrane structures after mechanical agitation thereby maintaining the native topology, as against the methods of organic acids, solvents and detergents (Goshe et al., 2003). The sample is incubated at 37°C for 3 h at a high pH (200 mM Na_2CO_3, pH 11) favoring the formation of membrane sheets and proteinase K (hpPK, 5 μg) cleaving the membrane protein hydrophilic domains. A MuDPIT analysis of the lysate identified 1610 proteins with the cellular proportion of membrane to soluble being close to 1:2.5.

Building on the initial analyses of membrane proteins from other tissues, recent studies have attempted to map the brain plasma membrane proteome. Nielsen et al. (2005) used the conventional plasma membrane isolation approach. A stepwise depletion of mouse brain cortex to remove the non-membrane proteins using high salt, carbonate and urea washes was performed. This was followed by treatment with sublytic concentrations of digitonin and density gradient fractionation. The enriched membrane fraction was lysed by endoproteinase LysC to identify about 1000 membrane proteins. Schindler et al. (2005) described a membrane fractionation protocol compatible with small amounts of brain tissue (e.g., cerebellum of a single rat) so that distinct functional and anatomical regions of brain from model animals can be studied (Schindler et al., 2005). Affinity-based partitioning of microsomes in an aqueous two-phase (polyethylene glycol (PEG) and dextran) system was used to enrich the plasma membrane fraction. Membranes from different subcellular fractions were separated based on their charge and hydrophobicity on the two-phase system. The plasma membranes have higher affinity for PEG phase and were separated on the top. Close to 500 proteins were characterized by LC-MS/MS, of which 197 (42%) proteins were bona fide plasma membrane proteins. Besides the proteins from plasma membrane such as transporters, channels and neurotransmitter receptors e.g., cerebellum-specific GABA receptor GABAR6, proteins belonging to the mitochondrial (1.2%) as well as endoplasmic reticulum (1%) membrane were also identified, signifying the cross-contamination between cellular compartments.

Nucleus

The nucleus has a high degree of organization, consisting of structurally and functionally distinct compartments; nucleolus, nuclear speckles, NPC and the nuclear envelope. The nucleus is a highly organized organelle consisting of domains fundamental for preserving the homeostasis of the cellular milieu.

Nuclear envelope

NE is a double-membrane system (inner and outer) continuous with the endoplasmic reticulum, perforated with NPCs and lined by nuclear lamina. Over the last few decades, various integral NE proteins have been characterized by biochemical and genetic approaches. However, a single proteomics study has been able to detect many of these along with numerous novel components (Dreger et al., 2001). The inner nuclear membrane (INM) comprising of distinct transmembrane proteins connects the INM to a polymer of intermediate filaments (lamins) to form the lamina. The outer nuclear membrane (ONM) is continuous with the endoplasmic reticulum and is functionally similar to it, making the proteomic analysis of NE a challenging task. There have been a few studies which have attempted to decipher the NE proteome exlusively. One of the studies undertaken by Dreger et al. (2001) assumed that NE proteins will have similar biochemical extraction characteristics as the proteins from lamina. With the above assumption, the nuclei from cultured neuroblastoma cells (5–8 mg of protein) were suspended in 40 ml of ice-cold TP buffer (10 mM Tris HCl (pH 7.4), 10 mM NaH_2PO_4, Na_2HPO_4 (pH 7.4), 1 mM PMSF, 10 μg/ml aprotinin, 10 μg/ml leupeptin) containing 250 μg/ml heparin, 1 mM Na_3VO_4, 10 mM NaF, and 400 units of Benzon Nuclease (Merck). After stirring for 90 min at 4°C, nuclear envelopes were sedimented by centrifugation ($10{,}000 \times g$) for 30 min at 4°C and resuspended in STM 0.25 buffer [20 mM Tris HCl (pH 7.4), 0.25 M sucrose, 5 mM $MgSO_4$, 1 mM Na_3VO_4, 1 mM PMSF, 10 μg/ml aprotinin, 10 μg/ml leupeptin). The NE proteins were extracted with 4 M urea in 0.1 M sodium carbonate generating an insoluble fraction rich in integral proteins. The known lamina-associated INM proteins from the non-ionic detergent-insoluble as well as salt-insoluble fractions were used for comparison. Proteins deemed likely candidates for novel INM proteins comprised the chaotrope pellet, which were also found in the detergent- or salt-extracted pellets. Most, but not all of the previously known INM were characherized along with four novel proteins. The second study was performed using subtractive approach. The proteins from the endoplasmic reticulum also present in the NE fraction were excluded (Schirmer et al., 2003).

Nuclear pore complex

The first comprehensive proteomic study of the yeast NPC by Rout et al. (2000) identified 40 proteins, comprising of previously known and unknown gene products, of which 11 were transport factors and 29 nucleoporins. More recently, Cronshaw et al. (2002) performed a proteomic analysis of rat liver NPC to identify ~30 proteins. Briefly, pelleted nuclei are resuspended with constant vortexing at a final concentration of 100 U/ml by drop-wise addition of buffer A (0.1 mM $MgCl_2$, 1 mM DTT, 0.5 mM PMSF, 1 μg/ml leupeptin/pepstatin/aprotinin) supplemented with 5 μg/ml DNase I and 5 μg/ml RNase A. After resuspension, nuclei are immediately diluted to 20 U/ml by addition of buffer B (buffer A + 10% sucrose, 20 mM triethanolamine, pH 8.5) with constant vortexing. After digestion at room temperature for 15 min, the suspension is underlayed with 4 ml ice-cold buffer C (buffer A + 30% sucrose, 20 mM triethanolamine, pH 7.5) and centrifuged at $3500 g$ for 10 min in a swinging bucket rotor (Sorvall SH-3000). The pellet is then resuspended in ice-cold buffer D (buffer A + 10% sucrose, 20 mM triethanolamine, pH 7.5) at a final concentration of 100 U/ml. The suspension is diluted to 67 U/ml with buffer D + 0.3 mg/ml heparin, and then immediately underlayed and pelleted as above. The heparin pellet is resuspended in ice-cold buffer D (100 U/ml), diluted to 67 U/ml with buffer D + 3% Triton X-100, 0.075% SDS, and then pelleted as above. The resultant pellet (the NPC-lamina fraction) is resuspended in buffer D + 0.3% Empigen BB (final concentration of 100 U/ml). After incubation on ice for 10 min, the insoluble lamina is separated from soluble nucleoporins by centrifugation in a microfuge at $16{,}000 g$ for 15 min. The NPC proteins in this study were separated by C4 reverse-phase chromatography followed by SDS-PAGE. The individual protein bands from SDS-PAGE were subjected for tandem MS analysis. The examination of the NPC has also been undertaken in some studies involving the brain; in hippocampus neurons of AD subjects and in the hypothalamic ventromedial neurons of rats post-exposure to high estrogen to name a few.

Nucleolus

The nucleolus coordinates the synthesis and assembly of ribosome's and has been implicated in cell growth, cell-cycle, apoptosis, senescence as well as the stress responses (Coute et al., 2005). Surprisingly not many studies have been conducted to document the changes in nucleolar proteome in neuroscience. The studies that will benefit the most by means of this approach will be the various studies on brain tumors. Before the year 2002, 121 human proteins were known to be located in the nucleolus based on publications, which used biochemical and subcellular localization techniques such as antibody staining and/or fluorescent tagging. Andersen et al. explored the proteome of human HeLa cells to identify 271 (of which 82 were novel nucleolar proteins) and 667 proteins in successive studies (Andersen et al., 2002, 2005). Proteins from important functional groups were revealed; ribosomal structural proteins as well as proteins involved in their synthesis and assembly; chromatin structural proteins; transcriptional and splicing factors; mRNA metabolism; translation factors; chaperones; DNA repair, replication; mitosis and cell-cycle regulation; ubiquitination and protein degradation; nucleocytoplasmic transport; kinases and phosphatases; enzymes; unpredictable function. Briefly, aliquots (250 µl) containing $\sim 1 \times 10^8$ nuclei were washed three times with PBS, resuspended in 5 ml buffer A (10 mM HEPES-KOH (pH 7.9), 1.5 mM $MgCl_2$, 10 mM KCl, 0.5 mM DTT), and Dounce homogenized 10 times using a tight pestle. Dounced nuclei were centrifuged at $228 \times g$ for 5 min at 4°C. The nuclear pellet was resuspended in 3 ml 0.25 M sucrose, 10 mM $MgCl_2$, and layered over 3 ml 0.35 M sucrose, 0.5 mM $MgCl_2$, and centrifuged at $1430 \times g$ for 5 min at 4°C. The clean, pelleted nuclei were resuspended in 3 ml of 0.35 M sucrose, 0.5 mM $MgCl_2$ and sonicated for 6×10 s using a microtip probe and a Misonix XL 2020 sonicator at power setting 5. The sonicate was checked using phase contrast microscopy, ensuring that there were no intact cells and that the nucleoli were readily observed as dense, refractile bodies. The sonicated sample was then layered on 3 ml 0.88 M sucrose, 0.5 mM $MgCl_2$ and centrifuged at $2800 \times g$ for 10 min at 4°C. The pellet contained the nucleoli, while the supernatant consisted of the nucleoplasmic fraction. The nucleoli were then washed by resuspension in 500 µl of 0.35 M sucrose, 0.5 mM $MgCl_2$, followed by centrifugation at $2000 \times g$ for 2 min at 4°C. The protein identification was performed by LC-MS/MS.

Expression proteomics

As mentioned previously, expression proteomics is the workhorse of current neuroproteomics initiative. Gel-based and MuDPIT approaches have enabled important advances in the measurement of protein expression alterations in normal and disease states. Important consideration for expression proteomics is the accuracy and reproducibility of each technique. It is widely believed that the two techniques offer complimentary data and it is critical to understand the advantages as well as the challenges inherent to them (Choe et al., 2005). The factors that are crucial for the successful implementation of proteomics technique are low detection limits, optimal signal-to-noise ratio and a wide dynamic range.

Two-dimensional gel electrophoresis

2-DGE is a widely used separation method in which the application of an electrical potential is applied across a solid-based gel whereby proteins are first separated by IEF and then by molecular mass in the second dimension. The basic principles of 2-DGE have remained the same since its introduction in 1972 by O'Farrell (1975) and Klose (1975). Recent technological improvements have enabled more reliable and reproducible IEF on strips that are gels that can be sued with the second-dimension slab gels and improved image analysis software that more accurately matches staining profiles across different gels. In general, approximately 1000–2000 proteins spots can be visualized on a gel depending upon the visualization technique, the pI range of the first dimension as well as the size of the 2-DGE. These usually represent the most-abundant housekeeping proteins and the more interesting proteins (signaling molecules and receptors) from a biological point of view, which are in lower abundance, remain obscured by the most-abundant proteins. Enrichment of such low-abundance proteins can be well achieved by subcellular fractionation (Fig. 1). Some of the inherent caveats/disadvantages associated with 2-DGE are (a) many protein spots are likely comprised of multiple proteins, (b) individual proteins may migrate as multiple spots based on differential digestion, (c) labor-intensive image analysis requires gel matching and manual removal of artifacts, (d) poor spot resolution at higher pIs, (e) difficulty in electrophoresing large and hydrophobic proteins in the first-dimension separation and (f) extreme acidic and basic proteins are not well represented (Van den Bergh et al., 2003). The technical variability seen with 2-DGE is due to sample preparation, sources of reagent, staining methods, image analysis software and individual experimenter variability accounting for a coefficient of variation of 20–30% (Molloy et al., 2003).

Sample preparation
Like any experiment, the quality of the results is dependent on the quality of the preparatory material. Adequate coverage of the intricacies of protein isolation would be difficult to address in current review; however, a few points of critical importance for proteomic experiments are discussed. Protein stability and purity, as well as prevention of protein degradation and modification, are of critical importance during sample preparation for proteomic approaches. Rapid removal of brain tissue, dissection and freezing are obvious imperatives for the maintenance of the proteome state in the animal. Human postmortem studies pose unique challenges, but can and should be undertaken with careful documentation of postmortem interval, brain pH and agonal state (Hynd et al., 2003). Specific proteins such as dihydropyrimidinase-related protein-2 has been putatively identified as a marker of *postmortem* interval and temperature (Franzen et al., 2003), highlighting the need for careful selection of controls in human brain postmortem studies (Fountoulakis, 2004). Protease and phosphatase inhibitors are used to help prevent degradation and dephosphorylation of proteins during protein preparation (Olivieri et al., 2001); however, care should be taken such that adducts and charge trains are not introduced by these inhibitors.

Purification of protein from other cellular substances is also necessary. Lipids and specific proteins (e.g., albumin and immunoglobulin isoform) are particularly abundant in the brain and, along with nucleic acids, must be eliminated from the protein sample. The most common methods of protein purification rely on selective precipitation. Acetone, TCA and other precipitation methods can be performed and a number of commercially available kits make this a routine procedure (Polson et al., 2003). In some instances though, pure protein may not be sufficient since proteins such as IgGs or albumin constitute the vast majority of protein concentrations in cells. Selective elimination of these proteins improves detection of less highly expressed proteins (Lollo et al., 1999).

Isoelectric focusing
IEF separates proteins according to their pI. The pI of a protein is primarily a function of its amino acid sequence although PTM can also contribute to the pI. Proteins are amphoteric molecules, capable of acting as either an acid or base. The side chains of the amino acids in proteins have acid or basic buffering groups that are protonated or deprotonated, depending on the pH of the solution in which the protein is present. IEF takes advantage of this property by placing proteins in a pH gradient and applying a potential such that the protein will migrate toward the anode or cathode, depending on the net charge. At the pI, the protein will reach the point in the pH gradient where the net charge of the protein is zero and stop migrating.

Initially, the preparation and use of the pH gradients needed for IEF was very difficult and inconsistent. pH gradients were often in the form of tube gels with carrier ampholytes, but the introduction of chemistries (Husi et al., 2000) to immobilize the pH gradient into the gel matrix and the popularization of this technique (Bjellqvist et al., 1982) was a significant step in making IEF more widely accessible. Most current applications of IEF use immobilized gradients in dedicated instruments which control both potential and temperature (Gorg et al., 1995). Exceedingly high potentials (e.g., 8000 V) are needed to focus proteins and consistent focusing requires close control of the temperature (Gorg et al., 2000). Commercial suppliers are producing IEF gels with narrow pH ranges, which when used in an overlapping fashion (pH 4–5, 4.5–5.5, 5–6, 5.5–6.7 and 6–9) enable the separation and detection of thousands of proteins (Gorg et al., 1991). IEF is rarely used on its own and is usually followed by applying the IEF strip to SDS-PAGE gel for electrophoresis in the second dimension based on molecular weight (described below).

Solution IEF

Solution IEF operates on the same principle as normal IEF except that proteins are separated into pI range bins, in solution (Wildgruber et al., 2000). Proteins can then be electrophoresed on standard IEF gels with a narrow pI range, the same as the pI range of the bin. One of the reasons for performing solution IEF is that when loading a whole-cell lysate onto a narrow pI range IEF gel, proteins outside the IEF range precipitate and can pull out proteins from within the range of the IEF gel. Solution IEF can increase the number of proteins observed, the amount of sample loaded and resolution. The drawback to this approach is the addition of another experimental step that can result in the sample loss; however, with further development, this technology has great potential.

SDS-PAGE

2-DGE along with MS are the two most commonly used techniques in proteomics, namely the separation of proteins by IEF (first dimension) followed by SDS-PAGE (second dimension) which involves the separation by molecular weight, of proteins that have already been separated by IEF. In general, the IEF gel or strip is equilibrated with SDS and placed on top of the SDS gel. The equilibration step is necessary to allow the SDS molecules to complex with the proteins and produce anionic complexes that have a net negative charge roughly equal to the molecular weight of the protein. The SDS gel is then electrophoresed and the proteins migrate out of the IEF gel and into the SDS gel, where they separate according to molecular weight. While most applications use denaturing SDS-PAGE, native approaches have also been used. Conventional SDS-PAGE instruments, such as those used for Western blotting and special purpose apparatuses can be used for this step.

Traditional stains. Coomassie brilliant blue (CBB), silver nitrate, and negative staining are common post-electrophoresis methods available for the 2-D gel-based proteomics analysis. The sensitivity of these stains ranges from 100 ng (e.g., Coomassie) down to 1 ng (e.g., silver) for individual protein spot detection (Neuhoff et al., 1990; Scheler et al., 1998). The organic dye CBB available in two modifications, Coomassie R-250 and Coomassie G-250, is one of the most widely used stains for expression analysis. In acidic medium, the dye binds to amino acids by electrostatic and hydrophobic interactions. However, it is not reproducible and reliable for quantitation as some of the proteins release the dye during the background destaining procedure. As a rule of thumb, naked-eye visualization of a spot by CBB stain infers adequate protein for mass spectrometric characterization. Colloidal CCB staining is more reproducible and has higher sensitivity, has quantitative protein binding and is lower in price. CCB is compatible with MS as complete destaining can be achieved using bicarbonate.

Silver staining method is the other traditional staining technique most widely followed for quantitative analysis because its sensitivity is as low as 1 ng per spot. The popularity of this stain can be gauged by extensive use of this method represented by 150 modifications of silver staining protocols published to date (Rabilloud et al., 1994). However, in principle, silver staining detects the proteins primarily at the gel surface. As it is not an endpoint procedure, there is a high degree of variability of staining intensities for particular spots and thus is unreliable for quantitation. A common glitch encountered with this method in the detection of abundant protein spots is the formation of yellow center, which result in concave peaks that are problematic for quantitative analysis. Despite its excellent sensitivity, silver staining lacks reproducibility, has a limited linear dynamic range, a subjective judgment of the staining end-point and interferes with the MS compatibility, resulting in a much lower sequence coverage compared to CBB (Mortz et al., 2001). Even though silver staining is still used currently, there has been an ever-increasing trend in the scientific community to use the new generation fluorescent stains especially for broad-scale proteomics analyses.

Other traditional staining, though less popular, methods employ copper and negative imidazol SDS zinc. The negative imidazol initially showed immense promise for further analysis of proteins after quantitation as it stains the background without modifying the protein. The protein spots were visualized against dark background and the detection sensitivity stayed between CBB and silver. Even though it was documented to show quantitative analysis, it was disputed because only the background was detected directly and not the protein (Ferreras et al., 1993).

Fluorescent stains. In general, the fluorescence-based detection methods are more sensitive than the absorbance-based methods because the detected wavelength is different to the incident wavelength. Prevailing over various issues limiting the traditional gel staining techniques, fluorescence-based gel stains are recently becoming widely accepted (White et al., 2004). SyproRuby™ dye (Molecular Probes, Eugene, OR), among the first of the fluorescent stains for proteins, is part of a stable organic complex composed of ruthenium which interacts non-covalently with the basic amino acids in proteins (Berggren et al., 1999). The stain can be visualized using a wide range of excitation sources commonly used in image analysis system and has a sensitivity which approximates silver staining while maintaining a broad linear dynamic range (three orders of magnitude). The fluorescent stain does not contain or require chemicals such as glutaraldehyde, formaldehyde and Tween-20 that normally impede with peptide mass fingerprinting (Berggren et al., 2000). A recent study showed an enhanced recovery of peptides from in-gel digests of SyproRuby™ stained proteins compared to silver staining using MALDI-TOF MS (Lopez et al., 2000); however, important drawbacks include the tendency to induce speckling and high background staining which blemish the visual appraisal of gel images, as well as concerns of stain disposal. Recently, DeepPurple™ (GE Healthcare, Piscataway, NJ), a fluorophore epicoccone from the fungus *Epicoccum nigrum* which interacts non-covalently with SDS and protein, has been introduced for protein gel staining. As sensitive as SyproRuby™, Deep Purple has a dynamic linear range over four orders of magnitude, shows no speckling and has a low background staining intensity (Mackintosh et al., 2003; Smejkal et al., 2004). Tannu et al. (2006a) in recent study concluded that the Deep-Purple™ stain results in an increased peptide recovery from in-gel digests compared to SyproRuby™ stain and also improves MALDI-TOF based identification of lower intensity proteins spots by increasing sequence coverage. The additional peptides seen from Deep Purple™ stained proteins were attributed to incomplete cleavage or modified (primarily with respect to methionine oxidation) forms of peptides already present in the spectrum. Incomplete cleavage by trypsin was attributed to the binding of epicoccone to the lysine residue, one of its primary cleavage sites (Coghlan et al., 2005). This study opens up the possibility of confident identification of low-abundance proteins for identification which have been evasive until now in spite of reliable quantitative data by the fluorescent dyes. With the availability of more sensitive stains, the challenge to acquire reasonable mass spectra for identification of lower-abundance proteins is crucial. It becomes important to confirm that dyes do not interfere with MS since dye interference can cause ion suppression with a resultant lower recovery of peptides or a reduction of signal intensities. In spite of this caveat, the majority of information currently available about protein gel staining involves comparing the efficacy of the staining techniques with little information regarding their comparative MS compatibilities.

2D-DIGE. One of the most-recent technical advances in 2-DGE has been the multiplexing fluorescent 2D-DIGE (Unlu et al., 1997). This method relies in direct labeling of the lysine groups on proteins with cyanine (Cy) dyes prior to IEF. The critical aspect of the use of 2D-DIGE technology is the ability to label 2–3 samples with different dyes and electorphorese all samples on the same 2-D gel. This ability reduces spot pattern variability and reduces the number of gels in an experiment making spot matching much more simple and accurate. The single positive charge of the CyDye replaces the single positive charge present in the lysine at neutral and acidic pH keeping the pI of the protein relatively unchanged. A mass of approximately 500 Da is also added by the CyDye to the labeled protein. The most popularized experimental design has been the use of a pooled internal standard (sample composed of equal aliquots of each sample in the experiment) labeled with the Cy2 dye and labeling the control and the diseased/treatment groups with either Cy3 or Cy5 dyes swapped equally across the samples, respectively (Fig. 2). Minimal labeling is performed to tag only one lysine residue in each protein to prevent the vertical train of spots due to the added mass of each fluorophores and prevent the protein precipitation due to increased hydrophobicity. The individual protein data from the control and diseased/treatment (Cy5 or Cy3) samples are normalized against the Cy2 dye-labeled sample, Cy5:Cy2 and Cy3:Cy2. These log abundance ratios are then compared between the control and diseased/treatment samples from all the gels using statistical analysis (t-test and ANOVA) (Tonge et al., 2001, Alban et al., 2003; Tannu).

This method has the advantage of being able to quantify the protein spots that are uniquely present in one group due to the presence of internal standard. The accuracy of quantitation as well as the statistical confidence obtained for the differentially regulated gene products is significantly higher using the experimental design of 2D-DIGE (Alban et al., 2003; Knowles et al., 2003; Tannu). CyDye-labeled proteins are scanned by Typhoon™ variable mode imager. Sequential scanning of Cy2, Cy3 and Cy5-labeled proteins is achieved by the following lasers/emission filters; 488/520 nm, 532/580 nm and 633/670 nm, respectively. Scanned images of fluorescence-labeled proteins are sequentially analyzed by differential in-gel analysis (DIA; performs Cy5/Cy3: Cy2 normalization) followed by biological variation analysis (BVA; performs inter-gel statistical analysis to provide relative abundance in various groups). The 2D-DIGE approach offers great promise and has been used increasingly by researchers to address a wide range of neuroscience questions from e.g. Alm et al. studying the neurodevelopment toxicity of PBDE-99 and Tannu et al. studying the altered phenotype of nucleus accumbens of human cocaine overdose victims (Prabakaran et al., 2004; Swatton et al., 2004; Beckner et al., 2005; Roelens et al., 2005; Sitek et al., 2005; Tannu). 2D-DIGE offers the most reliable quantitation of any 2-DGE method, is comparable in sensitivity to silver staining method and compatible with the downstream MS protein characterization (as majority of the lysine residues remain untagged and accessible for tryptic digestion).

The major drawback of this technique is that it is proprietary to GE Healthcare and requires expensive labeling dyes as well as specific equipment such as a three-laser fluorescent scanner and robotic spot picker including dedicated software. Also due to the prolonged scanning times, the protein diffusion affects the eventual protein spot quantitative analysis across the larger set of gels. However, Tannu et al. have recently documented the expediency of protein spot fixation prior to the scanning of gels in a large-scale 2D-DIGE experiment. This study circumvented the problem of protein spot diffusion at the same time maintaining the original protein spot quantitative analysis (Koichi Tanaka et al., 1988; Tannu and Scott, in press: 2006b). Also should be noted is that proteins with high percentage of lysine residues are possibly labeled more efficiently compared with the proteins with few/no lysines. Therefore, the possibility remains that a high-abundant protein spot in the conventional 2DGE can be a medium or even low-abundant protein in 2D-DIGE due to low lysine content. A modification of 2D-DIGE in which Cy dyes that label all of the cystine residues of proteins are labeled has recently been introduced. The detection limit for saturation labeling is 0.1 ng or protein per spot as opposed to 1 ng protein per spot thereby reducing the amount of protein sample required for analysis (Shaw et al., 2003).

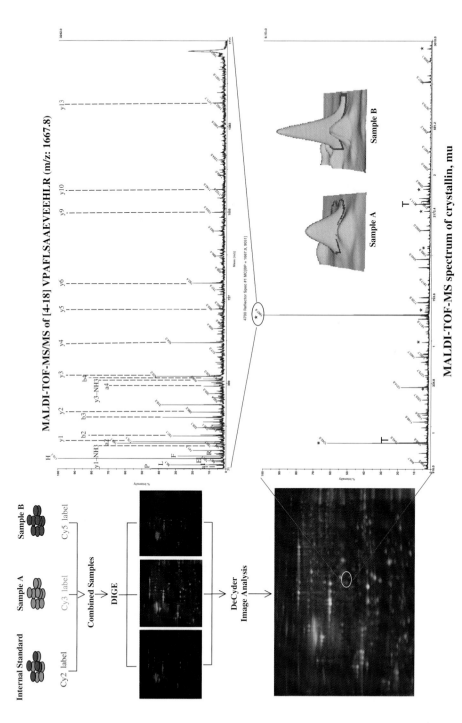

Fig. 2. Typical 2D-DIGE workflow for comparative expression proteomics. The samples are labeled by fluorescent dyes (Cy 2, 3 and 5) and combined together prior to the IEF. The scanned fluorescent images are analyzed by DeCyder™ analysis software using the normalization by Cy2-labeled pool sample. The protein spots (e.g., crystallin mu) with statistical significance that express different expression profiles between the two groups are further analyzed by MALDI-TOF-MS. The peptide mass fingerprint of crystallin (mu) shows the peptides marked with '*' matched to MASCOT search against the NCBInr *primate* database. The *x*- and *y*-axes show the mass to charge (*m/z*) ratio and the % abundance of the tryptic peptide fragments, respectively. Also depicted in the inset for crystallin (mu) mass spectrum is the representative 3-D protein-expression profiles for the two groups. Typically the most-abundant peptide fragment is sequenced by, additive series of the y- and b-ions generated as well as the immonium ions, MALDI-TOF-TOF (e.g., peptide fragment with *m/z* of 1667.8) to confirm the identification of protein and characterize the sites of PTM.

They provide a very attractive alternative for performing quantitative 2D-DIGE when dealing with low sample amounts, typical of neuroscience, even though only two saturation dyes are currently available (Cy3 and Cy5). A caveat, low protein amount complicates MS-based characterization of differentially regulated proteins analyzed by saturation labeling (Zhou et al., 2002a).

Quantitative image analysis
The image analysis in a classic 2-DGE-based proteomic approach encompasses the analysis of thousands of proteins simultaneously across different groups to reveal differentially regulated proteins. It becomes mandatory to use powerful algorithms to analyze these large data sets and a number of software packages are currently available (Marengo et al., 2005). These image analysis software packages are uniquely designed for either 2-DGE or 2D-DIGE, except Progenesis™ that has the capability of handling either method. The precision with which image analysis software performs spot detection, matching and normalization dictates the quality of data generated. The state-of-the-art commercial products have been automated with respect to these requirements; however, require some manual confirmations even with the higher-end products. Much software available has been compared to evaluate their individual pros and cons (Rosengren et al., 2003; Arora et al., 2005; Fodor et al., 2005; Marengo et al., 2005). PDQuest™, one of the most popular software for image analyses of 2-DGE ($\approx$ 70 studies), requires manual setting of appropriate selection parameters for spot detection. The sensitivity, size scale, minimum peak are determined based on manual selection of a faintest, smallest, large spot and a large representative section of the image containing spots, streaks, and background gradations to make corrections for noise filter. The typical normalization method used for each protein spot detected by PDQuest™ is a ratio of its raw spot intensity and the cumulative total intensity from each protein spot detected on the entire gel based on absorbance. Also to get the best-image matching between multiple gels, manual land-marking encompassing the entire gel is imperative. Progenesis™ Workstation ($\approx$ 8 studies) on the other hand is designed for automatic spot detection with no requirement for manual intervention and is by far the most high-throughput as well as the high-end image analysis software available for 2-DGE as well as the 2D-DIGE analysis. An iterative method is used in the spot detection, image warping and matching by the Progenesis™. DeCyder™ ($\approx$ 12 studies) spot detection is based on initial detection of the protein spots in the Cy2 image of pooled sample followed by application of similar spot boundaries to the remaining images (Cy3 and Cy5). The 3-D spot viewer which has been incorporated in most of the software program's recent version's have been very helpful for the purposes of manual land marking, confirming detection of true spots as well as the spot matching across the entire sub-set of gels in an experiment. A study comparing the softwares PDQuest™ and Progenesis™ for 2-DGE showed comparable accuracy of protein spot quantitation for well-resolved areas of gels (Arora et al., 2005). The integrated interpretation of results for studies using different image analysis softwares is a difficult task, and requires a common platform with integrated software addressing the weaknesses of individual programs and at the same time incorporating the strengths into a single user-friendly workstation.

Multi-dimensional separation of proteins

The coupling of efficient chromatographic and electrophoretic separation methods with high-performance MS hold great promise for qualitative and quantitative characterization of highly complex protein mixtures. The advances in chemical tagging and isotope labeling techniques have made possible the quantitative analysis of proteomes, and the specific isolation strategies have enabled the analysis of PTM. The multi-dimensional separation is typically based on using $\geqslant$ two physical properties of peptides (size, charge, hydrophobicity and affinity) to fractionate complex peptide mixture into individual components. The methods employed to fractionate peptides based on their corresponding physical and chemical properties are ultracentrifugation (density), capillary electrophoresis (size and charge), IEF (pI), size-exclusion chromatography (Stoke's radius), ion-exchange chromatography (charge), hydrophobic interaction chromatography (hydrophobicity), reverse-phase chromatography (hydrophobicity) and affinity chromatography (biomolecular interaction).

The drawback of 2-DGE to detect low-abundant proteins as well as the proteins with extreme pI, molecular weight and hydrophobicity has been the promise offered by the multi-dimensional chromatographic approach for proteomic analysis (Gygi et al., 2000; Washburn et al., 2001; Peng et al., 2003a). A caveat, no single chromatographic or electrophoretic method used in different combinations employed by multi-dimensional separations has been successfully devised to separate, detect and quantify all proteins in a given proteome. In most multi-dimensional approaches the proteins are digested into peptides prior to separation. The digestion of proteins produces complex mixture of peptides, however, at the same time it increases the overall solubility by eliminating non-soluble extremely hydrophobic peptides. This is extremely critical in neuroscience for studying the synaptic and PSD proteins, typically insoluble in aqueous buffers, involved in signal transduction (neurotransmitter receptors and G-proteins), molecular transport (carriers and voltage-gated ion channels) and cell–cell interactions. The peak capacity of multi-dimensional separation is the product of the peak capacities of its component one-dimensional methods. As mass spectrometer can perform mass measurements on several but not all coeluting peptides, fractionation is a critical aspect for mass spectral identification of peptides. The MS/MS cycle times of all conventional mass spectrometers are limited by the number of peptides that can be selected by collision-induced dissociation (CID).

Wolters et al. (2001) have initially showed that MuDPIT was reproducible within 0.5% between two analyses. Furthermore, a dynamic range of 10,000 to 1 between the most-abundant and least-abundant proteins/peptides in a complex peptide mixture was also demonstrated. The comprehensive proteome analysis requires the ability of a system to detect variation in protein abundance in $\geqslant$ six orders of magnitude to detect a potential biological significance (Corthals et al., 2000). The LC techniques currently used successfully in neuroscience have been ion-exchange (cation as well as anion), reverse-phase (RP) and affinity and will be elaborated here. Most RP-HPLC separations are carried out using acetonitrile (ACN) in combination with ion-pairing agent (formic acid or trifluoroacetic acid (TFA); depending on the downstream mass spectrometer to be used) to improve the selectivity. The ion-exchange chromatography is performed using salts (sodium chloride, potassium chloride, ammonium acetate/formate) at different concentrations at a gradient format. It is important to select the salt, its concentration and the buffer composition in such a way as to not affect the second-dimension separation in terms of resolution.

Quantitative analysis
Several strategies have been developed for relative quantitation of protein expression between samples. The labeling of proteins or peptides for quantitation followed by MS is currently rapidly advancing approach. The important steps (Fig. 3) in which this technique is practiced are: (1) isotopic labeling of separate protein mixtures, (2) combined digestion of the labeled proteins followed by multi-dimensional liquid chromatographic separation, (3) automated MS/MS of the separated peptides and (4) automated database search to identify the peptide sequences and quantify the relative protein abundance based on the MS/MS.

Isotope-coded affinity tags (ICAT and iTRAQ). This is the prototypical and the most popular method for quantitative proteome analysis based on stable isotope affinity tagging and MS (Gygi et al., 1999a). The ICAT reagent is a sulphydryl-directed alkylating agent composed of iodoacetate attached to biotin through a short oligomeric coupling arm (d0). The substitution of 8 deuterium atoms for hydrogen atoms in the coupling arm produces a heavy isotope version of the reagent (d8). Thus the reagent comprises of a cysteine-reactive group, a linker containing the heavy or light isotopes (d8/d0) and a biotin affinity

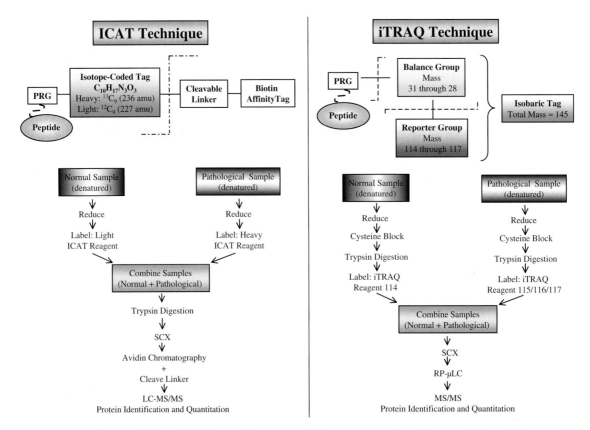

Fig. 3. Non-gel-based approach for analysis of expression proteomes. The most commonly used, commercially available, quantitation methods coupled with the MDLC approach of comparative proteomics are depicted. The figure shows the ICAT™ and iTRAQ™ reagent structures. The peptide-reactive group (PRG) covalently links iTRAQ™ as well as the ICAT™ reagent isobaric tag with the lysine side chain and N-terminal group of peptide. The fragmentation ( ) occurs as shown during MS/MS. The balance group of the iTRAQ™ reagent undergoes neutral loss, and the resultant reporter group (114–117) peaks in the low-mass region are used for relative quantitation. The elaborate description of the ICAT™ and iTRAQ™ workflow is in the text.

tag (Fig. 3). This method involves in vitro derivatization of cysteine residues in protein with d0 or d8 followed by enzymatic digestion of the combined sample. All the cysteine residues thus tagged with biotin are selectively separated by avidin column. The cysteine-containing peptides are further separated by reverse phase followed by MS analysis. In the process the complexity of the peptide mixture decreases significantly as only the cysteine-containing residues are enriched selectively. In humans, databases indicate that the occurrence of cysteine and the frequency of cysteine-containing peptides is ∼90% and 17%, respectively. The isotopically tagged peptides give quantitative MS analysis based on the relative peak intensities/areas of d0- and d8-labeled peptides (Gygi et al.,

1999a). Another advantage is the ability to analyze peptides with molecular weight more than 3000 Da easily because the mass difference between the coded isoforms is sufficiently large.

The major limitation of ICAT is that it can only be used to examine the concentration or structural changes in cysteine-containing peptides (10–20% of the peptides). The resolution is greatest in the case of smaller peptides where the d8/d0 ratio is higher and with peptides that have multiple cysteine residues (Regnier et al., 2002). Another limitation of the technique is that the biotin affinity tag remains linked to the peptides throughout the analysis causing shifts in chromatographic separation, shifts in m/z and changes to MS/MS spectra relative to the unlabelled peptides complicating the manual or computer-assisted interpretation (Gygi et al., 1999a; Ferguson and Smith, 2003). To address the issues of the first-generation ICAT technique, several second-generation chemistries for cysteine-specific isotope tagging have been developed. Zhou et al. used a solid-phase capture-and-release method and a photo- and acid-cleavable linker, an isotope-tag transfer group and a specific reactive group (Fig. 3) (Qiu et al., 2002; Zhou et al., 2002b). The major advantage provided by this method is that the isolation of cysteine-containing peptides and incorporation of the stable isotopes are achieved in single step, making it simpler, faster and easier to automate. There is also improved selective enrichment of cysteine-containing peptides as well as increased sensitivity by this method. The second-generation ICAT undergoes similar procedure as the original except that the biotin group is cleaved off before MS analysis of the tagged peptides. The isotope tag introduced with this reagent consists of $^{13}C_6$ instead of deuterium, which is relatively small. The peptides labeled with $^{13}C_6$ or $^{12}C_6$ have virtually the same retention time during the RP-HPLC (Zhang et al., 2001). The only chemical modification remaining on the peptides at the time of LC-MS/MS is an isotopically labeled leucine residue. The added advantage are, reduced sample handling and facilitation of extensive sample washing protocols prior to peptide elution (Zhou et al., 2002b).

Most of the analyses based on ICAT technology have coupled strong cation exchange (SCX) LC with reverse-phase microcapillary LC coupled on-line (RP-μLC) with MS and MS/MS (Link et al., 1999; Washburn et al., 2001; Gygi et al., 2002). Data-dependent software is used to select specific mass/charge (m/z) peptides for CID, alternating MS and MS/MS scans for collecting qualitative and quantitative data. The on-line LC-ESI-MS/MS has the drawback due to the requirement of continual sample consumption and the 'on the fly' selection of precursor ions for sequencing. The use of MALDI-MS/MS offers significant advantage that the temporal constraints of an on-line detection are eliminated as the peptides separated by the μLC are deposited on the MALDI sample plate before the MS analysis (Medzihradszky et al., 2000; Krutchinsky et al., 2001). Alternative strategies such as per-methyl esterification of carboxylic acid groups (Goodlett et al., 2001), specific labeling of lysine residues (Peters et al., 2001) and peptide N-termini (Munchbach et al., 2000) have also been probed recently. The quantification softwares such as XPRESS (http://www.systemsbiology.org/research/software/proteomics/) and ProICAT™ (Applied Biosystems, Foster City, CA) have been developed which can assemble a composite ratio for a protein based on the calculated expression ratio from all the peptides from a single protein. The data obtained from the above softwares can be analyzed collectively using INTERACT for multiple experiments (Han et al., 2001).

The iTRAQ™ technique capable of multiplexing samples is primarily based on the ICAT technique and compared in detail in Fig. 3. The iTRAQ™ technique uses four isobaric reagents (114, 115, 116 and 117) allowing the multiplexing of four different samples in a single LC-MS/MS experiment. The multiplexing capability of iTRAQ™ allows a control sample to be compared with different points in time of a disease state (e.g., acute, sub-acute, chronic and relapse) as well as with respect to different drug treatments. One of the major advantages of this technique over the ICAT™ is its ability to label multiple peptides per protein, which increases the confidence of identification as well as quantitation. As shown in the Fig. 3, each isobaric iTRAQ™ reagent constitutes of a reporter group, a balance group and a peptide-reactive group (PRG). Multiple peptides in a protein are labeled by covalent linking of the PRG with each lysine side and the N-terminal group of a peptide. The MS spectra of combined samples bear a resemblance to that of an individual sample. The balance group (31 to 28) makes it possible to display all the iTRAQ™ reagents (114–117) to be displayed at same mass. A neutral loss of the balance group occurs during the MS/MS, and the reporter group ion peaks appear in the low mass region. The area under the curve for each reporter ion peak represents the quantitation for that particular peptide. An average quantity can be assigned to a protein after incorporating the quantitative information from all the peptides that were identified for a particular protein. The relative amounts of protein from different samples are then the ratios of the average quantity obtained as above. A recent study compared the 2-DGE and iTRAQ™ technique using the *Escherichia coli* for consistent measurements showing an average CV of 0.24 for isobaric tagging to 0.31 for 2-DGE. Also a greater range of expression ratios was demonstrated by the proteins quantified by the isobaric tagging as compared with the 2-DGE (Choe et al., 2005). A more recent study by Wu et al. (2006) systematically compared the techniques of DIGE, ICAT™ and iTRAQ™. The DIGE technique was amenable for compromised quantitation due to partial/complete comigration of proteins. The global tagging iTRAQ™ was found to be more sensitive than the ICAT™ which was as sensitive as the DIGE. The complimentary nature of these techniques was confirmed by the limited overlap of the proteins characterized (Wu et al., 2006).

Peptide labeling with $H_2^{16}O/H_2^{18}O$. The samples to be compared are separately digested in either $H_2^{16}O$ or $H_2^{18}O$. The oxygen atom derived from the aqueous solvent is incorporated into the newly formed C-terminus acid functional group in each peptide, providing an effective isotope tag for relative quantitation (Mirgorodskaya et al., 2000; Stewart et al., 2001; Yao et al., 2001). The requirement of this technique to separately digest the protein samples to be compared has the potential to be imprecise due to separate sample handling. The difference of 4 Da between the ^{16}O and ^{18}O has limited usefulness for larger peptides, where mass spectral isotopes of labeled and unlabeled peptides separated by only 4 Da begin to overlap.

Tandem ion exchange /reverse-phase chromatography
The method of choice for multi-dimensional separation of peptides has been the SCX for the first-dimensional separation of peptides followed by the microcapillary RP chromatography. The negative charges at the carboxyl groups and the C-terminus are neutralized due to complete protonation at pH < 3. This leaves the arginine, lysine and histidine residues as well as the N-terminus contributing to a net positive charge of the peptide. The SCX chromatography fractionates the fully protonated peptides. On the other hand for anion exchange chromatography completely deprotonated basic residues by pH > 12 are required. A mixed mode effect is commonly exerted by most ionic exchangers, principal for the ionic interactions during tandem ion exchange (IEX), due to their hydrophobic influence (Zhu et al., 1991). Recently a biphasic column combining SCX and RP chromatography (direct analysis of large protein complexes: DALPC) has shown to have a improved resolution, loading capacity and the ability to detect low-abundance proteins as compared to a single-dimension column (Link et al., 1999). The SCX has four times greater loading capacity than the RP, greatly increasing the number of digested proteins that can be analyzed. DALPC has been shown to detect novel components of splicing, transcription and RNA processing as well as protein kinases which were not detected by earlier studies such as the 2-DGE analyses (Ohi et al., 2002; Sanders et al., 2002). Recently this method was optimized by the use of volatile salts to elute peptides, automated and combined to sensitive MS to be recoined as multi-dimensional protein identification technology identifying approximately 1500 proteins in single analysis (Washburn et al., 2001; Wolters et al., 2001; VerBerkmoes et al., 2002). The MuDPIT has been combined with ICAT™ for quantitative proteomic analyses in neuroscience as well (Li et al., 2004; Lovell et al., 2005; Schrimpf et al., 2005).

The peptides are separated by SCX using a LC system (Ultimate 3000, Dionex, Sunnyvale, CA, USA; Ettan, GE Healthcare, Uppsala, Sweden). Typically 4.6 × 200 mm Polysulphoethyl A™ column (PolyLC, Columbia, MD, USA), a silica-based column having a hydrophilic anionic polymer (poly(2-sulfoethly aspartamide)), is used for separation by 200 μl/min flow rate. The buffers generally used are 'A' 10 mM KH_2PO_4, ACN 30% (pH 3.0) and 'B' 10 mM KH_2PO_4, 350 mM KCl ACN 30% (pH 3.0). The gradient used is usually optimized for a particular tissue. A typical gradient run consists of 0–50% B over 30 min, 50–100% B from 30–31 min, remain at 100% B up to 36 min, return to 100% A and equilibrate for 20 min before the next run. The fractions collected vary from as low as 10 to as high as 62 depending on the sample complexity and the gradient used. The fractions are neutralized at this stage and loaded into 4 × 15 mm avidin column equilibrated in 2× phosphate-buffered saline. The ICAT™-labeled peptides are eluted with three column volumes of 30% ACN 0.4% TFA, dried and reconstituted in cleavable reagent to cleave the biotin portion of the tag from the labeled peptides. The fraction is dried and dissolved in the mobile phase for RP separation of peptides.

Since most of the multidimensional separation techniques are interfaced with MS, RP chromatography is the choice of second-dimension because the samples eluted from it are in most desirable form for injection into the mass spectrometer. The separation efficacy of the RP is dependent on the particle size, pore size, surface area, stationary phase as well as the chemistry of the substrate surface. The C_{18}-bound phase has been the most popular as it offers retention and selectivity for a wide range of compounds containing different polar and non-polar groups on their surface. To enhance mass transfer, silica monolith columns have been introduced recently (Minakuchi et al., 1997; Ishizuka et al., 2000; Premstaller et al., 2001). These columns are comprised of continuous rod of silica-based gel, which is made of highly interconnected network of large and small pores. The macropores (2 μm) allow fast flow of the eluent and the fine pores (13 nm) offer the surface area required for the separation process. The monolith material has a total porosity of over 80% that facilitates high permeability, good surface area. It also enhances mass transfer due to convection and not diffusion. The combination of effects results in practically no loss in peptide resolution, peak elution volume and concentration of analyte with flow rates (10 ml/min) 10 times higher than conventional rates (1 ml/min). Each salt fraction from the SCX is subjected to RP gradient of 60–90 min on a Ultimate 3000™ equipped with a Famos Micro Autosampler and Switchos Micro Column Switching Module (Dionex, Sunnyvale, CA, USA) using the buffers; A (0.1% TFA) and B (80% ACN, 0.08% TFA) at 0.4 μl/min. Typically the Ultimate 3000™ (Dionex, Sunnyvale, CA, USA) elutes the peptides through 5 mm C18 PepMap100 trapping column (300 μm i.d.) and a 15 cm C18 PepMap100 resolving column (75 μm i.d.) at 0.4 μl/min. A typical gradient run consists of 0–80% B over 60 min, 80–100% B from 60–61 min, remain at 100% B up to 71 min, return to 100% A and equilibrate for 20 min before the next run. The eluent is monitored at 214 nm and mixed with matrix (7 mg/ml CHCA in 70% ACN, 0.1% TFA spiked with 0.15 mg/ml dibasic ammonium citrate and 0.25 fmol/ml ACTH clip 18–39 (ratio of 1:2)) every 4 s via a micro-tee fitting of Probot Micro Fraction Collector (Dionex, Sunnyvale, CA, USA) on to a MALDI plate for MALDI-TOF-TOF analysis (Li et al., 2004; Peng et al., 2004; Lovell et al., 2005; Schrimpf et al., 2005; Trinidad et al., 2005).

Top–down proteomics

The above-described technique (bottom–up proteomics) is critically based on consistent enzymatic conversion of proteins to peptides. It is customary to accurately make mass measurements by a MS/MS of lower molecular weight peptides rather than higher molecular weight intact proteins. The bottom–down approach increases the sample complexity and the entire sequence coverage for proteins is rarely achieved. This seriously limits site-specific PTM analysis of proteins from a biological context of view. These limitations of the bottom–up approach have renewed interest in the top–down proteome characterization strategies. This technique characterizes the individual proteins by MS without prior enzymatic cleavage. Capillary isoelectric focusing (CIEF) coupled with Fourier transform-ion cyclotron resonance (FTICR) MS has been the first report to analyze complex protein mixture using top–down approach (Jensen et al., 1999; Valaskovic and Kelleher, 2002). A two-dimensional display, pI and molecular weight, similar to the conventional 2-DGE, however, with a higher resolution are seen on both the axes. The mass measurement accuracy can be enhanced by isotope depletion of proteins. The major limitation of this technique is that this level of information is not always sufficient for confident protein identification due to the possibilities of point mutations, PTM and the presence of ORFs having high sequence homology. This problem can be solved to some degree by incorporation of isotopically labeled amino acids into the cellular proteins of unicellular model organisms. The partial amino acid content information obtained combined with CIEF-FTICR, enables identification of proteins from genome databases without the MS/MS information (Jensen et al., 1999; Martinovic et al., 2002). The top–down approach using FTICR-MS/MS demonstrated localization of PTM and site-specific mutations in bovine carbonic anhydrase providing 100% sequence coverage for the protein. Simple protein mixtures can be analyzed by prior 1-D separation of proteins as well as 2-D separation followed by infrared multiphoton dissociation (IRMPD)-MS/MS by tandem quadrupole-FTICR. Besides requiring large amount of sample (1 g of yeast cells) the method is not high-throughput and not amenable for automation, much needed for analysis of complex protein mixtures.

Surface-enhanced laser desorption ionization

The surface-enhanced laser desorption ionization (SELDI) technique comprises of ProteinChip arrays, a mass analyzer and the data analysis software. ProteinChip array-based technology consists of spots with chromatographic surfaces. These surfaces are either preactivated for capture of protein molecules or have certain physiochemical properties such as hydrophobic, hydrophilic, cationic, etc. The technique requires an incubation of the sample (1–10 μg) on the spot, followed by washing of the unbound proteins as well as the salts. Matrix solution is added to the sample adsorbed on the spot to be analyzed by laser desorption ionization TOF-MS. The intensities of the different sample components plotted on y-axis against the m/z (x-axis) are used for differential mapping. The similar m/z components are clustered and the clusters are compared to give statistical significant p-value for a given profile.

SELDI-TOF-MS has shown application in proteomic profiling from pre-frontal cortex of schizophrenia and bipolar disorder subjects; CSF of frontotemporal dementia subjects; CSF of rat models of cerebral ischemia; CSF of glioma subjects; sera of neuroblastoma subjects and CSF of AD subjects. The technology was found to be well suited for generating differential maps of protein regulation; however, the major drawback is its inability to characterize the proteins of interest in succession. The identification needs subsequent purification and/or enrichment, followed by proteolytic digestion and peptide mass fingerprinting. The second major drawback is the inability to analyze all the proteins unlike most of the currently available proteomics techniques. It has also been noted that the higher molecular weight proteins (>30 kDa) are not well resolved. The technique also runs into problem due to a narrow dynamic linear range for purposes of quantitation. The diagnostic potential of SELDI technique-derived proteomic maps justifies further studies.

Functional proteomics

This field of proteomics monitors and analyzes the spatial and temporal properties of molecular network of proteins. The functional proteomics analyzes a large set of proteins for PTM critical to the function of proteins in signaling, their localization and turnover, and protein interactions. MS is a general method used to determine the PTM due to its ability to accurately measure the change in molecular weight. Some of the important PTM in neuroscience are: phosphorylation, glycosylation, acetylation, methylation, sulfation, ubiquitination and tyrosyl nitration. The most popular of the above PTM will be discussed with an emphasis on diverse ways to identify them on a proteomics scale.

Phosphorylation

Analysis of phosphorylation, conventionally regarded as the most imperative PTM, includes identification of phospho-proteins and localization of exact phosphorylated residue(s). Phosphorylation of serine, threonine and tyrosine residues is recognized as a key regulator for a wide range of biological functions and activities in eukaryotic cells, such as enzyme activity, signal transduction, transcriptional regulation, cell division, cytoskeletal rearrangement, cell movement, apoptosis and differentiation (Krebs, 1983; Hunter, 1998; Yan et al., 1998) affecting approximately one-third of all proteins at any given time (Zolnierowicz and Bollen, 2000). The various techniques currently available for phosphopeptide detection are schematized in Fig. 4. The classical method, detects phosphoproteins by autoradiography, by incorporating ^{32}P or ^{33}P by protein kinases into cultured cells or subcellular fractions. This approach is limited to specimens amenable to radio-labeling and poses certain safety and disposal problems (Guy et al., 1994; Wind et al., 2001). Immunoblotting is also used for phosphoprotein detection with antiphosphoserine, antiphosphotyrosine and antiphosphothreonine antibodies. In spite of the availability of high quality antibodies to phosphotyrosine residues, antiphosphoserine and antiphosphothreonine antibodies have inconsistent reproducibility. Immunoblotting also complicates subsequent use of the protein for sequencing by MS (Kaufmann et al., 2001).

Specific fluorescence-based detection methods such as Pro-Q Diamond stain devised recently have been rigorously established for staining phosphoproteins (Martin et al., 2003a; Schulenberg et al., 2003; Steinberg et al., 2003). Even though autoradiography is considered the most sensitive detection method for phosphorylation a recent study failed to show any significant differences in the number of proteins spots detected by autoradiography as compared to the Pro-Q Diamond stain (Wu et al., 2005). The Pro-Q Diamond staining method avoids the culture artifacts during culturing of cells with ^{32}P and is feasible with tissue from animal models as compared to the radioisotopes. This stain has a detection sensitivity of 1–16 ng of phosphoprotein and can be used to stain IEF gels, SDS-PAGE and 2-DGE for detection of phosphorylated serine, threonine and tyrosine residues (Martin et al., 2003b). This is a very convenient and less time consuming technique for preliminary assignment of phosphorylation status to hundreds of protein spots, from a total of thousands present on a 2-DGE map. This allows focusing on a small number of protein spots of interest stained by Pro-Q diamond to be later analyzed for protein content and characterization of the phosphorylation residues by MS/MS. Recently, Tannu et al. (Tannu) undertook the first proteomic scale phosphoproteome analysis of primate brain tissue, to throw light on some novel membrane-, receptor- and cytoskeletal-associated proteins, by coupling 2 DGE/Pro-Q Diamond staining with MS, to be involved after cocaine self-administration (Tannu). The use of both 2-DGE and LC-based approaches enables us to achieve greater proteome coverage (Collins et al., 2005a).

In spite of the importance of phosphorylation, identification of phosphorylation site(s) is still a challenge. There are several reasons which complicate phosphoprotein analysis: (1) only a small fraction of the intracellular proteins is phosphorylated at a given time, (2) the phosphorylated sites on proteins might vary (a protein can exist in many different phosphorylated forms), (3) most of the signaling molecules are present at low abundance intracellularly and (4) phosphotases can dephosphorylate a protein unless appropriate precautions are taken during the preparative stages of cell lysates. A general approach to overcome this challenge consists of phosphopeptide isolation from the enzymatic digest of proteins using the immobilized metal affinity chromatography (IMAC) columns, followed by derivatization of phosphoprotein and/or phosphopeptide prior to its analysis by MS/MS to confirm the identification as well as localizing of the phosphorylation site (Trinidad et al., 2005; Collins et al., 2005a). This method exploits the high affinity of phosphate groups toward a metal-chelated stationary phase such as Fe^{3+} (Posewitz and Tempst, 1999; Zhou et al., 2000; Stensballe et al., 2001; Trinidad et al., 2005). IMAC generally enriches for phosphoserine, phosphotyrosine and phosphothreonine residues as they are negatively charged groups. The affinity is extended over to aspartic acid and glutamic acid, and histidine (electron donor). Virtual elimination of the non-specific binding to IMAC column can be achieved (Ficarro et al., 2002). The phosphopeptides are first derivatized to corresponding peptide methyl esters rendering the IMAC selective for phosphopeptide and eliminating the confounding binding through

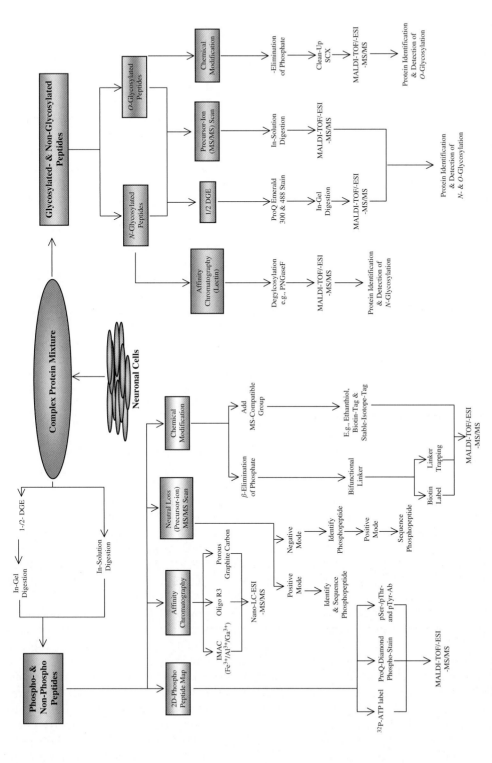

Fig. 4. Graphic description of the different detection methods routinely practiced for analysis of phosphorylation and glycosylation events at proteomics-scale (elaborate description in text). Adapted and reprinted with kind permission from Wiley-VCH Verlag GmbH & Co KG, and Elsevier (Mann et al., 2002; Reinders et al., 2004).

the carboxylate groups. The use of IMAC enriches for phosphopeptide, however, other alternatives such as purification on a polymer-based reverse-phase perfusion chromatography resin (oligo R3) (Matsumoto et al., 1997; Neubauer and Mann, 1999) or alternatively a porous graphite carbon (PGC) (Chin and Papac, 1999) are also beneficial. The chemical modifications derivatizing the phosphopeptides by 1% phosphoric acid (Kjellstrom and Jensen, 2004), ethanethiol (Resing et al., 1995), ethanedithiol (Oda et al., 2001), DTT (Amoresano et al., 2004) and diammonium citrate (Asara and Allison, 1999) have also shown to enhance the phosphopeptide ion signal in mass spectra.

There are some challenges which present during the MS stage of the phosphoproteome analysis: (1) the phosphopeptides are negatively charged whereas the MS is generally performed in the positive ion mode, (2) phosphopeptides being hydrophilic do not bind well to the columns which are routinely used for purification of peptides before analysis and (3) there is a strong possibility of ionic suppression by non-phosphopeptides making the phosphopeptides to be observed as less intense peaks in the mass spectra. The peptides containing phosphoserine or phosphothreonine residues when subjected to CID, undergo a gas-phase β-elimination reaction, resulting in neutral loss of phosphoric acid (-98 Da:H_3PO_4 loss) or are de-phosphorylated (-80 Da:HPO_3) (Bennett et al., 2002). However, phosphotyrosine is usually more resistant to this loss. As the mass spectrometer measures m/z values, doubly and triply charged peptide ions show an apparent loss of 49 and 32.66 Thompson (Th) in the mass spectrum (Covey et al., 1988; Schlosser et al., 2001). In the positive ion mode of MALDI-TOF-TOF, the serine and threonine residues show predominant neutral loss as compared to 80 Da loss, and are differentiated from tyrosine residue which generally shows only 80 Da loss (Annan and Carr, 1996). The suppression effect of the phosphopeptides in the mass spectra can be overcome by elimination of phosphate group to generate dehydroalanine from phosphoserine and dehydroamino-2-butyric acid from phosphothreonine; on the other hand, phosphotyrosine undergoes no elimination due to aromatic nature of the side chain. In the MS/MS spectrum, a spacing of 69 Da (dehydroalanine) or 83 Da (dehydroaminobutyric acid) indicates the exact location of phosphorylated serine and threonine residues respectively. There have also been reports where specific treatment with phosphatase was used to specifically identify phosphopeptides based on characteristic shift of mass due to the loss of phosphate after the treatment (Yip and Hutchens, 1992; Liao et al., 1994).

Specific MS scans methods such as parent ion scanning (precursor ion scanning operated in negative ion mode) as well as the neutral loss scanning have been used for analysis of phosphorylation sites (Carr et al., 1996; Schlosser et al., 2001; Steen et al., 2001). The negative mode of MALDI-MS has been recently shown to produce somewhat more intense signals from phosphopeptides as compared with the positive mode, an avenue which can be furthered (Ma et al., 2001). The fragments in the CID of triple quadrupole MS specifying for phosphate groups serve as reporter ions for precursor ion scanning by tandem MS. Quite a few large-scale analyses of synaptic phosphoproteome mapping have been undertaken using the above-discussed multiple complimentary approaches at the level of protein extraction, phosphopeptide enrichment and the analysis by MS and MS/MS (Collins et al., 2005b; Trinidad et al., 2005).

Glycosylation

Glycosylation, attachment of glycans/carbohydrates to proteins, is one of the most extensive and complex PTM. This modification is confounded by variable modifications such as phosphorylation, sulfation, methylation and acetylation of the glycans residues. It has been documented to play an important role in cell growth and development, cell–cell interaction, carcinogenesis, neurodegeneration and AD (Adamo et al., 1989; Lowe and Marth, 2003; Takeuchi et al., 2004; Lue et al., 2005; Moreira et al., 2005; Robbe et al., 2005). The brain is affected mostly to a severe degree in 10 of the 11 known congenital disorders of N-linked glycosylation (Jaeken and Matthijs, 2001). The branched glycans present as

N- and O-linked glycosylation have great linkage diversity as well as structural complexity presenting a significant challenge for characterization of glycosylation at global scale. Typically, O-linked glycosylation occurs on serine or threonine residues, while the N-linked glycosylation (most widely analyzed in the eukaryotes) occurs on asparagines residue. The protein glycosylation, particularly the N-linked glycosylation is encountered in extracellular proteins (Roth, 2002). The various techniques currently available for detection of glycosylation events in cellular proteins are represented in Fig. 4. Recently gel-based strategies have been developed, which do not require the breakdown of the glycoprotein, and can be integrated in the general proteomics workflow (Packer et al., 1999; Raju, 2000; Hart et al., 2003). The staining with Pro-Q Emerald 300 is based on reacting carbohydrate groups by a periodate/Schiff's base (PAS) mechanism. The oxidation of carbohydrate groups to aldehydes is followed by conjugation with chromogenic (acid fuschin, Alcian Blue) or fluorescent (Pro-Q Emerald 300 and 488) substrates. The glycoproteins segregated by 2 DGE and stained by Pro-Q Emerald 488 are scanned by fluorescent scanners and the images analyzed by image analysis softwares.

Another promising strategy is the coupling of lectin-affinity technology with LC-MS/MS (Hirabayashi et al., 2002). The general strategy used by this technique is the selective capture of glycopeptides from proteolytic protein products by lectin. This is followed by the determination of N-glycosylation sites using peptide-N-glycosidase F (PNGaseF) in presence of $H_2^{18}O$ (isotope-coded glycosylation site-specific tagging: IGOT) and identification of the protein by LC-MS/MS (Gonzalez et al., 1992; Hirabayashi, 2004). This technique has been able to successfully reveal 400 N-glycosylation sites in 250 proteins (Kaji et al., 2003). This technique is limited for its incapability to analyze O-glycopeptides by and large due to the unavailability of a universal glycosidase for release of O-glycans from glycoproteins. Typically, chemical modification of the O-linked glycosylated residues undergoing β-elimination (reducing conditions such as sodium borohydride) are used for assigning this modification for analysis by MS/MS.

A broader capture of glycoproteins based on the conjugation of glycoproteins to a solid support using hydrazide chemistry, stable isotope labeling of glycopeptides and the specific release of N-linked glycosylated peptides by PNGaseF has also been successfully applied to identify membrane and extracellular glycoproteins (Zhang et al., 2003). The above study also made a reasonable quantitative assessment of the glycopeptides by isotopically N-terminal labeling with d0 (light) and d4 (heavy) forms of succinic anhydride after C-terminal lysine residues were converted to homoarginines. The precursor ion scans in tandem mass spectrometers identify glycopeptides based on characterization of one or more of the daughter ions with m/z of 204 Da ($[HexNAc+H]^+$), 274 Da ($[NeuAc-H_2O+H]^+$), 292 Da ($[NeuAc+H]^+$) and 366 Da ($[Hex-HexNAc+H]^+$). A comprehensive N-glycoproteome analysis at global scale has been recently attempted by coupling the above technique with 2D LC-MS/MS (FTICR) to identify 303 non-redundant N-glycoproteins having 639 N-glycopeptides (Liu et al., 2005).

Ubiquitination

The 2004 Nobel Prize in chemistry was awarded to Hershko, Ciechanover and Rose for the central importance of ubiquitin in regulating protein degradation (Kirkpatrick et al., 2005). This PTM involves modification of protein substrates by a highly conserved 76 amino acid polypeptide, ubiquitin (Kaiser and Huang, 2005). The C-terminal glycine of ubiquitin covalently links through an isopeptide bond to the side chain of lysine(s) within the substrate as mono-, multi- or poly-ubiquitination. A family of ubiquitin-like (Ubl) proteins is also known to form similar covalent PTM. Ubiquitin typically is a degradation signal, while Ubl modulates exclusively non-proteasomal endpoints. Ubiquitination plays a central role in protein stabilization, localization, interactions as well as functional activity for many protein substrates (Finley et al., 2004). Mono-ubiquitination is linked to protein transport and poly-ubiquitination initiates the proteolysis of substrates; on the other hand, both regulate the protein function directly without affecting their stability (Hershko and Ciechanover, 1998; Hicke, 2001; Colledge et al., 2003; Pickart, 2004; Pickart and Eddins, 2004; Pickart and Fushman, 2004; Kato et al., 2005). Recent direct evidence has been established between oxidative damage to the neuronal ubiquitination/deubiquitination machinery and the pathogenesis of sporadic AD and Parkinson's disease (Choi et al., 2004). The enzymes involved in the multi-step transfer of ubiquitin are E1 enzymes (activating), E2 enzymes (conjugating) and E3 enzymes (ligases) (Kirkpatrick et al., 2005). The highly dynamic process of ubiquitination is balanced by the deubiquitinating enzymes.

The importance of ubiquitination in the cell physiology and disease necessitates a global approach to study cellular ubiquitination at large scale. This is difficult because the modification is large, 8 kDa, and due to the rapid turnover of ubiquitinated proteins the steady-state levels are characteristically low. A tryptic signature peptide at the ubiquitination site consists of two-residue remnant (glycine–glycine) derived from the C terminus of ubiquitin that remain attached by an isopeptide bond to the target lysine residue. This signature peptide shows a mass shift at the lysine residue of 114.1 Da. A caveat, because of the missed cleavage at the ubiquitin modification site the −GG signature peptides become too large for standard MS analyses and require alternate digestion strategy. The protein conjugates which have been detected by present technique are only a subset of all ubiquitin conjugates. Each ubiquitin-like protein modifier potentially will leave its own signature remnant peptide bound to its target and detected by MS. The substrates purified via an N-terminal epitope tag [DK 10] fused to ubiquitin, digested by trypsin and separated by LC-MS/MS identified as many as thousand proteins with ubiquitination along with 110 precise ubiquitination sites (Peng et al., 2003b; Kirkpatrick et al., 2005). The use of high mass accuracy Fourier transform-ion cyclotron resonance mass spectrometer (FTICR) have been beneficial for identifying the ubiquitination sites (Cooper et al., 2004).

The phosphorylation of some proteins is prerequisite for ubiquitination and subsequent substrate degradation (Ciechanover et al., 2000). Non-tagging strategies for enriching targets like ubiquitin-binding proteins as well as in vitro systems can be used effectively to identify targets for ubiquitin and Ubl proteins (Gocke et al., 2005; Mayor and Deshaies, 2005; Mayor et al., 2005). The amino-terminal labeling of −GG signature peptides by modification with fluorous affinity tags has also been used (Brittain et al., 2005). A double-affinity purification procedure is routinely used to increase the stringency of ubiquitin and Ubl substrates characterized by shotgun proteomics approach (Mayor and Deshaies, 2005; Mayor et al., 2005). Coupling the ICAT™ technique for quantitation of ubiquitinated proteins promises a unique analytical benefit for overcoming the huge excesses of peptides from ubiquitin for large-scale characterization (Kirkpatrick et al., 2005). Even though there has been a recent progress in global characterization of mono-ubiquitination, to develop a methodology for global analysis of poly-ubiquitination still remains a challenge.

Nitration

The chemical modifications exerted by nitric oxide having biological significance are through interactions with transition metals, free radicals, redox regulators and thiol groups such as in cysteine (Martinez-Ruiz and Lamas, 2004a, b). The S-nitrosylation of cysteine residues (addition of NO to sulfur atom to form S-NO bond) and nitration of tyrosine residues (addition of nitro group to position 3 of phenolic ring of tyrosine residue) are the primary PTMs by NO. The S-nitrosylation is implicated in cellular signal transduction pathways (Bolan et al., 2000; Lane et al., 2001; Martinez-Ruiz and Lamas, 2004b). The tyrosine nitration modifies protein function and causes irreversible protein damage due to oxidative stress as well as neuronal differentiation (Cappelletti et al., 2003). Neurodegenerative as well as inflammatory diseases viz. Parkinson's, Alzheimer's, familial amyotrophic lateral sclerosis and Huntington's have been associated with tyrosine nitration (Giannopoulou et al., 2002).

Tyrosine nitration is a relatively stable modification. The strategies used for separation, detection and quantitation of nitrotyrosine residues are, anti-3-NT antibodies, HPLC coupled to ESI-MS and GC-MS. Large scale nitro-tyrosine residues can be affinity tagged by reducing nitrotyrosine into aminotyrosine followed by biotinylation (Nikov et al., 2003). The affinity-tagged enrichment can be complemented to the 2-DGE and Western blotting proteomic methods for identification of nitrated proteins (Miyagi et al., 2002). Also chromatographic methods after tryptic digestion of proteins can be used for analysis of nitrotyrosine residues. Typically, the specific nitrotyrosine residues resolved by 2-DGE or HPLC are identified by N-terminal micro sequencing and mostly MS (Δ mass: +45 Da) (Marcondes et al., 2001; Haqqani et al., 2002). Large-scale proteomic approach for detection of global nitrotyrosine residues using the above techniques have been successfully employed recently (Miyagi et al., 2002; Zhan and Desiderio, 2004; Casoni et al., 2005). S-nitrosylation of protein cysteine residues to give nitrosothiols is a reversible (reduced by ascorbic acid, GSH and thioredoxin) modification. S-nitrosylated proteins can be detected by 2-DGE. This method derivatized SH with biotinylated thiol reagent and several derivatized proteins were identified by immunoblotting and/or immunoaffinity (Jaffrey et al., 2001). However, there are only few studies where techniques to detect nitrotyrosine and S-nitrosylation have been applied to large-scale proteomic analysis.

Multi-protein complex (protein–protein interactions)

The express and transitory associations in large protein complexes play an important role in modulating protein functions in various molecular mechanisms in neuron (Krapivinsky et al., 2004; Kim and Sheng, 2004; Soosairajah et al., 2005; Teng and Tang, 2005). The affinity-based techniques implementing the affinity tags and ligands; Poly-His (Ni2+), Biotin (Streptavidin), Calmodulin-binding peptide (Calmodulin), GST (Glutathione), and specific epitope such as FLAG, c-myc and HA (Monoclonal Ab) are commonly used for isolation of multi-protein complexes. The protein of interest is expressed with suitable tag to be used as bait, with the above antitag systems immobilized on agarose–sepharose supports, to isolate the entire multi-protein complex from cellular extract. The proteins in the multi-protein complex are further separated by SDS-PAGE followed by LC-MS/MS for protein characterization. The success of the affinity-based approach depends on non-specific binding which in turn depends on the specificity of the bait partners' recognition. The drawbacks of this technique include extensive pre-cleaning and the protein–protein interactions are essentially in vitro interactions (Monti et al., 2005). Most of the drawbacks of the affinity-based approaches can be overcome by immunoprecipitation strategies (Whetstone et al., 2004). The gene coding for the bait tagged with the epitope is expressed after transfection into appropriate cell line. The protein complexes formed in vivo are immunoprecipitated with antitag monoclonal antibodies. The multi-protein complex is further characterized by SDS-PAGE coupled with LC-MS/MS. The major disadvantage of this method is the cross-recognition of non-specific antigens and non-specific binding of proteins to antibodies and the peptide tags leads to false positives. The above problem has been largely overcome by tandem affinity purification (TAP) tag system (Anders et al., 1999; Puig et al., 2001). The TAP technique is based on combining two different tags (such as *Staphylococcus aureus* protein A (ProtA) and calmodulin-binding peptide (CBP)) on the same protein usually spaced by an enzyme-cleavable linker sequence (TEV protease cleavable site) (Wine et al., 2002; Hurst et al., 2004). The TAP technique has been recently employed in neuroscience (Borch et al., 2005; Davey et al., 2005; Gingras et al., 2005; Gottschalk et al., 2005; Swanson and Washburn, 2005).

Once the interacting partners are identified, to map the interaction site is usually desirable. Many techniques are currently available for studying protein–protein interactions; X-ray crystallography, multi-dimensional NMR, phage display, yeast two-hybrid screens and protein microarrays. Modern MS has increased the scope of these tools as well as improved the ability to investigate complex protein–protein interactions. The MS analysis can be simplified by digesting and isolating the cross-linked peptide, aided by incorporation of fluorescent probes or affinity handles into traditional cross-linkers, for determining the specific site of interaction (Trakselis et al., 2005). The most widely used fluorescent cross-linkers are bromobimanes and sulfosuccinimidyl-2-(7-azido-4-methylcoumarin-3-acetamido)-ethyl-1, 3′-dithiopropionate, and affinity handle as biotin specifically isolated using monomeric avidin or Streptavidin (Kim et al., 1995; Wine et al., 2002; Trester-Zedlitz et al., 2003). After tryptic digestion the fluorescent/affinity-purified protein fragments are separated by SDS-PAGE to be analyzed by tandem MS for interaction sites (Heck and Van Den Heuvel, 2004). The composition of the cross-linked peptide can be deciphered by comparison with the known sequences of the interacting proteins. The coupling of these techniques enables complete protein complex analysis in response to various physiological as well as pathological stimuli.

Mass spectrometry

The sequence analysis of peptides and proteins along with the PTM separated by electrophoresis and chromatography has been a major application of MS in proteomics (Aebersold and Mann, 2003). The mass spectrometer consists of three major units: ion source, mass analyzer and the ion-detection system (Fig. 5). A large variety of MSs are based on coupling of MALDI and ESI with different types of mass analyzers. MS is based on separating the ionized proteins or peptides based on mass to charge ratio (m/z). The tandem MS (MS/MS) on the other hand couples two MSs in time and space and has revolutionized the field of expression and functional proteomics (Smith, 2002). The MS/MS involves mass selection, fragmentation and mass analysis (in two stages) (Fig. 5). In the first stage of mass analysis (MS1) the precursor ion produced by the ion source gets selected for fragmentation in the CID. The fragmentation results in the production of product ions to be analyzed in the second stage of mass analysis (MS2). The inconvertible link between the precursor ion and the product ions is responsible for the unique molecular specificity of MS/MS.

All of these MS techniques can be applied to complex protein samples, i.e., mixtures of hundreds or thousands of proteins. It is important to separate the use of MS instruments to separate proteins from the MS used for protein identification as will be described later. For separation MS has great capabilities but also limitations. As described below, quantitative analysis by MS is limited to techniques like ICAT^{TM} and iTRAQ^{TM}. For researchers looking to profile the expression of proteins in a large number of samples, MS can be problematic and requires a great deal of time on expensive instruments.

Ion source

A number of ionization technologies exist including: fast ion bombardment (FAB) (Barber et al., 1981), MALDI (Karas and Hillenkamp, 1988), and ESI (Fenn et al., 1989; Wilm and Mann, 1996). MALDI and ESI are the techniques of choice for most proteomic applications of neuroscience research.

MALDI

MALDI operates based on irradiation by an intense laser beam of the protein mixture with matrix. There occurs a transfer of large amount of energy absorbed by the crystallized matrix to the sample molecules, desorbing and ionizing them into a gas phase plume by proton-transfer (Fig. 5). This is usually done on MALDI plates such as stainless steel and AnchorChip coated with a hydrophobic material (Tannu et al., 2004b). The most-popular fragmentation types utilized are, ion-source decay (source-accelerating region fast fragmentations) and post-source decay: field-free region metastable fragmentation process). The current proteomic applications overwhelmingly couple MALDI to TOF instrument as the pulsed nature of the laser beam matches well with the pulsed mode of TOF. MALDI-TOF MS is normally used for analysis of simple peptide mixture. Many matrix–laser combinations have been tried; however, for proteomics applications the two most

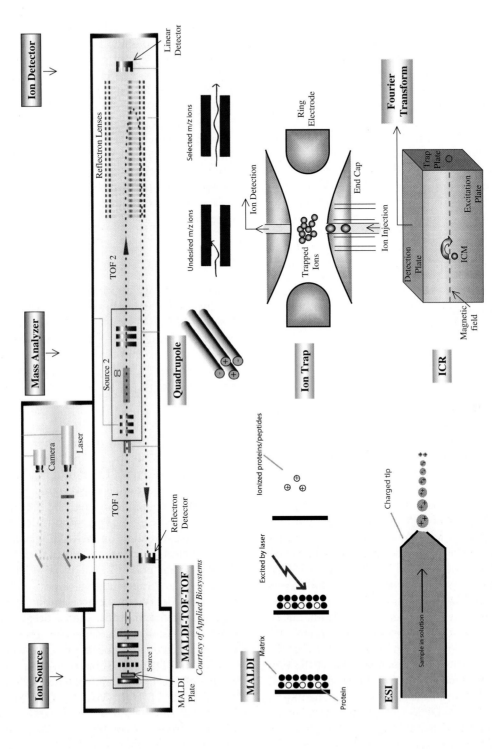

Fig. 5. Schematic representation of MS components (elaborate description in text). Adapted with kind permission from Springer Science and Business Media, and Applied Biosystems (Freeman and Hemby, 2004; Schrader and Klein, 2004; Biosystems, 2005).

commonly used matrices are α-cyano-4-hydroxycinnamic acid (CHCA) for peptides and small molecular-mass proteins (<10,000 Da), and sinapinic acid for high-mass proteins. Many techniques such as dried-drop, fast evaporation, sandwich matrix, spin-dry and seed-layer have been tested (solvents used: ACN, methanol, acetone, chloroform and propanol) to obtain a homogenous sample-matrix critical for obtaining good sample ion yield. A general theme for the formation of a homogenous sample-matrix, confirmed by Tannu et al. (2004b) among many studies undertaken so far, has been a hydrophobic MALDI plate surface. MALDI is capable of analyzing positive as well as negative ions. The MALDI mass spectra typically constitutes signals from singly protonated target molecules and their oligomeric ions ($[M+H]^+$, $[2M+H]^+$). The coupling of MALDI with quadrupole ion trap and FTICR instruments has also emerged, especially the MALDI-FTICR, contributing significantly to accurate mass measurements pre-requisite for assigning critical PTM to peptide residues.

ESI

ESI is an atmospheric pressure ionization (API) technique which can be performed in either positive or negative ionization mode to form $[M+nH]^{n+}$ and $[M-nH]^{n-}$-type ions, respectively. In ESI (Fig. 5) (and nanospray ionization), ions are produced in a liquid phase as three step process of droplet formation, droplet shrinkage, and gaseous ion formation (Wilm and Mann, 1996). The protein sample, in a solvent solution, is ejected as a mist of droplets from a charged capillary tip. As the solvent in the droplets evaporates the total charges of the proteins in the droplet remain but with a reduced surface area of the droplet. This continues to a point at which individual ions leave the droplet. Individual ions then pass on into the mass analyzer. The ESI has dominated bottom–up proteomics method to be coupled with the multidimensional chromatography. The wide popularity of interfacing ESI with HPLC and MS has been the continuous-flow function, acceptance of wide flow rates, tolerance to different types of solvents, and the ability to generate intact multiple-charged ions.

Mass analyzer

Which ever method of ionization is used, once the ions are created they must be separated before being detected in such a way as to provide information on the m/z ratio. Mass analyzers do not actually detect the ions or measure ion mass; they are only used to separate ions according to their m/z ratio. A number of mass analyzer types exist: time-of-flight (TOF), quadrupole, ion trap and FT-ICR. The coupling of HPLC with quadrupole and quadrupole ion trap (QIT) is widely practiced due to the high tolerance to low vacuum and absence of high potentials in the ion source. The coupling of LC with FTICR is known for high resolution and mass accuracy.

Time-of-flight

Time-of-flight (TOF) (Fig. 5) mass analyzers can be thought of as a tube. The ionized proteins/peptides enter the tube by passing through a high-voltage accelerator. The speed at which the ion travels is proportional to its mass (m). The TOF-MS is a velocity spectrometer separating the ions based on velocity differences. The principle of mass analysis is that after acceleration to constant kinetic energy (zV: where z = charge and V = accelerating potential) the ions travel at velocities, v which is an inverse function of the square root of m/z values. The short pulse of ions is dispersed as isomass aggregates such that the ones with lower mass travel faster and reach the detector early than the heavier ions. The time of arrival (t) is used for mass analysis of the ions for a particular length of the flight (L) and is given by

$$t = L(m/2 \cdot zV)^{1/2}$$

The measurement of flight times for two known mass ions is used for converting the time spectrum into the mass spectrum, typically displayed as m/z (Dass, 2000). MALDI ion source most commonly coupled with TOF mass analyzer is many a times referred to as the workhorse of proteomics due to its capability of detecting proteins of >300 kDa with the detection sensitivity in attomole–femtomole range. The MALDI-TOF-TOF (Fig. 5) has also found routine success in detection of peptide residues with PTM along with their locations.

Quadrupole

Quadrupole mass analyzer offers most of the desirable features necessary for MS viz. high scan speed, adequate mass range and resolution, high sensitivity and useful dynamic range. Also QIT is more sensitive and cheaper than quadrupole. Quadrupole mass analyzers (Fig. 5) also involve ions traveling down what can be thought of as a tube. In this case though, the tube consists of four parallel rods. The rods are two pairs of two that can be tuned to different currents and radio frequencies. The two pairs of rods have opposite currents and shifted radio frequencies allowing a form of tuning in which only ions of a particular m/z ratio pass though the tube. A range of m/z ratios can be scanned, generating an m/z profile of the sample. Quadrupole mass analyzers are often used with an ESI ion source.

Ion trap

Ion trap mass analyzers (Fig. 5) use the same principles as the quadrupole in that specific combinations of current and radio frequencies are used to select particular m/z ratios (Jonscher and Yates, 1997). The ion trap can be thought of as a small ball with one electrode around the equator and two more electrodes at the poles. Ions are introduced into the center of the ball and are kept in orbits within the trap. By changing current and radio frequency combinations particular m/z ratio ions are ejected from the ion trap through a port to the detector. By scanning through these voltages and radio frequencies a complete m/z profile can be made (Douglas et al., 2005).

FTICR

The major advantages of FTICR are high mass-resolution and -accuracy, and the ability to trap ions for extended periods of time along with performing multistage tandem MS. A cyclotron cell consists of three pairs of electrode plates assembled as a cube such that the front and end electrode act as trapping plates to trap the ions, the two excitation plates connected with a radio-frequency transmitter to excite the ions, and the detection plates detect the induced mirror current (Fig. 5). The basic principle of the ICR is to detect the ions in the cyclotron cell having a cyclotron motion (ICM) inside a uniform magnetic field produced by superconducting magnet with fixed field strength (4.7, 7, 9.4 and 12 T). A Fourier transform algorithm is used to digitize and process the transient signals of all the ions to give ion abundance for specific m/z (Marshall et al., 1998; Marshall et al., 2002; Schrader and Klein, 2004). The LC-FTICR (ESI ion source), LTQ FTICR and Qq FTICR are commercially available for proteomics applications.

Tandem mass spectrometry

A number of instrument designs strategies exist for MS/MS, each fulfilling special needs (Fig. 5). All of these were generally designed to increase the accuracy of m/z measurements and sensitivity to low abundance ions. Some of the instruments perform MS/MS in space i.e., the mass selection, fragmentation, and analysis is carried out in different regions of the tandem mass spectrometer (e.g., quadrupole and TOF instruments). In some of the instruments this is done tandem in the same region using temporal sequence (tandem-in-time: e.g., QIT and FTICR). Conceptually the operating principles of both designs MS/MS are practically similar. TOF analyzers can be placed in series (TOF/TOF) with a reflectron or collision cell (CID) between them (Fig. 5), quadrupoles and TOF can be placed in series (Q-TOF)(Morris et al., 1997) (Fig. 5), and extremely powerful magnets and Fourier transform algorithms (FT-ICR) (Bogdanov and Smith, 2005) can be used to determine the m/z ratios of all ions within an ion trap (Marshall et al., 1998) (Fig. 5). In the collision cell, an intermediate region of a tandem instrument is filled with neutral gas (helium and argon). The CID, a two-step process of collision activation and unimolecular dissociation, consists of selecting a product ion after dissociation of the precursor ions to be subjected for further activation and unimolecular dissociation (Fig. 5).

Most of the recently available tandem mass spectrometers are effective in identifying proteins in a complex mixture. Obtaining at least one MS/MS spectra by selecting couple of most-intense precursor ions from each constituent protein from a mixture can be used to confirm the identification of constituent proteins. With ions generated and separated, instruments like electron multipliers and scintillation counters detect the ions. Detectors change the kinetic energy of the ions into an electrical current that can be measured and passed along to a computer. While these detectors give information on abundance of ions, quantitation of protein abundance differences between samples by MS is limited unless samples are linked to isotopes (see ICAT^{TM}). Tandem mass spectrometry has enabled unprecedented sequence determination of large number of peptides. The various ion sources and mass analyzers which can be conceptually coupled to give a tandem mass spectrometer are schematized in Fig. 5. The most commonly used mass spectrometers today for quantitative analysis are the quadrupole ion trap (3D-IT), the triple quadrupole (QqQ) and the quadrupole TOF (QqTOF) (Chernushevich et al., 2001). The 3D-IT is relatively small and inexpensive instrument as compared to the standard triple quadrupole and had MS^n capability. Recently, the linear ion traps (LIT) combined with QTOF (QqLIT), having the uniqueness of Q3 to be run in two different modes, large ion storage capacity and higher trapping efficiency, permit very powerful scan combinations for information-dependent data acquisition (Hopfgartner et al., 2004, Douglas et al., 2005).

Protein identification

Peptide mass fingerprinting (PMF) and tandem mass spectrometry (MS/MS), are the main methods for determining protein identities (Rappsilber and Mann, 2002; Johnson et al., 2005). PMF was developed by a number of research groups and begins with digestion of a protein with sequence-specific endoproteinase, typically trypsin (Henzel et al., 1993; Mann et al., 1993; Pappin et al., 1993). PMF of spots from 2-DE gels is one very common application. Gel plugs are either excised by hand or robot and an in-gel-trypsin digestion performed. CBB, silver and negative staining are common post-electrophoresis methods available for the 2-D gel-based proteomics analysis. The gel plugs must often be destained, and some stains work better than others. The CBB stain remains the stain of choice for MS identification of proteins. Briefly, the gel plugs are washed for 20 min, twice in 50 mM ammonium bicarbonate/50% (v/v) methanol in water and once with 75% (v/v) acetonitrile in water. After drying, the gel plugs are incubated overnight at 37°C in 140 ng of sequencing grade trypsin (Promega, Madison, WI) resuspended in 20 mM ammonium bicarbonate. The peptide fragments are then extracted twice with 50% (v/v) acetonitrile, 0.1% (v/v) TFA in water and concentrated using a vacuum concentrator. The peptide fragments dissolved in 50% (v/v) acetonitrile, 0.1% (v/v) TFA in water are spotted on matrix-assisted laser desorption-ionization (MALDI) target plates, dried and mixed with a 50% saturated solution of CHCA in 50% (v/v) acetonitrile, 0.1% (v/v) TFA in water. Trypsin cleaves proteins at very specific locations (carboxy ends of arginine and lysine), resulting in a series of peptides. The experimentally obtained peptide masses are compared with the theoretical peptide masses of proteins stored in databases by means of search programs (Thiede et al., 2005); (1) Mascot: http://www.matrixscience.com/search_form_select.html, (2) ProFound: http://prowl.rockefeller.edu/profound_bin/WebProFound.exe, (3) MS-Fit: http://prospector.ucsf.edu/ucsfhtml4.0/msfit.htm and (4) Aldente: http://www.expasy.org/tools/aldente/ (Johnson et al., 2005).

A number of different strategies exist for MS/MS, in general the process entails the selection of one ion/protein generated during initial MS and then fragmenting this ion/protein into smaller pieces and measuring the mass of the resulting ions (Hernandez et al., 2005). These secondary ions can be decoded into protein sequence information which are searched against protein sequence databases to identify the protein (Perkins et al., 1999; Zhang and Chait, 2000). Almost all of the ionization and mass analyzer types can all be used in an automated mode for MS/MS as well peptide fragmentation identification provided that the instrument is appropriately configured (Fenyo and Beavis, 2002). One MS/MS method that is particularly suited for proteome determination, and recently become amenable to quantitation, is MuDPIT. In this method all the proteins in a sample are digested and loaded onto LC columns (see previous explanation). After fractionation of the peptides, the peptides are fed into a MS/MS instrument for protein identification (Hernandez et al., 2005). This method has identified thousands of proteins, can detect membrane proteins and is similar in concept to shotgun sequencing of DNA at the same time when coupled with iTRAQ is capable of quantitative analysis as well (Rappsilber and Mann, 2002).

Some of the more traditional methods for identifying proteins are still used for proteomic experiments. Edman protein sequencing can be performed on proteins or peptides extracted from gels or blotted from gels. This method is limited by low throughput and requires a comparatively large amount of protein. Another technique is the Far Western blot where a 2-DEG is blotted and the blot is probed with an antibody against a specific protein. This approach does not offer much progress over conventional immunoblotting.

Protein arrays

Because of some of the limitations of electrophoresis and MS methods, selected research groups are attempting to create proteomic chips/arrays. The basic approach is very similar to that of microarrays (Petricoin et al., 2002; Wilson and Nock, 2003; Lopez and Pluskal, 2003). Antibodies or other affinity reagents (e.g., aptamers and peptides) are spotted onto some sort of matrix. Hundreds to thousands of spots are on a single array. A labeled sample is then washed across the array and proteins bind to their specific antibody. The process can also be reversed whereby the protein samples of interest are spotted onto the matrix and then probed with different affinity reagents (Paweletz et al., 2001). While these array or chip approaches have potential for greatly increasing the throughput of proteomic experiments, the use of affinity reagents as the separation method is a severely limiting factor and cannot be ignored. A high-quality antibody is needed for each protein of interest and each modification of that protein. Generation of antibodies remains a laborious task which is almost as much art as science. Separate antibodies also have to be generated for different organisms. In order to generate quantitative data from antibody arrays, and because association kinetics between different antibodies and antigens can vary tremendously, relative concentrations of each antibody and antigen have to be optimized for each protein in order to have quantitative information (Haab et al., 2001). Lastly, it should not be forgotten that sequence/structure knowledge is needed of any protein to be analyzed by protein microarrays in order to generate the affinity reagent, limiting this approach to known protein sequences and modifications. Though there seem to be a number of pitfalls to proteomic chips/arrays as an open-screen technique they do hold promise for routine examination of a small group of proteins. Well-known pathways or gene families could be easily examined by such an approach.

Conclusion

The immense potential for neuroproteomic has increased its recognition over years. It becomes crucial for neuroscientist to address some preliminary questions; such as the completeness of the database for the animal model to be used for study, the amount of tissue available for experiment, the subcellular location showing most promise for scouting and the precedence of quantitative or functional evaluation. The long-term success will most certainly depend on defining the questions to be answered and thereafter integrating the advanced technology with sound and relevant experimental design. Ever-increasing neuroscientists have adopted proteomic approaches to their research; the true advances will come from those that use these new tools to create not just data but discoveries.

References

Adamo, M., Raizada, M.K. and LeRoith, D.. (1989). Mol. Neurobiol., 3: 71–100.
Aebersold, R. and Mann, M.. (2003). Nature, 422: 198–207.

Alban, A., David, S.O., Bjorkesten, L., Andersson, C., Sloge, E., Lewis, S. and Currie, I.. (2003). Proteomics, 3: 36–44.

Amoresano, A., Marino, G., Cirulli, C. and Quemeneur, E.. (2004). Eur. J. Mass Spectrom. (Chichester, Eng), 10: 401–412.

Anders, J., Bluggel, M., Meyer, H.E., Kuhne, R., ter Laak, A.M., Kojro, E. and Fahrenholz, F.. (1999). Biochemistry, 38: 6043–6055.

Andersen, J.S., Lam, Y.W., Leung, A.K., Ong, S.E., Lyon, C.E., Lamond, A.I. and Mann, M.. (2005). Nature, 433: 77–83.

Andersen, J.S., Lyon, C.E., Fox, A.H., Leung, A.K., Lam, Y.W., Steen, H., Mann, M. and Lamond, A.I.. (2002). Curr. Biol., 12: 1–11.

Anderson, L. and Seilhamer, J.. (1997). Electrophoresis, 18: 533–537.

Anderson, N.G., Matheson, A. and Anderson, N.L.. (2001). Proteomics, 1: 3–12.

Annan, R.S. and Carr, S.A.. (1996). Anal. Chem., 68: 3413–3421.

Arora, P.S., Yamagiwa, H., Srivastava, A., Bolander, M.E. and Sarkar, G.. (2005). J. Orthop. Sci., 10: 160–166.

Asara, J.M. and Allison, J.. (1999). J. Am. Soc. Mass Spectrom., 10: 35–44.

Barber, M., Bordoli, R.S., Sedgwick, R.D., Tyler, A.N. and Bycroft, B.W.. (1981). Biochem. Biophys. Res. Commun., 101: 632–638.

Beckner, M.E., Chen, X., An, J., Day, B.W. and Pollack, I.F.. (2005). Lab. Invest., 85: 316–327.

Bennett, K.L., Stensballe, A., Podtelejnikov, A.V., Moniatte, M. and Jensen, O.N.. (2002). J. Mass Spectrom., 37: 179–190.

Berggren, K., Chernokalskaya, E., Steinberg, T.H., Kemper, C., Lopez, M.F., Diwu, Z., Haugland, R.P. and Patton, W.F.. (2000). Electrophoresis, 21: 2509–2521.

Berggren, K., Steinberg, T.H., Lauber, W.M., Carroll, J.A., Lopez, M.F., Chernokalskaya, E., Zieske, L., Diwu, Z., Haugland, R.P. and Patton, W.F.. (1999). Anal. Biochem., 276: 129–143.

Biosystems, A. (2005). Applera Corporation and MDS Inc.

Bjellqvist, B., Ek, K., Righetti, P.G., Gianazza, E., Gorg, A., Westermeier, R. and Postel, W.. (1982). J. Biochem. Biophys. Methods, 6: 317–339.

Blonder, J., Goshe, M.B., Moore, R.J., Pasa-Tolic, L., Masselon, C.D., Lipton, M.S. and Smith, R.D.. (2002). J. Proteome Res., 1: 351–360.

Bogdanov, B. and Smith, R.D.. (2005). Mass Spectrom. Rev., 24: 168–200.

Bolan, E.A., Gracy, K.N., Chan, J., Trifiletti, R.R. and Pickel, V.M.. (2000). J. Neurosci., 20: 4798–4808.

Borch, J., Jorgensen, T.J. and Roepstorff, P.. (2005). Curr. Opin. Chem. Biol., 9: 509–516.

Brittain, S.M., Ficarro, S.B., Brock, A. and Peters, E.C.. (2005). Nat. Biotechnol., 23: 463–468.

Cappelletti, G., Maggioni, M.G., Tedeschi, G. and Maci, R.. (2003). Exp. Cell Res., 288: 9–20.

Carr, S.A., Huddleston, M.J. and Annan, R.S.. (1996). Anal. Biochem., 239: 180–192.

Casoni, F., Basso, M., Massignan, T., Gianazza, E., Cheroni, C., Salmona, M., Bendotti, C. and Bonetto, V.. (2005). J. Biol. Chem., 280: 16295–16304.

Chernushevich, I.V., Loboda, A.V. and Thomson, B.A.. (2001). J. Mass Spectrom., 36: 849–865.

Chin, E.T. and Papac, D.I.. (1999). Anal. Biochem., 273: 179–185.

Choe, L.H., Aggarwal, K., Franck, Z. and Lee, K.H.. (2005). Electrophoresis, 26: 2437–2449.

Choi, J., Levey, A.I., Weintraub, S.T., Rees, H.D., Gearing, M., Chin, L.S. and Li, L.. (2004). J. Biol. Chem., 279: 13256–13264.

Churchward, M.A., Butt, R.H., Lang, J.C., Hsu, K.K. and Coorssen, J.R.. (2005). Proteome Sci., 3: 5.

Ciechanover, A., Orian, A. and Schwartz, A.L.. (2000). Bioessays, 22: 442–451.

Coghlan, D.R., Mackintosh, J.A. and Karuso, P.. (2005). Org. Lett., 7: 2401–2404.

Colledge, M., Snyder, E.M., Crozier, R.A., Soderling, J.A., Jin, Y., Langeberg, L.K., Lu, H., Bear, M.F. and Scott, J.D.. (2003). Neuron, 40: 595–607.

Collins, M.O., Yu, L., Coba, M.P., Husi, H., Campuzano, I., Blackstock, W.P., Choudhary, J.S. and Grant, S.G.. (2005a). J. Biol. Chem., 280: 5972–5982.

Collins, M.O., Yu, L., Husi, H., Blackstock, W.P., Choudhary, J.S. and Grant, S.G.. (2005b). Sci. STKE, 298: l6.

Cooper, H.J., Heath, J.K., Jaffray, E., Hay, R.T., Lam, T.T. and Marshall, A.G.. (2004). Anal. Chem., 76: 6982–6988.

Corthals, G.L., Wasinger, V.C., Hochstrasser, D.F. and Sanchez, J.C.. (2000). Electrophoresis, 21: 1104–1115.

Coute, Y., Burgess, J.A., Diaz, J.J., Chichester, C., Lisacek, F., Greco, A. and Sanchez, J.C.. (2006) Mass Spectrom. Rev, 25(2): 215–234.

Covey, T.R., Bonner, R.F., Shushan, B.I. and Henion, J.. (1988). Rapid Commun. Mass Spectrom., 25(2): 249–256.

Cronshaw, J.M., Krutchinsky, A.N., Zhang, W., Chait, B.T. and Matunis, M.J.. (2002). J. Cell Biol., 158: 915–927.

Dass, C. (2000). Principles and Practice of Biological Mass Spectrometry. Wiley.

Davey, F., Hill, M., Falk, J., Sans, N. and Gunn-Moore, F.J.. (2005). J. Neurochem., 94: 1243–1253.

Davidsson, P., Folkesson, S., Christiansson, M., Lindbjer, M., Dellheden, B., Blennow, K. and Westman-Brinkmalm, A.. (2002). Rapid Commun. Mass Spectrom., 16: 2083–2088.

Douglas, D.J., Frank, A.J. and Mao, D.. (2005). Mass Spectrom. Rev., 24: 1–29.

Dreger, M.. (2003a). Eur. J. Biochem., 270: 589–599.

Dreger, M.. (2003b). Mass Spectrom. Rev., 22: 27–56.

Dreger, M., Bengtsson, L., Schoneberg, T., Otto, H. and Hucho, F.. (2001). Proc. Natl. Acad. Sci. USA, 98: 11943–11948.

Fenn, J.B., Mann, M., Meng, C.K., Wong, S.F. and Whitehouse, C.M.. (1989). Science, 246: 64–71.

Fenyo, D. and Beavis, R.C.. (2002). Trends Biotechnol., 20: S35–S38.

Ferguson, P.L. and Smith, R.D.. (2003). Annu. Rev. Biophys. Biomol. Struct., 32: 399–424.

Ferreras, M., Gavilanes, J.G. and Garcia-Segura, J.M.. (1993). Anal. Biochem., 213: 206–212.

Ferro, M., Salvi, D., Riviere-Rolland, H., Vermat, T., Seigneurin-Berny, D., Grunwald, D., Garin, J., Joyard, J. and Rolland, N.. (2002). Proc. Natl. Acad. Sci. USA, 99: 11487–11492.

Ficarro, S.B., McCleland, M.L., Stukenberg, P.T., Burke, D.J., Ross, M.M., Shabanowitz, J., Hunt, D.F. and White, F.M.. (2002). Nat. Biotechnol., 20: 301–305.

Finley, D., Ciechanover, A. and Varshavsky, A.. (2004). Cell, 116: S29–S32+.

Fodor, I.K., Nelson, D.O., Alegria-Hartman, M., Robbins, K., Langlois, R.G., Turteltaub, K.W., Corzett, T.H. and McCutchen-Maloney, S.L.. (2005). Bioinformatics, 21: 3733–3740.

Fountoulakis, M.. (2004). Mass Spectrom. Rev., 23: 231–258.

Franzen, B., Yang, Y., Sunnemark, D., Wickman, M., Ottervald, J., Oppermann, M. and Sandberg, K.. (2003). Proteomics, 3: 1920–1929.

Freeman, W.M. and Hemby, S.E.. (2004). Neurochem. Res., 29: 1065–1081.

Galeva, N. and Altermann, M.. (2002). Proteomics, 2: 713–722.

Giannopoulou, E., Katsoris, P., Polytarchou, C. and Papadimitriou, E.. (2002). Arch. Biochem. Biophys., 400: 188–198.

Gingras, A.C., Aebersold, R. and Raught, B.. (2005). J. Physiol., 563: 11–21.

Gocke, C.B., Yu, H. and Kang, J.. (2005). J. Biol. Chem., 280: 5004–5012.

Godovac-Zimmermann, J., Kleiner, O., Brown, L.R. and Drukier, A.K.. (2005). Proteomics, 5: 699–709.

Gonzalez, J., Takao, T., Hori, H., Besada, V., Rodriguez, R., Padron, G. and Shimonishi, Y.. (1992). Anal. Biochem., 205: 151–158.

Goodlett, D.R., Keller, A., Watts, J.D., Newitt, R., Yi, E.C., Purvine, S., Eng, J.K., von Haller, P., Aebersold, R. and Kolker, E.. (2001). Rapid Commun. Mass Spectrom., 15: 1214–1221.

Gorg, A., Boguth, G., Obermaier, C., Posch, A. and Weiss, W.. (1995). Electrophoresis, 16: 1079–1086.

Gorg, A., Obermaier, C., Boguth, G., Harder, A., Scheibe, B., Wildgruber, R. and Weiss, W.. (2000). Electrophoresis, 21: 1037–1053.

Gorg, A., Postel, W., Friedrich, C., Kuick, R., Strahler, J.R. and Hanash, S.M.. (1991). Electrophoresis, 12: 653–658.

Goshe, M.B., Blonder, J. and Smith, R.D.. (2003). J. Proteome Res., 2: 153–161.

Gottschalk, A., Almedom, R.B., Schedletzky, T., Anderson, S.D., Yates III, J.R. and Schafer, W.R.. (2005). EMBO J., 24: 2566–2578.

Guy, G.R., Philip, R. and Tan, Y.H.. (1994). Electrophoresis, 15: 417–440.

Gygi, S.P., Corthals, G.L., Zhang, Y., Rochon, Y. and Aebersold, R.. (2000). Proc. Natl. Acad. Sci. USA, 97: 9390–9395.

Gygi, S.P., Rist, B., Gerber, S.A., Turecek, F., Gelb, M.H. and Aebersold, R.. (1999a). Nat. Biotechnol., 17: 994–999.

Gygi, S.P., Rist, B., Griffin, T.J., Eng, J. and Aebersold, R.. (2002). J. Proteome Res., 1: 47–54.

Gygi, S.P., Rochon, Y., Franza, B.R. and Aebersold, R.. (1999b). Mol. Cell Biol., 19: 1720–1730.

Haab, B.B., Dunham, M.J. and Brown, P.O. (2001). Genome Biol. 2 RESEARCH0004.

Hammack, B.N., Owens, G.P., Burgoon, M.P. and Gilden, D.H.. (2003). Multiple Sclerosis, 9: 472–475.

Han, D.K., Eng, J., Zhou, H. and Aebersold, R.. (2001). Nat. Biotechnol., 19: 946–951.

Hansson, S.F., Puchades, M., Blennow, K., Sjogren, M. and Davidsson, P.. (2004). Proteome Sci., 2: 7.

Haqqani, A.S., Kelly, J.F. and Birnboim, H.C.. (2002). J. Biol. Chem., 277: 3614–3621.

Hart, C., Schulenberg, B., Steinberg, T.H., Leung, W.Y. and Patton, W.F.. (2003). Electrophoresis, 24: 588–598.

Heck, A.J. and Van Den Heuvel, R.H.. (2004). Mass Spectrom. Rev., 23: 368–389.

Henzel, W.J., Billeci, T.M., Stults, J.T., Wong, S.C., Grimley, C. and Watanabe, C.. (1993). Proc. Natl. Acad. Sci. USA, 90: 5011–5015.

Hernandez, P., Muller, M. and Appel, R.D.. (2005) Mass Spectrom. Rev, 25(2): 235–254.

Hershko, A. and Ciechanover, A.. (1998). Annu. Rev. Biochem., 67: 425–479.

Hesse, C., Rosengren, L., Andreasen, N., Davidsson, P., Vanderstichele, H., Vanmechelen, E. and Blennow, K.. (2001). Neurosci. Lett., 297: 187–190.

Hicke, L.. (2001). Nat. Rev. Mol. Cell Biol., 2: 195–201.

Hirabayashi, J.. (2004). Glycoconj. J., 21: 35–40.

Hirabayashi, J., Hayama, K., Kaji, H., Isobe, T. and Kasai, K., (2002). J. Biochem. (Tokyo), 132: 103–114.

Holland, M.J.. (2002). J. Biol. Chem., 277: 14363–14366.

Hopfgartner, G., Varesio, E., Tschappat, V., Grivet, C., Bourgogne, E. and Leuthold, L.A.. (2004). J. Mass Spectrom., 39: 845–855.

Hunter, T.. (1998). Philos. Trans. R. Soc. Lond. B Biol. Sci., 353: 583–605.

Hurst, G.B., Lankford, T.K. and Kennel, S.J.. (2004). J. Am. Soc. Mass Spectrom., 15: 832–839.

Husi, H., Ward, M.A., Choudhary, J.S., Blackstock, W.P. and Grant, S.G.. (2000). Nat. Neurosci., 3: 661–669.

Hynd, M.R., Lewohl, J.M., Scott, H.L. and Dodd, P.R.. (2003). J. Neurochem., 85: 543–562.

Ishizuka, N., Minakuchi, H., Nakanishi, K., Soga, N., Nagayama, H., Hosoya, K. and Tanaka, N.. (2000). Anal. Chem., 72: 1275–1280.

Jaeken, J. and Matthijs, G.. (2001). Annu. Rev. Genomics Hum. Genet., 2: 129–151.

Jaffrey, S.R., Erdjument-Bromage, H., Ferris, C.D., Tempst, P. and Snyder, S.H.. (2001). Nat. Cell. Biol., 3: 193–197.

Jensen, P.K., Pasa-Tolic, L., Anderson, G.A., Horner, J.A., Lipton, M.S., Bruce, J.E. and Smith, R.D.. (1999). Anal. Chem., 71: 2076–2084.

Johnson, R.S., Davis, M.T., Taylor, J.A. and Patterson, S.D.. (2005). Methods, 35: 223–236.

Jonscher, K.R. and Yates III, J.R.. (1997). Anal. Biochem., 244: 1–15.

Kaiser, P. and Huang, L.. (2005). Genome Biol., 6: 233.

Kaji, H., Saito, H., Yamauchi, Y., Shinkawa, T., Taoka, M., Hirabayashi, J., Kasai, K., Takahashi, N. and Isobe, T.. (2003). Nat. Biotechnol., 21: 667–672.

Karas, M. and Hillenkamp, F.. (1988). Anal. Chem., 60: 2299–2301.

Kato, A., Rouach, N., Nicoll, R.A. and Bredt, D.S.. (2005). Proc. Natl. Acad. Sci. USA, 102: 5600–5605.

Kaufmann, H., Bailey, J.E. and Fussenegger, M.. (2001). Proteomics, 1: 194–199.

Kim, E. and Sheng, M.. (2004). Nat. Rev. Neurosci., 5: 771–781.

Kim, M.J., Kim, H.J., Kim, J.M., Kim, B., Han, S.H. and Cha, G.S.. (1995). Anal. Biochem., 231: 400–406.

Kirkpatrick, D.S., Denison, C. and Gygi, S.P.. (2005). Nat. Cell Biol., 7: 750–757.

Kjellstrom, S. and Jensen, O.N.. (2004). Anal. Chem., 76: 5109–5117.

Klose, J.. (1975). Humangenetik, 26: 231–243.

Knowles, M.R., Cervino, S., Skynner, H.A., Hunt, S.P., de Felipe, C., Salim, K., Meneses-Lorente, G., McAllister, G. and Guest, P.C.. (2003). Proteomics, 3: 1162–1171.

Koichi Tanaka, H.W., Yutaka Ido, Satoshi Akita, Yoshikazu Yoshida and Tamio Yoshida (1988). Rapid Commun. Mass Spectrom., 2: 151–153.

Krapivinsky, G., Medina, I., Krapivinsky, L., Gapon, S. and Clapham, D.E.. (2004). Neuron, 43: 563–574.

Krebs, E.G.. (1983). Philos. Trans. R. Soc. Lond B Biol. Sci., 302: 3–11.

Krutchinsky, A.N., Kalkum, M. and Chait, B.T.. (2001). Anal. Chem., 73: 5066–5077.

Lane, P., Hao, G. and Gross, S.S. (2001). Sci STKE, RE1, 86.

Li, K.W., Hornshaw, M.P., Van Der Schors, R.C., Watson, R., Tate, S., Casetta, B., Jimenez, C.R., Gouwenberg, Y., Gundelfinger, E.D., Smalla, K.H. and Smit, A.B.. (2004). J. Biol. Chem., 279: 987–1002.

Liao, P.C., Leykam, J., Andrews, P.C., Gage, D.A. and Allison, J.. (1994). Anal. Biochem., 219: 9–20.

Link, A.J., Eng, J., Schieltz, D.M., Carmack, E., Mize, G.J., Morris, D.R., Garvik, B.M. and Yates III, J.R.. (1999). Nat. Biotechnol., 17: 676–682.

Liu, T., Qian, W.J., Gritsenko, M.A., Camp II, D.G., Monroe, M.E., Moore, R.J. and Smith, R.D.. (2005). J. Proteome Res., 4: 2070–2080.

Lollo, B.A., Harvey, S., Liao, J., Stevens, A.C., Wagenknecht, R., Sayen, R., Whaley, J. and Sajjadi, F.G.. (1999). Electrophoresis, 20: 854–859.

Lopez, M.F., Berggren, K., Chernokalskaya, E., Lazarev, A., Robinson, M. and Patton, W.F.. (2000). Electrophoresis, 21: 3673–3683.

Lopez, M.F. and Pluskal, M.G.. (2003). J. Chromatogr. B Analyt. Technol. Biomed. Life Sci., 787: 19–27.

Lovell, M.A., Xiong, S., Markesbery, W.R. and Lynn, B.C.. (2005). Neurochem. Res., 30: 113–122.

Lowe, J.B. and Marth, J.D.. (2003). Annu. Rev. Biochem., 72: 643–691.

Lue, L.F., Yan, S.D., Stern, D.M. and Walker, D.G.. (2005). Curr. Drug. Targets CNS Neurol. Disord., 4: 249–266.

Ma, Y., Lu, Y., Zeng, H., Ron, D., Mo, W. and Neubert, T.A.. (2001). Rapid Commun. Mass Spectrom., 15: 1693–1700.

Mackintosh, J.A., Choi, H.Y., Bae, S.H., Veal, D.A., Bell, P.J., Ferrari, B.C., Van Dyk, D.D., Verrills, N.M., Paik, Y.K. and Karuso, P.. (2003). Proteomics, 3: 2273–2288.

Malinow, R. and Malenka, R.C.. (2002). Annu. Rev. Neurosci., 25: 103–126.

Mann, M., Hojrup, P. and Roepstorff, P.. (1993). Biol. Mass Spectrom., 22: 338–345.

Mann, M., Ong, S.E., Gronborg, M., Steen, H., Jensen, O.N. and Pandey, A.. (2002). Trends Biotechnol., 20: 261–268.

Marcondes, S., Turko, I.V. and Murad, F.. (2001). Proc. Natl. Acad. Sci. USA, 98: 7146–7151.

Marengo, E., Robotti, E., Antonucci, F., Cecconi, D., Campostrini, N. and Righetti, P.G.. (2005). Proteomics, 5: 654–666.

Marshall, A.G., Hendrickson, C.L. and Jackson, G.S.. (1998). Mass Spectrom. Rev., 17: 1–35.

Marshall, A.G., Hendrickson, C.L. and Shi, S.D.. (2002). Anal. Chem., 74: 252A–259A.

Martin, K., Steinberg, T.H., Cooley, L.A., Gee, K.R., Beechem, J.M. and Patton, W.F.. (2003a). Proteomics, 3: 1244–1255.

Martin, K., Steinberg, T.H., Goodman, T., Schulenberg, B., Kilgore, J.A., Gee, K.R., Beechem, J.M. and Patton, W.F.. (2003b). Comb. Chem. High Throughput Screen, 6: 331–339.

Martinez-Ruiz, A. and Lamas, S.. (2004a). Arch. Biochem. Biophys., 423: 192–199.

Martinez-Ruiz, A. and Lamas, S.. (2004b). Cardiovasc. Res., 62: 43–52.

Martinovic, S., Veenstra, T.D., Anderson, G.A., Pasa-Tolic, L. and Smith, R.D.. (2002). J. Mass Spectrom., 37: 99–107.

Matsumoto, M., Fu, Y.X., Molina, H., Huang, G., Kim, J., Thomas, D.A., Nahm, M.H. and Chaplin, D.D.. (1997). J. Exp. Med., 186: 1997–2004.

Mayor, T. and Deshaies, R.J.. (2005). Methods Enzymol., 399: 385–392.

Mayor, T., Lipford, J.R., Graumann, J., Smith, G.T. and Deshaies, R.J.. (2005). Mol. Cell Proteomics, 4: 741–751.

Medzihradszky, K.F., Campbell, J.M., Baldwin, M.A., Falick, A.M., Juhasz, P., Vestal, M.L. and Burlingame, A.L.. (2000). Anal. Chem., 72: 552–558.

Minakuchi, H., Nakanishi, K., Soga, N., Ishizuka, N. and Tanaka, N.. (1997). J. Chromatogr. A, 762: 135–146.

Mirgorodskaya, O.A., Kozmin, Y.P., Titov, M.I., Korner, R., Sonksen, C.P. and Roepstorff, P.. (2000). Rapid Commun. Mass Spectrom., 14: 1226–1232.

Miyagi, M., Sakaguchi, H., Darrow, R.M., Yan, L., West, K.A., Aulak, K.S., Stuehr, D.J., Hollyfield, J.G., Organisciak, D.T. and Crabb, J.W.. (2002). Mol. Cell Proteomics, 1: 293–303.

Molloy, M.P., Brzezinski, E.E., Hang, J., McDowell, M.T. and VanBogelen, R.A.. (2003). Proteomics, 3: 1912–1919.

Monti, M., Orru, S., Pagnozzi, D. and Pucci, P.. (2005). Clin. Chim. Acta, 357: 140–150.

Mootha, V.K., Bunkenborg, J., Olsen, J.V., Hjerrild, M., Wisniewski, J.R., Stahl, E., Bolouri, M.S., Ray, H.N., Sihag,

S., Kamal, M., Patterson, N., Lander, E.S. and Mann, M.. (2003). Cell, 115: 629–640.

Moreira, P.I., Smith, M.A., Zhu, X., Nunomura, A., Castellani, R.J. and Perry, G.. (2005). Ann. N Y Acad. Sci., 1043: 545–552.

Morris, H.R., Paxton, T., Panico, M., McDowell, R. and Dell, A.. (1997). J. Protein Chem., 16: 469–479.

Mortz, E., Krogh, T.N., Vorum, H. and Gorg, A.. (2001). Proteomics, 1: 1359–1363.

Mullis, K.B.. (1990). Ann. Biol. Clin. (Paris), 48: 579–582.

Munchbach, M., Quadroni, M., Miotto, G. and James, P.. (2000). Anal. Chem., 72: 4047–4057.

Navarre, C., Degand, H., Bennett, K.L., Crawford, J.S., Mortz, E. and Boutry, M.. (2002). Proteomics, 2: 1706–1714.

Neubauer, G. and Mann, M.. (1999). Anal. Chem., 71: 235–242.

Neuhoff, V., Stamm, R., Pardowitz, I., Arold, N., Ehrhardt, W. and Taube, D.. (1990). Electrophoresis, 11: 101–117.

Nielsen, P.A., Olsen, J.V., Podtelejnikov, A.V., Andersen, J.R., Mann, M. and Wisniewski, J.R.. (2005). Mol. Cell Proteom., 4: 402–408.

Nikov, G., Bhat, V., Wishnok, J.S. and Tannenbaum, S.R.. (2003). Anal. Biochem., 320: 214–222.

Oda, Y., Nagasu, T. and Chait, B.T.. (2001). Nat. Biotechnol., 19: 379–382.

O'Farrell, P.H.. (1975). J. Biol. Chem., 250: 4007–4021.

Ohi, M.D., Link, A.J., Ren, L., Jennings, J.L., McDonald, W.H. and Gould, K.L.. (2002). Mol. Cell Biol., 22: 2011–2024.

Olivieri, E., Herbert, B. and Righetti, P.G.. (2001). Electrophoresis, 22: 560–565.

Packer, N.H., Ball, M.S. and Devine, P.L.. (1999). Methods Mol. Biol., 112: 341–352.

Pappin, D.J., Hojrup, P. and Bleasby, A.J.. (1993). Curr. Biol., 3: 327–332.

Paweletz, C.P., Charboneau, L., Bichsel, V.E., Simone, N.L., Chen, T., Gillespie, J.W., Emmert-Buck, M.R., Roth, M.J., Petricoin, I.E. and Liotta, L.A.. (2001). Oncogene, 20: 1981–1989.

Peng, J., Elias, J.E., Thoreen, C.C., Licklider, L.J. and Gygi, S.P.. (2003a). J. Proteome Res., 2: 43–50.

Peng, J., Kim, M.J., Cheng, D., Duong, D.M., Gygi, S.P. and Sheng, M.. (2004). J. Biol. Chem., 279: 21003–21011.

Peng, J., Schwartz, D., Elias, J.E., Thoreen, C.C., Cheng, D., Marsischky, G., Roelofs, J., Finley, D. and Gygi, S.P.. (2003b). Nat. Biotechnol., 21: 921–926.

Perkins, D.N., Pappin, D.J., Creasy, D.M. and Cottrell, J.S.. (1999). Electrophoresis, 20: 3551–3567.

Peters, E.C., Horn, D.M., Tully, D.C. and Brock, A.. (2001). Rapid Commun. Mass Spectrom., 15: 2387–2392.

Petricoin, E.F., Zoon, K.C., Kohn, E.C., Barrett, J.C. and Liotta, L.A.. (2002). Nat. Rev. Drug Discov., 1: 683–695.

Pickart, C.M.. (2004). Cell, 116: 181–190.

Pickart, C.M. and Eddins, M.J.. (2004). Biochim. Biophys. Acta, 1695: 55–72.

Pickart, C.M. and Fushman, D.. (2004). Curr Opin Chem Biol, 8: 610–616.

Polson, C., Sarkar, P., Incledon, B., Raguvaran, V. and Grant, R.. (2003). J Chromatogr B Anal. Technol Biomed Life Sci, 785: 263–275.

Posewitz, M.C. and Tempst, P.. (1999). Anal Chem, 71: 2883–2892.

Prabakaran, S., Swatton, J.E., Ryan, M.M., Huffaker, S.J., Huang, J.T., Griffin, J.L., Wayland, M., Freeman, T., Dudbridge, F., Lilley, K.S., Karp, N.A., Hester, S., Tkachev, D., Mimmack, M.L., Yolken, R.H., Webster, M.J., Torrey, E.F. and Bahn, S. (2004). Mol. Psychiatry, 9: 684–697, 643.

Premstaller, A., Oberacher, H., Walcher, W., Timperio, A.M., Zolla, L., Chervet, J.P., Cavusoglu, N., van Dorsselaer, A. and Huber, C.G.. (2001). Anal. Chem., 73: 2390–2396.

Puig, O., Caspary, F., Rigaut, G., Rutz, B., Bouveret, E., Bragado-Nilsson, E., Wilm, M. and Seraphin, B.. (2001). Methods, 24: 218–229.

Qiu, Y., Sousa, E.A., Hewick, R.M. and Wang, J.H.. (2002). Anal. Chem., 74: 4969–4979.

Rabilloud, T., Vuillard, L., Gilly, C. and Lawrence, J.J.. (1994). Cell Mol. Biol. (Noisy-le-grand), 40: 57–75.

Raju, T.S.. (2000). Anal. Biochem., 283: 125–132.

Rappsilber, J. and Mann, M.. (2002). Trends. Biochem. Sci., 27: 74–78.

Raymackers, J., Daniels, A., De Brabandere, V., Missiaen, C., Dauwe, M., Verhaert, P., Vanmechelen, E. and Meheus, L.. (2000). Electrophoresis, 21: 2266–2283.

Regnier, F.E., Riggs, L., Zhang, R., Xiong, L., Liu, P., Chakraborty, A., Seeley, E., Sioma, C. and Thompson, R.A.. (2002). J. Mass Spectrom., 37: 133–145.

Reinders, J., Lewandrowski, U., Moebius, J., Wagner, Y. and Sickmann, A.. (2004). Proteomics, 4: 3686–3703.

Resing, K.A., Johnson, R.S. and Walsh, K.A.. (1995). Biochemistry, 34: 9477–9487.

Righetti, P.G., Castagna, A., Antonioli, P. and Boschetti, E.. (2005). Electrophoresis, 26: 297–319.

Robbe, C., Paraskeva, C., Mollenhauer, J., Michalski, J.C., Sergi, C. and Corfield, A.. (2005). Biochem. Soc. Trans., 33: 730–732.

Roberts, G.C. and Smith, C.W.. (2002). Curr. Opin. Chem. Biol., 6: 375–383.

Roelens, S.A., Beck, V., Aerts, G., Clerens, S., Vanden Bergh, G., Arckens, L., Darras, V.M. and VAN DER Geyten, S.. (2005). Ann. N Y Acad. Sci., 1040: 454–456.

Rosengren, A.T., Salmi, J.M., Aittokallio, T., Westerholm, J., Lahesmaa, R., Nyman, T.A. and Nevalainen, O.S.. (2003). Proteomics, 3: 1936–1946.

Roth, J.. (2002). Chem. Rev., 102: 285–303.

Rout, M.P., Aitchison, J.D., Suprapto, A., Hjertaas, K., Zhao, Y. and Chait, B.T.. (2000). J. Cell Biol., 148: 635–651.

Sanders, S.L., Jennings, J., Canutescu, A., Link, A.J. and Weil, P.A.. (2002). Mol. Cell Biol., 22: 4723–4738.

Scheler, C., Lamer, S., Pan, Z., Li, X.P., Salnikow, J. and Jungblut, P.. (1998). Electrophoresis, 19: 918–927.

Schindler, J., Lewandrowski, U., Sickmann, A., Friauf, E. and Nothwang, H.G.. (2006). Mol. Cell Proteom, 5(2): 390–400.

Schirmer, E.C., Florens, L., Guan, T., Yates III, J.R. and Gerace, L.. (2003). Science, 301: 1380–1382.

Schirmer, E.C. and Gerace L. (2002). Genome Biol., 3 REVIEWS1008.

Schirmer, E.C. and Gerace, L.. (2005). Trends Biochem. Sci., 30: 551–558.

Schlosser, A., Pipkorn, R., Bossemeyer, D. and Lehmann, W.D.. (2001). Anal. Chem., 73: 170–176.

Schrader, W. and Klein, H.W.. (2004). Anal. Bioanal. Chem., 379: 1013–1024.

Schrimpf, S.P., Meskenaite, V., Brunner, E., Rutishauser, D., Walther, P., Eng, J., Aebersold, R. and Sonderegger, P.. (2005). Proteomics, 5: 2531–2541.

Schulenberg, B., Aggeler, R., Beechem, J.M., Capaldi, R.A. and Patton, W.F.. (2003). J. Biol. Chem., 278: 27251–27255.

Shaw, J., Rowlinson, R., Nickson, J., Stone, T., Sweet, A., Williams, K. and Tonge, R.. (2003). Proteomics, 3: 1181–1195.

Sickmann, A., Dormeyer, W., Wortelkamp, S., Woitalla, D., Kuhn, W. and Meyer, H.E.. (2000). Electrophoresis, 21: 2721–2728.

Sickmann, A., Reinders, J., Wagner, Y., Joppich, C., Zahedi, R., Meyer, H.E., Schonfisch, B., Perschil, I., Chacinska, A., Guiard, B., Rehling, P., Pfanner, N. and Meisinger, C.. (2003). Proc. Natl. Acad. Sci. USA, 100: 13207–13212.

Sitek, B., Apostolov, O., Stuhler, K., Pfeiffer, K., Meyer, H.E., Eggert, A. and Schramm, A.. (2005). Mol. Cell Proteom., 4: 291–299.

Smejkal, G.B., Robinson, M.H. and Lazarev, A.. (2004). Electrophoresis, 25: 2511–2519.

Smith, R.D.. (2002). Trends Biotechnol., 20: S3–S7.

Soosairajah, J., Maiti, S., Wiggan, O., Sarmiere, P., Moussi, N., Sarcevic, B., Sampath, R., Bamburg, J.R. and Bernard, O.. (2005). EMBO J., 24: 473–486.

Southan, C.. (2004). Proteomics, 4: 1712–1726.

Steen, H., Kuster, B., Fernandez, M., Pandey, A. and Mann, M.. (2001). Anal. Chem., 73: 1440–1448.

Steinberg, T.H., Agnew, B.J., Gee, K.R., Leung, W.Y., Goodman, T., Schulenberg, B., Hendrickson, J., Beechem, J.M., Haugland, R.P. and Patton, W.F.. (2003). Proteomics, 3: 1128–1144.

Stensballe, A., Andersen, S. and Jensen, O.N.. (2001). Proteomics, 1: 207–222.

Stevens, T., Garcia, J.G., Shasby, D.M., Bhattacharya, J. and Malik, A.B.. (2000). Am. J. Physiol. Lung Cell Mol. Physiol., 279: L419–L422.

Stewart, Ian. I., Thomson, T. and Figeys, D.. (2001). Rapid Commun. Mass Spectrom., 15: 2456–2465.

Swanson, S.K. and Washburn, M.P.. (2005). Drug Discov. Today, 10: 719–725.

Swatton, J.E., Prabakaran, S., Karp, N.A., Lilley, K.S. and Bahn, S.. (2004). Mol. Psychiatry, 9: 128–143.

Takeuchi, M., Kikuchi, S., Sasaki, N., Suzuki, T., Watai, T., Iwaki, M., Bucala, R. and Yamagishi, S.. (2004). Curr. Alzheimer Res., 1: 39–46.

Tang, W.X., Fasulo, W.H., Mash, D.C. and Hemby, S.E.. (2003). J. Neurochem., 85: 911–924.

Tannu, N., Leonard, H and Scott E.H. (Manuscript Submitted) J. Neurochem.

Tannu, N.S., Deborah M and Hemby S.E. (Manuscript in Preparation).

Tannu, N.S., Gabriela S.B., Pam K. and Tracy M.A. (2006a). Electrophoresis.

Tannu, N.S., Rao, V.K., Chaudhary, R.M., Giorgianni, F., Saeed, A.E., Gao, Y. and Raghow, R.. (2004a). Mol. Cell Proteom., 3: 1065–1082.

Tannu, N.S., Scott E.H. (In Press: 2006b). *Electrophoresis*.

Tannu, N.S., Wu, J., Rao, V.K., Gadgil, H.S., Pabst, M.J., Gerling, I.C. and Raghow, R.. (2004b). Anal. Biochem., 327: 222–232.

Tapiola, T., Overmyer, M., Lehtovirta, M., Helisalmi, S., Ramberg, J., Alafuzoff, I., Riekkinen Sr., P. and Soininen, H.. (1997). Neuroreport, 8: 3961–3963.

Taylor, S.W., Fahy, E., Zhang, B., Glenn, G.M., Warnock, D.E., Wiley, S., Murphy, A.N., Gaucher, S.P., Capaldi, R.A., Gibson, B.W. and Ghosh, S.S.. (2003). Nat. Biotechnol., 21: 281–286.

Teng, F.Y. and Tang, B.L.. (2005). Cell Mol Life Sci, 62: 1571–1578.

Terry, D.E. and Desiderio, D.M.. (2003). Proteomics, 3: 1962–1979.

Thiede, B., Hohenwarter, W., Krah, A., Mattow, J., Schmid, M., Schmidt, F. and Jungblut, P.R. (2005). Methods, 35: 237–247.

Tonge, R., Shaw, J., Middleton, B., Rowlinson, R., Rayner, S., Young, J., Pognan, F., Hawkins, E., Currie, I. and Davison, M. (2001). Proteomics, 1: 377–396.

Trakselis, M.A., Alley, S.C. and Ishmael, F.T. (2005). Bioconjug. Chem., 16: 741–750.

Trester-Zedlitz, M., Kamada, K., Burley, S.K., Fenyo, D., Chait, B.T. and Muir, T.W. (2003). J. Am. Chem. Soc., 125: 2416–2425.

Trinidad, J.C., Thalhammer, A., Specht, C.G., Schoepfer, R. and Burlingame, A.L. (2005). J. Neurochem., 92: 1306–1316.

Unlu, M., Morgan, M.E. and Minden, J.S. (1997). Electrophoresis, 18: 2071–2077.

Valaskovic, G.A. and Kelleher, N.L. (2002). Curr. Top. Med. Chem., 2: 1–12.

Van den Bergh, G., Clerens, S., Vandesande, F. and Arckens, L. (2003). Electrophoresis, 24: 1471–1481.

Van Gelder, R.N., von Zastrow, M.E., Yool, A., Dement, W.C., Barchas, J.D. and Eberwine, J.H. (1990). Proc. Natl. Acad. Sci. USA, 87: 1663–1667.

Venter, J.C., Adams, M.D., Myers, E.W., Li, P.W., Mural, R.J., Sutton, G.G., Smith, H.O., Yandell, M., Evans, C.A., Holt, R.A., Gocayne, J.D., Amanatides, P., Ballew, R.M., Huson, D.H., Wortman, J.R., Zhang, Q., Kodira, C.D., Zheng, X.H., Chen, L., Skupski, M., Subramanian, G., Thomas, P.D., Zhang, J., Gabor Miklos, G.L., Nelson, C., Broder, S., Clark, A.G., Nadeau, J., McKusick, V.A., Zinder, N., Levine, A.J., Roberts, R.J., Simon, M., Slayman, C., Hunkapiller, M., Bolanos, R., Delcher, A., Dew, I., Fasulo, D., Flanigan, M., Florea, L., Halpern, A., Hannenhalli, S., Kravitz, S., Levy, S., Mobarry, C., Reinert, K., Remington, K., Abu-Threideh, J., Beasley, E., Biddick, K., Bonazzi, V., Brandon, R., Cargill, M., Chandramouliswaran,

I., Charlab, R., Chaturvedi, K., Deng, Z., Di Francesco, V., Dunn, P., Eilbeck, K., Evangelista, C., Gabrielian, A.E., Gan, W., Ge, W., Gong, F., Gu, Z., Guan, P., Heiman, T.J., Higgins, M.E., Ji, R.R., Ke, Z., Ketchum, K.A., Lai, Z., Lei, Y., Li, Z., Li, J., Liang, Y., Lin, X., Lu, F., Merkulov, G.V., Milshina, N., Moore, H.M., Naik, A.K., Narayan, V.A., Neelam, B., Nusskern, D., Rusch, D.B., Salzberg, S., Shao, W., Shue, B., Sun, J., Wang, Z., Wang, A., Wang, X., Wang, J., Wei, M., Wides, R., Xiao, C., Yan, C., et al. (2001). Science, 291: 1304–1351.

VerBerkmoes, N.C., Bundy, J.L., Hauser, L., Asano, K.G., Razumovskaya, J., Larimer, F., Hettich, R.L. and Stephenson, J.L. (2002). J. Proteome Res., 1: 239–252.

Wang, H. and Hanash, S. (2005). Mass Spectrom. Rev., 24: 413–426.

Washburn, M.P., Wolters, D. and Yates III, J.R. (2001). Nat. Biotechnol., 19: 242–247.

Whetstone, P.A., Butlin, N.G., Corneillie, T.M. and Meares, C.F. (2004). Bioconjug. Chem., 15: 3–6.

White, I.R., Pickford, R., Wood, J., Skehel, J.M., Gangadharan, B. and Cutler, P. (2004). Electrophoresis, 25: 3048–3054.

Wildenauer, D.B., Korschenhausen, D., Hoechtlen, W., Ackenheil, M., Kehl, M. and Lottspeich, F. (1991). Electrophoresis, 12: 487–492.

Wildgruber, R., Harder, A., Obermaier, C., Boguth, G., Weiss, W., Fey, S.J., Larsen, P.M. and Gorg, A. (2000). Electrophoresis, 21: 2610–2616.

Wilkins, M.R., Gasteiger, E., Sanchez, J.C., Bairoch, A. and Hochstrasser, D.F. (1998). Electrophoresis, 19: 1501–1505.

Wilm, M. and Mann, M. (1996). Anal. Chem., 68: 1–8.

Wilson, D.S. and Nock, S. (2003). Angew. Chem. Int. Ed. Engl., 42: 494–500.

Wind, M., Edler, M., Jakubowski, N., Linscheid, M., Wesch, H. and Lehmann, W.D. (2001). Anal. Chem., 73: 29–35.

Wine, R.N., Dial, J.M., Tomer, K.B. and Borchers, C.H. (2002). Anal. Chem., 74: 1939–1945.

Wolters, D.A., Washburn, M.P. and Yates III, J.R. (2001). Anal. Chem., 73: 5683–5690.

Wu, C.C., MacCoss, M.J., Howell, K.E. and Yates III, J.R. (2003). Nat. Biotechnol., 21: 532–538.

Wu, J., Lenchik, N.J., Pabst, M.J., Solomon, S.S., Shull, J. and Gerling, I.C. (2005). Electrophoresis, 26: 225–237.

Wu, W.W., Wang, G., Baek, S.J. and Shen, R.F. (2006). J. Proteome Res., 5: 651–658.

Yan, J.X., Packer, N.H., Gooley, A.A. and Williams, K.L. (1998). J. Chromatogr. A, 808: 23–41.

Yao, X., Freas, A., Ramirez, J., Demirev, P.A. and Fenselau, C. (2001). Anal. Chem., 73: 2836–2842.

Yates III, J.R., Gilchrist, A., Howell, K.E. and Bergeron, J.J. (2005). Nat. Rev. Mol. Cell Biol., 6: 702–714.

Yip, T.T. and Hutchens, T.W. (1992). FEBS Lett., 308: 149–153.

Yuan, X. and Desiderio, D.M. (2005). Proteomics, 5: 541–550.

Yuan, X., Russell, T., Wood, G. and Desiderio, D.M. (2002). Electrophoresis, 23: 1185–1196.

Zhan, X. and Desiderio, D.M. (2004). Biochem. Biophys. Res. Commun., 325: 1180–1186.

Zhang, H., Li, X.J., Martin, D.B. and Aebersold, R. (2003). Nat. Biotechnol., 21: 660–666.

Zhang, R., Sioma, C.S., Wang, S. and Regnier, F.E. (2001). Anal. Chem., 73: 5142–5149.

Zhang, S., Fu, J. and Zhou, Z. (2005). Toxicol. Appl. Pharmacol., 202: 13–17.

Zhang, W. and Chait, B.T. (2000). Anal. Chem., 72: 2482–2489.

Zheng, P.P., Luider, T.M., Pieters, R., Avezaat, C.J., van den Bent, M.J., Sillevis Smitt, P.A. and Kros, J.M. (2003). J. Neuropathol. Exp. Neurol., 62: 855–862.

Zhou, G., Li, H., DeCamp, D., Chen, S., Shu, H., Gong, Y., Flaig, M., Gillespie, J.W., Hu, N., Taylor, P.R., Emmert-Buck, M.R., Liotta, L.A., Petricoin III, E.F. and Zhao, Y. (2002a). Mol. Cell Proteom., 1: 117–124.

Zhou, H., Ranish, J.A., Watts, J.D. and Aebersold, R. (2002b). Nat. Biotechnol., 20: 512–515.

Zhou, W., Merrick, B.A., Khaledi, M.G. and Tomer, K.B. (2000). J. Am. Soc. Mass Spectrom., 11: 273–282.

Zhu, B.Y., Mant, C.T. and Hodges, R.S. (1991). J. Chromatogr., 548: 13–24.

Zolnierowicz, S. and Bollen, M. (2000). EMBO J., 19: 483–488.

CHAPTER 4

Functional genomics and proteomics in the clinical neurosciences: data mining and bioinformatics

John H. Phan, Chang-Feng Quo and May D. Wang*

The Wallace H. Coulter Department of Biomedical Engineering, Georgia Institute of Technology and Emory University, Atlanta, GA 30322, USA

Abstract: The goal of this chapter is to introduce some of the available computational methods for expression analysis. Genomic and proteomic experimental techniques are briefly discussed to help the reader understand these methods and results better in context with the biological significance. Furthermore, a case study is presented that will illustrate the use of these analytical methods to extract significant biomarkers from high-throughput microarray data.

Genomic and proteomic data analysis is essential for understanding the underlying factors that are involved in human disease. Currently, such experimental data are generally obtained by high-throughput microarray or mass spectrometry technologies among others. The sheer amount of raw data obtained using these methods warrants specialized computational methods for data analysis.

Biomarker discovery for neurological diagnosis and prognosis is one such example. By extracting significant genomic and proteomic biomarkers in controlled experiments, we come closer to understanding how biological mechanisms contribute to neural degenerative diseases such as Alzheimers' and how drug treatments interact with the nervous system.

In the biomarker discovery process, there are several computational methods that must be carefully considered to accurately analyze genomic or proteomic data. These methods include quality control, clustering, classification, feature ranking, and validation.

Data quality control and normalization methods reduce technical variability and ensure that discovered biomarkers are statistically significant. Preprocessing steps must be carefully selected since they may adversely affect the results of the following expression analysis steps, which generally fall into two categories: unsupervised and supervised.

Unsupervised or clustering methods can be used to group similar genomic or proteomic profiles and therefore can elucidate relationships within sample groups. These methods can also assign biomarkers to sub-groups based on their expression profiles across patient samples. Although clustering is useful for exploratory analysis, it is limited due to its inability to incorporate expert knowledge.

On the other hand, classification and feature ranking are supervised, knowledge-based machine learning methods that estimate the distribution of biological expression data and, in doing so, can extract important information about these experiments. Classification is closely coupled with feature ranking, which is essentially a data reduction method that uses classification error estimation or other statistical tests to score features. Biomarkers can subsequently be extracted by eliminating insignificantly ranked features.

These analytical methods may be equally applied to genetic and proteomic data. However, because of both biological differences between the data sources and technical differences between the experimental

*Corresponding author. E-mail: maywang@bme.gatech.edu

methods used to obtain these data, it is important to have a firm understanding of the data sources and experimental methods.

At the same time, regardless of the data quality, it is inevitable that some discovered biomarkers are false positives. Thus, it is important to validate discovered biomarkers. The validation process may be slow; yet, the overall biomarker discovery process is significantly accelerated due to initial feature ranking and data reduction steps. Information obtained from the validation process may also be used to refine data analysis procedures for future iteration. Biomarker validation may be performed in a number of ways — bench-side in traditional labs, web-based electronic resources such as gene ontology and literature databases, and clinical trials.

Introduction

Genomics, the study of gene expression, is fundamental for two reasons. First, genes encode information that is the basic blueprint of life. Second, genes are the vehicles responsible for the transmission of hereditary material from one generation to the next. Gene sequences, or DNA, are relatively simple to analyze in the sense that gene function is neatly contained and determined by the primary structure, which is the order of base pairs. In addition, the flavor of base pairs is constrained to four basic nucleotides, with the exception of some special cases. Interaction between base pairs is readily described by a simple complementary pairing rule.

On the other hand, proteins are the active agents in the cell that ultimately determine the cellular characteristics. Briefly, proteins carry out intra- and intercellular processes and facilitate communication within cells as well as between cells and larger systems. In comparison with genes, protein function may be determined by up to four levels — primary, secondary, tertiary, and quaternary structures. The primary structure of a protein is the amino acid sequence directly encoded from DNA base pairs, while secondary, tertiary, and quaternary structures are derived from various degrees of protein folding and aggregation. There is also greater variety in the flavor of amino acids, as evident by the 20 natural amino acids apart from other rare species. Interaction between amino acids arises as a function of multiple chemical factors.

The completion of the Human Genome Project created new inroads into the study of complex biological systems, such as the human body. Given the functional complexity of the human body, it is somewhat surprising that only a relatively small number of genes are required to encode this information. Expectedly, it turns out that among other classes of compounds, proteins play a leading role in expressing the wide array of functions within the human body. Consequently, the study of protein expression, or proteomics, refers to any procedure that characterizes large sets of proteins; this includes composition, modification, quantification, localization, and functional interaction (Pandey and Mann, 2000; Fields, 2001; Aebersol and Mann, 2003; Glish and Vachet, 2003; Steen and Mann, 2004).

Experimentally, because of the ease at which genes may be sequenced as well as the relative simplicity of experiments that assay genes compared to proteins, genomics has been more widely exploited as a high-throughput means for screening. Furthermore, genomics is essential for detecting and possibly treating disease conditions in their primary manifestations in DNA resulting from inheritance and mutation. At the same time, while we may obtain a vague idea of the quality and quantity of expressed gene products by studying gene expression profiles, the level of assessment is insufficient to determine the dynamic state of cellular processes. In many instances such as post-translational modifications, gene expression is not easily correlated to protein expression. Thus, genomics and proteomics are complementary efforts that must be coupled to reveal the secrets of complex biological systems, in particular, the human body.

Microarrays and mass spectrometry (MS) are current common genomic and proteomic technologies that play a vital role in clinical neurosciences, providing the tools to describe, quantify and ultimately predict the behavior of neurological systems. We may be interested in the signaling pathways involved in disease pathologies or the response of the neurological system to therapeutic agents. To meet the demands of current clinical interests however, microarrays and MS are used to assay genomic and proteomic expression levels on a large scale, enabling a birds' eye view of neurological systems that is, at the same time, voluminous. Consequently, to utilize experimental data fruitfully, we have a need for data mining and in a broader context, bioinformatics.

The objective of bioinformatics is to discover significant biomarkers. Here, we are presented with several challenges in bioinformatics. As with all biological experiments, we need to deal with technical and biological variability, or noise, in the data. Because multiple platforms exist for microarrays, these experiments are highly variable across platforms. In addition, the complexity of the multi-step process contributes to technical variability. This problem is also inherent in MS. Thus, it is important to understand the construct of the experiments, including concrete details, so that we are aware of the limits of the data obtained as well as kinks in the process where unwanted variability is introduced. Furthermore, within a single patient, or clinical source, pathologies are non-uniform leading to biological variability. Quality control, both experimentally and analytically, coupled with normalization methods are steps to reducing technical and biological noise.

At the same time, the relatively high cost of obtaining patient or clinical samples is a prohibitive factor that contributes to the common phenomenon of 'ill-posed' problems, especially in bioinformatics. While microarrays and MS may assay samples in multiple dimensions (to the order of 10^4 in genomic studies), the small number of patient samples leads to statistical problems.

Subsequently, while robust algorithms may produce mathematically valid results, these results must be also biologically relevant to be useful for clinical applications. The problem of false discovery is non-trivial given that there is no unique solution to ill-posed problems. Various tools exist for the purpose of validating analytical results with independent sources such as literature annotation or clinical experiments. The results of validation also provide performance measures and feedback for further iteration of data-mining methods. Finally, validated findings from bioinformatics then lead to clinical tools and applications. These steps are combined to provide accurate analyses of genomic and proteomic data with each step informing other steps in the process.

In the data-mining process, each step — quality control, normalization, biomarker discovery, interpretation, and validation — informs other steps. These challenges will be addressed in detail in later sections so that we may achieve accurate analyses of genomic and proteomic data for potential clinical applications.

Experimental methods

Genomic technology

The production of mRNA in cells is the first step in the expression of genes to functional proteins. By quantifying mRNA expression, microarray technology is able to roughly measure genetic processes. The concept that underscores genomic microarray technology is the specific and complementary hybridization of nucleotide sequences. Generally, mRNA can be isolated from cells and exposed to an array of complementary sequences to which the mRNA sequences of interest have high affinity. Hybridization can then be measured by fluorescence. The two primary microarray technologies based on this premise that are widely used today are cDNA (complementary DNA) and oligonucleotide microarrays (Liao, 2005).

cDNA microarrays

cDNA microarrays, developed by Schena et al. (1995), are based on long sequences (0.6–2.4 kb) of cDNA fixed, or printed, onto a substrate (usually a glass slide) in a spotted matrix such that each spot on the array corresponds to a specific gene or transcript. cDNA sequences are selected from libraries of gene sequences and amplified using polymerase chain reaction (PCR).

RNA is extracted and isolated from separate control and test cells, reverse transcribed, then amplified with PCR. During the process of PCR, special fluorescent base pairs, Cy5 and Cy3 are incorporated into the cDNA for tagging purposes. Tags are incorporated into the control and test products in order to distinguish different cases; for example, the Cy3 in control and Cy5 in test cases represent green and red fluorescence, respectively. cDNA from both control and test cases are mixed and allowed to hybridize to the glass slide. After the microarray is washed, each cDNA spot on the array, consisting of many similar sequences, will have hybridized with sequences from both the control and test cases.

Theoretically, the amount of hybridization from each case, control or test, is proportional to the amount of cDNA, which is proportional to the amount of original mRNA expression. Each spot on the microarray can then be quantified by analyzing the ratio of fluorescence of each of the two colors, red or green. Typically, the log ratio of the two fluorescence colors is used for analysis such that, for example, a positive number would indicate over expression of the test case relative to the control case and a negative number would indicate under expression.

Oligonucleotide microarrays

Oligonucleotide microarrays consist of short nucleotide sequences (on the order of 20–60 base pairs) rather than the long sequences in cDNA microarrays. A photo-lithographic technique developed by Affymetrix (Lipshutz et al., 1999) enables production of high-density microarrays by building nucleotide sequences directly onto a substrate. Unfortunately, this method limits the length of sequences to 25 nucleotides, reducing the sensitivity and specificity of hybridization. Careful design of the microarray and selection of sequences, however, can overcome this problem. Because of the length limitation, sequences must be selected that are unique to the transcript or target gene, called perfect match sequences. For real biological experiments, different target genes may have many regions with similar sequences, therefore several different perfect match sequences may be required for high specificity to a single target.

In addition to perfect match sequences, mismatch sequences are included on the chip. Mismatch sequences are the same as perfect match sequences with the exception of a single nucleotide. These sequences are used to quantify the amount of specificity. Sequences that normally would bind to perfect match sequences have a small chance of binding to mismatch sequences, and the amount of non-specific binding is subtracted from the amount of specific, perfect binding. Statistically, this technique corrects any bias caused by non-specific binding. As with cDNA arrays, mRNA from test cells are extracted and hybridized to oligonucleotide arrays, except that only a single fluorescence channel is necessary.

Quality control of microarrays

Variance and bias in microarray experiments can be introduced through a number of steps, including the extraction and amplification of mRNA, design of chips to maximize hybridization specificity, dye intensity imbalances (depending on chip type), and quantification of signals with image processing. Variability can be reduced by increasing the accuracy and precision of hardware (Zien et al., 2001), by removing outlier samples, or by increasing the number of replicates. This section will focus on the removal of outliers and various normalization methods to improve overall data quality.

Outlier removal

Model et al. (2002) describe methods of handling variations between single microarray slides and between batches of slides. They propose the use of multivariate statistical process control to detect deviations from normal working conditions. Once these deviations are detected, samples are either removed or replicates produced so that confident average values can be obtained. The simplest method of detecting outliers would be able to measure the deviation of a gene on an array from the mean expression of that gene over all arrays. This is also known as the sample variance. The threshold of tolerance for variance can be defined depending on the desired t-distribution significance level, assuming that the data are normally distributed. If the number of outlier genes on an array reaches a threshold, the entire array is deemed as an outlier. It is usually the case, however, that genes are highly correlated and single-dimensional tests will not account for this. In such cases, the use of Hotelling's T^2 statistic in combination with robust PCA is better suited for multi-dimensional tests (Model et al., 2002). However, removing samples in an already small pool of samples may be problematic. An alternative would be to increase the sample size so that the effect of outliers is reduced. Unfortunately, increasing sample size may also be difficult due to the cost of microarray experiments and, depending on the study, a lack of test subjects. Many data normalization techniques have been explored to clean data without removing or adding samples.

Multi-channel microarrays

Multi-channel microarray chips, such as cDNA chips, use different colored dyes and enable the measuring of relative expression levels, indicated by the amount of fluorescence of each dye. Although these dyes have very similar properties, slight differences exist that affect hybridization or amount of fluorescence and, ultimately, the observed gene-expression level. The most common dyes for two channel cDNA microarrays are Cy3 (green) and Cy5 (red). A method often recommended for correcting dye intensity imbalances is to fit the data on a transformed MA scatter plot, in which the axes are

$$m_i = \log(R_i) - \log(G_i) \qquad (1)$$

$$a_i = \frac{1}{2}(\log(R_i) + \log(G_i)) \qquad (2)$$

In the ideal case, values along the M axis, which is the log ratio of red over green fluorescence, is approximately constant as values along the A axis, average intensity, increases. For some dyes, however, the intensity of one dye increases more quickly than the intensity of the other dye, resulting in a slight positive slope in the graph. The relationship may also be non-linear. The data can be smoothed after transformation to the MA axis using either locally weighted linear regression (LOWESS) or smoothing splines and adjusted by the smoothing residuals to obtain normalized data (Cleveland and Devlin, 1988; Yang et al., 2002).

Normalization

Normalization has a significant effect on the detection of differentially expressed genes (Hoffmann et al., 2002). The combination of normalization and gene selection algorithms should be selected carefully so that the number of relevant genes is maximized while reducing the number of false positives. This may be a daunting task, since the process of interpreting and validating results can be very tedious. Nevertheless, many normalization algorithms have been widely used without fully understanding their effects. Some of these methods include background correction, dye-, global-, and quantile normalization.

Fluorescent signals representing hybridization on microarrays may include some background signal that is present regardless of hybridization. These noisy signals, which may be caused by non-specific hybridization of dye or tagged transcripts to the array, tend to reduce the signal-to-noise ratio but may also provide some information regarding dye intensities (for two channel arrays). Assuming that the background signal is constant over the entire array and is additive, it may be subtracted from all spot intensities to normalize the array (Wolkenhauer et al., 2002). For instances in which the background signal may be variable over the array, signals can be normalized by subtracting a multiple of the standard deviations of the background or a fraction of the background to avoid negative signals. Background correction is often applied during the image acquisition step when fluorescence of each microarray spot is quantified, but may also be applied at a later step.

Global normalization methods adjust overall intensity of each microarray by assuming that the total amount of mRNA is consistent across most cells. Gene expression signals can be divided by the sum of gene expression over the entire chip, resulting in normalized fractions of total mRNA expression (Zien et al., 2001). This method may also be used with housekeeping genes (genes that are consistently expressed) by dividing or subtracting other gene expression signals by the expression values of housekeeping genes. However, the assumption of total gene expression and constant expression of housekeeping genes may not be accurate (Suzuki et al., 2000). An alternative is to use invariant or control genes, which are not necessarily housekeeping genes, as a basis for normalization. Control genes may be spiked into experiments and in cases where controls are not available, invariant genes must be estimated from the given data. These estimations can be erroneous, however, and the use of a handful of genes estimated to be invariant for normalization of an entire chip may bias the results (Reilly et al., 2003).

Other more general normalization techniques focus on ensuring similarity of distributions across microarrays. For example, the most basic normalization of several microarrays is mean or median centering to ensure that all chips have a similar baseline of expression. Similarly, each sample can be scaled so that all spots are expressed within the same range. Both of these methods are relatively simple, but incorrectly assume that inconsistencies in microarray experiments are linear. A more sophisticated method of distribution adjustment is quantile normalization. Quantile normalization forces all samples into identical distributions and can even be applied across conditions if invariant genes are taken into account (Bolstad et al., 2003). It is important that the effects of a normalization technique on a dataset are well known before drawing conclusions from the results. The best method may be to iteratively apply these methods to a dataset and interpret or validate results from differential expression analysis before selecting the best method. The following sections discuss several differential expression analysis methods.

Proteomic technology

Proteomics can be crudely classified into three areas (Pandey and Mann, 2000). First, peptide characterization for large-scale protein identification and post-translational modifications focuses on extracting sequence and structure information leading to further functional studies. Second, 'differential display' for the comparison of protein levels with potential clinical applications is similar to comparative genomic microarray studies. Third, the study of protein aggregates and protein–protein interactions focuses on cellular reaction mechanisms, also leading to further functional studies.

Technologies that have been used for protein analysis include two-dimensional (2D) polyacrylamide gel electrophoresis (2D-PAGE), MS, and protein microarrays. 2D-PAGE has been the common method for protein analysis; however, for the analysis of a large number of proteins, it is difficult for several reasons. Although it can accurately identify a large amount of proteins, it is very labor-intensive and requires large quantities of protein (Li et al., 2002). MS and protein microarrays are high-throughput methods that have been able to overcome the limitations of 2D-PAGE. Many of the algorithms used for genomic microarray analysis can be applied to MS and protein microarrays. Like genomic microarrays, experiments from such methods often produce data with small sample sizes and large dimensions.

2 Dimensional gel electrophoresis

2D gel electrophoresis (GE) is a highly popular technique among proteomics researchers. Besides other reasons, GE is not equipment intensive and relies simply on fundamental molecular properties of the sample such as the isoelectric point (pI), molecular weight, and charge. These properties can be measured without disrupting the sample significantly. The resulting 2D gel maps can be used for 'differential display' of protein levels between control and treated samples; specific proteins may also be extracted for further analysis. It is common for 2D GE to be coupled with MS for this latter part of experimental analysis.

The first dimension in 2D GE is isoelectric focusing. The nature of the amino acid backbone inherent in all proteins allows for the formation of zwitterions — ions that possess both positive and negative charges at different sites. The amino group readily accepts a proton to become positively charged, while the carboxyl group loses a proton to become negatively charged. Thus, given an applied electric field along a simple pH gradient mounted on a gel strip, these zwitterions will migrate, or be 'focused', to their respective isoelectric points known as the pI.

The second dimension is separation by size and molecular weight through a poly-acrylamide gel mesh. The mesh is prepared using sodium dodecyl sulfate (SDS) that confers a uniform negative charge density, based on molecular size, to the proteins present. Furthermore, mercaptans are used to disrupt the secondary disulfide bonds between amino acids so that the proteins acquire a large rod-like conformation. Because of the materials used, this procedure is also termed as SDS-PAGE. The size of the pores in this gel mesh is dependent on the concentration of polyacrylamide used. An electric field is applied across the gel with the positive electrode as the end point to attract and motivate the migration of the proteins. Given the uniform charge density, proteins move through the gel mesh based only on size and molecular weight. A variety of staining methods may be applied to visualize the location of the proteins as spots. The resulting 2D map gives us information about the isoelectric point and molecular weight of the sample proteins.

In general, a typical gel map may contain up to 5000 spots. Detection sensitivity is heavily conditioned on the staining method applied. Even then, the number of spots visualized is practically limited and is potentially less than the number of possible gene products. This is a consequence of constraints from the separation steps such as a limited spectrum of isoelectric points and limited dimensions of the polyacrylamide gel, resulting in a limited molecular weights range to less than 250 kDa in general. Furthermore, there may be problems with loading the sample in that the loading volume is restricted and hydrophobic proteins are problematic. In addition, issues of quantification and reproducibility do arise as this is a labor-intensive procedure.

Mass spectrometry
MS is a popular high-throughput experimental technique that has quickly become an integral component of proteomics. The experimental procedure is conceptually simple where sample particles are ionized; the motion of these charged particles in applied electric and magnetic fields are then captured and represented in a spectrum of intensities (peaks). Depending on the sample composition, the spectrum generated contains large volumes of data previously embedded in the sample. Consequently, information extraction depends on high-precision spectrum analysis.

The mass spectrometer is comprised of three essential components: an ion source, analyzer, and detector. The ion source is the key to ionizing the sample — it should ideally only provide the sample particles with charges but not interfere with the motion of these particles. Achieving stable and abundant ions from the sample of interest is essential in MS. Subsequently, the analyzer separates the charged sample particles based on the mass-to-charge (m/z) ratio — analyzers can be classified into beam and trapping analyzers. Finally, the detector reports the intensities of sample particles based on the m/z ratio. Understanding the construct of the equipment will make it easier to refine or troubleshoot equipment specifications to achieve better spectra.

Importantly, while evaluating the appeal of MS, we consider the following factors: sample preparation, sample consumption, accuracy, and precision. The sample state is critical in determining if MS is viable. Sample preparation is required for obtaining a sufficiently pure sample — a clean sample results in a spectrum with the appearance of junk peaks significantly reduced. Furthermore, in many clinical applications today, sample volume is generally low. Thus, it is important that sample consumption is optimized to extract the most information. The accuracy of the spectrum peaks in locating and reporting the true m/z ratios in the relative intensities give us both qualitative and quantitative information about the sample composition. A high-precision mass spectrometer is desired to allow the resolution of closely located peaks.

MS can be coupled with other techniques such as liquid chromatography and 2D-GE to achieve more information about the sample. Furthermore, the process of MS can be repeated sequentially — this procedure is known as tandem MS.

Tandem mass spectrometry
Tandem mass spectrometry (MS/MS) is a natural extension of MS to further derive more

information about isolated spectrum components from a previous MS stage. Theoretically, multiple stages of MS can be involved, leading to an MS^n experiment where n is the number of stages. Here we discuss MS/MS with two stages. The extension to higher orders is similar.

The parent ion undergoes the first stage of MS; selected components from the spectra are then isolated and may be subject to a variety of induced dissociation methods and one more stage of MS to produce daughter ions and neutral masses. This process is represented by a simple concept equation:

$$m^+_{\text{parent}} \rightarrow m^+_{\text{daughter}} + m_{\text{neutral}} \qquad (3)$$

The signal-to-noise ratio is greatly improved in the second stage of MS because the cloud of ions, released from the ion source, that contribute to 'chemical noise' is filtered by the first stage of MS. Furthermore, technological advances have made it possible to select to scan either the first or second stage of MS — this becomes the independent stage — and investigate the spectra from the other stage — the dependent stage. This is the primary advantage of MS/MS.

Protein microarrays

Protein microarrays are a relatively new technology similar to genomic microarrays, and exist in two types: forward and reverse phase. Forward phase protein microarrays, like genomic microarrays, immobilize bait molecules on the surface of a chip and detect proteins from a sample by washing the chip with a solution of several analytes extracted from that sample. The result is that each chip can assay a large number of proteins for a single sample. Reverse phase arrays immobilize several analytes per spot with each spot representing an entire sample. The entire chip is then probed with a single molecule of interest, such as a labeled antibody, so that several samples can be analyzed at once under the same conditions. A unique advantage of using protein microarrays is that they can detect whether proteins are in a phosphorylated state, a switching mechanism often found in protein networks. Protein microarrays are also efficient, requiring very little cell lysate to produce several microarrays (Espina et al., 2003).

Data analysis

Microarray analysis

Microarray technology is most commonly used for either identifying patterns in gene expression by clustering or detecting differentially expressed genes between two or more classes of samples. Gene clustering and differential expression analysis fall into the realm of unsupervised and supervised algorithms that have been widely explored in microarray applications. Unsupervised methods comprise algorithms that do not require prior knowledge of the microarray samples. On the other hand, supervised methods require that the treatment conditions of samples are known.

Gene clustering can be used to identify patterns in microarray data based on similarity of expression (Eisen et al., 1998). Samples can be clustered to identify groups of samples that have similar expression profiles or that have been exposed to similar treatments. In addition, genes can be clustered to identify groups of genes that may be biologically correlated.

Differential gene expression experiments are typically carried out with one set of microarray samples comprising mRNA extracted from test subjects or tissue under certain treatment conditions and a control set of samples comprising mRNA from independent, normal subjects or tissue. Ideally, all microarrays in the experiment are identical with regard to the types and number of spotted gene or DNA transcripts. The fluorescent signal from each transcript in the set of microarrays of the same condition is representative of a single population. Since the intensities of the fluorescent signals are correlated with mRNA transcription, statistical testing between multiple populations can be performed to derive conclusions about gene expression. Gene transcripts that have significant differential expression between the two conditions may be used to infer the biological mechanisms altered by treatment of the test subjects or tissue.

Although these analytical techniques have been used for microarray analysis, they are not unique to genomics and can easily be applied to proteomics. The section on 'Statistical analysis and pattern classification' will describe some supervised and unsupervised statistical and pattern classification techniques in detail.

Mass spectrometry analysis

MS is primarily used for identification of proteins based on cleavage of the amino acid backbone. Analytical methods can be broadly classified as database search and de novo analysis algorithms. Nonetheless, all identification methods depend heavily on the specificity of the enzyme proteases used for cleavage during sample preparation. Fortunately, with the exception of some proteases, the majority of experimentally viable proteases are sufficiently specific and consistent.

Mass spectra obtained can be searched against theoretical peptides derived from genomic databases to establish an identity. Clearly, such approaches fail in two cases: first, when there is insufficient genomic information, which is the prevailing scenario, taking into account possible sequencing errors and second, in general when extensive post-translational modifications occur. In addition, because there is only limited mass information available, the problem of false-positive matches returned from such database searches is non-trivial. By requiring more stringent match criteria, the number of false positives will decrease, however at the same time, possible candidates may be eliminated. It is important to note that all these methods return a score, or probability, that the match is true. Matches are then ranked in order of the assigned scores. The best matches are not necessarily unique; furthermore, by varying the combination of match criteria, different 'best' matches may be obtained. Consequently, the rational approach will be to seek peptide matches that consistently achieve high scores and to validate analysis results using other sources such as published literature or ontological approaches.

There are three common database methods for protein identification. The first, Peptide Sequence Tags, relies on the observation that most spectra usually contain a small series of easily interpretable sequence by mass. It is assumed that the lowest and highest masses in the series contain information about the distance to either ends of the peptide chain in mass units. The easily interpretable sequence forms the link between these two masses — these three pieces combine to form a peptide sequence tag that is matched against sequences in the genomic database. This method is most notably implemented over the Internet by the EMBL Bioanalytical Research Group in the PeptideSearch program (2005).

The second method is a correlation method that compares theoretically derived spectra, based on information returned from the genomic database search, to empirical data. This method is derived from signal-processing techniques common in communications and works better for low-resolution data. The computational cost for processing high-resolution data this way is too high for practical considerations although this may change with technological advances. Another advantage of this method is that it is also robust for low signal-to-noise spectra.

The third approach utilizes the intensity information by matching the masses of the theoretical fragments with the experimental peaks beginning with the most intense peak. The probability of random fragment matches, i.e. the probability that the fragments match by chance, is also determined as the basis for comparison. This is known as probability-based matching.

De novo analysis

Apart from database search methods, another different approach toward spectrum analysis is de novo analysis, literally 'a new analysis'. This approach derives theoretical peptide sequences solely based on the distribution of intensities in spectra. The true identity of the sample peptide is then determined by verification from other independent sources. Because of the limitations of database search-based methods and the potential of machine-learning algorithms, de novo analysis must not be lightly dismissed as an academic approach.

The first step in de novo analysis is to assess the quality of the spectra (Bern et al., 2004). There is little purpose in investing significant effort in analyzing spectra of inferior quality. At the same time, the concept of spectra quality is not a well-defined one. Good quality spectra may be defined in terms of resolution or signal-to-noise ratio among the other criteria. Specific criteria are determined with regard to the experimental focus.

Spectra peaks are the result of the cleavage of the amino acid backbone at specific sites based on the specificity of the proteases. In general, peptide fragments attached with the N-terminus are known as b-ions, while peptide fragments attached with the C-terminus is known as y-ions. There have been multiple attempts to reconstruct the sample peptide using graph methods as well as probabilistic network approaches (Bruni et al., 2005; Yan et al., 2005). Typically, peaks are classified as b- and y-ions and represented as nodes in a graph; these nodes are linked by weighted edges. Subsequently, the best possible peptide sequence is obtained from the path through the graph that has an optimized score along the edges.

Besides proteases, the tertiary structure of the sample peptide influences the tendency for fragmentation at specific sites, affecting directly the locations and intensities of peaks within the spectra. Hence, there are efforts to correlate peak intensities with peptide properties such as hydrophobicity and helicity (Gay et al., 2002; Elias et al., 2004). However, such intensity-based approaches are still in its infancy because of unresolved issues with the reproducibility of absolute intensity information across replicate MS runs.

Tandem mass spectrometry analysis

In a parent-ion scan, the product (daughter) ion is fixed as the independent variable. The first spectrum is scanned for parent ions that give rise to specified spectrum component in the second stage. In other words, we are searching for a class of parent ions that dissociate to give target product ions. The product-ion scan is analogous to the parent-ion scan. In addition, the neutral-loss scan refers to scanning both stages of MS for a specified mass loss between the two stages.

The parent-ion and neutral-loss scans can be useful for drug discovery. Homologous compounds, such as variations of a likely therapeutic compound or metabolites of a target compound, will exhibit similar m/z ratios in the second stage of MS; these homologues may also be lost as neutral masses. The product-ion scan is useful for further revealing the structure and sequence information of complex protein mixtures passed into the first stage of MS.

2Dimensional gel electrophoresis

Because the primary mode of comparing individual gel maps is visual, there is a variety of software that is integrated with laboratory equipment to process the gel images. The huge number of spots on a single gel has also necessitated the automation of this process for high-throughput screening.

Gel maps can first be compared visually for 'differential display', i.e. a comparison of protein expression levels between control and treated samples. 'Interesting' proteins may be identified by location on the map or by quantifying the expression levels in identical spots. Furthermore, the spots can be excised and prepared for MS for further analysis.

Statistical analysis and pattern classification

Once the quality of the genomic or proteomic experiment has been verified, the next step is to employ any of the number of mathematical techniques have been applied to or developed for genomic or proteomic data. As mentioned previously, these techniques can be categorized into either unsupervised or supervised methods. The purpose of these algorithms is to identify underlying relationships between gene or protein expression or experimental samples and to identify significant genes or proteins in a controlled experiment.

Unsupervised methods

Unsupervised methods can provide insights to the natural organization of data. Typically for clustering algorithms, samples, genes, or protein expression levels can be arranged based on a metric of similarity. There are several methods for determining feature similarity: these include Euclidian distance, correlation, and dot product. Euclidian distance is the simplest metric. For example, given n samples, the distance between gene expression values $X \in R^n$ and $Y \in R^n$ can be calculated as

$$D(X, Y) = \sqrt{\sum_{i=1}^{n}(X_i - Y_i)^2} \quad (4)$$

where X and Y represent two different genes across all n samples. For the purpose of computational efficiency the square root may be removed with no adverse effect on clustering.

Pearson's correlation coefficient can also be used as a distance metric and can be computed as

$$D(X, Y) = \frac{1}{n}\sum_{i=1}^{n}\left(\frac{X_i - X_{\text{offset}}}{\Phi_X}\right)\left(\frac{Y_i - Y_{\text{offset}}}{\Phi_Y}\right) \quad (5)$$

in which Φ is the standard deviation of gene or protein expression over n samples. It has been suggested that the standard deviation for this metric be modified to account for a reference state that is not necessarily based on sample mean expression (Eisen et al., 1998). The modified standard deviation is

$$\Phi_G = \sqrt{\sum_{i=1}^{n}\frac{(G_i - G_{\text{offset}})^2}{n}} \quad (6)$$

where G_{offset} is a reference state (perhaps median or another reference state not apparent in the data). It may also be necessary to divide by $n-1$ instead of n to obtain the sample standard deviation if the sample size is small.

The dot product of two expression vectors may also be used as a measure of similarity. Vectors that are similar will have a large inner product, whereas orthogonal vectors will have a zero inner product. A larger result implies that genes are more closely related.

$$D(X, Y) = \langle \overline{X} \cdot \overline{Y} \rangle \quad (7)$$

The unsupervised clustering methods described here are based on similarity measures and include hierarchical clustering, self-organizing maps, and principal component analysis.

Hierarchical clustering

Once the distance metric has been defined, hierarchical clustering can be performed using one of several algorithms. Hierarchical clustering is an agglomerative technique, meaning that initially, there are several single member clusters which are gradually combined based on similarity until a single cluster remains (Quackenbush, 2001). When clusters are formed, the distance between two clusters may be calculated in different ways. Single-linkage clustering uses the minimum distance between a member of one cluster and a member of the other cluster, for all members. Complete-linkage clustering uses the maximum distance, while average linkage uses the average distance. Average distance can be computed as the average of the distance of each point in one cluster to all other points in the other cluster. Averages may also be weighted to account for unbalanced cluster sizes using a method called weighted pair-group average.

Hierarchical clustering is very useful for microarray analysis because it can be easily visualized on all dimensions. Gene expression can be represented with a heat map and ordered with hierarchical clustering so that patterns in expression become visually accessible. Figure 1 is an example of hierarchical clustering using two distance metrics. Both distance metrics result in the same two groups consisting of genes 1, 2, 4, 6, 9 and genes 3, 5, 7, 8, 10, however, the ordering of genes within these groups differs slightly between Euclidian and correlation distance metrics.

There are drawbacks in the use of hierarchical clustering for gene/protein expression analysis. The hierarchical structure imposes a strict ordering of groups of genes that may not be representative of all possible relationships. For multi-dimensional clustering, most metrics can result in non-unique distances, therefore failing to capture important differences between groups (Tamayo et al., 1999). More sophisticated clustering techniques such as self-organizing maps have been shown to outperform hierarchical clustering when applied to noisy biological data (Mangiameli et al., 1996).

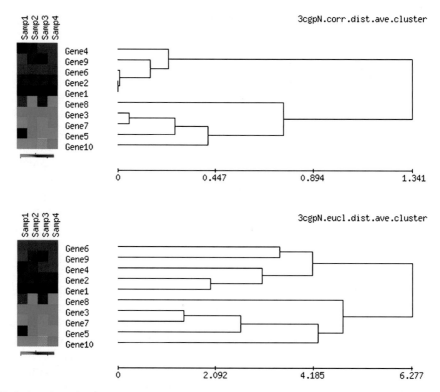

Fig. 1. Hierarchical clustering of 10 genes and four samples using correlations (top) and Euclidian distance (bottom). Clustering groups change slightly depending on the distance metric.

Self-organizing maps

The self-organizing map (SOM) is an unsupervised clustering neural network in which a single neuron, or node, is activated for each input vector, or sample. A SOM is trained by first selecting the number of nodes (equal to the number of samples) and arranging them in a 1- or 2dimensional grid. Each node is then mapped to a d-dimensional space in which d is the number of features (genes or proteins). Nodes are randomly compared to samples and iteratively moved, in d-dimensional space, toward the sample to which it is closest. The nodes in the grid surrounding the node that is closest to a sample are also moved slightly in the direction of the sample. After a number of iterations, the nodes in the grid become organized in a manner that represents the topological structure of the input samples. The nodes will form clusters that represent clusters of the underlying data.

Mathematically, this is applied as follows (Ham and Kostanic, 2001). Each node is associated with a synaptic weight vector $w_i \in R^d$ where $i = 1, 2, \ldots, m$ and m is the number of nodes. For each input data point $X_i \in R^d$ where d is the number of features or genes, the closest match to a node is computed as

$$q(X) = \min \|X - w_i\|_2 \quad (8)$$

where $q(X)$ indicates the winning node of the associated weight vector. The synaptic weight vector of the winning node is adjusted by

$$w_i(k+1) = w_i(k) + \mu(k)[X(k) - w_i(k)] \quad (9)$$

where $\mu(k)$ is the learning rate parameter that is decreased at each iteration. Nodes in the vicinity of the winning node are also adjusted, but with a different, smaller $\mu(k)$. The number of neighboring nodes affected by the winning node may also be reduced over time. The number of nodes, representing clusters, can be defined so that the input data can be optimally clustered. SOMs have been successfully applied to microarray gene expression data (Tamayo et al., 1999). A fundamental difference between SOMs and hierarchical clustering is the ability to select the number of nodes, or clusters, in SOMs. This is equivalent to the clustering of data by the most significant features, the number of which is equal to the number of nodes.

Principal component analysis

Although both hierarchical clustering and SOMs have been applied to large dimensional datasets, it may be desirable to reduce dimensionality without discarding any important features. Principal component analysis (PCA) has been used as a statistical tool for dimensional reduction. Given a set of features, or genes in a microarray, the principal components retain most of the information in the original features (Wang and Gehan, 2005). PCA compresses the input sample x with an optimal transformation matrix W according to

$$y = Wx \qquad (10)$$

where $y \in R^m$, $x \in R^n$, and $m << n$ (n is the original number of features). y retains most of the information in the input samples x by combining dimensions that have high covariance or correlation. W is a matrix of eigenvectors

$$W = [w_1, w_2, ..., w_m]^T \qquad (11)$$

in which w_i is an n dimensional eigenvector. PCA can be solved by computing the eigenvalue decomposition of the covariance matrix of the input data:

$$C_x = W^T \Lambda W \qquad (12)$$

in which C_x is the covariance matrix, defined as $E[xx^T]$, and Λ the diagonal matrix of eigenvalues (Ham and Kostanic, 2001).

Transformation with the W matrix can serve two purposes. First, it reduces the dimensionality of each sample so that the complexity of other clustering methods is reduced. Second, PCA can be a preliminary step to feature reduction in that the components of each eigenvector can be examined to extract the original dimensions that are most significant. For example, there are typically only a handful of eigenvalues that are much larger than the rest, corresponding to a small number of eigenvectors and components in the reduced space.

Each principal eigenvector in turn has a number of components that correspond to the original data dimensions. These components can be sorted by magnitude to rank the original dimensions that mostly contribute to the reduced dimension.

A drawback of PCA is that ranking or clustering by principal components is not intuitive; each value does not directly represent gene expression, instead it is a weighted linear combination of the original dimensions. Classical PCA has also been shown to be highly sensitive to outliers in the data, consequently, Hubert and Engelen (2004) have developed a robust PCA method to avoid this problem.

Supervised methods

In many genomic and proteomic experiments, the underlying grouping or clustering of samples is already known, in which case it would be more informative to identify the genes or proteins most affected by a controlled treatment of the samples, also known as the differentiating genes or proteins. These genes or proteins are called significant biomarkers and are actively sought after because of their potential in helping to understand and control biological mechanisms. They also help to reduce the complexity and increase the accuracy of classification rules that can be used for disease diagnosis and prognosis. Further applications of supervised methods are discussed by Olshen and Jain (2002).

Supervised algorithms continually being developed and refined to search for these biomarkers are not trivial for several reasons. First, the algorithm must be able to distinguish truly significant biomarkers from non-significant biomarkers in noisy data. Because of inherent variations in biological data, algorithms are seldom perfect. Second, algorithm parameters depend on properties of the data such that a set of parameters that works well for one dataset may not be correct for another dataset. Finally, identified biomarkers must be validated, which can be a tedious process given the large number of gene and protein features assayed in a single experiment. Significant biomarker identification is an iterative process that must be mathematically and biologically sound.

Statistical testing and pattern classification have been used for identifying significant biomarkers and building predictive models for diagnosis and prognosis. Statistical testing can be coupled with pattern classification as a method of feature reduction by ranking and filtering insignificant features before building a predictive classifier. Determination of the best statistical test to use on a particular dataset depends on the properties of the data distribution, such as mean, variance, and normality.

Pattern classifiers can be used to for both feature ranking and model building. The problem of pattern recognition is one of predicting future behavior or output of a system based on past input for which the output is already known. In general, this can be thought of as interpolating or extrapolating the output of a system. Output behavior of past input or training data is known, thus pattern recognition is a supervised learning method.

Given a set of data points X in the space R^d with known class labels $C \in \{1, 2, ..., l\}$, a pattern recognition algorithm, or classifier, defines a function that maps the data points to their correct labels in C. This classifier can then attempt to map an additional set of test data points, which were not used in the training phase, to their correct labels. Correct mapping of test data points is not guaranteed and the primary problem in designing the classifier is to minimize classification error. Determining true classification error is difficult, especially with small sample size problems, and is discussed in the error estimation and cross-validation section. An accurate error estimation technique can also be used as a feature ranking method.

Several pattern classifiers have been applied to both genomic and proteomic experiments for both feature ranking and predictive model building. Some commonly used supervised methods — k-nearest neighbors (knn), linear discriminant analysis and support vector machines (SVM) — are discussed in the following sections. Statistical hypothesis testing is also discussed as a method of feature ranking.

Feature ranking with hypothesis testing
Microarray or MS data can be represented mathematically as a set of samples $X_i \in R^d$ where d is the number of features, e.g. genes for microarrays or peptides for MS, and i the sample number. The number of features for both experiments can be very large, on the order of tens of thousands and it is possible that only a handful of these features significantly differentiate samples into defined classes. Feature space reduction can minimize the computational complexity of the classifier and provide useful information about underlying properties of the data. For example, a small set of features that can be used to build an accurate classifier may include features that are very important in the biological mechanisms under examination.

The problem of producing an accurate classifier, either by selecting the appropriate algorithm or adjusting parameters and feature selection are often closely related. In order to select features, a metric must be defined which can evaluate the overall contribution of a feature to classification.

The simplest metric that has often been used in microarray analysis is average fold change between classes:

$$F(X) = \frac{G_1}{G_2} \quad (13)$$

in which G_1 and G_2 are the average expression of a single feature in conditions 1 and 2, respectively. With a small number of samples, however, the difference in mean expression may not be significant. Even with an adequate number of samples, it may be useful to quantify the significance of mean difference using a statistical hypothesis test. Depending on properties of the data, a z- or t-test can be used to test the confidence of difference between two population means. The two population means, in this case, are the mean gene expression for each condition.

The z-test can be used for data in which both populations are known to be normally distributed and population variances are also known. The null hypothesis for this two-tailed test is that mean feature expression is zero.

$$H_0 : \mu_1 - \mu_2 = 0 \quad (14)$$

$$H_a : \mu_1 - \mu_2 \neq 0 \quad (15)$$

in which μ_1 and μ_2 are the population means of the feature in conditions 1 and 2, respectively. The z-statistic is computed as

$$z = \frac{\overline{G_1} - \overline{G_2} - 0}{\sqrt{\frac{\sigma_1^2}{M} + \frac{\sigma_2^2}{N}}} \quad (16)$$

in which M and N are sample sizes in conditions 1 and 2, respectively and σ^2 is the known population variance of each condition. The null hypothesis is rejected if $z \geq z_{\alpha/2}$ or $z \leq -z_{\alpha/2}$, where α is selected based on desired confidence. Even when the populations cannot be assumed to be normal and the population variances are not known, the z-statistic can still be used if samples sizes are sufficiently large. The z-statistic in this case is slightly altered

$$z = \frac{\overline{G_1} - \overline{G_2} - 0}{\sqrt{\frac{S_1^2}{M} + \frac{S_2^2}{N}}} \quad (17)$$

where S is the sample standard deviation.

For the typical microarray case in which the sample size is small and population variances are not known, the t-test can be applied if both populations are assumed to be normal. The t-statistic is very similar to the z-statistic:

$$T = \frac{\overline{G_1} - \overline{G_2} - 0}{\sqrt{\frac{S_1^2}{M} + \frac{S_2^2}{N}}} \quad (18)$$

T is approximately a t-distribution with degrees of freedom, v, estimated as

$$v = \frac{\left(\frac{S_1^2}{M} + \frac{S_2^2}{N}\right)^2}{\frac{(S_1^2/M)^2}{M-1} + \frac{(S_2^2/N)^2}{N-1}} \quad (19)$$

The assumption that populations are normal may be problematic and can be handled by methods such as significance analysis of microarrays (SAM) or the Wilcoxon rank-sum test (Tusher et al., 2001).

The Wilcoxon rank-sum test is effective on data that is neither normally distributed nor has a large sample size. This test is more robust on data with outliers. If the distributions of the two conditions are similar, the test statistic can be computed as

$$w = \sum_{i=1}^{m} r_i \quad (20)$$

where r_i is the rank of sample i in all samples, with 1 being the lowest ranked and $M+N$ being the highest ranked. For gene expression, we are interested in a two-tailed test, therefore the rejection region of the test is $w > c$ or $w <= M(M+N+1)-c$ where c can be obtained from a table of Wilcoxon critical values (Devore, 2004).

Using the methods described above, features can be ranked by ordering p-values of the hypothesis tests or ordering fold changes.

k-Nearest neighbors
k-nearest neighbors (kNN) is a simple supervised classification method that is used to predict the class of an unknown sample based on surrounding samples with known labels (Cover and Hart, 1967). Given an unknown sample, $X \in R^d$, the nearest k known samples are found by computing and sorting all distances to adjacent samples. Each of the k samples is associated with a label y and the predicted class of the unknown sample is determined by a majority vote. The distance can be computed using any number of metrics, including the simple Euclidian distance. Variations on kNN have been proposed that include studies of distance weighting (Dudani, 1976) and selection of k (Kulkarni et al., 1998).

kNN has the advantage of being very simple to implement yet competitive with other pattern classifiers. On the other hand, it can become computationally intensive as the number of samples increases. The distance metric must be computed for a large number of sample combinations, which would be compounded when using different cross validation or feature selection methods. The kNN algorithm has been applied to many gene selection problems, ranging from bone disease (Theilhaber et al., 2002) to breast cancer (Modlich and Bojar, 2005).

Linear discriminant analysis
Linear discriminant analysis (LDA) is similar to PCA except that it considers class labels when reducing dimensionality. The LDA solution is the projection of data points onto an optimized feature space which maximizes the ratio of between-class to within-class variance. In other words, the

data points are mapped to a space in which classes are maximally separated. Given microarray samples $\{X_1, X_2, ..., X_N\}$ belonging to classes $\{C_1, C_2, ..., C_C\}$ the between-class scatter matrix can be computed as

$$S_b = \frac{1}{N}\sum_{i=1}^{C} N_i(\mu_i - \mu)(\mu_i - \mu)^T \qquad (21)$$

where $\mu_i = 1/N\sum_{X \in C_i} X$ is the mean of samples in class C_i and $\mu = 1/N\sum_{i=1}^{N} X_i$ is the mean of all samples. The within-class scatter matrix is computed as

$$S_w = \frac{1}{N}\sum_{i=1}^{C}\sum_{X_k \in C_i} (X_k - \mu_i)(X_k - \mu_i)^T \qquad (22)$$

The solution is obtained by solving the optimization problem

$$\arg\max_W \frac{|W^T S_b W|}{|W^T S_w W|} = [w_1, w_2, ..., w_m] \qquad (23)$$

where W_i are eigenvectors of S_b and S_w that correspond to the largest eigenvalues, λ_i. The problem is essentially an eigenvalue decomposition

$$S_b w_i = \lambda_i S_w w_i \qquad (24)$$

Once W has been solved, a data point, or sample can be mapped to a predicted class label by determining the samples position relative to the hyperplane defined by w (Nhat and Lee, 2005).

Support vector machines
SVM, developed by Vapnik (1995), are a class of optimization problems that partitions a set of data points according to class label with a maximal margin hyperplane. Given a set of data points $X \in R^d$ and class labels $y \in \{+1, -1\}$, a linear classifier defines the hyperplane

$$w \cdot X + b = 0 \qquad (25)$$

where w is also a vector in R^d. The optimization problem searches for w and b that satisfy the conditions

$$y_i(X_i \cdot w + b) \geq 1 - \xi_i \qquad (26)$$

in which ξ_i is a slack variable that allows for small errors in classification. The primal Lagrangian of this problem is

$$L_P = \frac{1}{2}\|w\|^2 + C\sum_i \xi_i - \sum_i \alpha_i\{y_i(X_i \cdot w + b) - 1 + \xi_i\} - \sum_i \mu_i \xi_i \qquad (27)$$

where α_i and μ_i are Lagrange multipliers. Equation (26) can be converted into the Lagrange dual, maximizing

$$L_D = \sum_i \alpha_i - \frac{1}{2}\sum_{i,j} \alpha_i \alpha_j y_i y_j K(X_i, X_j) \qquad (28)$$

subject to:

$$0 \leq \alpha_i \leq C \qquad (29)$$

$$\sum_i \alpha_i y_i = 0 \qquad (30)$$

then the solution can be obtained by

$$w = \sum_i \alpha_i y_i X_i \qquad (31)$$

The function $K(X_i, X_j)$ in Eq. (27) is known as the kernel function and enables the SVM to handle non-linear separations by mapping each data point X_i to an alternate space $\Phi(X_i)$. The function $\Phi(X_i)$ does not have to be explicit, keeping computational complexity of the SVM at a minimum even though the alternate space, usually high dimension, can be very complex. For the linear case, the kernel function is a dot product

$$K(X_i, X_j) = \langle X_i \cdot X_j \rangle \qquad (32)$$

Mapping can be very complex, to an infinite dimensional space, using the Gaussian kernel function

$$K(X_i, X_j) = e^{-\|X_i - X_j\|^2/2\sigma^2} \qquad (33)$$

Once the SVM has been trained, it can be evaluated with the function

$$f(X) = \sum_{i=1} \alpha_i y_i K(X_i, X) + b \qquad (34)$$

in which the classification of sample X is determined by the sign of $f(X)$. Note that when evaluating the classifier, the mapping function $\Phi(X)$ does not need to be known (Cristianini and Shawe-Taylor, 2000).

Performance evaluation

Cross validation and error-estimation methods
A pattern classifier is typically able to correctly classify all the samples on which it was trained. Of course, this depends on the type of classifier used. For example, a linear discriminant classifier will have 100% accuracy on a set of data points only if those data points can be linearly separated. A non-linear problem is more difficult for a linear classifier but can be easily handled by the SVM using a non-linear kernel. However, testing a classifier on the same set of points on which the classifier was trained is only useful for determining the classifier's ability to map input samples to output class labels, also known as the resubstitution or training error. If, for instance, the classifier was then used to test an independent set of samples, the accuracy may be significantly lower. This phenomenon of high accuracy on training data is called overfitting and is a difficult problem that arises in ill-posed problems such as the classification of microarray and MS in which the number of features is much larger than the number of samples. The accuracy of a classifier when tested with an independent set of samples corresponds to the classifier's ability to generalize. The combination of accuracy and generalization ability of a classifier can be used as a measure of overall performance. Methods used to deal with small sample problems are cross validation, bootstrap, and bolstering.

Typically, cross validation consists of training a classifier on a subset of samples, then testing the resulting classifier on an independent set of test samples. This method in its simplest form is also called holdout cross validation. This method is adequate for datasets that have a large number of samples. However, for small sample sizes, as is often the case with microarray experiments, alternative methods must be used (Goutte, 1997). k-fold cross validation is the process of dividing a dataset into k subsets and performing the holdout cross validation k times, each time with a different training and test set. The training set in this case would be $k-1$ subsets, while the test set would be the remaining subset. The extreme case of k-fold cross validation is complete leave-one-out cross validation, in which $k = n$, the total number of samples.

Correct prediction rate or error of the classifier can be computed as the average over all k-tests. k-fold cross validation reduces the variance in error estimation by increasing the number of tests.

Error estimation using cross validation for small sample microarray studies may still be biased and highly variable (Fu et al., 2005). Therefore bootstrapping has been proposed to reduce biases in error estimation for microarray analysis (Braga-Neto and Dougherty, 2004b). Bootstrapping is the process of randomly selecting a number of samples for training and testing on the remaining samples. By randomly selecting samples, training sets may not be unique, but by repeating this process a sufficient number of times, the variance and bias of the error can be improved. Generally, bootstrap error estimation iteratively selects B random samples of n points each for training a classifier. The n points of sample are selected with replacement, thus the total sample size in a set B is usually less than n. For each B sample, the classifier is trained then tested on the remaining points, resulting in an error value. The B error values are then averaged to produce the error estimation.

The 0.632 bootstrap method is similar to bootstrap cross validation except that the cross validation error rate is weighted by 0.632, the average fraction of samples that are in a training set if samples are selected randomly with replacement. For n total samples, the training set consists of n draws with replacement, therefore some samples are duplicated, while others are left out (Efron, 1983). Therefore, the total 0.632 bootstrap error is computed as

$$E_{0.632} = 0.632 E_0 + (1 - 0.632) E_{\text{resub}} \qquad (35)$$

where E_0 is the bootstrap zero estimator (Efron, 1983; Braga-Neto and Dougherty, 2004b) and E_{resub} the resubstitution error.

The 0.632 bootstrap method has been improved to alleviate the case in which resubstitution error, E_{resub}, is 0 (sometimes this means that overfitting has occurred). With this method, a larger weight (>0.632) is placed on the bootstrap zero estimator when overfitting occurs, detected when E_0 is much greater than E_{resub}. This improved bootstrap method is known as the 0.632+ estimator (Efron, 1997).

Braga-Neto and Dougherty, 2004b found that the 0.632 bootstrap error estimator shows the least bias and variability compared to resubstitution, leave-one-out, and 5- and 10-fold cross validation. Fu et al., however, compared bootstrap, cross validation, leave-one-out bootstrap, and 0.632 bootstrap methods in a comprehensive simulated study and showed that the simpler bootstrap performs better than leave-one-out bootstrap, 0.632 bootstrap, and the improved 0.632 + bootstrap.

Another method for error estimation in which the original data distribution is 'bolstered' using a kernel function performs, in most cases, as well as 0.632 bootstrap error estimation but can be computed faster. This method is essentially a density estimation for a set of samples using a mixture of Gaussians. The area around a data point, defined by the kernel function, is part of a Gaussian distribution and is used to compute a smooth error. If the area around any point is misclassified, the total performance of the classifier is deducted by an amount equal to the fraction of misclassified area. The error can be quickly estimated using Monte-Carlo simulation (Braga-Neto and Dougherty, 2004a).

A set of features used for training a classifier may be an optimal set of features depending on the performance of the classifier using various cross-validation techniques for error estimation. If the estimated error is accurate and small, then these features may be significant contributors to the natural partitioning of the conditions of interest. This is the basis of feature ranking using error estimation and classifiers.

Feature ranking with classifiers

The most intuitive method of ranking features with a classifier is to perform one of the above mentioned error estimation techniques on individual features and rank these features by increasing error or decreasing prediction rate. Once again, the total number of features governs the number of identified significant features, which depends on the desired threshold of significance. This threshold of significance, as with the *p*-values in hypothesis testing, should be selected carefully and the process may have to undergo validation through an iterative process in order to select the optimal number of significant features.

An alternate method, recursive feature elimination (RFE), is a process of iteratively reducing the feature space of a dataset by ranking (Guyon et al., 2002). At each iteration, the classifier is trained with the remaining features (which, for the first iteration, will be all features), and the genes are ranked based on properties of the trained classifier. For a SVM, the ranking may be based on the absolute value or square of the weights, w, with a small weight indicating the lowest ranked feature. The lowest ranked feature is then removed and the process is repeated. For computational efficiency, more than one feature can be removed at each iteration, however, removing only one feature results in a complete ranking. Ranking produced by RFE may differ from individual gene ranking since the performance of a feature set is the result of a combination of several genes.

Although single feature ranking is the simplest method that has been proven in several applications, it assumes that features are expressed independently. Yet the ultimate mechanisms that govern a biological process most certainly depend on combinations of features. Because of the combinatorial complexity of ranking groups of features, single-feature ranking remains an indispensable tool that may be used to reduce the search space for multiple-feature ranking. Methods for feature combination selection are introduced in the next section.

ROC curves

The receiver operating characteristic (ROC) curve is often used as a tool for measuring the performance of a machine-learning algorithm (Bradley, 1997). When designing a classifier with optimal generalization properties, there is always some inherent error involved. For the simplest two class problems, these errors can be categorized as either false positives or false negatives. The overall performance of a classifier can be summarized with a confusion matrix (Table 1).

The confusion matrix represents all information about a classifier at a specific operating point. The only information needed to produce an ROC curve is the sensitivity (Eq. (36)) and specificity (Eq. (37)).

Table 1. Confusion matrix

True class	Predicted class		True total
	−	+	
−	T_n	F_p	C_n
+	F_n	T_p	C_p
Predicted total	R_n	R_p	

$$\text{Sensitivity} = P(T_p) = \frac{T_p}{C_P} \quad (36)$$

$$\text{Specificity} = P(T_n) = \frac{T_n}{C_n} \quad (37)$$

As the decision threshold of the classifier is varied, $P(T_p)$ and $P(T_n)$ also vary. Several ROC points can be plotted in this manner, while varying the decision threshold to produce the ROC curve. ROC points are plotted as false positives versus true positives, or $1-P(T_n)$ versus $P(T_p)$. For a good classifier, a slight increase in false positives would result in a faster increase in true positives, whereas for random data, these values would increase at approximately the same rate. Therefore, to measure the performance of the classifier, all points on the curve should be considered. The area under the ROC curve (AUC) is a good measure of performance since, for an optimal classifier, this area would be approximately 1, whereas for random data, this area would be 0.5.

Feature combinations and global search methods

Feature selection and ranking methods that analyze microarray genes or MS proteomic expression a single dimension at a time ignore the fact that many of these features are highly correlated. Correlation information in microarray data can provide insights to gene regulatory networks and improve the effectiveness of pattern classifiers. Selection of significant feature combinations may produce classification rules with higher predictive ability than any single biomarker. Furthermore, these combinations may include genes or proteins that may not be detectable by conventional univariate methods (Szabo et al., 2002).

Szabo et al. discuss multivariate statistical methods for comparing gene combinations. The application of these statistical methods to microarray data is not easy for two reasons. First, some of these methods rely on estimating covariance, such as the Mahalanobis distance (Mahalanobis, 1936), which is a generalized statistical distance. Covariance estimation is usually unreliable because of the small number of samples often associated with microarray experiments. Second, the number of dimensions in the dataset is usually very large, rendering a full search of all gene combinations computationally impossible (Chilingaryan et al., 2002).

Algorithms such as the Monte-Carlo and genetic algorithm (GA) have been applied to overcome the problem of combinatorial complexity in feature selection searches. The Monte-Carlo, a random search algorithm, by no means searches the entire space but has been shown to be a good estimator for optimization problems. Chilingaryan et al. (2002) proposed a multi-start random search algorithm which can be implemented in parallel. This algorithm searches local maximum regions and avoids finding global maximums by stopping the algorithm at an iteration limit. The reasoning being that a global optimum may be overfitting, especially in discontinuous or discrete data.

GAs are similar to random searches except the solution is directed, or evolved, toward the optimal solution by producing new combinations from previously good combinations, as evaluated by an objective function. In other words, it uses survival of the fittest to retain variations only if they are beneficial. Random permutations are periodically introduced into the population in the form of mutations that prevent the algorithm from stalling in a local optimum. The GA has been applied to many biological problems (Kim et al., 2004; Ni and Liu, 2004; Paul and Iba, 2004; Liu et al., 2005).

In addition to using multivariate statistical tests such as the Mahalanobis distance, cross-validation methods can be used as the objective function for Monte-Carlo and GA searches. Pattern classifiers such as kNN, linear discriminant, and SVM can easily handle multi-dimensional data. Furthermore, the use of an effective cross validation or error estimation method as the objective function may alleviate the problem of overfitting with global search algorithms.

While the search for single biomarkers can provide some information about underlying biological mechanisms, the search for multiple feature combinations is essential for discovering complex interactions that govern gene and protein networks. The inference of genetic and proteomic networks, however, is an extension of the methods described here and will not be covered.

Microarray case study

As a case study, a set of microarray data is analyzed according to some of the methods outlined above. CodeLink bioarrays were used to measure gene expression for two sample classes, cocaine overdose (OD) and control (C) (Hemby, S.). Seven cocaine OD and seven C samples for a total of 14 samples were analyzed after four samples (two from each class) were removed based on quality.

CodeLink bioarrays have a very large dynamic range, on the order of thousands, which is good for accurately detecting small changes in expression. However, for the numerical stability of some computational algorithms, this dynamic range can be reduced to a more manageable range. The dataset was normalized with a $\log_{10}$ transformation, which reduces the data's range as well as variances across samples (Fig. 2).

Dataset features were ranked individually using fold change, SAM, linear SVM, radial basis SVM, and the polynomial SVM with degrees 2 and 3. For each SVM, a full parameter analysis was performed to determine the optimal SVM cost or gamma (in the case of the radial basis and polynomial SVMs) that reduced the average error over all features. The SVM (implemented using LIBSVM (Cheng and Lin, 2001)) was used to estimate resubstitution, leave-one-out cross validation and bootstrap error estimation for each feature. Bootstrap error estimation with the 0.632 method (Braga-Neto and Dougherty, 2004b) was performed on each feature with 100 sampling iterations.

For the linear SVM, the cost parameter was ranged from 0.01 to 10,000 on the log scale to minimize resubstitution, cross validation, and bootstrap error. In all cases, the error tended to decrease with increasing cost, however, computational time also increased with increasing cost. Cost must be selected based on problem size and computing power, but does not significantly improve error after a problem-specific threshold (Fig. 3). The optimal cost was selected to be 1000.

In addition to the SVM cost parameter, the radial basis kernel expects an additional parameter, gamma (inversely proportional to sigma in the radial basis kernel equation), which affects the size of resulting classification regions. Both the cost and gamma must be selected in conjunction since they can have a direct effect on each other. Values for each parameter are selected by evaluating the performance of each feature across a 2D grid of parameter values (Fig. 4). For the cocaine overdose dataset, cost was varied from 1 to 1000 and gamma was varied from 100 to 10,0000 on the log scale. Each gene is evaluated independently; however, it is not feasible to select parameters for each individual gene. Therefore, the performance of all genes for a single parameter grid point is averaged, creating a smooth surface representing overall performance. The performance measures used were resubstitution, leave-one-out cross validation and 0.632 bootstrap. Resubstitution results (Fig. 4, top left) show that as gamma and cost are increased, the accuracy of classification on training data increases. Cross-validation results (Fig. 4, top right) show that the generalization ability of the classifier decreases as gamma increases. It is desirable to have both training accuracy and good generalization ability; therefore, a combination of resubstitution and cross validation should be used to determine the optimal parameters. The bootstrap error-estimation method achieves this, selecting an intermediate gamma (Fig. 4, bottom). The optimal gamma and cost for the radial basis SVM are 800 and 1000, respectively.

The genes of the dataset were ranked using the optimal selected parameters for linear and radial basis SVM as well as with fold change and student's t-test. Fold change, t-test, and linear SVM rankings were somewhat correlated, while radial basis SVM rankings were significantly different (Fig. 5). The gene ranking method must therefore

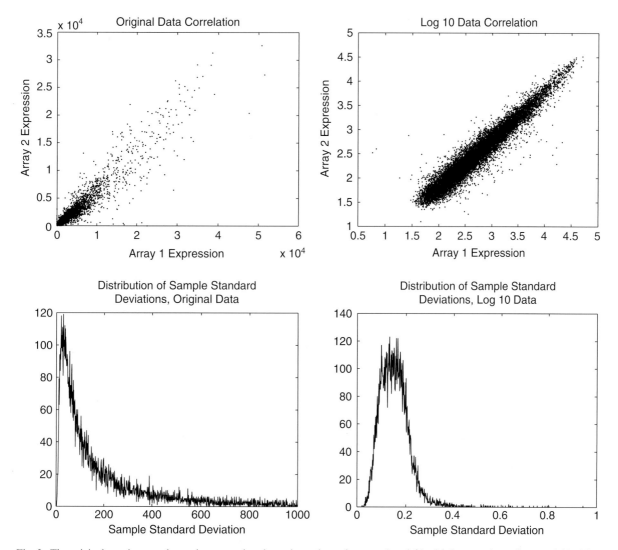

Fig. 2. The original cocaine overdose microarray data has a large dynamic range (top left) with large variance (bottom left). After normalization with a $\log_{10}$ transformation, the data's range (top right) and variance (bottom right) have been significantly reduced.

be selected based on validation and interpretation results.

Once genes have been ranked, the threshold for selection of significant genes should be determined to maximize the number of truly significant genes and minimize the false-discovery rate. To determine if a gene is truly significant, that gene must be understood in context of the disease in question. In many cases, this information is incomplete, which is why gene ranking and selection methods are employed in the first place. Full literature surveys on the top genes are possible but can be very time-consuming. In addition, different ranking and normalization methods can drastically change the selected top genes, therefore increasing the number of possible significant genes.

The simplest method of selecting top genes from the ranking results is to set the threshold at a statistically significant level. For our test case, the threshold for selecting genes from the radial basis

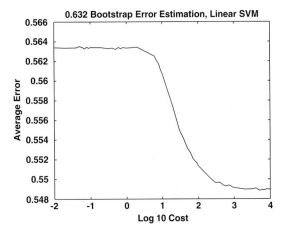

Fig. 3. Parameter-selection curve for the linear SVM. The average error over all features decreases as the SVM cost is increased.

and linear SVM was set to three standard deviations under the mean error. For normally distributed data, this means that there is about a 0.15% chance that any samples would fall below that threshold. However, true errors are only approximately normal, resulting in slightly different percentages. For the fold change data, genes were selected if their fold change fell outside of three standard deviations from the mean, both greater than and less than, for an approximately 0.3% probability.

Correlation of significant markers is not significant among the three ranking methods (Table 2). There were 14 common genes that were significant in both linear SVM ranking and absolute fold change ranking. Only two were in common between radial basis and linear SVM and none in common between radial basis and absolute fold

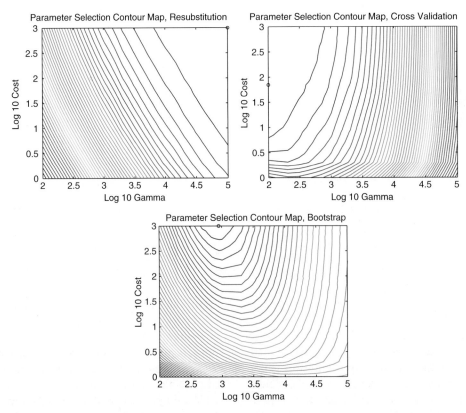

Fig. 4. Parameter selection contour maps for several error-estimation methods: resubstitution (top left), cross validation (top right), bootstrap (bottom).

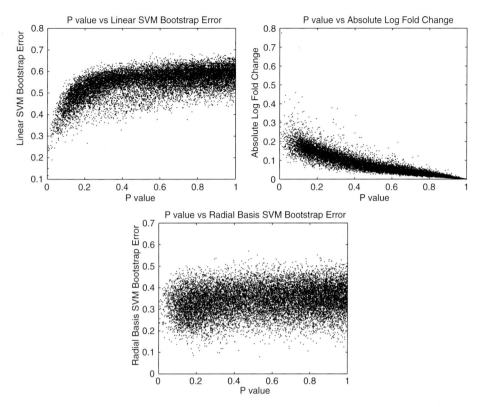

Fig. 5. Comparison of gene-ranking methods. Bootstrap error estimation for the linear SVM tends to increase as the t-test p-value increases (top left), fold change between class means decreases as t-test p-value increases (top right) and bootstrap error estimation for the radial basis SVM has no correlation with p-value.

Table 2. Correlation of gene ranking methods after establishing a statistically significant threshold

	RBF SVM	Linear SVM	Abs. FC
RBF SVM	31	2	0
Linear SVM	2	185	14
Abs. FC	0	14	73

Note: The radial basis function SVM (RBF SVM) and linear SVM thresholds were set to three standard deviations below the mean error. The absolute fold change (Abs. FC) threshold was set to three standard deviations from the mean.

change. The threshold for t-test ranking is more difficult to select since the distribution of p-values is not normal. However, p-values were shown to be correlated with linear SVM and absolute fold change rankings (Fig. 5).

ROC curves were generated for the radial basis SVM, linear SVM, and absolute fold change ranking methods using top-, mid-, and low-ranked genes. ROC curves for the absolute fold change method were generated using a linear SVM. Using optimal parameters, the radial basis SVM performs well with most genes, showing only a slight decrease in AUC as lower-ranked genes are used. AUC of the linear SVM, however, decreases significantly as the classifier is trained with lower-ranked genes (Fig. 6).

Interpretation and validation

Following the biomarker discovery process, efforts must be undertaken to interpret and validate the results in a clinically meaningful way. Solving the problems posed by biomarker discovery, as we have discussed, may be an academic challenge and will remain at that until we verify our findings with independent sources.

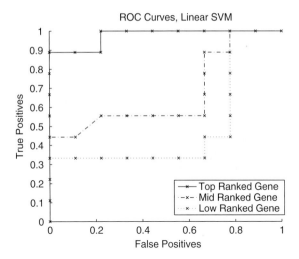

Fig. 6. Linear ROC curves generated using a top-, mid-, and low-ranked gene. AUC of the ROC curves decrease as gene rank is decreased.

There are a variety of approaches to interpreting and validating selected biomarkers. Gene ontologies are shared vocabularies that classify gene products by three major classes: molecular function, biological process, and cellular component. These vocabularies are by no means exhaustive; however, they represent significant progress toward a unified standard for biological annotation. The Gene Ontology Project is a collaborative effort toward this means — to achieve a consistent annotation for gene products across various databases (Ashburner et al., 2000). By locating the discovered biomarkers in the biological context of an ontological tree, we gather more evidence supporting or refuting the significance of these biomarkers.

Besides ontologies, considerable research has been done on cellular pathways, in particular, pathways that are potentially involved in pathological expression. Coupled with high-throughput screening methods, such as microarrays and MS, extensive networks of cellular pathways may be sampled and then visualized simultaneously to achieve a systems-based perspective. For instance, the Pathway Explorer developed by the Institute of Genomics and Bioinformatics, Graz University of Technology, Austria (Mlecnik et al., 2005) is a popular web-based tool for visualizing biological pathways derived from databases such as KEGG, BioCarta, and GenMapp. On the other hand, because the field of pathway annotation and visualization is still in infancy, the number of curated pathways remains at a minimum and may not necessarily be consistently annotated. For a more comprehensive list of bioinformatics tools, refer to the Bioinformatics Links Directory: Protein Interaction et al., 2005, (http://bioinformatics.ubc.ca/resources/links directory/) provided by the University of British Columbia Bioinformatics Centre. Similarly, as is the case with ontologies, discovered biomarkers may be verified by their location within the visualized pathways.

Furthermore, comprehensive tools have been developed to provide simultaneous verification of discovered biomarkers across multiple platforms and sources. For instance, GoMiner (Zeeberg et al., 2003) provides a variety of representations of biomarker information such as trees, statistical tables, molecular structures, and pathway visualizations derived from multiple databases. The strength of such cross-platform validation lies in the fact that the sources are independent compartmentalized views of cellular processes and that agreement throughout these sources is a robust verification of the validity of the discovered biomarkers.

Although microarray technology is fairly well understood, it may be desirable to validate the mRNA expression in a more focused experiment. For this reason, real-time RT-PCR (qPCR) has been used to corroborate the expression of selected microarray transcripts. By eliminating the many variables involved in a microarray experiment, a qPCR experiment can serve to increase the confidence of expression level before drawing conclusions.

Undoubtedly, the golden standard for interpretation and validation of biomarkers is empirical and clinical studies. However, besides cost, practical, and ethical considerations dictate that clinical studies must be fully justified before trials can be conducted. Consequently, the importance of interpretation and validation tools, as we have just discussed, cannot be slighted.

Acknowledgments

The authors want to thank Georgia Cancer Coalition, Georgia Research Alliance, and NIH for Research support for this work.

References

Aebersol, R. and Mann, M. (2003) Mass spectrometry-based proteomics. Nature, 422: 198–207.

Ashburner, M., et al. (2000) Gene Ontology: tool for the unification of biology. The Gene Ontology Consortium. Nat. Genet., 25: 25–29.

Bern, M., Goldberg, D., McDonald, W.H. and Yates, J.R. (2004) Automatic quality assessment of peptide tandem mass spectra. Bioinformatics, 20: i49–i54.

Bioinformatics Links Directory: Protein Interaction, Pathways, Enzymes. (2005).

Bolstad, B.M., Irizarry, R.A., Astrand, M. and Speed, T.P. (2003) A comparison of normalization methods for high density oligonucleotide array data based on variance and bias. Bioinformatics, 19(2): 185–193.

Bradley, A.P. (1997) The use of area under the ROC curve in the evaluation of machine learning algorithms. Pattern Recog., 30(7): 1145–1159.

Braga-Neto, U.M. and Dougherty, E.R. (2004a) Bolstered error estimation. Pattern Recog., 37: 1267–1281.

Braga-Neto, U.M. and Dougherty, E.R. (2004b) Is cross-validation valid for small-sample microarray classification? Bioinformatics, 20(3): 374–380.

Bruni, R., Gianfranceschi, G. and Koch, G. (2005) On peptide de novo sequencing: a new approach. J. Pept. Sci., 11: 225–234.

Cheng, C.-C., Lin, C.-J. (2001). LIBSVM: a library for support vector machines.

Chilingaryan, A., Gevorgyan, N., Vardanyan, A., Jones, D. and Szabo, A. (2002) Multivariate approach for selecting sets of differentially expressed genes. Math. Biosci., 176: 59–69.

Cleveland, W.S. and Devlin, S.J. (1988) Locally weighted regression: an approach to regression analysis by local fitting. J. Am. Stat. Assoc., 83(403): 596–610.

Cover, T.M. and Hart, P.E. (1967) Nearest neighbor pattern classification. IEEE Trans. Inform. Theory, IT-13(1): 21–27.

Cristianini, N. and Shawe-Taylor, J. (2000) An Introduction to Support Vector Machines and Other Kernel-based Learning Methods. Cambridge University Press, Cambridge.

Devore, J.L. (2004) Probability and Statistics for Engineering and the Sciences. Thomson Brooks/Cole, Toronto.

Dudani, S. (1976) The distance weighted K-nearest-neighbor rule. IEEE Trans. Systems, Man, Cybernet., 6: 325–327.

Efron, B. (1983) Estimating the error rate of a prediction rule: some improvements on cross-validation. J. Am. Stat. Assoc., 78: 316–331.

Efron, B. (1997) Improvements on cross-validation: the .632+ bootstrap method. J. Am. Stat. Assoc., 92(438): 548–560.

Eisen, M.B., Spellman, P.T., Brown, P.O. and Botstein, D. (1998) Cluster analysis and display of genome-wide expression patterns. Proc. Natl. Acad. Sci., 95: 14863–14868.

Elias, J., Gibbons, F., King, O., Roth, F. and Gygi, S. (2004) Intensity-based protein identification by machine learning from a library of tandem mass spectra. Nat. Biotechnol., 22(2): 214–219.

Espina, V., Mehta, A.I., Winters, M.E., Calvert, V., Wulfkuhle, J., Petricoin, E.F. and Liotta, L.A. (2003) Protein microarrays: molecular profiling technologies for clinical specimens. Proteomics, 3: 2091–2100.

Fields, S. (2001) Proteomics in genomeland. Science, 291: 1221–1223.

Fu, W.J., Carroll, R.J. and Wang, S. (2005) Estimating misclassification error with small samples via bootstrap cross-validation. Bioinformatics, 21(9): 1979–1986.

Gay, S., Binz, P.-A., Hochstrasser, D. and Appel, R. (2002) Peptide mass fingerprinting peak intensity prediction: extracting knowledge from spectra. Proteomics, 2: 1374–1391.

Glish, G. and Vachet, R. (2003) The basics of mass spectrometry in the twenty-first century. Nat. Rev. Drug Discov., 2: 140–150.

Goutte, C. (1997) Note on free lunches and cross-validation. Neural Comput., 9: 1245–1249.

Guyon, I., Weston, J., Barnhill, S. and Vapnik, V. (2002) Gene selection for cancer classification using support vector machines. Mach. Learning, 46: 389–422.

Ham, F.M. and Kostanic, I. (2001) Principles of Neurocomputing for Science and Engineering. McGraw-Hill, New York.

Hoffmann, R., Seidl, T. and Dugas, M. (2002) Profound effect of normalization on detection of differentially expressed genes in oligonucleotide microarray data analysis. Bioinformatics, 21(8): 1509–1515.

Hubert, M. and Engelen, S. (2004) Robust PCA and classification in biosciences. Bioinformatics, 20: 1728–1736.

Kim, Y.H., Lee, S.Y. and Moon, B.R. (2004) A genetic approach for gene selection on microarray expression data. Lecture Notes Comput. Sci., 3102: 346–355.

Kulkarni, S.R., Lugosi, G. and Venkatesh, S.S. (1998) Learing pattern classification — a survey. IEEE Trans. Inform. Theory, 44(6): 2178–2206.

Li, J., Zhang, Z., Rosenzweig, J., Wang, Y.Y. and Chan, D.W. (2002) Proteomics and bioinformatics approaches for identification of serum biomarkers to detect breast cancer. Clin. Chem., 48(8): 1296–1304.

Liao, M. (2005) Baysian Models and Machine Learning with Gene Expression Analysis Applications. Institute of Statistics and Decision Sciences, Duke University.

Lipshutz, R., Fodor, S., Gingeras, T. and Lockhart, D. (1999) High density synthetic oligonucleotide arrays. Nat. Genet. Suppl., 21: 20–24.

Liu, J.J., Cutler, G., Li, W., Pan, Z., Peng, S., Hoey, T., Chen, L. and Ling, X.B. (2005) Multiclass cancer classification and biomarker discovery using GA-based algorithms. Bioinformatics, 21(11): 2691–2697.

Mahalanobis, P.C. (1936) On the generalized distance in statistics. Proc. Natl. Inst. India, 12: 49.

Mangiameli, P., Chen, S. and West, D.A. (1996) A comparison of SOM neural network and hierarchical clustering methods. Eur. J. Oper. Res., 93: 402–417.

Model, F., Konig, T., Piepenbrock, C. and Adorjan, P. (2002) Statistical process control for large scale microarray experiments. Bioinformatics, 18: S155–S163.

Modlich, O., Prisack, H.B., Munnes, M., Audretsch, W., and Bojar, H. (2005) Predictors of primary breast cancers responsiveness to preoperative epirubicin/cyclophosphamide-based chemotherapy: translation of microarray data into clincally useful predictive signatures. J. Transl. Med., 32.

Nhat, V.D.M. and Lee, S. (2005) Block LDA for face recognition. Lecture Notes Comput. Sci., 3512: 899.

Ni, B. and Liu, J. (2004) A novel method of searching the microarray data for the best gene subsets by using a genetic algorithm. Lecture Notes in Comput. Sci., 3242: 1153–1162.

Olshen, A. and Jain, A. (2002) Deriving quantitative conclusions from microarray expression data. Bioinformatics, 18: 961–970.

Pandey, A. and Mann, M. (2000) Proteomics to study genes and genomes. Nature, 405: 837–846.

Mlecnik, B., Scheideler, M., Hackl, H., Hartler, J., Sanchez-Cabo, F. and Trajanoski, Z. (2005) Pathway Explorer: web service for visualizing high-throughput expression data on biological pathways. Nucleic Acids Res, 33: W633–W637.

Paul, T.K. and Iba, H. (2004) Identification of informative genes for molecular classification using probabilistic model building genetic algorithm. Lecture Notes Comput. Sci., 3102: 414–425.

PeptideSearch: FingerPrint, Bioanalytical Research Group. (2005).

Quackenbush, J. (2001) Computational analysis of microarray data. Nat. Rev.: Genet., 2: 418–427.

Reilly, C., Wang, C. and Rutherford, M. (2003) A method for normalizing microarrays using genes that are not differentially expressed. J. Am. Stat. Assoc., 98(464): 868–878.

Schena, M., Shalon, D., Davis, R. and Brown, P. (1995) Quantitative monitoring of gene expression patterns with a complementary DNA microarray. Science, 270: 467–470.

Steen, H. and Mann, M. (2004) The abc's (and xyz's) of peptide sequencing. Nat. Rev. Mol. Cell Biol., 5: 699–711.

Suzuki, T., Higgins, P.J. and Crawford, D.R. (2000) Control selection for RNA quantitation. Biotechniques, 29(2): 332.

Szabo, A., Boucher, K., Carroll, W.L., Klebanov, L.B., Tsodikov, A.D. and Yakovlev, A.Y. (2002) Variable selection and pattern recognition with gene expression data generated by the microarray technology. Math. Biosci., 176: 71–98.

Tamayo, P., Slonim, D., Mesirov, J., Zhu, Q., Kitareewan, S., Dmitrovsky, E., Lander, E.S. and Golub, T.R. (1999) Interpreting patterns of gene expression with self-organizing maps: methods and application to hematopoietic differentiation. Proc. Natl. Acad. Sci., 96: 2907–2912.

Theilhaber, J., Connolly, T., Roman-Roman, S., Bushnell, S., Jackson, A., Call, K., Garcia, T. and Baron, R. (2002) Finding genes in the C2C12 osteogenic pathway by k-nearest-neighbor classification of expression data. Genome Res., 12: 165–176.

Tusher, V.G., Tibshirani, R. and Chu, G. (2001) Significance analysis of microarrays applied to the ionizing radiation response. Proc. Natl. Acad. Sci., 98(9): 5116–5121.

Vapnik, V. (1995) The Nature of Statistical Learning Theory. Springer, New York.

Wang, A. and Gehan, E. (2005) Gene selection for microarray data analysis using principal component analysis. Stat. Med., 24: 2069–2087.

Wolkenhauer, O., Moller-Levet, C. and Sanchez-Cabo, F. (2002) The curse of normalization. Comp. Funct. Genom., 3: 375–379.

Yan, B., Pan, C., Olman, V., Hettich, R. and Xu, Y. (2005) A graph-theoretic approach for the separation of b and y ions in tandem mass spectra. Bioinformatics, 21(5): 563–574.

Yang, Y.H., Dudoit, S., Luu, P., Lin, D.M., Peng, V., Ngai, J. and Speed, T.P. (2002) Normalization of cDNA microarray data: a robust composite method addressing single and multiple slide systematic variation. Nucleic Acids Res., 30(4): e15.

Zeeberg, B., Feng, W., Wang, G., Wang, M., Fojo, A., Sunshine, M., Narasimhan, S., Kane, D., Reinhold, W., Lababidi, S., Bussey, K., Riss, J., Barrett, J. and Weinstein, J. (2003) oMiner: a resource for biological interpretation of genomic and proteomic data. Genome Biol., 4: R28.

Zien, A., Aigner, T., Zimmer, R. and Lengauer, T. (2001) Centralization: a new method for the normalization of gene expression data. Bioinformatics, 17: S323–S331.

CHAPTER 5

Reproducibility of microarray studies: concordance of current analysis methods

Matt T. Wayland and Sabine Bahn*

Cambridge Centre for Neuropsychiatric Research, Institute of Biotechnology, University of Cambridge, Tennis Court Road, Cambridge CB2 1QT, UK

Introduction

Many studies have addressed the reproducibility of transcriptomics studies with respect to platform (Shippy et al., 2004; Hollingshead et al., 2005; Petersen et al., 2005), sample quality (Ryan et al., 2004), sample storage method (Mutter et al., 2004), RNA amplification (Jeanette et al., 2003; Petalidis et al., 2003; Wilson et al., 2004) and assay variation (Zakharkin et al., 2005). Fewer have investigated systematic bias resulting from the experimenters choice of analysis protocol, and the majority of these have focused on one step in the analysis protocol, such as computation of expression measures (Shedden et al., 2005). Variation between analysis protocols is an important issue, since numerous tools have been developed and applied to the problem of measuring the global transcriptome using microarrays. Many of these tools differ radically in their theoretical underpinning (e.g. hypothesis tests may be based on frequentist or Bayesian statistics). Furthermore, since the gene expression changes observed in many neuropsychiatric disorders are subtle (Prabakaran et al., 2004), their detection is likely to be especially dependent on choice of appropriate analysis methods. It is instructive to pose the question: if multiple analysis tools are applied to the same microarray data set, do they yield similar results?

Conventionally, microarray studies incorporate the following sequential steps in the data analysis pipeline: pre-processing of data, detection of differentially expressed genes, functional profiling of the lists of differentially expressed genes and validation. These core tasks are the subject of this chapter. The aim here is not to define the optimum analysis protocol (a contentious problem which may only be resolved empirically using wholly defined data sets generated from "spike-in" experiments (Choe et al., 2005)), but rather to explore the range of conclusions which might be drawn from a single data set when different analysis methods are used.

We have selected data from Blalock's study of incipient Alzheimer's disease (AD) (Blalock et al., 2004) for our comparison of analysis tools, because of its high quality, associated patient metadata and relevance to neuroscientists. This microarray study was performed on the Affymetrix human GeneChip (HG-U133A) platform and the raw data are available from the Gene Expression Omnibus (GEO; www.ncbi.nlm.nih.gov/geo) under accession number GSE1297. A total of 31 subjects were expression-profiled (9 control and 22 with AD); a single hippocampal specimen from each subject hybridized to each array. Data for two clinical variables are available from GEO: MiniMental State Examination (MMSE, an index of AD-related cognitive

*Corresponding author. E-mail: sb209@cam.ac.uk

decline) and neurofibrillary tangle count (NFT). MMSE and NFT are markers used to assess the severity of AD. Based on MMSE score, AD progression was categorized as "incipient", "moderate" or "severe" (Blalock et al., 2004).

In this chapter we start by describing the microarray data analysis pipeline, making special reference to the methods appropriate to our example data set. We then make a systematic comparison of the outcomes of using different analysis tools. Finally, we discuss the concordance of analysis methods in the context of experiment validation and data mining.

The data analysis pipeline

Pre-processing

Probe set definitions

Many Affymetrix GeneChips were designed at a time when genome and transcriptome annotation was not as comprehensive as it is today. For example, the HG-U133A GeneChip used in the AD study was designed in 2001 using information from build 133 of the UniGene database (April 20, 2001) and other publicly available databases. At this time, only 25% of the human genome had been sequenced (Dai et al., 2005). As the genome and transcriptome annotation has evolved, it has become apparent that many Affymetrix probe sets do not accurately measure the expression of the transcripts they are supposed to target. A number of informatics-related problems have been identified in the original Affymetrix probe set definition and annotation, including: unreliable representative accession numbers, probe set redundancy, non-specific probes, deleted target sequence, genomic location issues and allele-specific probes (Dai et al., 2005). More specifically, in the case of the current Affymetrix annotation for the HG-U133A GeneChip, 14.4% of probe sets have an unreliable public ID, 34.2% show UniGene redundancy, 36% contain one or more probes which match to multiple UniGene clusters, 10.1% have genomic location or strand issues and 3.6% include one or more probes with no known target (Dai et al., 2005).

These problems have been addressed by Dai et al. (2005), who have used the latest genome and transcriptome information to pool all probes targeting the same gene and so define gene-specific probe sets. The one-to-one correspondence of probe sets to genes is an appealing feature of gene-based probe sets, since most researchers are interested in the overall transcription activity of a gene. Furthermore, gene-specific probe sets tend to contain more probes than Affymetrix probe sets, and thus may provide increased statistical power for detecting the subtle expression changes characteristic of many neuropsychiatric disorders. In the current Affymetrix probe set definitions, multiple probe sets may map to a single gene, yet each may yield a different expression value, leading to problems with the interpretation of experiments.

Quality control

Affymetrix provide guidelines for assessing the quality of data generated using their GeneChip arrays (Affymetrix, 2004). Here, we describe alternative approaches to the quality control of Affymetrix data using tools developed by the BioConductor project (Gentleman et al., 2004).

Simple boxplots and histograms of raw signal data will often reveal data quality problems. For example, histograms of raw, probe level, Affymetrix data typically display a positively skewed distribution with a long right tail. Small peaks near the far end of this tail may be indicative of image saturation.

A more refined approach is to fit a robust linear model to the probe level data, with an effect estimated for each chip (Bolstad, 2004). Summary statistics derived from the fitted model can then be used to diagnose problems with specific chips. Boxplots of the model standard errors (each probe set standardized to have median 1) can be used to assess inter-assay variability and reveal outlier chips. Generating pseudo images of either weights or residuals from the robust linear model fit is an effective means of visualizing artefacts. Such artefacts are not always appararent on the raw scanned image.

If the pseudo images of residuals reveal that a large proportion of the probes on a chip have signal values deviating from what would be predicted by the robust linear model, then it may be advisable to repeat the assay. Such a chip will typically be flagged as an outlier in boxplots of model standard errors. Conversely, small, localized artefacts are not a serious cause for concern. Discordant probes in a probe set can usually be adequately controlled for if a robust model is used to generate the gene expression measures (see the next section).

The Affy package (Gautier et al., 2004) from the BioConductor project provides RNA digestion plots for the assessment of RNA sample and assay quality. In the RNA digestion plots, individual probes in a probe set are ordered by location relative to the 5′ end of the targeted RNA molecule. For each chip, probe intensities are averaged by their location within the probe set, with the average taken over all probe sets. Mean probe intensities are then standardized and plotted against the probe's position within the probe set, allowing any 5′–3′ trend to be visualized. Plots demonstrating a 3′-signal bias are indicative of degraded RNA or inefficient transcription of double-stranded cDNA or biotinylated cRNA. Assessment of RNA degradation is particularly important in the context of expression profiling of postmortem tissue samples.

Computation of expression measures

Computation of expression measures for Affymetrix probe sets may be viewed as a three stage process comprising background correction (adjustment of probe intensities to correct for optical noise and non-specific binding), normalization (adjustment of probe intensities to remove systematic error and make measurements from different arrays comparable) and probe summary (summarization of probe level data to produce one expression measure per probe set). Many different procedures for computing expression measures have been developed (see Irizarry et al., 2005). This study will feature five methods which are representatives of the field, namely: microarray suite 5 (MAS5) (Affymetrix, 2002), probe logarithmic intensity error estimation (PLIER) (Affymetrix, 2005), robust multichip average (RMA) (Irizarry et al., 2003),

GCRMA (Wu, and Irizarry, 2004; Wu et al., 2003) and variance stabilizing normalization (VSN) (Huber et al., 2002).

Detection of differentially expressed genes

Variable types

Choice of a statistical test for differential gene expression depends on whether the condition of interest is described by a nominal (categorical), ordinal (ranked) or metric (interval scale) variable. Nominal variables classify data into orderless, non-numerical categories, such as presence/absence of disease or sex. Ordinal variables are ordered (or ranked) categories, such as disease-severity (incipient/moderate/severe). Ordinal variables do not establish the numeric difference between data points, they indicate only that one data point is ranked higher or lower than another. Unlike nominal and ordinal variables, metric variables have numerical values (e.g. MMSE, NFT and gene expression).

The association of gene expression data with a nominal variable can be assessed with simple univariate statistics such as the t-test (comparison of the means of two groups) and ANOVA (comparison of more than two groups) or by uisng more powerful methods designed specifically for microarray data analysis, including, those found in the software packages LIMMA (Smyth, 2004) and MAANOVA (Wu, 2005). For testing the association of gene expression with ordinal or metric variables, correlation analysis is an appropriate approach.

Correlation analysis

The correlation coefficient, usually denoted as r, is a measure of the degree of association between two variables and can take values from -1 to $+1$. A value of $+1$ indicates maximal positive association, while a value of -1 indicates maximal negative association. An r-value of zero signifies no correlation. Various formulae exist for computing correlation coefficients, each applicable to a particular type of variable and/or relationship.

Pearson's product–moment correlation coefficient (r_p) is perhaps the most widely used statistic and measures the strength of the linear relationship between two metric variables. It is computed using the following formula

$$r_p = \frac{\sum_{i=1}^{n}(x_i - \bar{x})(y_i - \bar{y})}{\sqrt{\sum_{i=1}^{n}(x_i - \bar{x})^2}\sqrt{\sum_{i=1}^{n}(y_i - \bar{y})^2}} \quad (1)$$

where x and y are metric variables. Linear correlation may be descriptively inadequate if the two variables display curvilinearity or if the data set contains outliers. Such deviations from bivariate normality can usually be detected by eye in scatterplots.

For ordinal variables, a rank-based measure of correlation, such as Spearman's rank–order correlation coefficient (r_s) or Kendall's tau (τ) should be used. These two non-parametric methods measure the strength of any monotonic relationship (linear or non-linear) between two variables. Spearman's rank–order correlation coefficient is computed using

$$r_s = 1 - \frac{6\sum_{i=1}^{n}d_i^2}{(n^3 - n)} \quad (2)$$

where d_i is the difference between the ranks of the paired variables and n the number of pairs of ranks.

The formula for Kendall's τ is

$$\tau = \frac{n_c - n_d}{n(n-1)/2} \quad (3)$$

where n_c is the number of concordant pairs of ranks and n_d the number of discordant pairs of ranks.

In the case of r_s, the weight accorded to a pair of ranks is proportional to the distance between them, whereas τ weighs each disagreement in ranks equally. Therefore, if there is uncertainty about the reliability of close ranks, r_s is the correlation coefficient of choice.

Significance tests determine whether or not a sample correlation coefficient could have come from a population with a correlation coefficient of zero. For Pearson's product moment correlation coefficient and Spearman's rank–order correlation coefficient the t-statistic can be expressed in terms of r:

$$t_s = r\sqrt{\frac{n-2}{1-r^2}} \quad (4)$$

In the case of Kendall's τ, the t-statistic can be derived from a normal approximation:

$$t_s = \frac{\tau}{\sqrt{2(2n+5)/9n(n-1)}} \quad (5)$$

Caution should be exercised when interpreting correlation coefficients and tests of significance. Even strongly correlated variables may be causally unrelated. Moreover, using the term *significant* to describe the results of testing a correlation coefficient can be misleading. *Significance* does not necessarily imply strong correlation, it merely indicates that the observed pattern cannot be explained away as a chance observation.

All three of the statistics described above are useful measures of correlation in microarray studies. However, the choice of a particular correlation coefficient should be based on the types of variables to be analysed, the distribution of the data and the type of relationship anticipated. Pearson's r_p is only applicable to the measurement of linear correlation between two metric variables. However, the non-parametric methods can be applied to both ordinal and metric variables, which have been converted into ordinal variables. Substituting metric variables for their ranks results in some loss of information in the original data, but enables us to use the rank-based correlation coefficients (r_s or τ), which may be more appropriate in some instances. For example, erroneous conclusions can be drawn if Pearson's r_p is used to measure correlation between two metric variables in the presence of outliers or other deviations from normality. In such a situation, it may be sensible to convert the metric variables into their ranks and then use r_s or τ as the measure of correlation, since they are robust to the effects of outliers. Similarly, if metric variables are converted into their ranks, r_s and τ can be used to detect any monotonic relationship, whether linear or non-linear. A comprehensive treatment of correlation and association analysis in the context of microarray studies is provided by Lin and Johnson (2003).

Control of multiplicity

Microarray experiments simultaneously measure the expression of thousands to tens-of-thousands of genes, depending on platform. If a hypothesis test is applied to each gene to determine whether or not it is differentially expressed, then the problem of multiple comparisons arises with an unprecedented severity. In the context of transcriptomics studies, a false positive is a gene, which is called differentially expressed by the statistical test applied, but in reality it is not. In statistical terms, a false positive is referred to as a Type I error; that is rejection of the null hypothesis when in fact it is true.

In a microarray experiment where hypothesis tests have been used to test for differential expression in each of m genes, the overall risk of detecting one false positive is given by:

$$p' = 1 - (1-p)^m \tag{6}$$

The family-wise error rate (FWER) is the rate at which a statistical test would be expected to yield one or more false positives among a collection (family) of tests and can be conservatively controlled using the Bonferroni corretion. To ensure the overall risk of a Type I error does not exceed the conventional threshold of 0.05, each of the separate tests is conducted at a significance level of $\alpha = 0.05/m$. Therefore, if a statistical test for differential expression is applied to 10,000 genes, the individual p-value for each gene must be $\leq 0.05/10,000 = 0.000005$ if it is to be called significant at a FWER of $\leq 5\%$. Protection against any single false positive is considered too stringent a criterion for microarray studies as it may result in many genuinely differentially expressed genes not passing the significance threshold.

In transcriptomics studies, the goal is to identify as many differentially expressed genes as possible, while keeping the number of false positives at a relatively low level. For this type of problem, Benjamini and Hochberg (1995) proposed the concept of a false discovery rate (FDR), defined as the ratio of the number of false-positive features to the number of significant features.

Benjamini and Hochberg provided a step-wise procedure for controlling the FDR. First, the p-values from the hypothesis tests are ordered

$$p_1 \leq p_2 \leq \cdots p_k \cdots \leq p_m \tag{7}$$

p_1 being the smallest and thus most significant, p-value and p_m being the least significant. Next, the largest k is found for an FDR level α such that

$$p_k \leq \alpha k/m \tag{8}$$

Finally, if k exists, the rejected null hypotheses (significant test statistics) correspond to

$$p_1 \leq \cdots \leq p_k \tag{9}$$

Otherwise, no null hypotheses are rejected. For independent tests, the expected proportion of false positives is $\leq \alpha$.

Alternatives to Benjamini and Hochberg's frequentist algorithm for FDR control include the empirical Bayes approaches, exemplified by the q-value. The q-value gives a measure of significance to each individual statistical test in terms of the FDR (Storey and Tibshirani, 2003). If all genes with a q-value of ≤ 0.05 are called significant, then up to 5% of these significant genes are expected to be false discoveries. A full description of the computation of q-values is beyond the scope of this chapter (see instead Storey, 2003). In brief, q-values are dependent on the proportion, p_0, of tests in which the null hypothesis is true (i.e. there is no actual differential gene expression), and on the distribution of p-values for which the null hypothesis is false. Given that p-values are uniformly distributed under the null hypothesis, these parameters can be estimated from the data.

Functional interpretation

An important interpretive step in most transcriptomics studies is the functional profiling of candidate gene lists. More than a dozen different tools have been developed for this secondary analysis (Khatri and Draghici, 2005). The aim of these tools is to characterize the list of differentially expressed genes in terms of Gene Ontology (GO) categories. A variety of statistical models are employed by these tools to test for the over-representation of a GO term in the candidate gene list, relative to its occurrence on the microarray. These models include the χ^2, binomial, hypergeometric and Fisher's exact test. The problem of multiplicity also arises here, necessitating an appropriate procedure to control the FDR.

Assessment of data quality

Before starting a comparison of analysis tools it is important to make a thorough evaluation of the quality of the example data set and also to select an appropriate disease marker from our choice of MMSE and NFT. The procedures described here are applicable to many microarray studies of clinical samples.

Outlier removal

The original analysis of this microarray experiment was based on all 31 chips. However, a recent re-analysis of this data set (Jackson et al., 2005), employing the quality control methods described above, revealed two outlier chips. If a robust linear model is fitted to the probe level data, the two outlier chips can be readily visualized in a plot of model standard errors (Fig. 1). Elevated standard errors result from factors including low signal from perfect match probes and suggest that the hybridization assay has not performed well when compared to other chips in the experiment. For a more detailed analysis of the quality of the data generated in this experiment, see Jackson et al. (2005). Since the discordance of these chips is apparently due to technical variation rather than any biological phenomenon, they should be excluded from the analysis.

In order to get an indication of the noise these two discordant chips were introducing into the data set, the numbers of differentially expressed genes detected before and after outlier removal were compared. Five different methods (MAS5, PLIER, RMA, GCRMA and VSN) were used to compute expression measures for the probe sets defined by Affymetrix. Pearson's r was used to measure correlation between gene expression and the two markers of disease progression (MMSE and NFT) and Benjamini and Hochberg's method used to control the FDR. Outlier removal had a marked effect on the number of differentially expressed genes, which could be detected at a given FDR. At the conventional FDR threshold of $q< =0.05$, outlier removal resulted in dramatic gains in the sensitivity of the MMSE correlation analysis, no matter which method had been used to generate expression measures (Fig. 2). In the case of VSN, the number of genes whose expression was significantly ($q< =0.05$) correlated with MMSE increased by two orders of magnitude after the elimination of the discordant chips. Similar results were obtained for the analysis of correlation with NFT. This observation underscores the value of a rigorous approach to quality control in the early stages of a microarray study. If poor quality chips are not detected, not only is noise introduced into the data set, but also an opportunity to repeat failed assays is missed.

Choice of disease marker

AD is a degenerative disease, thus to understand the gene expression changes associated with it (either causal or otherwise), we need a reliable measure of disease progression. For the example data set, we have a cognitive measure of AD severity (MMSE) as well as a histopathological measure (NFT). Using a set of 30 questions, the MMSE test assesses various cognitive functions, including: orientation in time, orientation in place, registration, arithmetic, memory, language and spacial insight (Folstein et al., 1975). MMSE scores decline with AD severity. NFT quantifies the pathological protein aggregates found within neurons and is positively correlated with AD. In the sampled patients, MMSE and NFT display moderate negative correlation ($r_p = -0.51$; $r = -0.75$).

Blalock et al. (2004) defined their candidate list of "AD-related genes" as those which showed a statistically significant correlation with both MMSE and NFT. Following removal of the two outlier chips, correlation between the expression of unigene probe sets (as summarized by RMA) and the two disease markers was measured using both Spearman and Pearson's statistics. Figure 3 shows the number of genes whose expression is significantly correlated with MMSE, NFT or both at a range of FDR values. In this analysis, it can be seen that the expression of relatively few genes is

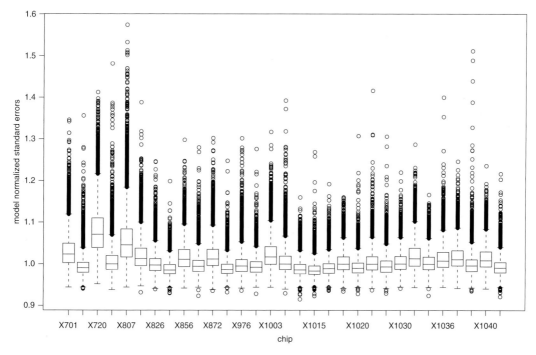

Fig. 1. Boxplots of normalized standard errors from the probe level model for each of the 31 chips in the study. Standard errors have been normalized across probe sets so each has median one. The median interquartile ranges of the normalized standard errors are particularly high for samples 720 and 807, relative to other chips.

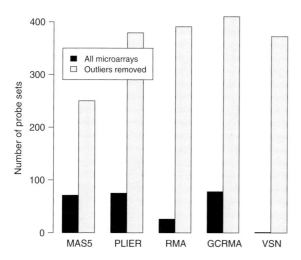

Fig. 2. Effect of outlier removal on candidate gene lists. Histogram shows a number of Affymetrix probe sets which are significantly (Benjamini and Hochberg $q <\, = 0.05$) correlated (Pearson's r) with MMSE, before and after outlier removal, for five different expression measures.

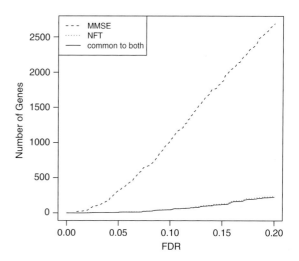

Fig. 3. Number of genes whose expression is significantly correlated (Pearson's r) with MMSE, NFT or both at a range of FDR values. The FDR was estimated using the method of Benjamini and Hochberg.

significantly correlated with NFT compared to MMSE. Furthermore, those genes whose expression is significantly correlated with NFT are a subset of the genes whose expression is significantly correlated with MMSE. Similar results are obtained if Pearson's r is used as the measure of correlation. Thus, in the analysis that follows, MMSE will be used as the measure of AD severity. This suffices for the current demonstration, but in a more comprehensive study it would be interesting to pay special attention to that subset of genes whose expression exhibits significant correlation with both MMSE and NFT. Clearly, the choice of a measure of disease progression has a profound influence on the interpretation of microarray studies of neuropsychiatric disorders.

Performance comparison

Aim and scope

The aim of this comparison of analysis tools was solely to address reproducibility. Identification of optimal strategies for the detection of differentially expressed genes is beyond the scope of this chapter and in any case could not be established using a single data set. Here we compare methods available at four key stages in the data analysis pipeline, namely: (i) definition of probe sets; (ii) computation of expression measures; (iii) correlation analysis; and (iv) FDR control (see Fig. 4).

Approach

In the first stage of the data analysis pipeline, probes were grouped according to two different probe set definitions, the original Affymetrix system and the alternative UniGene system (Dai et al., 2005) (CDF version 6 based on UniGene build 186 and available from http://brainarray.mhri.med.umich.edu/Brainarray/Database/CustomCDF/). Next, expression values were computed for each type of probe set using five different methods: MAS5, PLIER, RMA, GCRMA and VSN. This step was followed by an analysis of correlation of gene expression with MMSE using the methods of Pearson, Spearman and Kendall. Finally, two different FDR-controlling procedures were applied to the results of the correlation analyses.

All paths through the flow diagram depicted in Fig. 4 were followed, resulting in 60 different combinations of analyses (2 probe set definitions × 5 methods of computing expression measures × 3 correlation coefficients × 2 methods of FDR control).

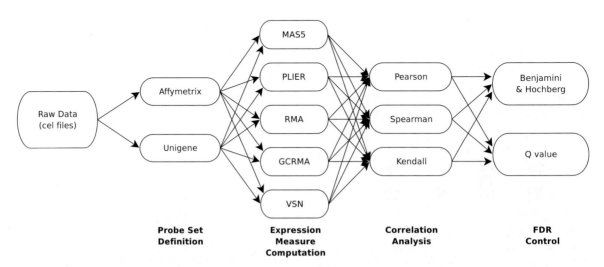

Fig. 4. Flow diagram showing methods available at four key stages in the microarray data analysis pipeline which was used to study the AD data set. Starting with raw data, there are 60 combinations of methods which can be used to arrive at a candidate gene list.

Where a particular method could optionally be modified by adjusting various parameters (as was the case with the tools for computing expression measures and Q-value), the default settings were used. The intention here was that the use of these tools in this study should reflect the way they are used by most researchers.

For each of the 60 protocols (i.e. combinations of methods), a "candidate gene list" was formed from those genes, which passed the conventional FDR threshold of 5%. Concordance between any pair of candidate gene lists was defined as the ratio of the number of genes in the intersect of the two lists to the number of genes in the union of the two lists, and expressed as a percentage. Affymetrix probe set identifiers were mapped to UniGene cluster IDs, so that the results of the analyses based on the two different probe set definitions could be compared.

Computational setup

All data analysis was performed using R 2.2.0 (R Development Core Team., 2005) and BioConductor 1.7 (Gentleman et al., 2004) on a PC with an AMD 64 bit processor and 2 GB of RAM, running SUSE Linux Professional 9.1 (Linux kernel 2.6). The R language/environment and the BioConductor libraries are open source software and can be downloaded from www.r-project.org and www.bioconductor.org, respectively.

Results

Probe set definitions

Expression data based on each system of probe set definition were processed by 30 combinations of analysis methods (5 methods of computing expression measures × 3 methods of measuring correlation × 2 methods of controlling the FDR). Thus, the concordance between the two systems of probe set definitions was measured from 30 pairs of candidate gene lists and found to be typically around 30% (Fig. 5). In a previous study a number of HG-U133A data sets were used to compare the performance of Affymetrix probe set definitions with UniGene and other alternative probe set

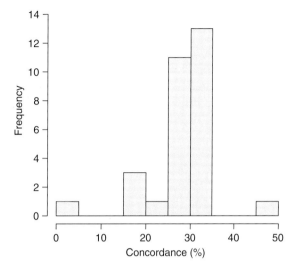

Fig. 5. Concordance between analyses based on Affymetrix probe set definitions and those based on Unigene probe set definitions. Expression data based on each system of probe set definition were processed by 30 combinations of analysis methods (five methods of computing expression measures, three methods of measuring correlation and two methods of controlling the FDR). Concordance for each of the 30 pairwise analyses was defined as the number of significant ($q < = 0.05$) genes common to both probe set definitions as a percentage of the number of significant genes found with either probe set definition.

definitions (Dai et al., 2005). In these analyses, use of the alternative probe set definitions only resulted in a 30–40% change in the composition of lists of differentially expressed genes at a range of FDR values, including 5%. The cause of the greater discordance observed in the current study is far from clear, but may be related to the data set or methods of differential gene detection.

Methods for computing expression measures

Concordance of the five methods for computing expression measures could be compared under 12 combinations of the other analysis methods. Table 1 shows the concordance values for one such comparison, which is representative of the other 11. MAS5, the method used in the original study, displays relatively low concordance (<30%) with the other methods. However, even the highest concordance values fall below 50%.

Table 1. Concordance of candidate gene lists derived using different methods for computation of expression measures. Concordance values are from an analysis based on UniGene probe sets, using r_p as the correlation coefficient and the Benjamini and Hochberg method to control the FDR

	MAS5	PLIER	RMA	GCRMA
PLIER	22.4			
RMA	24.4	37.0		
GCRMA	27.7	38.2	45.3	
VSN	24.1	43.4	49.5	37.2

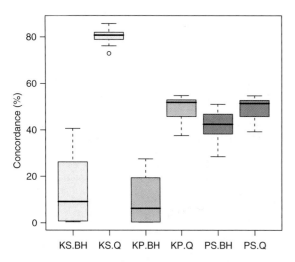

Fig. 6. Concordance between each pair of correlation coefficients, stratified by procedure for FDR control. Key: K; τ; S; r_s; P; r_p; BH, Benjamini and Hochberg FDR control; Q, Q value FDR control. KS.BH represents the concordance between the 10 pairs of candidate gene lists derived using τ or r_s and the Benjamini and Hochberg method for FDR control (2 probe set definitions × 5 methods of computing expression measures).

Correlation coefficients

Figure 6 shows the concordance between each pair of correlation coefficients, under different methods of FDR control. Candidate gene lists derived using Kendall's τ and the Benjamini and Hochberg method for FDR control were short; in some cases containing just a few genes. Consequently, comparisons involving these gene lists yielded low concordance values. The apparent low power of Kendall's τ in this study may be due to close ranks in the data set (see the section on Correlation analysis). However, when Q-value was used to control the FDR, τ and r_s exhibited strong concordance of around 80%, as might be anticipated for two non-parametric correlation coefficients.

The comparisons involving the parametric correlation coefficient r_p and each of the non-parametric correlation coefficients, r_s and τ demonstrate only partial concordance ($\approx 50\%$ when Q-value is used to control the FDR). This is not surprising, because the two types of methods are sensitive to different correlation patterns. As discussed earlier, the parametric r_s is a statistically powerful method for measuring the strength of the linear relationship between two variables, but may give misleading results in the presence of outliers. By contrast, the rank-based, non-parametric correlation coefficients are sensitive to any monotonic relationship (linear or non-linear) and are robust to outliers. Given the different strengths of parametric and rank-based correlation coefficients, and the marked differences in candidate gene lists detected by each, it would be advisable to use both types to quantify the association of gene expression with metric variables. If correlation is measured using only a parametric correlation coefficient (as it was in the original study), some interesting patterns of gene expression may fail to be detected.

Procedures for FDR control

Using Q-value to control the FDR resulted in considerably longer candidate gene lists than when the Benjamini and Hochberg method was used (Fig. 7). Candidate gene lists generated with the aid of the Benjamini and Hochberg FDR control procedure numbered hundreds of genes, but were invariably subsets of the Q-value candidate gene lists, which contained up to several thousand genes.

Protocols: combinations of analysis methods

The preceding sections have highlighted the variability to be found at each stage in the data analysis pipeline. Such variability is presumably compounded when divergent paths are taken through the flow diagram in Fig. 4. To examine this further, concordance was measured for all 1770 pairwise comparisons of the 60 candidate gene lists generated in this study and the results are summarized in Fig. 8. Median concordance was found to be just

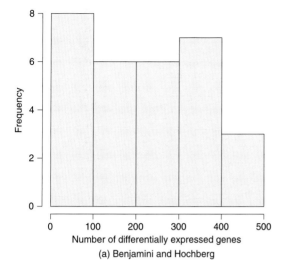

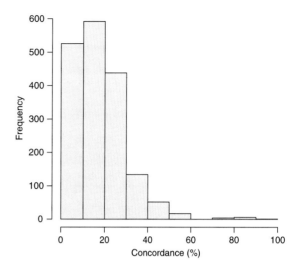

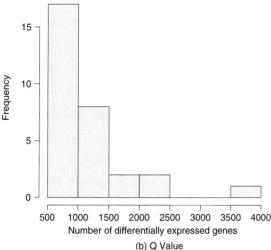

Fig. 7. Histograms of the length of candidate gene lists when (a) the Benjamini and Hochberg and (b) the Q-value methods are used to control the FDR at a level of $q< = 0.05$. In each histogram, 30 different combinations of analysis methods are summarized.

16.5%, with an interquartile range of 8.1–23.8%. The R-function hclust was used (at its default setting) to create a dendrogram to show the similarity of the 60 candidate gene lists, based on pairwise concordance values (Fig. 9). In this dendrogram, it can be seen that candidate gene lists are primarily grouped by the method used to control the FDR. It should also be noted that candidate gene lists derived from different probe set definitions do not generally co-cluster.

Fig. 8. Histogram of concordance found in each of the 1770 pairwise comparisons of the 60 candidate gene lists generated in this study.

Impact of analysis protocol on functional interpretation

In the above example, the candidate gene list derived from a microarray experiment has been shown to be dependent on the methods selected at each step in the data analysis pipeline. While the significance of expression changes in individual genes should not be ignored, the functional interpretation of a transcriptomics study relies on a secondary analysis in which the candidate gene list(s) is summarized by its ontological profile. Gene lists do not need to be identical to have similar functional properties, therefore it is of interest to examine the agreement between ontological profiles based on candidate gene lists which have been generated from the same data set, but using different analytical approaches.

Since comparison of gene ontological profiles from the 60 candidate gene lists derived from the above analysis cannot be easily automated, a single list was selected and its ontological profile compared with that of the candidate gene list from the original study of Blalock et al. (2004). The selected candidate gene list, totaling 1279 unique UniGene identifiers (573 up-regulated and 706 down-regulated), was chosen because the protocol used to generate it differed as much as possible from that

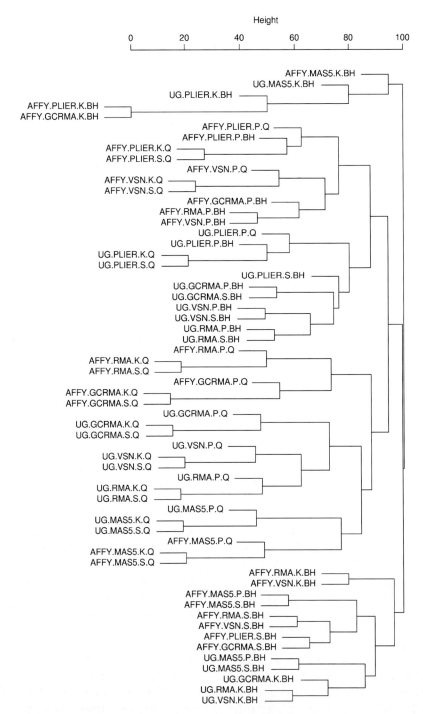

Fig. 9. Hierarchical clustering of candidate gene list similarity. Leaves are labeled using the following notation: [probe set definition]. [method used to compute expression measures]. [correlation coefficient]. [FDR correction].

Table 2. Derivation of candidate gene lists for ontological profiling

Method	Original study	This study
Number of samples	31	29
Probe set definition	Affymetrix	UniGene
Expression measure	MAS5	GCRMA
Correlation coefficient	Pearson's r_p	Spearman's r_s
FDR correction	Benjamini and Hochberg	Q-value
FDR rate	≈ 0.2	0.05
Up-regulated genes	1977	573
Down-regulated genes	1436	706
Ontological tool	EASE[a]	Onto-express

[a]Expression Analysis Systematic Explorer (http://david.niaid.nih.gov/david/ease.htm).

Table 3. GO biological process categories over-represented in list of genes up-regulated in AD

GO.ID	Biological process	n	Corrected p-value
GO:0006355	Regulation of transcription, DNA-dependent	84	< 0.0000
GO:0006350	Transcription	59	0.0038
GO:0000004	Biological process unknown	22	0.0236
GO:0007186	G-protein coupled receptor protein signalling pathway	5	0.0246
GO:0001501	Skeletal development	8	0.0409
GO:0043123	Positive regulation of I-kappaB kinase/NF-kappaB cascade	7	0.0561
GO:0007160	Cell–matrix adhesion	6	0.0624
GO:0008380	RNA splicing	6	0.0637
GO:0007267	Cell–cell signalling	5	0.0657
GO:0006915	Apoptosis	17	0.0841
GO:0045449	Regulation of transcription	6	0.0842
GO:0007169	Transmembrane receptor protein tyrosine kinase singalling pathway	6	0.0848
GO:0006817	Phosphate transport	6	0.0893
GO:0006357	Regulation of transcription from RNA polymerase II promoter	13	0.0918
GO:0007596	Blood coagulation	6	0.0958

employed in the original study. Furthermore, in this study Onto-Express (Khatri et al., 2002) (http://vortex.cs.wayne.edu/ontoexpress/) was used to perform the ontological analysis, whereas the Expression Analysis Systematic Explorer (EASE; http://david.niaid.nih.gov/david/ease.htm) tool was used by the authors of the original study (Blalock et al., 2004). The main points of difference between the two approaches to detection of differentially expressed genes are summarized in Table 2. It is also important to note that the two analyses were based on different versions of the GO database. The original study was conducted sometime in mid-2003, whereas the re-analysis presented here was performed in December 2005; the GO database may have undergone considerable expansion and refinement in this time.

The candidate gene list from this study was divided into those genes, which are up-regulated in AD (i.e. negatively correlated with MMSE) and those which are down-regulated in AD (i.e. positively correlated with MMSE). These two subsets of genes were then profiled using Onto-Express. Tables 3 and 4 show the biological process categories, which are over-represented in the lists of up- and down-regulated genes, respectively. Results of an equivalent analysis from the original study can be found in Table 2 of Blalock et al. (2004). A comparison of the results of the two studies reveals a number of striking similarities. Among the biological processes over-represented in the lists of down-regulated genes, "generation of precursor metabolites and energy" (synonym of "energy pathways"), "tricarboxylic acid cycle", "ATP synthesis", "ion transport", "microtubule-based movement", "protein folding", "glycolysis", "ubiquitin-dependent protein catabolism" and "amino acid metabolism" are common to both studies. The same concordance was not found for the ontological profiles of the up-regulated gene lists, but "regulation of transcription" and "apoptosis"

Table 4. GO biological process categories over-represented in list of genes down-regulated in AD

GO.ID	Biological process	N	Corrected p-value
GO:0006120	Mitochondrial electron transport, NADH to ubiquinone	12	< 0.0000
GO:0015992	Proton transport	18	< 0.0000
GO:0006099	Tricarboxylic acid cycle	11	< 0.0000
GO:0015986	ATP synthesis coupled proton transport	13	< 0.0000
GO:0051258	Protein polymerization	7	< 0.0000
GO:0006355	Regulation of transcription, DNA-dependent	33	< 0.0000
GO:0006811	Ion transport	31	0.0002
GO:0007018	Microtubule-based movement	11	0.0010
GO:0006118	Electron transport	29	0.0016
GO:0006350	Transcription	27	0.0060
GO:0006457	Protein folding	20	0.0063
GO:0006091	Generation of precursor metabolites and energy	10	0.0066
GO:0006096	Glycolysis	7	0.0095
GO:0006367	Transcription initiation from RNA polymerase II promoter	5	0.0359
GO:0007218	Neuropeptide signalling pathway	8	0.0403
GO:0007186	G-protein coupled receptor protein signalling pathway	8	0.0406
GO:0006810	Transport	35	0.0505
GO:0007267	Cell–cell signalling	7	0.0548
GO:0006511	Ubiquitin-dependent protein catabolism	10	0.0551
GO:0006520	Amino acid metabolism	6	0.0554
GO:0006139	Uncleobase, nucleoside, nucleotide and nucleic acid metabolism	5	0.0969
GO:0007155	Cell adhesion	10	0.0978

were flagged in both. Overall these are encouraging observations because they suggest that even disparate approaches to differential gene detection can converge, to some extent, on a similar functional interpretation.

Validation

Although untested, it is possible that transcriptomics studies of neuropsychiatric disorders are particularly sensitive to the choice of analysis protocol because the gene expression changes observed are generally subtle. While the field of microarray data analysis is maturing, there is as yet no consensus on optimal protocols. However, a number of groups are starting to address this issue (Cope et al., 2004; Choe et al., 2005; Irizarry et al., 2005; Shedden et al., 2005).

As shown here and in other studies, the method used to compute expression measures from Affymetrix GeneChips has a considerable impact on the size and composition of lists of differentially expressed genes. The Affycomp project has been set up to provide a benchmark for Affymetrix GeneChip expression measures (Cope et al., 2004; Irizarry et al., 2005). Assessment data comprises dilution and spike-in experiments performed on HGU95 and HGU133 GeneChips, for which the true signal in the data is known. The benchmark uses a range of summary statistics to evaluate performance of probe processing methods in terms of accuracy and overall detection ability. A webtool implementing the benchmark enables developers to compare their algorithms with existing methods using the same assessment data and criteria. This is an ongoing project, the current results of which can be found at http://affycomp.biostat.jhsph.edu/ . Investigators may use this resource to guide their selection of an appropriate measure of gene expression for their particular study. At present there is no single best method, choice will depend on factors including the tradeoff between bias (lack of accuracy) and variance (precision).

Empirical evaluation of analysis methodologies will aid experimenters in their choice of tools, but confirmation that a sensible choice of protocol has been employed will come from the ability of the experimenter to validate her or his findings using a different platform. Indeed, a multi-tiered approach, encompassing transcriptomics, proteomics and metabonomics, is the strategy of choice for addressing the complex questions posed by those working in the neurosciences, as illustrated by recent work (Prabakaran et al., 2004).

Implications for data mining

While this study has shown broad concordance between analysis tools (at least in terms of functional profile, if not in composition of candidate gene lists) the considerable differences identified suggest that the dedicated data miner might extract further biological signal from published microarray data sets, simply by applying novel methods. As we have seen, a rigorous approach to the detection and removal of outlier chips alone can result in substantial gains in statistical power. Transcriptomics studies are a rich source of data and it is impossible to distill the full information content of a microarray experiment into a single scientific paper; typically the most striking and easily interpretable observations are reported. Therefore, revisiting existing microarray data sets may reveal previously unrecognized patterns, as illustrated in the following example.

While the above analysis explored a number of methodological issues, it was by no means an exhaustive examination of the data set. By focusing on the detection of genes whose expression was correlated with MMSE, it ignored those genes with an expression pattern that did not show a monotonic relationship with AD severity. Similarly, one of the aims of the original study was to identify transcriptional changes during incipient AD. The authors of the original study addressed this by first using Pearson's r_p to find a set of 3413 genes whose expression was significantly correlated with MMSE or NFT. Next they used Pearson's r_p to establish which of this subset of 3413 AD-related genes were significantly correlated with MMSE or NFT within the incipient AD samples only. This final step in their gene detection algorithm yielded a total of 609 incipient AD-related genes. A limitation of this approach is that the expression of all 609 of these incipient AD-related genes must show a linear relationship with AD severity across the full spectrum of the disorder. While this candidate gene list is undoubtedly relevant to understanding the pathology of the disorder, it is pertinent to also consider gene expression changes which are nonlinear with respect to disease severity.

"Template matching" is a simple method of finding genes which follow an expression profile defined by the experimenter (Pavlidis, 2003). The expression profile of genes exhibiting a change in expression profile in incipient AD can be represented by a binary vector in which incipient AD samples are coded as 1 and all other samples are coded as 0. Genes whose expression profile displays a peak or a trough in incipient AD will respectively show positive or negative correlation with this template. We compared this template to

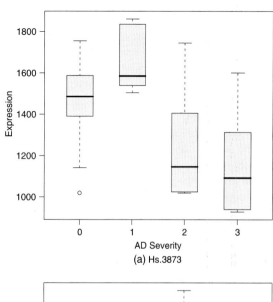

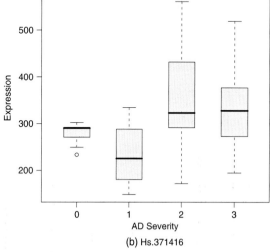

Fig. 10. Expression of genes, which do not show a monotonic relationship with AD severity. Disease progression is measured on a three-point scale: 0, healthy control; 1, incipient; 2, moderate; 3, severe.

the GCRMA expression profile of UniGene probe sets using Spearman's r_s. Figure 10 shows the expression profiles of two of the genes showing the strongest positive and negative correlation with this template, respectively. Hs.3873, which appears to be up-regulated in incipient AD (Fig. 10(a)), represents palmitoyl-protein thioesterase 1 (PPT1). This small glycoprotein is thought to be associated with the maintenance of synaptic function (Lehtovirta et al., 2001) and so a peak of expression in incipient AD, followed by a decline in expression in later stages of the disorder is noteworthy. Coactivator-associated arginine methyltransferase 1 (CARM1), represented by Hs.371416 and possibly down-regulated in incipient AD (Fig. 10(b)), is probably involved in transcriptional regulation (Chen et al., 1999).

An alternative method of finding genes which show interesting patterns of expression across the sample set, is the unsupervised, exploratory approach of cluster analysis. With this technique genes are partitioned into groups with similar expression profiles, but the expression profiles are "discovered" by the clustering algorithm rather than being predefined by the experimenter (for a recent evaluation of clustering methods in the context of microarray studies, see Garge et al., 2005). Cluster analysis of the AD data set may prove to be a fruitful adjunct to the methods of investigation employed so far.

Summary and conclusions

This chapter has demonstrated considerable discordance between the content of candidate gene lists output from different microarray data analysis protocols performed on the same microarray data set. Conversely, it has shown that a similar functional signal (as expressed in terms of the Gene Ontology) may emerge from these diverse approaches to the detection of differentially expressed genes. The robustness of the functional signal is reassuring. Nevertheless, given the plethora of analysis tools available and the divergent results they produce, there is a pressing need for an optimal data analysis protocol to be established for Affymetrix GeneChip and other microarray studies.

References

Affymetrix. (2002). Statistical algorithms description document. Technical report. www.affymetrix.com.

Affymetrix. (2004). Affymetrix genechip expression analysis: data analysis fundamentals. Technical report. www.affymetrix.com.

Affymetrix. (2005). Guide to probe logarithmic intensity error (plier) estimation. Technical report. www.affymetrix.com.

Benjamini, Y. and Hochberg, Y. (1995) Controlling the false discovery rate: a practical and powerful approach to multiple testing. J. R. Stat. Soc. B, 57: 289–300.

Blalock, E.M., Geddes, J.W., Chen, K.C., Porter, N.M., Markesbery, W.R. and Landfield, P.W. (2004) Incipient Alzheimer's disease: microarray correlation analyses reveal major transcriptional and tumour suppressor responses. Proc. Natl. Acad. Sci. USA, 101(7): 2173–2178.

Bolstad, B. (2004) affyplm: methods for fitting probe level models to affy data. Technical report. www.bioconductor.org.

Chen, D., Ma, H., Hong, H., Koh, S.S., Huang, S.M., Schurter, B.T., Aswad, D.W. and Stallcup, M.R. (1999) Regulation of transcription by a protein methyltransferase. Science, 284: 2174–2177.

Choe, S.E., Boutros, M., Michelson, A.M., Church, G.M. and Halfon, M.S. (2005) Preferred analysis methods for affymetrix genechips revealed by a wholly defined control dataset. Genome Biol., 6: R16.

Cope, L.M., Irizarry, R.A., Jaffee, H.A., Wu, Z. and Speed, T.P. (2004) A benchmark for affymetrix genechip expression measures. Bioinformatics, 20: 323–331.

Dai, M., Wang, P., Boyd, A.D., Kostov, G., Athey, B., Jones, E.G., Bunney, W.E., Myers, R.M., Speed, T.P., Akil, H., Watson, S.J. and Meng, F. (2005) Evolving gene/transcript definitions significantly alter the interpretation of genechip data. Nucleic Acids Res., 33: e175.

Folstein, M., Folstein, S. and McHugh, P. (1975) Minimental state a practical method for grading the cognitive state of patients for the clinician. J. Psychiatr. Res., 12: 189–198.

Garge, N.R., Page, G.P., Sprague, A.P., Gorman, B.S. and Allison, D.B. (2005) Reproducible clusters from microarray research: whither? BMC Bioinform., 6: S10.

Gautier, L., Cope, L., Bolstad, B.M. and Irizarry, R.A. (2004) affy — analysis of affymetrix genechip data at the probe level. Bioinformatics, 20(3): 307–315.

Gentleman, R.C., Carey, V.J., Bates, D.M., Bolstad, B., Dettling, M., Dudoit, S., Ellis, B., Gautier, L., Ge, Y., Gentry, J., Hornik, K., Hothorn, T., Huber, W., Iacus, S., Irizarry, R., Li, F.L.C., Maechler, M., Rossini, A.J., Sawitzki, G., Smith, C., Smyth, G., Tierney, L., Yang, J.Y.H. and Zhang, J. (2004) Bioconductor: open software development for computational biology and bioinformatics. Genome Biol., 5: R80.

Hollingshead, D., Lewis, D.A. and Mirnics, K. (2005) Platform influence on DNA microarray data in postmortem brain research. Neurobiol. Dis., 18(3): 649–655.

Huber, W., von Heydebreck, A., Sultmann, H., Poustka, A. and Vingron, M. (2002) Variance stabilization applied to microarray data calibration and to the quantification of differential expression. Bioformatics, 18: S96–S104.

Irizarry, R.A., Hobbs, B., Collin, F., Beazer-Barday, Y.D., Antonellis, K.J., Schery, U. and Speed, T.P. (2003) Exploration, normalization, and summaries of high density oligonucleotide array probe level data. Biostatistics, 4(2): 249–264.

Irizarry, R.A., Wu, Z. and Jaffee, H.A. (2005) Comparison of affymetrix genechip expression measures. Johns Hopkins University, Dept. of Biostatistics Working Papers, Working Paper 86: 1–7.

Jackson, E.S., Wayland, M.T., Fitzgerald, W. and Bahn, S. (2005) A microarray data analysis framework for postmortem tissues. Methods, 37: 247–260.

Khatri, P., Draghici, S., Ostermeier, C. and Krawetz, S. (2002) Profiling gene expression utilizing onto-express. Genomics, 79: 266–270.

Khatri, P. and Draghici, S. (2005) Ontological analysis of gene exression data: current tools, limitations, and open problems. Bioinformatics, 21(18): 3587–3595.

Lehtovirta, M., Kyttala, A., Eskelinen, E.L., Hess, M., Heinonen, O. and Jalanko, A. (2001) Palmitoyl protein thioesterase (ppt) localizes into synaptosomes and synaptic vesicles in neurons: implications for infantile neuronal ceroid lipofuscinosis (incl). Hum. Mol. Genet., 10: 69–75.

Lin, S.M. and Johnson. K.F. (2003) Correlation and Association Analysis, Chap. 17. Kluwer, Dordrecht, The Netherlands, pp. 289–305.

Jeanette, N.M., Jerome, R.E., Nicholson, C.R., Crabb, D.W. and Edenberg, H.J. (2003) Reproducibility of oligonucleotide arrays using small samples. BMC Genom., 4(1): 4.

Mutter, G.L., Zahrieh, D., Liu, C., Neuberg, D., Finkelstein, D., Baker, H.E. and Warrington, J.A. (2004) Comparison of frozen and rnalater solid tissue storage methods for use in rna expression microarrays. BMC Genom., 5: 88.

Pavlidis, P. (2003) Using ANOVA for gene selection from microarray studies of the nervous system. Methods, 31: 282–289.

Petalidis, L., Bhattacharyya, S., Morris, G.A., Collins, V.P., Freeman, T.C. and Lyons, P.A. (2003) Global amplification of mRNA by template-switching pcr: linearity and application to microarray analysis. Nucleic Acids Res., 31(22): e142.

Petersen, D., Chandramouli, G.V.R., Geoghegan, J., Hilburn, J., Paarlberg, J., Kim, C.H., Munroe, D., Gangi, L., Han, J., Puri, R., Staudt, L., Weinstein, J., Barrett, J.C., Green, J. and Kawasaki, E.S. (2005) Three microarray platforms: an analysis of their concordance in profiling gene expression. BMC Genom., 6: 63.

Prabakaran, S., Swatton, J.E., Ryan, M.M., Huffaker, S.J., Huang, J.T.-J., Griffin, J.L., Wayland, M., Freeman, T., Dudbridge, F., Lilley, K.S., Karp, N.A., Hester, S., Tkachev, D., Mimmack, M.L., Yolken, R.H., Webster, M.J., Torrey, E.F. and Bahn, S. (2004) Mitochondrial dysfunction in schizophrenia: evidence for compromised brain metabolism and oxidative stress. Mol. Psychiatry, 9(7): 684–697.

R Development Core Team. (2005) R: A Language and Environment for Statistical Computing. R Foundation for Statistical Computing, Vienna, Austria. ISBN 3-900051-07-0.

Ryan, M.M., Huffaker, S.J., Webster, M.J., Wayland, M., Freeman, T. and Bahn, S. (2004) Application and optimization of microarray technologies for human postmortem brain studies. Biol. Psychiatry, 55(4): 329–336.

Shedden, K., Chen, W., Kuick, R., Ghosh, D., Macdonald, J., Cho, K.R., Giordano, T.J., Gruber, S.B., Fearon, E.R., Taylor, J.M.G. and Hanash, S. (2005) Comparison of seven methods for producing affymetrix expression scores based on false discovery rates in disease profiling data. BMC Bioinform., 6: 26.

Shippy, R., Sendera, T.J., Lockner, R., Palaniappan, C., Kaysser-Kranich, T., Watts, G. and Alsobrook, J. (2004) Performance evaluation of commercial short-oligonucleotide microarrays and the impact of noise in making cross-platform correlations. BMC Genom., 5: 61.

Smyth, G.K. (2004) Linear models and empirical Bayes methods for assessing differential expression in microarray experiments, Stat. Appl. Genet. Mol. Biol., 3(1): Article 3. Available at: http://www.bepress.com/Sagmb/vol3/issi/art3.

Storey, J.D. (2003) The positive false discovery rate: a Bayesian interpretation and the q-value. Ann. Stat., 31(6): 2013–2035.

Storey, J.D. and Tibshirani, R. (2003) Statistical significance for genomewide studies. Proc. Natl. Acad. Sci., 100(16): 9440–9445.

Wilson, C.L., Pepper, S.D., Hey, Y. and Miller, C.J. (2004) Amplification protocols introduce systematic but reproducible errors into gene expression studies. Biotechniques, 36(3): 498–506.

Wu, H. (2005). with ideas from Gary Churchill, Katie Kerr, and Xiangqin Cui. maanova: tools for analyzing Micro Array experiments. R package version 0.98.8.

Wu, Z. and Irizarry, R.A. (2005) Stochastic models inspired by hybridisation theory for short oligonucleotide arrays. J. Comput Biol., 12(6): 882–893.

Wu, Z., Irizarry, R.A. et al. (2004) A model based background adjustment for oligonucleotide expression arrays. J. Am. Stat. Assoc. To appear

Wu, Z., Irizarry, R.A., Gentleman, R., Martinez Murillo, F., Spencer, F. (2003). A Model Based Background Adjustement for Oligonucleotide Expression Arrays. To appear in JASA.

Zakharkin, S.O., Kim, K., Mehta, T., Chen, L., Barnes, S., Parrish, R.S., Scheirer, K.E., Allison, D.B. and Page, G.P. (2005) Sources of variation in affymetrix microarray experiments. BMC Bioinform., 6: 214.

SECTION II

Applications of Genomics and Proteomic Technologies to Clinical Neuroscience

CHAPTER 6

The genomics of mood disorders

Salvatore Alesci[1], Michelle Rodak[1], Ioannis Ilias[2], Rulun Zhou[3,4] and Husseini K. Manji[4,*]

[1]Clinical Neuroendocrinology Branch, National Institute of Mental Health, National Institutes of Health, Bethesda, MD 20892, USA
[2]Department of Pharmacology, Faculty of Medicine, University of Patras, Rion-Patras, Greece
[3]Department of Psychiatry, Uniformed Services, University of the Health Sciences, Bethesda, MD, USA
[4]Laboratory of Molecular Pathophysiology, National Institute of Mental Health, NIH, Bethesda, MD 20892-3711, USA

Introduction

Attempts to comprehend the brain's role in severe mood disorders began in earnest as clinically effective mood altering drugs began to appear in the late 1950s and early 1960s. The psychopharmacological revolution fortuitously coincided with the arrival of new techniques that were making it possible to characterize neurotransmitter function in the central nervous system (CNS). Over the next three decades, clinical studies attempted to uncover the biological factors mediating the pathophysiology of manic-depressive illness utilizing a variety of biochemical strategies. Studies were, by and large, designed to detect relative excess or deficiency associated with pathological states; not surprisingly, progress in unraveling the unique neurobiology of this disorder was slow using such strategies in isolation. However, the last decade has truly been a remarkable one for biomedical research, with the "molecular medicine revolution" bringing to bear the power of sophisticated cellular and molecular biology methodologies to tackle many of society's most devastating illnesses. Psychiatry, like much of the rest of medicine, has entered a new and exciting age demarcated by current rapid advances and the future promises of genetics, molecular and cellular biology, and improving technologies. Unfortunately, clinical translation of these findings vis-à-vis a direct benefit to patients who suffer from severe psychiatric diseases has not been as rapid.

Complete sequencing of the human genome was announced in April 2003 — a truly landmark milestone in biomedical research that undoubtedly marks the beginning of a new era in molecular medicine research. While knowledge of the full human genetic sequence is a major step forward, there are additionally many other advances of significant importance in our efforts to elucidate the pathophysiology of severe psychiatric illnesses. Recent years have witnessed a more wide-range understanding of the neural circuits and the various mechanisms of synaptic transmission, the molecular mechanisms of receptor and postreceptor signaling, a finer understanding of the process by which genes encode for specific functional proteins, the identification of causative genes in many neurological disorders (e.g. Huntington's disease, early onset Alzheimer's disease, and numerous seizure disorders) that *in toto* reduce the complexity in gene-to-behavior pathways (Goodwin and Guze, 1996). These advances have generated considerable excitement among the clinical neuroscience community, and are reshaping views about the neurobiological underpinnings of these disorders. It is our firm belief that the impact of

*Corresponding author.; E-mail: manjih@mail.nih.gov

molecular and cellular biology — which has been felt in every corner of clinical medicine — will have major repercussions for our understanding about the fundamental core pathophysiology of severe mood disorders in the new millennium, and that we will see the development of markedly improved treatments for this devastating illness.

We now turn to a brief description of the clinical disorders in question, before critiquing and integrating the advances in our understanding of the genomics of the disorders.

Bipolar disorder (BPD) and unipolar depression are classified as mood disorders. They are common, severe, and chronic illnesses. Depression is typified by a depressed mood, anhedonia (inability to experience pleasure), feelings of worthlessness or excessive guilt, impaired sleep (either insomnia or hypersomnia), cognitive and concentration deficits, psychomotor changes, recurrent thoughts of death or suicide, and a variety of neurovegetative symptoms. In BPD, patients typically alternate (albeit not in a one-to-one manner) between episodes of depression (mostly indistinguishable from unipolar depression) and episodes of mania, which is characterized by a heightened mood, hyperaroused state, racing thoughts, increased speed and volume of speech, quicker thought, brisker physical and mental activity levels, inflated self-esteem, grandiosity, increased energy (with a corresponding decreased need for sleep), irritability, impaired judgment, heightened sexuality, and sometimes frank psychotic symptoms such as hallucinations and delusions. BPD is further classified into bipolar I or II (BP1 or BP2), based on clinical presentation. Bipolar I disorder is defined by a history of one or more manic episodes. Bipolar II is defined by a history of one or more hypomanic episodes and at least one depressive episode. In hypomania, a less severe form of mania, the changes noted for the description of mania above are generally moderate and do not result in hospitalization. Both diseases commonly first appear in young adulthood, though early- and late-onset of both diseases is not uncommon. They are often comorbid with other conditions, most commonly anxiety disorders with depression and drug and alcohol abuse with BPD.

The courses of both depressions, bipolar I, or II are characterized by episodes of mood changes, separated by periods of euthymia (normal mood). However, their episodic courses with intervening periods of recovery belie the severe impact of these diseases. The accumulating effect of recurring bouts of depression and/or mania leads to an increased rate of marital and family breakdown, unemployment, impaired career progress, and consequent financial difficulties. Both diseases are pervasive, being associated with increased morbidity and mortality. This arises not only from a rate of suicide approaching 15% for both disorders, but additionally the existence of significant medical co-morbidities, often limited social and economic functioning, and poor inter-episode recovery. Recent studies indicate that for a large percentage of patients, outcome is quite poor. Patients afflicted with BPD generally experience high rates of relapse, chronicity, lingering residual symptoms, cognitive and functional impairment, psychosocial disability, and diminished well-being. In addition, many deleterious health-related effects are increasingly being recognized. BPD is increasingly being viewed as an illness not only with purely psychological manifestations, but as a systemic disease frequently associated with cardiovascular disease, diabetes mellitus, obesity, and thyroid disease.

Genetics of mood disorders: the progress

Twin and adoption studies, conducted throughout United States and Europe since the early 1950s, offered some insights on the role of genetic factors in mental disorders (Kallman, 1953). The vast majority of these studies focused on BPD. The weighted average rate of BPD among first-degree relatives of bipolar probands is 5.5%, as opposed to a mere 0.6% for the relatives of control subjects (Merikangas and Risch, 2003). Furthermore, first-degree relatives of BP1 and BP2 probands are at higher risk of developing unipolar depression (Tsuang and Faraone, 1994). For unipolar depression, studies involving monozygotic and dizygotic twins estimated an overall heritability (i.e., the proportion of variance attributable to genetic factors) of 40–70% (Sullivan et al., 2000). Compared with the general population, first-degree relatives of major depressive disorder (MDD) patients have a 2–3 times increased risk for MDD. Two clinical features that predict a greater risk for MDD in first-degree relatives of depressed patients are recurrent episodes and early age of onset. However, relatives of MDD patients are not at an increased risk of BPD, thus suggesting partial genetic independence of the two disorders (Gershon et al., 1982).

Linkage studies are primarily concerned with testing anonymous DNA markers (usually in the context of a genome-wide scan) for cosegregation with the disease in families (Baron, 2002). Several chromosomal regions report significant evidence of linkage, such as 1q31–32, 4p16, 6pter-p24, 10p14, 10q25–26, 12q23–24, 13q31–32, 18p11, 18q21–23, 21q22, 22q11–13, and Xq24-28 (Baron, 2002). Early linkage studies mainly used a sample of extended multiplex pedigrees, which critics contend: (a) reduced power for linkage detection because of intrafamilial heterogeneity caused by extraneous genes (introduced by marrying-in spouses) and (b) created a bias in favor of single-gene, dominant-like inheritance (Baron, 2002). However, others believe that small sib-pair families are also vulnerable to intra-familial heterogeneity and that extended pedigrees contain more genetic information than small families, which may protect against potential loss of power due to heterogeneity. Subsequent research has not shown a definite advantage of one method of sampling over the other.

Association studies involve candidate genes, dynamic mutations, mitochondrial mutations, and chromosomal aberrations (Baron, 2002). Association studies involving candidate genes are becoming increasingly popular due to the fact that linkage studies have produced inconsistent results and are costly, both time-wise and financially (Baron, 2002). The candidate-gene strategy tests gene markers with presumed functional relevance for the disease. Using either a case–control design (unrelated cases versus population-based controls) or a family-based design, these studies are able to examine the co-occurrence of a marker and a disease. Because of their role in regulating mood and behavior, most association studies include genes implicated in serotonergic, dopaminergic, and noradrenergic systems. An insertion/deletion polymorphism in the promoter of the serotonin gene has been reported to influence the expression of the serotonin transporter, targeted by many antidepressants (Johansson et al., 2001). Other functional polymorphisms have been reported in the catechol-O-methyltransferase (COMT) gene (which is located in a region that has been linked to BPD); this finding has not always been replicated (Johansson et al., 2001).

Dynamic mutations (i.e., expanded trinucleotide repeat sequences) are implicated in diseases that demonstrate both "anticipation" (i.e., increase in disease severity, or decrease in age-at-onset, in succeeding generations) and imprinting (i.e., parent-of-origin effect on disease transmission), such as fragile-X syndrome and Huntington's disease (Baron, 2002). An association between BPD and unstable repeat sequences was found in several studies, although not all the results are conclusive and larger data sets might be needed (O'Donovan et al., 1995).

Consistent with mitochondrial inheritance, which is maternal, preferential transmission of BPD appears to occur through the maternal line. There are several reports of bipolar subjects with mitochondrial abnormalities, though much more study is warranted in this area (Kato et al., 1997).

Chromosomal aberrations may aid in gene discovery by marking candidate regions for further examination. Numerous reports of associations between BPD and chromosomal abnormalities exist, such as translocations, deletions, insertions, trisomies, and fragile sites (Baron, 2002). Craddock and Owen identified four genomic regions of importance: chromosomes 11q21–25 (balanced translocation), 15q11–13 (deletions), 21q22 (Down's syndrome trisomy), and Xq27–28 (fragile X) (Craddock and Owen, 1994). 21q22 and Xq27–28 are possibly linked to BPD, though once again, due to the high prevalence of the disease and the fact that the aberrations were observed in isolated cases, further study is warranted.

Neurobiological and neuroanatomical substrates of severe mood disorders

Positron emission tomography (PET) imaging studies have revealed multiple abnormalities of regional cerebral blood flow (CBF) and glucose metabolism in limbic and prefrontal cortical (PFC) structures in mood disorders. Although disagreement exists regarding the specific locations and the direction of some of these abnormalities, in unmedicated subjects with familial major depression, regional CBF and metabolism are consistently increased in the amygdala, orbital cortex, and medial thalamus, and decreased in the dorsomedial/dorsal anterolateral PFC, and anterior cingulate cortex ventral to the genu of the corpus callosum (in other words, subgenual PFC) relative to healthy controls (Drevets, 2001). These abnormalities implicate limbic–thalamic–cortical and limbic–cortical–striatal–pallidal–thalamic circuits, involving the amygdala, orbital and medial PFC, and anatomically related parts of the striatum and thalamus in the pathophysiology of major depression. These circuits have also been implicated more generally in emotional behavior by the results of electrophysiological, lesion analysis, and brain mapping studies of humans and experimental animals.

During symptom remission some of these abnormalities reverse, implicating areas where neurophysiological activity may increase or decrease to mediate or respond to the emotional and cognitive manifestations of depression. Nevertheless, in many of these areas CBF and metabolism do not entirely normalize during effective antidepressant treatment. In the latter regions, recent morphometric magnetic resonance imaging (MRI) and postmortem investigations have also demonstrated abnormalities of brain structure that persist independently of mood state and may contribute to the corresponding abnormalities of metabolic activity (Drevets, 2001; Manji et al., 2001).

Structural imaging studies have demonstrated reduced gray matter volumes in areas of the orbital and medial PFC, ventral striatum, and hippocampus, and enlargement of third ventricles in mood-disordered individuals relative to healthy control samples. Complementary postmortem neuropathological studies have shown abnormal reductions in cortex volume, and either glial cell counts, neuron size, or both in the subgenual PFC, orbital cortex, dorsal anterolateral PFC, and amygdala (Rajkowska et al., 2005). It is not known whether these deficits constitute developmental abnormalities that may confer vulnerability to abnormal mood episodes, compensatory changes to other pathogenic processes, or the sequelae of recurrent affective episodes *per se*. Understanding these issues will partly depend upon experiments that delineate the onset of such abnormalities within the illness course and determine whether they antedate depressive episodes in individuals at high familial risk for mood disorders. Nevertheless, the marked reduction in glial cells in these regions has been particularly intriguing in view of the growing appreciation that glia play critical roles in regulating synaptic glutamate concentrations and CNS energy homeostasis, and in releasing trophic factors that participate in the development and maintenance of synaptic networks formed by neuronal and glial processes (Rajkowska et al., 2005). Abnormalities of glial function could thus prove integral to the impairment of structural plasticity and overall pathophysiology of major depression.

Taken together with other clinical and preclinical data regarding these structures' specific roles in emotional processing, the neuroimaging and neuropathological abnormalities in major depression suggest that major depression is associated with activation of regions that putatively mediate emotional and stress responses (for example, amygdala), whereas areas that appear to inhibit emotional expression (such as posterior orbital cortex) contain histological abnormalities that may interfere with the modulation of emotional or stress responses. For example, in major depression the elevation of CBF and metabolism in the amygdala is positively correlated with depression severity, consistent with this structure's role in organizing the autonomic, neuroendocrine, and behavioral manifestations of some types of emotional responses.

In contrast, some of the medial and orbital PFC areas where metabolism is abnormal in major depression appear to play roles in reducing autonomic and endocrine responses to stressors or threats and in extinguishing behavioral responses to fear-conditioned stimuli that are no longer reinforced (Drevets, 2001). Activation of the orbital cortex during depressive episodes may thus reflect endogenous attempts to interrupt unreinforced, aversive thought and emotion. However, the histopathological abnormalities identified in these areas in major depression postmortem suggest that the ability to mediate these functions may be impaired. The hypothesis that dysfunction of these regions may contribute to the development of depression is consistent with evidence that lesions (such as strokes or tumors) involving either the PFC or the striatum (a major target of efferent projections from the PFC), and degenerative diseases affecting the striatum (for instance, Parkinson's and Huntington's diseases) are associated with increased risk for developing the depressive syndrome.

The pathophysiology of severe mood disorders: insights from recent gene profiling studies

The systems of the brain that have thus far received the greatest attention in neurobiologic studies of these illnesses have been the monoaminergic neurotransmitter systems (Manji et al., 2001). Although such investigations have had heuristic value over the years, they have been of limited value in (a) elucidating the unique biology of mood disorders, and in (b) discovering the underlying basis for the tendency of the illness to become episodic and progressively worse over time. Thus, mood disorders likely arise from the complex interaction of multiple susceptibility (and protective) genotypic and environmental factors, and the phenotypic expression of the disease, including episodic and often profound mood disturbance, and a constellation of cognitive, motor, autonomic, endocrine, and sleep/wake abnormalities. Furthermore, whereas most antidepressants exert their initial effects by increasing the intrasynaptic levels of serotonin and/or norepinephrine, their clinical antidepressant effects are only observed after chronic administration (days to weeks), suggesting that a neurobiological cascade of downstream interactions are eventually responsible for the therapeutic effects of antidepressants.

These observations have led to the appreciation that although dysfunction within the monoaminergic neurotransmitter systems is likely to play an important role in mediating some facets of the pathophysiology of mood disorders, they likely represent the downstream effects of other, more primary abnormalities (Manji and Lenox, 2000a, b). A true understanding of the pathophysiology of mood disorders must address its neurobiology at different physiological levels, i.e., at molecular, cellular, systems, and behavioral levels. Abnormalities in gene expression undoubtedly underlie the neurobiology of the disorder at the molecular level, and these abnormalities will become evident as we identify the susceptibility and protective genes for recurrent unipolar and BPDs in the coming years. Once this has been accomplished, however, an even more difficult task of examining the impact of the faulty expression of these gene-products (proteins) on integrated cell function must begin.

It is at these different levels that investigators have identified several protein candidates using the psychopharmacological strategies more fully elucidated later in this chapter (Gould and Husseini, 2004). The precise manner in which these candidate molecular targets may or may not relate to the faulty expression of gene-product susceptibility has yet to be determined. This task becomes even more difficult when considering that a major component of the pathophysiology of BPD may originate from discordant biological rhythms ranging from ultradian (weeks to months) to infradian (<24 h) that ultimately drive the periodic recurrent nature of the disorder (Rosenthal et al., 1983; Russo and Mandell, 1984; Ikonomov and Manji, 1999; Bunney and Bunney, 2000). The subsequent challenge for the basic and clinical neuroscientist will be the integration of these molecular and cellular changes into the systems level and ultimately into the behavioral level, wherein the clinical expression of mood disorders becomes manifest.

Despite these formidable obstacles, there has been considerable progress in our understanding of the underlying molecular and cellular basis of affective disorders in recent years. Recognition of a clear need for better treatments, and the lack of significant advances in our ability to develop novel, improved, therapeutics for these devastating illnesses has led to investigation of the putative roles of intracellular signaling cascades and synaptic plasticity. Recent evidence demonstrating that impairments of neuroplasticity may underlie the pathophysiology of mood disorders, and that antidepressants and mood stabilizers exert major effects on signaling pathways that regulate cellular plasticity, has generated considerable excitement among the clinical neuroscience community, reshaping views about the neurobiological underpinnings of these disorder (Manji and Lenox, 2000a; Manji et al., 2001; D'Sa and Duman, 2002; Nestler et al., 2002; Young et al., 2002). We will now review these data, and discuss their implications for changing existing conceptualizations regarding the pathophysiology of mood disorders, and for the strategic development of improved therapeutics.

Genomics encompass the sphere of methods used to analyze gene transcription (Wilson et al., 2004). The molecular revolution has led to the development of novel strategies to profile gene expression in various cells and tissues. Today, microarray technologies allow examining the expression of tens of thousands of genes simultaneously at one time (Zhou et al., 2005b). With the advent of microarray technology, there have been tremendous advances in the last decade in our ability to study an entire transcriptome (all the genes transcribed at one time). The methodology is not only providing an important new lead in our understanding of the molecular and cellular pathophysiology of severe psychiatric disorders, but also showing insight into ways for developing novel treatments. Microarray has illustrated its power in hypothesis generating (aimed at identifying novel gene target(s) and then fishing out novel mechanism(s) that may not be noticed other ways, it is as opposed to hypothesis-dependent, which limits the research within some candidate genes selected in a knowledge-based way), particularly when dealing with disorders whose pathophysiology remains largely unknown. The power of microarray is self-evident; this approach has largely become the method of choice to interrogate the whole transcriptome.

With the advances of genome projects of many organisms including human, mouse and rat, and also with fast-developing molecular biological analysis technologies, the application of microarray has expended from human samples to samples from many different kinds of animals. Trends show microarray is becoming more consistent, more comprehensive, less expensive than before and can tolerate much smaller amount of sample (RNA) quantity which makes its application easier and affordable (Xiang et al., 2002; Wang et al., 2005). Recent microarray studies in mood disorders can roughly be divided into three categories of gene profile comparisons: (1) Patients versus control subjects. Representative gene profiling researches for mood disorders in this field are those studies using human postmortem brains. Most of the studies have provided hitherto unnoticed genes or/and mechanisms for the pathophysiology of mood disorders (Evans et al., 2004; Ogden et al., 2004). (2) Drug-treated samples versus non-treated controls. Generally in these microarray studies, clinically proven effective drugs (such as mood stabilizers and antidepressants) are tested by gene expression profiling (Detera-Wadleigh, 2001; Drigues et al., 2003; Bosetti et al., 2005). For those differentially expressed genes picked out through comparison between treated and non-treated, some potential therapeutically relevant molecular targets can be selected for evaluation at protein levels and further subjected to functional tests. With rigorous validation steps, some genes may come into focus as novel molecular targets for development of new drugs and treatments (Zhou et al., 2005b). (3) Specially manipulated samples (such as knock-out mice) versus the corresponding controls. This kind of studies can evaluate the pathway changes caused by a specific candidate gene or its wide-range downstream effects (Poulsen et al., 2005).

Comparison between levels of gene transcripts in the healthy and malfunctioning brain can be quite useful in psychiatric research, potentially leading to discovery of novel diagnostic markers and targets for treatment (Evans et al., 2004; Konradi et al., 2004). Indeed, recent research in post-mortem brain tissue illustrates how microarray technology can be applied to psychiatry. Microarray studies in postmortem brains from BPD patients have shown a downregulation of oligodendrocytes messenger ribonucleic acid (mRNA) in the cortex (Tkachev et al., 2003), and of transcripts for the mitochondrial respiratory chain and ubiquitine/proteasome subunits in the hippocampus (Konradi et al., 2004). These findings support the hypothesis of abnormal energy regulation in BPD and with the proposed role played by maternal inheritance in BPD (Konradi, 2005). Consistently with these findings, spectroscopic studies have recently provided evidence of decreased pH and high-energy phosphates in the frontal and temporal lobes of bipolar patients (Konradi, 2005).

Overall though, results from microarray studies on BPD using postmortem brains seem to vary widely with regard to the expression profiles of mitochondria-related genes, perhaps because of the difficulty in controlling experimental conditions (Iwamoto et al., 2005). As patients tend to have more complex causes of death than controls, a downregulation of mitochondrial genes could occur because of a low pH, rather than a disease (Iwamoto et al., 2005). Medication might also play an important role, as the majority of drugs used to treat psychiatric disorders have mitochondrial toxicity (Haas et al., 1981).

Potential confounders are factors affecting the quality of RNA in postmortem studies, such as the subject's age, gender, postmortem interval, sample storage time in the freezer, mode of sample freezing, brain pH, and pre-mortem events (Konradi, 2005). If the variability in a sample group is too large, there might be insufficient power to detect significant changes in a poorly matched design. It is therefore crucial to match the samples *a priori* to the best possible level for age, gender, postmortem interval, freezer storage time, brain pH, and brain hemisphere (Konradi, 2005). Because of the limited sample supply, compromises are likely to be made when matching patients to controls, especially when it comes to disease-specific treatments or mode of death (Konradi, 2005).

The microarray technology has also its own set of disadvantages and should not necessarily be assumed as superior to other methods, including mRNA differential display (Zhou et al., 2005b). A drawback of gene microarrays for example is that they only measure levels of RNA, intermediate product between gene and protein, and not of the protein itself. Furthermore, microarrays only allow for the identification of known transcripts, and cannot distinguish between splice variants (Ryan et al., 2004). In addition, especially for transcripts of low abundance, the test-retest reliability of microarray is often poor. The high cost of microarrays prevents researchers from sufficiently replicating their results, thus jeopardizing to some extent the value of their conclusions (Lee et al., 2000). Therefore, microarray should be thought of as a screening technique to be used in conjunction with other methods of independent validation (Zhou et al., 2005b).

In order to enhance the stability and specificity of microarray, several strategies have been introduced for both the experimental design and data analysis. In experimental design for studies of mood disorder, the convergent method has been adopted to increase the specificity (Ogden et al., 2004). In our own studies, we profiled gene expressions of two chemically distinct mood stabilizers lithium (a monovalent cation) and valproate (a small branched fatty acid) using microarray. It was thought that those common targets shared by lithium and valproate are more likely — although not necessarily — to be attributed to their similarity as mood stabilizing agents. With a series of post-microarray functional studies, one gene — BCL2 associated athanogene or BAG1, which was similarly regulated by two chemically far different mood stabilizers and play major roles in at least three pathways (ERK MAP kinase pathway, BCL2 mediated antiapoptotic pathway and glucocorticoid receptor pathway), showed up as a novel target for mood stabilizing (Zhou et al., 2005a).

In microarray study, consistency of experimental data has been strongly suggested by many researchers. In a report by Michael Waddell et. al. from 4 universities and institutes of United States and United Kingdom, there are two conclusions about the consistency and magnitude based on their analysis, first, "consistency is a more important predictor than magnitude" and second, "considering consistency alone is far better than considering magnitude alone"(Waddell et al., 2002). One method to be addressed for consistency in microarray is replication (Lee et al., 2000).

In addition, for two-dye hybridization system, dye-swap (two fluorescent dyes Cy3 and Cy5 are used to label treated and control respectively, to counterbalance the possible bias of the two dyes, they are switched as labeling dyes between treated and control in paired tests) is also very important. According to the study by Yao et al. (2004), 32 differentially expressed genes were screened using microarray, but when the dye-swap method was adopted, this number dropped to nine, suggesting a substantial bias exists. Analyses based on both mathematical models and actual experiments supported dye swap as an effective method for quality control (Yang et al., 2002).

Consistency of microarray may also be enhanced in data analysis. A study showed error probabilities (as measured using the t-test) could be used as important indicators of consistency (Bilke et al., 2003) Thus, student's t-test is suggested for picking out positive targets in microarray. Comprehensive analyses of gene expression studies, looking at biological, molecular and structural commonalities of genes with altered regulation should be performed to yield valuable clues on underlying pathophysiology as well as to hone future research strategies.

Clues from animal models

Although not identical, being able to identify a rodent genome sequence for a disease might similarly improve the efficiency of identifying disease vulnerability in humans. Unfortunately, the lack of well-validated animal models for most psychiatric disorders has delayed the understanding of neurobiology and the development of new medications. Animal models of psychiatric disorders can be classified into two groups: pharmacological animal models and genetically altered animal models (Niculescu and Kelsoe, 2002). To mimic mania in rodents, a number of pharmacological substances such as ouabain and 6-hydroxydopamine, have been utilized. The most popular pharmacological substance used in research though is methamphetamine, which produces the core symptoms of mania (increased energy, decreased need for sleep, and psychomotor hyperactivity) with just a single dose (Niculescu and Kelsoe, 2002). In rodents, expression in the caudate-putamen and the prefrontal cortex of selected genes (PFC-TAC1, PENK, DARPP-32, MEF2C, CP-CCK and TBR1) is altered following methamphetamine administration (Ogden et al., 2004). In humans, chronic use of amphetamine also produces delusions and hallucinations, similar to psychotic mania or the positive symptoms of schizophrenia (Ogden et al., 2004). Amphetamine psychosis tends to be paranoid in nature, while that of mania tends to be grandiose, which therefore jeopardizes the face validity of such a study. Another leading limitation is that the primary symptoms in humans involve thought, affect, and speech; these cannot be observed in rodents.

Genetically engineered mutations in rodents, especially in mice provided new opportunities in behavioral, biochemical, and gene expression research. Mutations in mice have generally been targeted to specific genes, like deleting a gene (knock-out) or inserting an altered copy (knock-in) (Niculescu and Kelsoe, 2002). The dopamine transporter knock-out mouse has provided an interesting model for hyperactivity states, like mania, which has been applied to the study of BPD (Niculescu and Kelsoe, 2002).

It is a well-known fact antidepressant therapy needs at least two to six weeks to show any clinical efficacy. Thus, the presence of long-term regulation of molecules activated by drug treatment appears to be necessary for this effect (Carlezon et al., 2005; Malberg and Blendy, 2005). Interestingly, in animal models more than 300 fragments of complementary DNA (cDNA; single-stranded DNA synthesized in the laboratory using mRNA as a template and the enzyme reverse transcriptase) that have been cloned appear to be affected by chronic antidepressant therapy (Yamada and Higuchi, 2002). Current research on antidepressants has focused on long-term transcriptional regulators such as cAMP response element-binding protein (CREB) and trophic factors such as brain-derived nerve growth factor and insulin-like growth factor and how they are enmeshed with the clinical manifestations of depression and the effects of therapy (Carlezon et al., 2005; Malberg and Blendy, 2005a).

Concluding remarks

A substantial body of evidence suggests that impairments in cellular plasticity and resilience may play a central role in the underlying biology of mood disorders. Additionally, there is a growing appreciation that new medications that simply imitate "traditional" drugs, those aiming to directly or indirectly alter monoaminergic throughput, may be of limited benefit to those patients with refractory depression. Those strategies assume that the target circuits are functionally intact and that changes in synaptic activity will alter the postsynaptic throughput of the system. The evidence discussed here indicates that, in addition to neurochemical changes, many patients suffering from mood disorders also have marked structural alterations in crucial neuronal circuits. Therefore in order to obtain an optimal treatment response, it will likely be crucial to provide both trophic and neurochemical support. The aim of the trophic support would be to enhance and maintain normal synaptic connectivity, therefore permitting the chemical signal to restore maximum functioning of vital circuits essential for normal affective functioning. In fact, preliminary studies suggest that regional structural changes in the brains of patients with mood disorders may be related with, not only severity and duration of the illness, but also with altered treatment response to pharmacotherapy and electroconvulsive therapy.

The evidence also suggests that, somewhat similar to the treatment of other chronic medical conditions, such as hypertension and diabetes, prompt and sustained treatment may be necessary to prevent many of the injurious long-term sequellae associated with mood disorders. Although the evidence hints at an association between hippocampal atrophy and illness duration in depressed patients, it remains unclear whether the volumetric and cellular changes observed in other brain areas are related to affective episodes. In fact, some studies have described reduced gray matter volumes and increase ventricle size in patients with mood disorders at the time of their first episode and in early onset of the disease.

In conclusion, relevant genotypes for mood disorders are being identified, and clinical research techniques are now capable of defining neurobiological phenotypes. Similarly, results from transcription and proteomic studies which identified neurotrophic signaling as targets for the long-term actions of antidepressants and mood stabilizers have played a role (along with neuroimaging and postmortem brain studies) in a reconceptualization about the pathophysiology, course and optimal long-term treatment of severe mood disorders. These data suggest that, while mood disorders are clearly not classical neurodegenerative diseases, they are in fact, associated with impairments of cellular plasticity and resilience. As a consequence, there is a growing appreciation that optimal long-term treatment will likely be achieved by attempting to prevent the underlying disease progression and its attendant cellular dysfunction, rather than exclusively focusing on the treatment of signs and symptoms. We are optimistic that a new generation of research will clarify the relation among environmental and genetic risk factors, to quantify the risk for the development of depression more precisely. These advances will result in a dramatically different diagnostic system based upon etiology, and ultimately in the discovery of new approaches to the prevention and treatment of some of mankind's most devastating and least-understood illnesses.

Acknowledgments

This work was supported by the NIMH Intramural Research Program, NARSAD, and the Stanley Medical Research Institute. The authors thank Alex Noury for his outstanding assistance.

References

Baron, M. (2002) Manic-depression genes and the new millennium: poised for discovery. Mol. Psychiatry, 7: 342–358.

Bilke, S., Breslin, T. and Sigvardsson, M. (2003) Probabilistic estimation of microarray data reliability and underlying gene expression. BMC Bioinform., 4: 40.

Bosetti, F., Bell, J.M. and Manickam, P. (2005) Microarray analysis of rat brain gene expression after chronic administration of sodium valproate. Brain Res. Bull., 65: 331–338.

Bunney, W.E. and Bunney, B.G. (2000) Molecular clock genes in man and lower animals: possible implications for circadian abnormalities in depression. Neuropsychopharmacology, 22: 335–345.

Carlezon Jr., W.A., Duman, R.S. and Nestler, E.J. (2005) The many faces of CREB. Trends Neurosci., 28: 436–445.

Craddock, N. and Owen, M. (1994) Chromosomal aberrations and bipolar affective disorder. Br. J. Psychiatr., 164: 507–512.

Detera-Wadleigh, S.D. (2001) Lithium-related genetics of bipolar disorder. Ann. Med., 33: 272–285.

Drevets, W.C. (2001) Neuroimaging and neuropathological studies of depression: implications for the cognitive-emotional features of mood disorders. Curr. Opin. Neurobiol., 11: 240–249.

Drigues, N., Poltyrev, T., Bejar, C., Weinstock, M. and Youdim, M.B. (2003) cDNA gene expression profile of rat hippocampus after chronic treatment with antidepressant drugs. J. Neural Transm., 110: 1413–1436.

D'Sa, C. and Duman, R.S. (2002) Antidepressants and neuroplasticity. Bipolar Disord., 4: 183–194.

Evans, S.J., Choudary, P.V., Neal, C.R., Li, J.Z., Vawter, M.P., Tomita, H., Lopez, J.F., Thompson, R.C., Meng, F., Stead, J.D., Walsh, D.M., Myers, R.M., Bunney, W.E., Watson, S.J., Jones, E.G. and Akil, H. (2004) Dysregulation of the fibroblast growth factor system in major depression. Proc. Nat. Acad. Sci. USA,, 101: 15506–15511.

Gershon, E.S., Hamovit, J., Guroff, J.J., Dibble, E., Leckman, J.F., Sceery, W., Targum, S.D., Nurnberger Jr., J.I., Goldin, L.R. and Bunney Jr., W.E. (1982) A family study of schizoaffective bipolar I, bipolar II, unipolar and normal control probands. Arch. Gen. Psychiatr., 39: 1157–1167.

Goodwin, D.W. and Guze, S.B. (1996) Psychiatric Diagnosised. Oxford University Press, New York.

Gould, T.D. and Husseini, K.M. (2004) The molecular medicine revolution and psychiatry: bridging the gap between basic neuroscience research and clinical psychiatry. J. Clin. Psychiatr., 65: 598–604.

Haas, R., Stumpf, D.A., Parks, J.K. and Eguren, L. (1981) Inhibitory effects of sodium valproate on oxidative phoshorylation. Neurology, 31: 1473–1476.

Ikonomov, O.C. and Manji, H.K. (1999) Molecular mechanisms underlying mood stabilization in manic-depressive illness: the phenotype challenge. Am. J. Psychiatry, 156: 1506–1514.

Iwamoto, K., Bundo, M. and Kato, T. (2005) Altered expression of mitochondria-related genes in postmortem brains of patients with bipolar disorder or schizophrenia, as revealed by large-scale DNA microarray analysis. Hum. Mol. Genet., 14: 241–253.

Johansson, C., Jansson, M., Linner, L., Yuan, Q.P., Pedersen, N.L., Blackwood, D., Barden, N., Kelsoe, J. and Schalling, M. (2001) Genetics of affective disorders. Eur. Neuropsychopharmacol., 11: 385–394.

Kallman, F.J. (1953) Review of psychiatric progress. Am. J. Psychiatr., 110: 489–492.

Kato, T., Stine, O.C., McMahon, F.J. and Crowe, R.R. (1997) Increased levels of a mitochondrial DNA deletion in the brain of patients with bipolar disorder. Biol. Psychiatr., 42: 871–875.

Konradi, C. (2005) Gene expression microarray studies in polygenic psychiatric disorders: applications and data analysis. Brain. Res. Rev., 50: 142–155.

Konradi, C., Eaton, M., MacDonald, M.L., Walsh, J., Benes, F.M. and Heckers, S. (2004) Molecular evidence for mitochondrial dysfuction in bipolar disorder. Arch. Gen. Psychiatr., 61: 300–308.

Lee, M.L., Kuo, F.C., Whitmore, G.A. and Sklar, J. (2000) Importance of replication in microarray gene expression studies: statistical methods and evidence from repetitive cDNA hybridizations (2000) Proc. Natl. Acad. Sci. USA, 97: 9834–9839.

Malberg, J.E. and Blendy, J.A. (2005) Antidepressant action: to the nucleus and beyond. Trends Pharmacol. Sci., 26: 631–638.

Manji, H.K., Drevets, W.C. and Charney, D.S. (2001) The cellular neurobiology of depression. Nat. Med., 7: 541–547.

Manji, H.K. and Lenox, R.H. (2000a) The nature of bipolar disorder. J. Clin. Psychiatry, 61(Suppl 13): 42–57.

Manji, H.K. and Lenox, R.H. (2000b) Signaling: cellular insights into the pathophysiology of bipolar disorder. Biol Psychiatry, 48: 518–530.

Manji, H.K., Moore, G.J. and Chen, G. (2001) Bipolar disorder: leads from the molecular and cellular mechanisms of action of mood stabilizers. Br. J. Psychiatry. Suppl., 41: s107–s119.

Merikangas, K.R. and Risch, N. (2003) Will the genomics revolution revolutionize psychiatry? Am. J. Psychol., 160: 625–635.

Nestler, E.J., Barrot, M., DiLeone, R.J., Eisch, A.J., Gold, S.J. and Monteggia, L.M. (2002) Neurobiology of depression. Neuron, 34: 13–25.

Niculescu, A.B. and Kelsoe, J.R. (2002) Finding genes for bipolar disorder in the functional genomics era: from convergent functional genomics to phenomics and back. CNS Spectr., 7: 215–226.

O'Donovan, M.C., Guy, C., Craddock, N., Murphy, K.C., Cardno, A.G., Jones, L.A., Owen, M.J. and McGuffin, P. (1995) Expanded CAG repeats in schizophrenia and bipolar disorder. Nature Genet., 10: 380–381.

Ogden, C.A., Rich, M.E., Schork, N.J., Paulus, M.P., Geyer, M.A., Lohr, J.B., Kuczenski, R. and Niculescu, A.B. (2004) Candiate genes, pathways and mechanisms for bipolar (manic-depressive) and related disorders: an expanded convergent functional genomics approach. Mol. Psychiarty., 9: 1007–1029.

Poulsen, C.B., Penkowa, M., Borup, R., Nielsen, F.C., Caceres, M., Quintana, A., Molinero, A., Carrasco, J., Giralt, M. and Hidalgo, J. (2005) Brain response to traumatic brain injury in wild-type and interleukin-6 knockout mice: a microarray analysis. J. Neurochem., 92: 417–432.

Rajkowska, G., Miguel-Hidalgo, J.J., Dubey, P., Stockmeier, C.A. and Krishanan, K.R. (2005) Prominent reduction in pyramidal neurons density in the orbitofrontal cortex of elderly depressed patients. Biol. Psychiatry., 58: 297–306.

Rosenthal, N.E., Lewy, A.J., Wehr, T.A., Kern, H.E. and Goodwin, F.K. (1983) Seasonal cycling in a bipolar patient. Psychiatry Res., 8: 25–31.

Russo, P.V. and Mandell, A.J. (1984) A kinetic scattering approach to the non-linear instabilities of rat brain tyrosine hydroxylase preparations at several levels of tetrahydrobiopterin cofactor demonstrates evolutionary behavior characteristic of global dynamical systems. Brain Res., 299: 313–322.

Ryan, M.M., Huffaker, S.J., Webster, M.J., Wayland, M., Freeman, T. and Bahn, S. (2004) Application and optimization of microarray technologies for human postmortem brain studies. Biol. Psychitry, 55: 329–336.

Sullivan, P.F., Neale, M.C. and Kendler, K.S. (2000) Genetic epidemiology of major depression: review and meta-analysis. Am. J. Psychiatr., 157: 1552–1562.

Tkachev, D., Mimmack, M.L., Ryan, M.M., Wayland, M., Freeman, T., Jones, P.B., Starkey, M., Webster, M.J., Yolken, R.H. and Bahn, S. (2003) Oligodendrocyte dysfunction in schizophrenia and bipolar disorder. Lancet, 362: 798–805.

Tsuang, M.T. and Faraone, S.V. (1994) The genetic epidemiology of schizophrenia. Compr. Ther., 20: 130–135.

Waddell, M., Page, D., Zhan, F., Barlogie, B., Shaughnessy, J., Hardin, J., and Cussen, J., (2002). Comparative Data Mining for Microarrays. 10th International Conference on Intelligent Systems for Molecular Biology, Edmonton, Canada.

Wang, H., He, X., Band, M., Wilson, C. and Liu, L. (2005) A study of inter-lab and inter-platform agreement of DNA microarray data. BMC Genomics, 6: 71.

Wilson, K.E., Ryan, M.M., Prime, J.E., Pashby, D.P., Orange, P.R., O'Beirne, G., Whateley, J.G., Bahn, S. and Morris, C.M. (2004) Functional genomics and proteomics: application in neurosciences. J. Neurol. Neurosurg. Psychiatr., 75: 529–538.

Xiang, C.C., Kozhich, O.A., Chen, M., Inman, J.M., Phan, Q.N., Chen, Y. and Brownstein, M.J. (2002) Amine-modified random primers to lebel probes for DNA microarrays. Nat. Biotechnol., 20: 738–742.

Yamada, M. and Higuchi, T. (2002) Functional genomics and depression research. Beyond the monoamine hypothesis. Eur. Neuropsychopharmacol., 12: 235–244.

Yang, Y.H., Dudoit, S., Luu, P., Lin, D.M., Peng, V., Ngai, J. and Speed, T.P. (2002) Normalization for cDNA microarray data: a robust composite method addressing single and multiple slide systematic variation. Nucleic Acids Res., 30: e15.

Yao, B., Rakhade, S.N., Li, Q., Ahmed, S., Krauss, R., Draghici, S. and Loeb, J.A. (2004) O1_MRKO1_MRKAccuracy of cDNA microarray methods to detect small gene expression changes induced by neuregulin on breast epithelial cells. BMC Bioinformatics, 5: 99.

Young, A.H., Hughes, J.H., Marsh, V.R. and Ashton, C.H. (2002) Acute tryptophan depletion attenuates auditory event related potentials in bipolar disorder: a preliminary study. J. Affect. Disord., 69: 83–92.

Zhou, R., Gray, N.A., Yuan, P., Chen, J., Damschroder-Williams, P., Du, J., Zhang, L. and Manji, H.K. (2005a) The anti-apoptotic, glucocorticoid receptor cochaperone protein BAG-1 is a long-term target for the actions of mood stabilizers. J. Neurosci., 25: 4493–4502.

Zhou, R., Zarate, C.A. and Manji, H.K. (2005b) Identification of molecular mechanisms underlying mood stabilization through genome-wide gene expression profiling. Int. J. Neuropsychopharmacol., 1–4.

CHAPTER 7

Transcriptome alterations in schizophrenia: disturbing the functional architecture of the dorsolateral prefrontal cortex

David A. Lewis[1,*] and Karoly Mirnics[1,2]

[1]*Department of Psychiatry, and Department of Neuroscience, University of Pittsburgh, 3811 O'Hara Street, W1652 BST, Pittsburgh, PA 15213-2593, USA*
[2]*Department of Neurobiology, University of Pittsburgh, 3811 O'Hara Street, W1652 BST, Pittsburgh, PA 15213-2593, USA*

Abstract: The availability of methods for quantifying tissue concentrations of messenger RNAs in the postmortem of the human brain has provided a number of new findings in schizophrenia. However, understanding how these findings actually relate to the disease process of schizophrenia requires knowledge both of the factors that might give rise to such changes in gene expression and of the impact of these changes on the function of the affected neural circuits. Consequently, this chapter provides a review of the potential causes and consequences of some of the schizophrenia-related transcriptome changes in the dorsolateral prefrontal cortex, a brain region implicated in the pathophysiology of certain core cognitive deficits in this illness.

Keywords: GABA; glutamic acid decarboxylase; prefrontal cortex; regulator of G-protein signaling 4; schizophrenia

As described in other chapters in this volume, the advent of a range of new methods for quantifying messenger RNA (mRNA), the transcription product of genomic DNA, has made it possible to assess tissue levels of virtually every transcript expressed in the human brain. The application of these approaches to the study of postmortem brain specimens from individuals with schizophrenia has produced a number of interesting, and potentially very important, findings. The altered tissue concentration of a given mRNA in schizophrenia has generally been interpreted to be the result of a change in the active regulation of that transcript. The same assumption will be followed in this chapter, although with the recognition that differences in transcript tissue levels could also reflect changes in mRNA stability and/or degradation. It is also generally assumed, but much less frequently demonstrated, that a disease-related difference in the tissue concentration of a given transcript is associated with a similar change in the cognate protein. This assumption will also be followed in this chapter, but with the caveat that exceptions to this rule have been reported.

In the context of the disease process of schizophrenia, the significance of transcriptome alterations depends upon their impact on the neural circuitry or functional architecture of the brain region in which the gene expression changes arise. The purpose of this chapter is to consider some of the potential causes and consequences of the

*Corresponding author. Tel.: +1-412-624-3934; Fax: +1-412-624-9910; E-mail: lewisda@upmc.edu

schizophrenia-related transcript alterations in the dorsolateral prefrontal cortex (DLPFC), a brain region frequently implicated in the pathophysiology of core cognitive features of the illness. Thus, we first consider the nature of the functional disturbances in the DLPFC in schizophrenia.

Dysfunction of the DLPFC in schizophrenia

Although a plethora of clinical signs and symptoms may be manifest by individuals with schizophrenia, disturbances in certain cognitive processes (e.g., attention, memory, and other higher-order processes such as the ability to plan, initiate, and regulate goal-directed behavior) appear to represent the core features of the illness (Elvevåg and Goldberg, 2000). For example, cognitive abnormalities have been found throughout the life span of affected individuals, including during childhood and adolescence (Davidson et al., 1999), at the initial onset of psychosis (Saykin et al., 1994), and during later stages of the illness (Heaton et al., 1994). In addition, the unaffected relatives of individuals with schizophrenia also exhibit similar, if less severe, cognitive deficits (Sitskoorn et al., 2004). Thus, these cognitive deficits are not the result of treatment with antipsychotic medications or a consequence of the chronic nature of the illness. In addition, the cognitive disturbances of schizophrenia are more debilitating and have a greater negative impact on daily activities than do psychotic symptoms (Green, 1996). Finally, the degree of cognitive dysfunction is inversely related to, and may be the best predictor of, long-term functional outcome in schizophrenia (Green, 1996).

Many of these cognitive deficits reflect alterations in executive control, the processes that facilitate complex information processing and behavior. Executive control includes context representation and maintenance functions, such as working memory, that involve the active retention of limited amounts of information for a short period of time in order to guide thought processes or sequences of behavior. Working memory is dependent on the circuitry of the DLPFC (Miller and Cohen, 2001). Subjects with schizophrenia tend to perform poorly on various working memory tasks and to demonstrate reduced activation of the DLPFC when attempting to perform such tasks (Weinberger et al., 1986; Perlstein et al., 2001). In addition, under conditions in which subjects with schizophrenia perform normally on working memory tasks, activity in the DLPFC may actually be increased, suggesting a reduced efficiency of DLPFC function in schizophrenia (Callicott et al., 2003).

The relevance of these alterations in DLPFC function to the disease process of schizophrenia is supported by several findings. First, subjects with major depression show normal activation of the DLPFC when performing working memory tasks, suggesting that the abnormalities observed in schizophrenia might be specific to the disease process of schizophrenia (Barch et al., 2003). Indeed, although first-episode, never-medicated subjects with schizophrenia exhibit DLPFC dysfunction during such tasks, first-episode, never-medicated subjects with other psychotic disorders do not (MacDonald III et al., 2005). Second, deficits in activation of the DLPFC, but not in other cortical regions, during working memory tasks predict the severity of cognitive disorganization symptoms observed clinically in subjects with schizophrenia (Perlstein et al., 2001). Third, reduced working memory capacity has been suggested to be rate limiting in the performance of other cognitive functions in schizophrenia (Silver et al., 2003). Because working memory deficits appear to be a central feature of schizophrenia, identifying the gene expression alterations in the DLPFC that might contribute to these functional alterations is essential for understanding the underlying disease process.

Types of transcriptome alterations in the DLPFC in schizophrenia

A number of transcript abnormalities have been observed in the DLPFC of subjects with schizophrenia. The first microarray study of postmortem brain tissue from subjects with schizophrenia (Mirnics et al., 2000) found a marked reduction in the expression of a number of transcripts that encode proteins involved in the machinery required for presynaptic neurotransmitter release. Many of the same transcripts, such as N-ethylmaleimide-sensitive factor, γ-SNAP, synaptogyrin 1, synaptobrevin, clathrin, and 42 kDa vacuolar proton pump, were also identified as underexpressed in a microarray study conducted using a different platform in an independent subject cohort (Vawter et al., 2002). These findings were interpreted as an evidence that schizophrenia is a disease of the synapse (Mirnics et al., 2001a), an interpretation supported by reduced levels of some synaptic proteins (Glantz and Lewis, 1997; Karson et al., 1999) and synaptic structural elements (Garey et al., 1998; Selemon and Goldman-Rakic, 1999; Glantz and Lewis, 2000) in the DLPFC of subjects with schizophrenia.

Several microarray studies have also found evidence of altered expression of genes involved in mitochondrial function and energy generation. For example, Middleton et al. (2002) reported evidence of expression abnormalities in genes-encoding proteins involved in ornithine and polyamine metabolism, the mitochondrial malate shuttle system, the transcarboxylic acid cycle, aspartate and alanine metabolism, and ubiquitin metabolism. Related abnormalities in the expression of mitochondrial genes were also reported by other research groups (Prabakaran et al., 2004; Altar et al., 2005; Iwamoto et al., 2005). Together, these findings suggest that markers of altered energy metabolism are a characteristic feature of the transcriptome of schizophrenia, although these changes may be neither disease-specific nor part of the primary disease process (Mirnics et al., 2006).

A consistent down-regulation of oligodendrocyte transcripts has also been observed in the DLPFC of subjects with schizophrenia. For example, Hakak et al. (2001) identified decreased tissue levels of a number of mRNAs-encoding proteins related to oligodendrocyte function and myelination in schizophrenia. The same or related transcripts were also reported to be reduced in the DLPFC of subjects with schizophrenia by other investigators using microarray techniques (Pongrac et al., 2002; Tkachev et al., 2003; Sugai et al., 2004). The idea that these transcript alterations are associated with other evidence of white matter changes in schizophrenia has been recently reviewed (Davis et al., 2003; Katsel et al., 2005). However, it remains unclear to what extent the transcript alterations represent a reduction in gene expression or the consequence of a decrease in the number of oligodendrocytes (Hof et al., 2003).

Causes of transcriptome alterations in the DLPFC in schizophrenia

Because the substantial heritability of schizophrenia appears to be polygenic in nature (Owen et al., 2004), some transcript differences might be due to DNA sequence variants that confer disease susceptibility through changes in the expression level of the encoded mRNA. Alternatively, certain transcript abnormalities might represent secondary responses of the genome that are either deleterious or compensatory. Finally, some differences in mRNA levels observed in the DLPFC might reflect changes in gene expression induced by the treatment of the illness. Clearly, distinguishing among these alternatives is critical for assessing whether a given transcript alteration represents a cause, consequence, compensation, or confound of the disease process.

In an initial microarray study of the DLPFC in schizophrenia, the expression of the transcript for Regulator of G-protein signaling 4 (RGS4) was found to be markedly and consistently reduced (Mirnics et al., 2001b). RGS proteins play critical roles in controlling the duration of signaling through G-protein-coupled receptors (GPCR), which mediate the effects of a number of neurotransmitters (e.g., dopamine, serotonin, and glutamate) that are involved in the pathophysiology or pharmacotherapy of schizophrenia. The cytoplasmic domain of GPCRs is tightly bound to a heterotrimeric G protein composed of subtypes of G_α, G_β, and G_γ subunits until the binding of a specific ligand produces a conformational change in the structure of the receptor, resulting in the dissociation of the G protein and the rapid replacement of guanine diphosphate (GDP) with guanine triphosphate (GTP) at the G_α subunit. The low affinity of GTP-bound G_α for the $G_{\beta\gamma}$ subunit causes them to dissociate, and the independent subunits then interact with a number of cellular effectors, activating or inhibiting a variety of signaling cascades. Once the G_α-bound GTP is hydrolyzed to GDP, the G_α–GDP complex regains its high affinity to the $G_{\beta\gamma}$ and the heterotrimeric G protein re-associates. This effectively ends signaling via the GPCR, and prepares the receptor for a new round of ligand binding. The spontaneous hydrolysis of $G_{\alpha i}$-bound GTP to GDP is slow, and thus limits dynamic signaling via the receptor. RGS proteins are GTPase-activating proteins that markedly facilitate the hydrolysis of G_α-bound GTP, and thus they play a critical role in determining the duration and timing of signaling through GPCRs (De Vries et al., 2000).

Interestingly, none of the other 11 RGS family members represented on the microarrays showed altered expression levels in the subjects with schizophrenia, suggesting that specific aspects of GPCR signaling are altered in the illness. Furthermore, the decreased RGS mRNA levels detected in the DLPFC by microarray were verified by in situ hybridization in additional subjects, and similar findings have been reported by other investigators (Erdely et al., 2003). In addition, a similar magnitude of reduction in RGS4 mRNA levels was also found in motor and visual cortices of the same subjects with schizophrenia, regions that differ substantially from the DLPFC in structure, connectivity and function. In contrast, no changes in RGS4 expression were observed in the DLPFC of subjects with major depression or in monkeys exposed chronically to haloperidol in a manner that mimicked the clinical treatment of schizophrenia. Thus, these findings suggested that decreased RGS4 expression in schizophrenia (1) is selective for this particular regulator of G-protein signaling, (2) is a common feature of the illness, (3) is not restricted to the DLPFC, (4) might be specific to the clinical syndrome of schizophrenia, and (5) is part of the disease process of schizophrenia, and not merely a confound of factors such as treatment with antipsychotic medications.

Together, these findings suggested the hypothesis that reductions in RGS4 mRNA expression in schizophrenia reflect the effects of allelic variants in the *RGS4* gene that confer increased risk for the illness. Interestingly, *RGS4* is present at chromosomal locus 1q23.3, a region identified as linked to schizophrenia (Brzustowicz et al., 2000). Consistent with this observation, a meta-analysis of 20 independent genome-wide linkage scans also found evidence suggestive of a susceptibility locus at 1q21–22 (Lewis et al., 2003). However, other large number of studies have not detected linkage at this locus (Levinson et al., 2002). Interestingly, of the 70 transcripts that map to this cytogenetic region and that were represented on the microarray, only the mRNA for *RGS4* was consistently under-expressed in subjects with schizophrenia. However, it should be noted that another gene (*CAPON*), with increased expression in the DLPFC (Xu et al., 2005), has been suggested to account for the signal from this locus (Brzustowicz et al., 2004).

Thus, *RGS4* appeared to meet three criteria as a candidate susceptibility gene in schizophrenia. First, *RGS4* meets the criterion of being functionally relevant since its protein product is involved in neurotransmitter systems implicated in the pathophysiology and treatment of schizophrenia. Second, *RGS4* meets the criterion of position since it is present at a chromosomal locus previously linked to schizophrenia. Finally, *RGS4* meets the criterion of altered expression since tissue concentrations of its mRNA are reduced across cortical regions with apparent specificity for the clinical syndrome of schizophrenia. Consistent with this hypothesis, allelic variations in the 5' region of *RGS4* were found to be associated with schizophrenia in independent samples from Pittsburgh, the National Institutes on Mental Health (NIMH) Collaborative Genetics Initiative and New Delhi (Chowdari et al., 2002). Evidence of transmission distortion of individual alleles and haplotypes was found at four SNPs (designated SNPs 1, 4, 7, and 18) located in the promoter region (SNPs 1, 4, and 7) or first intron (SNP 18) of the gene. However, the associated alleles and haploptypes differed across the samples. For example, the four marker risk haplotype in the Pittsburgh sample was the mirror image of the risk haplotype in the NIMH sample. Four subsequent reports from independent groups have found evidence of an association between RGS4 and schizophrenia, although the specific findings differed across samples. Significant associations with alleles at SNPs 4 and 18, but not with the four SNP haplotypes, were found in a large case–control study by the Cardiff group (Williams et al., 2004). Significant associations were found at SNPs 1 and 7, and a trend at SNP 4, in a case–control study from Dublin, but only after subjects with a diagnosis of schizoaffective disorder were removed (Morris et al., 2004). A family-based analysis of multiply affected pedigrees from Ireland revealed an association at SNP 18 (Chen et al., 2004). Although significant case–control differences were not found in a Brazilian sample, evidence of modest over-transmission at SNP 18 was observed (Cordeiro et al., 2005). In addition, a novel meta-analysis of over 13,000 individuals from 13 independent samples (including those studies summarized above as well as unpublished data) suggested a complex association between *RGS4* variants and schizophrenia, potentially consistent with either disease risk due to two common haplotypes or clinical/genetic heterogeneity (Talkowski et al., 2006).

Together, these findings suggest that genetic variants in the promoter region of *RGS4* could contribute to its altered expression in schizophrenia, but as is the case for other putative susceptibility genes in schizophrenia (Owen et al., 2004), the genetic findings regarding the specific risk alleles are inconsistent across subject cohorts. In addition, the frequency of subjects carrying the putative risk alleles is inconsistent with the very high percentage (<90%) of postmortem subjects that show decreased RGS4 mRNA expression. Finally, it remains to be determined how (and if) the putative risk alleles actually contribute to the reduced levels of RGS4 mRNA in schizophrenia. Nonetheless, gray matter volume in the DLPFC of first-episode, never-mediated subjects with schizophrenia was associated with certain *RGS4* alleles at SNPs 4 and 18 (Prasad et al., 2005). Together, these findings suggest a dual origin for the observed RGS4 transcript deficiencies (and perhaps those of other transcripts as well). That is, some subjects with schizophrenia might show a direct RGS4 down-regulation as a result of allelic variants in the *RGS4* gene that influence its transcription, whereas other individuals exhibit decreased RGS4 mRNA levels as a result of adaptations to upstream disease processes. This interpretation implies that RGS4 protein represents a critical convergence point (or "intracellular hub") in neurons (Mirnics et al., 2006).

Consequences of transcriptome alterations in the DLPFC in schizophrenia

Perhaps the most widely replicated finding in postmortem studies of schizophrenia (Knable et al., 2002) is a reduced tissue concentration in the DLPFC of the mRNA for the 67 kDa isoform of glutamic acid decarboxylase (GAD_{67}), an enzyme responsible for the synthesis of the inhibitory neurotransmitter GABA. Indeed, GAD_{67} mRNA expression was found to be reduced in the DLPFC of subjects with schizophrenia by several research groups in a number of subject cohorts using several different techniques (Akbarian et al., 1995; Guidotti et al., 2000; Mirnics et al., 2000; Volk et al., 2000; Vawter et al., 2002; Hashimoto et al., 2005). The only published exception to this findings was in one cohort of elderly, chronically hospitalized individuals with schizophrenia (Dracheva et al., 2004). Although less-extensively studied, the deficit in GAD_{67} mRNA appears to be accompanied by a corresponding decrease in the cognate protein (Guidotti et al., 2000). In contrast, overall protein and mRNA expression levels of another synthesizing enzyme for GABA, GAD_{65} have been reported to be unchanged in the DLPFC in schizophrenia (Guidotti et al., 2000), as it has the density of GAD_{65}–IR puncta (Benes et al., 2000). Interestingly, elimination of the GAD_{65} gene in mice results in minor reductions of cortical GABA levels, whereas genetically engineered reductions in GAD_{67} are associated with profound decreases in cortical GAD activity and GABA content (Asada et al., 1997).

Thus, these findings suggest that reduced expression of GAD_{67} in the DLPFC could contribute to the substrate for the working memory disturbances present in individuals with schizophrenia. Indeed, although working memory is clearly an emergent property of the coordinated activation of a neuronal network distributed across a number of brain regions, it depends upon the coordinated and sustained firing of subsets of

DLPFC pyramidal neurons between the temporary presentation of a stimulus cue and the later initiation of a behavioral response (Goldman-Rakic, 1995). Although other neurotransmitter systems are also involved, DLPFC neurons that utilize the inhibitory neurotransmitter γ-aminobutyric acid (GABA) appear to be critical for such synchronization of pyramidal neuron activity during working memory processes. For example, fast-spiking GABA neurons in monkey DLPFC are active during the delay period of working memory tasks (Wilson et al., 1994), and are necessary for task-related neuronal firing and the spatial tuning of neuronal responses during working memory (Rao et al., 2000). In addition, the injection of GABA antagonists in the DLPFC disrupts working memory performance (Sawaguchi et al., 1989). Indeed, Constantinidis et al. (2002) suggested that during working memory tasks inhibition serves both a spatial role (in which DLPFC pyramidal neurons are activated during working memory) and a temporal role (when they are active during the different phases of working memory).

However, interpreting the significance of these alterations in GAD_{67} expression for the function of DLPFC circuitry in schizophrenia depends upon knowing whether all or a subset of GABA neurons are affected (Lewis, 1995). Indeed, any observed difference in the tissue level of a given transcript could reflect a change in the number of the cells that normally express that mRNA, a uniform expression change in all cells that express the transcript, or an alteration in expression restricted to a subpopulation of neurons (Mirnics et al., 2004). Clearly, each of these scenarios has markedly different implications for the circuitry in which the affected neurons participate.

Cortical GABA neurons are quite heterogeneous, with a number of different subpopulations distinguishable by a combination of morphological, physiological, and molecular attributes, with each subpopulation playing specialized roles in regulating pyramidal neuron activity (DeFelipe, 1997; Gupta et al., 2000; McBain & Fisahn, 2001). For example, the chandelier (or axoaxonic) subpopulation of GABA neurons express the calcium-binding protein parvalbumin (PV) (Lewis and Lund, 1990), exhibit a fast-spiking, non-adapting firing pattern (Kawaguchi, 1995), and furnish a linear array of axon terminals (termed cartridges) that exclusively target the axon initial segment of pyramidal neurons (Somogyi, 1977). Wide arbor (basket) cells in the monkey DLPFC also contain PV (Condé et al., 1994) and have fast-spiking electrophysiological features that are indistinguishable from those of chandelier neurons (González-Burgos et al., 2005). However, the axons of wide arbor neurons have a much larger spread than those of chandelier cells and their axon terminals principally target the somata and proximal dendrites of pyramidal neurons (Lewis and Lund, 1990). The proximity of the perisomatic inhibitory synapses formed by PV-containing chandelier and wide arbor neurons to the site of action potential generation in pyramidal neurons suggests that these GABA neurons are specialized to powerfully regulate the output of pyramidal neurons. The calcium-binding protein calbindin is present primarily in double-bouquet cells which exhibit a regular spiking, adapting firing pattern, and provide axon terminals that synapse on the distal dendrites of pyramidal cells (Kawaguchi and Kubota, 1998; Zaitsev et al., 2005). Finally, about 50% of GABA neurons in the monkey DLPFC express the calcium-binding protein calretinin (Condé et al., 1994; Gabbott and Bacon, 1996), exhibit a regular-spiking, adaptive firing pattern (Kawaguchi, 1995; Zaitsev et al., 2005) and provide axon terminals that target distal dendritic spines or shafts on pyramidal cells or other GABA neurons (Melchitzky et al., 2005).

Several findings at the cellular level suggest that only certain subsets of GABA neurons exhibit altered gene expression in schizophrenia. For example, although the total number of prefrontal neurons was not reduced in schizophrenia (Thune et al., 2001), the density of neurons with detectable levels of GAD_{67} mRNA was decreased in DLPFC layers 2–5, but not in the layer 6, of subjects with schizophrenia (Akbarian et al., 1995; Volk et al., 2000). In contrast, in neurons with detectable levels of GAD_{67} mRNA, the expression level per neuron did not differ from controls (Volk et al., 2000). Together, these observations suggest that in subjects with schizophrenia, about two-thirds of DLPFC GABA neurons express normal levels of

GAD$_{67}$ mRNA, but the remainder lack detectable amounts of this transcript. Similar changes in expression were found for the mRNA for the GABA membrane transporter (GAT1), which is responsible for the re-uptake of released GABA (Volk et al., 2001). Thus, both the synthesis and re-uptake of GABA appear to be greatly reduced in a subset of DLPFC inhibitory neurons in schizophrenia.

Furthermore, the mRNA level of PV, but not of calretinin, was decreased, whereas the density of neurons with detectable levels of PV mRNA was not changed in the DLPFC subjects with schizophrenia (Hashimoto et al., 2003), consistent with the results of immunocytochemical studies which failed to find decreased densities of PV-immunoreactive neurons in the DLPFC of subjects with schizophrenia (Woo et al., 1997; Beasley et al., 2002). However, the expression level of PV mRNA per neuron was significantly decreased, and the expression level of PV mRNA per neuron was strongly correlated with the change in density of GAD$_{67}$ mRNA-positive neurons. Together these findings suggest that GAD$_{67}$ mRNA expression was markedly reduced in PV-containing neurons that also had reduced, but still detectable, levels of PV mRNA. Consistent with this interpretation, dual label in situ hybridization studies demonstrated that 50% of PV mRNA-positive neurons in subjects with schizophrenia lacked detectable levels of GAD$_{67}$ mRNA (Hashimoto et al., 2003).

The distinctive developmental trajectories of chandelier neurons in the monkey DLPFC, especially during adolescence (Cruz et al., 2003), suggested that these neurons might be particularly vulnerable in schizophrenia (Lewis, 1997). Consistent with this hypothesis, the density of chandelier neuron axon cartridges immunoreactive for GAT1 was significantly reduced in the middle cortical layers in the DLPFC of subjects with schizophrenia (Woo et al., 1998; Pierri et al., 1999). In contrast, GAT1 immunoreactivity in other populations of axon terminals was unchanged (Woo et al., 1998). These alterations appear to be specific to the disease process of schizophrenia since similar alterations were not present in subjects with other psychiatric disorders or in monkeys exposed chronically to antipsychotic medications in a fashion that mimics the clinical treatment of schizophrenia (Pierri et al., 1999).

The combination of gene expression alterations and cell type-specific protein changes suggest that in schizophrenia, chandelier neurons in the DLPFC express decreased levels of PV mRNA and undetectable levels of GAD$_{67}$ and GAT1 mRNAs, with the latter resulting in reduced GAT1 protein in chandelier neuron axon cartridges. This combination of findings appears to reflect deficient inhibition, resulting from a primary reduction in GABA synthesis, because GABA$_A$ receptors that contain α_2 subunits, which are preferentially located in pyramidal neuron axon initial segments (Loup et al., 1998), appear to be up-regulated in schizophrenia. For example, in DLPFC of subjects with schizophrenia, the density of pyramidal neuron axon initial segments immunoreactive for the GABA$_A$ α_2 subunit was significantly increased by over 100% compared to control subjects (Volk et al., 2002). The increased density of α_2-immunoreactive axon initial segments must reflect higher levels of α_2 subunits at the axon initial segment because neither the density of pyramidal neurons (Pierri et al., 2003; Maldonado-Aviles et al., 2006) nor their axon initial segments (Cruz et al., 2004) were increased in these subjects. Consistent with this interpretation, in a preliminary study the level of GABA$_A$ α_2 mRNA was reported to be increased in pyramidal cell bodies in subjects with schizophrenia (Kim et al., 2005). Thus, in the DLPFC cortex of subjects with schizophrenia, GABA$_A$ receptors appear to be up-regulated at pyramidal neuron axon initial segments in response to deficient GABA synthesis and release from chandelier axon terminals.

This interpretation also suggests that the expression changes in GAT1 and PV mRNAs represent compensatory, but inadequate, responses to the deficit in GABA synthesis in PV-containing neurons. For example, although GAT1 levels do not affect single inhibitory postsynaptic currents (IPSCs), the blockade of GABA re-uptake prolongs the duration of IPSCs when synapses located close to each other are activated synchronously (Overstreet and Westbrook, 2003), as is the case for the inputs of chandelier axon cartridges to pyramidal cell axon initial segments.

Thus, a presynaptic reduction in GAT1 would result in the prolongation of IPSCs, leading to increased probability of IPSC summation, and therefore enhanced efficacy of IPSC trains. Similarly, by buffering presynaptic Ca^{2+} transients, PV is thought to decrease the Ca^{2+}-dependent facilitation of GABA release (Vreugdenhil et al., 2003). That is, during periods of repetitive firing, PV binds, Ca^{2+} and reduces residual intra-terminal Ca^{2+} levels, resulting in decreased GABA release. Thus, decreased expression of GAT1 and PV transcripts, with corresponding reductions of GAT1 and PV proteins in chandelier axon cartridges, may act synergistically as compensatory responses designed to increase the efficacy of GABA neurotransmission at pyramidal cell axon initial segments in the DLPFC of individuals with schizophrenia.

Despite these compensations, the detrimental functional consequences of reduced GABA activity at pyramidal neuron axon initial segments is revealed by the role that fast-spiking, PV-positive GABA neurons play in regulating working memory function. For example, these neurons appear to be specialized to coordinate the firing of pyramidal neurons at gamma frequency oscillations (see Lewis et al. (2005) for review). Interestingly, the power of gamma frequency oscillations in the human frontal cortex increases in proportion to working memory load (Howard et al., 2003), and subjects with schizophrenia demonstrate reduced prefrontal gamma band activity when performing working memory tasks (Cho et al., 2004). Thus, these findings suggest that impairments in GABAergic inhibition, resulting from a conserved down-regulation in the expression of GAD_{67} mRNA, might disrupt the functional architecture of the DLPFC and underlie the working memory disturbances characteristic of schizophrenia.

Conclusions

The current ability to characterize the disease-related transcriptome(s) in specific brain regions, like the DLPFC, offers great potential for increasing our knowledge of the neurobiology of certain clinical features, such as working memory, in schizophrenia. However, understanding whether these gene expression changes represent causes, consequences, or compensations in the disease process requires the parallel study of the mechanisms that can give rise to these gene changes and of the impact they have on the functional architecture of the DLPFC. The results of such investigations will help determine the viability of a given gene product as a novel target for pharmacological intervention in the illness.

Acknowledgments

The work by the authors cited in this article was supported by National Institutes of Health grants MH045156, MH043784 and MH070786.

References

Akbarian, S., Kim, J.J., Potkin, S.G., Hagman, J.O., Tafazzoli, A., Bunney Jr., W.E. and Jones, E.G. (1995) Gene expression for glutamic acid decarboxylase is reduced without loss of neurons in prefrontal cortex of schizophrenics. Arch. Gen. Psychiatry, 52: 258–266.

Altar, C.A., Jurata, L.W., Charles, V., Lemire, A., Liu, P., Bukhman, Y., Young, T.A., Bullard, J., Yokoe, H., Webster, M.J., Knable, M.B. and Brockman, J.A. (2005) Deficient hippocampal neuron expression of proteasome, ubiquitin, and mitochondrial genes in multiple schizophrenia cohorts. Biol. Psychiatry, 58: 85–96.

Asada, H., Kawamura, Y., Maruyama, K., Kume, H., Ding, R., Kanbara, N., Kuzume, H., Sanbo, M., Yagi, T. and Obata, K. (1997) Cleft palate and decreased brain γ-aminobutyric acid in mice lacking the 67-kDa isoform of glutamic acid decarboxylase. Proc. Natl. Acad. Sci. USA, 94: 6496–6499.

Barch, D.M., Sheline, Y.I., Csernansky, J.G. and Snyder, A.Z. (2003) Working memory and prefrontal cortex dysfunction: specificity to schizophrenia compared with major depression. Biol. Psychiatry, 53: 376–384.

Beasley, C.L., Zhang, Z.J., Patten, I. and Reynolds, G.P. (2002) Selective deficits in prefrontal cortical GABAergic neurons in schizophrenia defined by the presence of calcium-binding proteins. Biol. Psychiatry, 52: 708–715.

Benes, F.M., Todtenkopf, M.S., Logiotatos, P. and Williams, M. (2000) Glutamate decarboxylase(65)-immunoreactive terminals in cingulate and prefrontal cortices of schizophrenic and bipolar brain. J. Chem. Neuroanat., 20: 259–269.

Brzustowicz, L.M., Hodgkinson, K.A., Chow, E.W.C., Honer, W.G. and Bassett, A.S. (2000) Location of a major susceptibility locus for familial schizophrenia on chromosome 1q21-q22. Science, 288: 678–682.

Brzustowicz, L.M., Simone, J., Mohseni, P., Hayter, J.E., Hodgkinson, K.A., Chow, E.W. and Bassett, A.S. (2004) Linkage disequilibrium mapping of schizophrenia susceptibility to the CAPON region of chromosome 1q22. Am. J Hum. Genet., 74: 1057–1063.

Callicott, J.H., Mattay, V.S., Verchinski, B.A., Marenco, S., Egan, M.F. and Weinberger, D.R. (2003) Complexity of prefrontal cortical dysfunction in schizophrenia: more than up or down. Am. J. Psychiatry, 160: 2209–2215.

Chen, X., Dunham, C., Kendler, S., Wang, X., O'Neill, F.A., Walsh, D. and Kendler, K.S. (2004) Regulator of G-protein signaling 4 (RGS4) gene is associated with schizophrenia in Irish high density families. Am. J. Med. Genet., 129B: 23–26.

Cho, R.Y., Konecky, R.O. and Carter, C.S. (2004) Impaired task-set maintenance and frontal cortical gamma-band synchrony in schizophrenia. *Cognitive Neuroscience Society Annual Meeting.*

Chowdari, K.V., Mirnics, K., Semwal, P., Wood, J., Lawrence, E., Bhatia, T., Deshpande, S.N., BK, T., Ferrell, R.E., Middleton, F.A., Devlin, B., Levitt, P., Lewis, D.A. and Nimgaonkar, V.L. (2002) Association and linkage analyses of RGS4 polymorphisms in schizophrenia. Hum. Mol. Genet., 11: 1373–1380.

Condé, F., Lund, J.S., Jacobowitz, D.M., Baimbridge, K.G. and Lewis, D.A. (1994) Local circuit neurons immunoreactive for calretinin, calbindin D-28k, or parvalbumin in monkey prefrontal cortex: distribution and morphology. J. Comp. Neurol., 341: 95–116.

Constantinidis, C., Williams, G.V. and Goldman-Rakic, P.S. (2002) A role for inhibition in shaping the temporal flow of information in prefrontal cortex. Nat. Neurosci., 5: 175–180.

Cordeiro, Q., Talkowski, M.E., Chowdari, K.V., Wood, J., Nimgaonkar, V. and Vallada, H. (2005) Association and linkage analysis of RGS4 polymorphisms with schizophrenia and bipolar disorder in Brazil. Genes Brain Behav., 4: 45–50.

Cruz, D.A., Eggan, S.M., Azmitia, E.C. and Lewis, D.A. (2004) Serotonin1A receptors at the axon initial segment of prefrontal pyramidal neurons in schizophrenia. Am. J. Psychiatry, 161: 739–742.

Cruz, D.A., Eggan, S.M. and Lewis, D.A. (2003) Postnatal development of pre- and post-synaptic GABA markers at chandelier cell inputs to pyramidal neurons in monkey prefrontal cortex. J. Comp. Neurol., 465: 385–400.

Davidson, M., Reichenberg, A., Rabinowitz, J., Weiser, M., Kaplan, Z. and Mark, M. (1999) Behavioral and intellectual markers for schizophrenia in apparently healthy male adolescents. Am. J. Psychiatry, 156: 1328–1335.

Davis, K.L., Stewart, D.G., Friedman, J.I., Buchsbaum, M., Harvey, P.D., Hof, P.R., Buxbaum, J. and Haroutunian, V. (2003) White matter changes in schizophrenia: evidence for myelin-related dysfunction. Arch Gen. Psychiatry, 60: 443–456.

De Vries, L., Zheng, B., Fischer, T., Elenko, E. and Farquhar, M.G. (2000) The regulator of G protein signaling family. Annu. Rev. Pharmacol. Toxicol., 40: 235–271.

DeFelipe, J. (1997) Types of neurons, synaptic connections and chemical characteristics of cells immunoreactive for calbindin-D28 K, parvalbumin and calretinin in the neocortex. J. Chem. Neuroanat., 14: 1–19.

Dracheva, S., Elhakem, S.L., McGurk, S.R., Davis, K.L. and Haroutunian, V. (2004) GAD67 and GAD65 mRNA and protein expression in cerebrocortical regions of elderly patients with schizophrenia. J. Neurosci. Res., 76: 581–592.

Elvevåg, B. and Goldberg, T.E. (2000) Cognitive impairment in schizophrenia is the core of the disorder. Crit. Rev. Neurobiol., 14: 1–21.

Erdely, H.A., Lahti, R.A., Roberts, R.C., Vogel, M.W. and Tamminga, C.A. (2003) Reduced levels of RGS4 mRNA and protein in schizophrenia. Soc. Neurosci. Abstr., 33: 317.8.

Gabbott, P.L.A. and Bacon, S.J. (1996) Local circuit neurons in the medial prefrontal cortex (areas 24a,b,c, 25 and 32) in the monkey: II. Quantitative areal and laminar distributions. J. Comp. Neurol., 364: 609–636.

Garey, L.J., Ong, W.Y., Patel, T.S., Kanani, M., Davis, A., Mortimer, A.M., Barnes, T.R.E. and Hirsch, S.R. (1998) Reduced dendritic spine density on cerebral cortical pyramidal neurons in schizophrenia. J. Neurol. Neurosurg. Psychiatry, 65: 446–453.

Glantz, L.A. and Lewis, D.A. (1997) Reduction of synaptophysin immunoreactivity in the prefrontal cortex of subjects with schizophrenia: regional and diagnostic specificity. Arch. Gen. Psychiatry, 54: 943–952.

Glantz, L.A. and Lewis, D.A. (2000) Decreased dendritic spine density on prefrontal cortical pyramidal neurons in schizophrenia. Arch. Gen. Psychiatry, 57: 65–73.

Goldman-Rakic, P.S. (1995) Cellular basis of working memory. Neuron, 14: 477–485.

González-Burgos, G., Krimer, L.S., Povysheva, N.V., Barrionuevo, G. and Lewis, D.A. (2005) Functional properties of fast spiking interneurons and their synaptic connections with pyramidal cells in primate dorsolateral prefrontal cortex. J. Neurophysiol., 93: 942–953.

Green, M.F. (1996) What are the functional consequences of neurocognitive deficits in schizophrenia? Am. J. Psychiatry, 153: 321–330.

Guidotti, A., Auta, J., Davis, J.M., Gerevini, V.D., Dwivedi, Y., Grayson, D.R., Impagnatiello, F., Pandey, G., Pesold, C., Sharma, R., Uzunov, D. and Costa, E. (2000) Decrease in reelin and glutamic acid decarboxylase67 (GAD67) expression in schizophrenia and bipolar disorder. Arch. Gen. Psychiatry, 57: 1061–1069.

Gupta, A., Wang, Y. and Markram, H. (2000) Organizing principles for a diversity of GABAergic interneurons and synapses in the neocortex. Science, 287: 273–278.

Hakak, Y., Walker, J.R., Li, C., Wong, W.H., Davis, K.L., Buxbaum, J.D., Haroutunian, V. and Fienberg, A.A. (2001) Genome-wide expression analysis reveals dysregulation of myelination-related genes in chronic schizophrenia. Proc. Natl. Acad. Sci. USA, 98: 4746–4751.

Hashimoto, T., Bergen, S.E., Nguyen, Q.L., Xu, B., Monteggia, L.M., Pierri, J.N., Sun, Z., Sampson, A.R. and Lewis, D.A. (2005) Relationship of brain-derived neurotrophic factor and its receptor TrkB to altered inhibitory prefrontal circuitry in schizophrenia. J. Neurosci., 25: 372–383.

Hashimoto, T., Volk, D.W., Eggan, S.M., Mirnics, K., Pierri, J.N., Sun, Z., Sampson, A.R. and Lewis, D.A. (2003) Gene expression deficits in a subclass of GABA neurons in the prefrontal cortex of subjects with schizophrenia. J. Neurosci., 23: 6315–6326.

Heaton, R., Paulsen, J.S., McAdams, L.A., Kuck, J., Zisook, S., Braff, D., Harris, J. and Jeste, D.V. (1994) Neuropsychological deficits in schizophrenics: relationship to age, chronicity, and dementia. Arch. Gen. Psychiatry, 51: 469–476.

Hof, P.R., Haroutunian, V., Friedrich Jr., V.L., Byne, W., Buitron, C., Perl, D.P. and Davis, K.L. (2003) Loss and altered spatial distribution of oligodendrocytes in the superior frontal gyrus in schizophrenia. Biol. Psychiatry, 53: 1075–1085.

Howard, M.W., Rizzuto, D.S., Caplan, J.B., Madsen, J.R., Lisman, J., Aschenbrenner-Scheibe, R., Schulze-Bonhage, A. and Kahana, M.J. (2003) Gamma oscillations correlate with working memory load in humans. Cereb. Cortex, 13: 1369–1374.

Iwamoto, K., Bundo, M. and Kato, T. (2005) Altered expression of mitochondria-related genes in postmortem brains of patients with bipolar disorder or schizophrenia, as revealed by large-scale DNA microarray analysis. Hum. Mol. Genet., 14: 241–253.

Karson, C.N., Mrak, R.E., Schluterman, K.O., Sturner, W.Q., Sheng, J.G. and Griffin, W.S.T. (1999) Alterations in synaptic proteins and their encoding mRNAs in prefrontal cortex in schizophrenia: a possible neurochemical basis for 'hypofrontality'. Mol. Psychiatry, 4: 39–45.

Katsel, P.L., Davis, K.L. and Haroutunian, V. (2005) Large-scale microarray studies of gene expression in multiple regions of the brain in schizophrenia and Alzheimer's disease. Int. Rev Neurobiol, 63: 41–82.

Kawaguchi, Y. (1995) Physiological subgroups of nonpyramidal cells with specific morphological characteristics in layer II/III of rat frontal cortex. J. Neurosci., 15: 2638–2655.

Kawaguchi, Y. and Kubota, Y. (1998) Neurochemical features and synaptic connections of large physiologically-identified GABAergic cells in the rat frontal cortex. Neuroscience, 85: 677–701.

Kim, A.M., Matzilevich, D.A., Walsh, J.P., Benes, F.M. and Woo, T.W. (2005) Parvalbumin-containing neurons and disturbances of prefrontal cortical circuitry in schizophrenia. Soc. Neurosci. Abstr., 35: 912.1.

Knable, M.B., Barci, B.M., Bartko, J.J., Webster, M.J. and Torrey, E.F. (2002) Molecular abnormalities in the major psychiatric illnesses: classification and regression tree (CRT) analysis of post-mortem prefrontal markers. Mol. Psychiatry, 7: 392–404.

Levinson, D.F., Holmans, P.A., Laurent, C., Riley, B., Pulver, A.E., Gejman, P.V., Schwab, S.G., Williams, N.M., Owen, M.J., Wildenauer, D.B., Sanders, A.R., Nestadt, G., Mowry, B.J., Wormley, B., Bauché, S., Soubigou, S., Ribble, R., Nertney, D.A., Liang, K.Y., Martinolich, L., Maier, W., Norton, N., Williams, H., Albus, M., Carpenter, E.B., deMarchi, N., Ewen-White, K.R., Walsh, D., Jay, M., Deleuze, J.-F., O'Neill, F.A., Papadimitriou, G., Weilbaecher, A., Lerer, B., O'Donovan, M.C., Dikeos, D., Silverman, J.M., Kendler, K.S., Mallet, J., Crowe, R.R. and Walters, M. (2002) No major schizophrenia locus detected on chromosome 1q in a large multicenter sample. Science, 296: 739–741.

Lewis, C.M., Levinson, D.F., Wise, L.H., DeLisi, L.E., Straub, R.E., Hovatta, I., Williams, N.M., Schwab, S.G., Pulver, A.E., Faraone, S.V., Brzustowicz, L.M., Kaufmann, C.A., Garver, D.L., Gurling, H.M., Lindholm, E., Coon, H., Moises, H.W., Byerley, W., Shaw, S.H., Mesen, A., Sherrington, R., O'Neill, F.A., Walsh, D., Kendler, K.S., Ekelund, J., Paunio, T., Lonnqvist, J., Peltonen, L., O'Donovan, M.C., Owen, M.J., Wildenauer, D.B., Maier, W., Nestadt, G., Blouin, J.L., Antonarakis, S.E., Mowry, B.J., Silverman, J.M., Crowe, R.R., Cloninger, C.R., Tsuang, M.T., Malaspina, D., Harkavy-Friedman, J.M., Svrakic, D.M., Bassett, A.S., Holcomb, J., Kalsi, G., McQuillin, A., Brynjolfson, J., Sigmundsson, T., Petursson, H., Jazin, E., Zoega, T. and Helgason, T. (2003) Genome scan meta-analysis of schizophrenia and bipolar disorder, part II: schizophrenia. Am. J. Hum. Genet., 73: 34–48.

Lewis, D.A. (1995) Neural circuitry of the prefrontal cortex in schizophrenia. Arch. Gen. Psychiatry, 52: 269–273.

Lewis, D.A. (1997) Development of the prefrontal cortex during adolescence: insights into vulnerable neural circuits in schizophrenia. Neuropsychopharmacology, 16: 385–398.

Lewis, D.A., Hashimoto, T. and Volk, D.W. (2005) Cortical inhibitory neurons and schizophrenia. Nat. Rev. Neurosci., 6: 312–324.

Lewis, D.A. and Lund, J.S. (1990) Heterogeneity of chandelier neurons in monkey neocortex: corticotropin-releasing factor and parvalbumin immunoreactive populations. J. Comp. Neurol., 293: 599–615.

Loup, F., Weinmann, O., Yonekawa, Y., Aguzzi, A., Wieser, H.-G. and Fritschy, J.-M. (1998) A highly sensitive immunoflourescence procedure for analyzing the subcellular distribution of GABAA receptor subunits in the human brain. J. Histochem. Cytochem., 46: 1129–1139.

MacDonald III., A.W., Carter, C.S., Kerns, J.G., Ursu, S., Barch, D.M., Holmes, A.J., Stenger, V.A. and Cohen, J.D. (2005) Specificity of prefrontal dysfunction and context processing deficits to schizophrenia in never-medicated patients with first-episode psychosis. Am. J. Psychiatry, 162: 475–484.

Maldonado-Aviles, J.G., Wu, Q., Sampson, A.R. and Lewis, D.A. (2006) Somal size of immunolabeled pyramidal cells in the prefrontal cortex of subjects with schizophrenia. Biol. Psychiatry, in press.

McBain, C.J. and Fisahn, A. (2001) Interneurons unbound. Nat. Rev. Neurosci., 2: 11–23.

Melchitzky, D.S., Eggan, S.M. and Lewis, D.A. (2005) Synaptic targets of calretinin-containing axon terminals in macaque monkey prefrontal cortex. Neuroscience, 130: 185–195.

Middleton, F.A., Mirnics, K., Pierri, J.N., Lewis, D.A. and Levitt, P. (2002) Gene expression profiling reveals alterations of specific metabolic pathways in schizophrenia. J. Neurosci., 22: 2718–2729.

Miller, E.K. and Cohen, J.D. (2001) An integrative theory of prefrontal cortex function. Annu. Rev. Neurosci., 24: 167–202.

Mirnics, K., Levitt, P. and Lewis, D.A. (2004) DNA microarray analysis of postmortem brain tissue. Int. Rev. Neurobiol., 60: 153–181.

Mirnics, K., Levitt, P. and Lewis, D.A. (2006) Critical appraisal of DNA microarrays in psychiatric genomics. Biol. Psychiatry, in press.

Mirnics, K., Middleton, F.A., Lewis, D.A. and Levitt, P. (2001a) Analysis of complex brain disorders with gene expression microarrays: schizophrenia as a disease of the synapse. Trends Neurosci., 24: 479–486.

Mirnics, K., Middleton, F.A., Marquez, A., Lewis, D.A. and Levitt, P. (2000) Molecular characterization of schizophrenia viewed by microarray analysis of gene expression in prefrontal cortex. Neuron, 28: 53–67.

Mirnics, K., Middleton, F.A., Stanwood, G.D., Lewis, D.A. and Levitt, P. (2001b) Disease-specific changes in regulator of G-protein signaling 4 (RGS4) expression in schizophrenia. Mol. Psychiatry, 6: 293–301.

Morris, D.W., Rodgers, A., McGhee, K.A., Schwaiger, S., Scully, P., Quinn, J., Meagher, D., Waddington, J.L., Gill, M. and Corvin, A.P. (2004) Confirming RGS4 as a susceptibility gene for schizophrenia. Am. J. Med. Genet., 125B: 50–53.

Overstreet, L.S. and Westbrook, G.L. (2003) Synapse density regulates independence at unitary inhibitory synapses. J. Neurosci., 23: 2618–2626.

Owen, M.J., Williams, N.M. and O'Donovan, M.C. (2004) The molecular genetics of schizophrenia: new findings promise new insights. Mol. Psychiatry, 9: 14–27.

Perlstein, W.M., Carter, C.S., Noll, D.C. and Cohen, J.D. (2001) Relation of prefrontal cortex dysfunction to working memory and symptoms in schizophrenia. Am. J. Psychiatry, 158: 1105–1113.

Pierri, J.N., Chaudry, A.S., Woo, T.-U. and Lewis, D.A. (1999) Alterations in chandelier neuron axon terminals in the prefrontal cortex of schizophrenic subjects. Am. J. Psychiatry, 156: 1709–1719.

Pierri, J.N., Volk, C.L., Auh, S., Sampson, A. and Lewis, D.A. (2003) Somal size of prefrontal cortical pyramidal neurons in schizophrenia: differential effects across neuronal subpopulations. Biol. Psychiatry, 54: 111–120.

Pongrac, J., Middleton, F.A., Lewis, D.A., Levitt, P. and Mirnics, K. (2002) Gene expression profiling with DNA microarrays: advancing our understanding of psychiatric disorders. Neurochem. Res., 27: 1049–1063.

Prabakaran, S., Swatton, J.E., Ryan, M.M., Huffaker, S.J., Huang, J.T., Griffin, J.L., Wayland, M., Freeman, T., Dudbridge, F., Lilley, K.S., Karp, N.A., Hester, S., Tkachev, D., Mimmack, M.L., Yolken, R.H., Webster, M.J., Torrey, E.F. and Bahn, S. (2004) Mitochondrial dysfunction in schizophrenia: evidence for compromised brain metabolism and oxidative stress. Mol. Psychiatry, 9: 684–697 643.

Prasad, K.M., Chowdari, K.V., Nimgaonkar, V.L., Talkowski, M.E., Lewis, D.A. and Keshavan, M.S. (2005) Genetic polymorphisms of the RGS4 and dorsolateral prefrontal cortex morphometry among first episode schizophrenia patients. Mol. Psychiatry, 10: 213–219.

Rao, S.G., Williams, G.V. and Goldman-Rakic, P.S. (2000) Destruction and creation of spatial tuning by disinhibition: GABAA blockade of prefrontal cortical neurons engaged by working memory. J. Neurosci., 20: 485–494.

Sawaguchi, T., Matsumura, M. and Kubota, K. (1989) Delayed response deficits produced by local injection of bicuculline into the dorsolateral prefrontal cortex in Japanese macaque monkeys. Exp. Brain Res., 75: 457–469.

Saykin, A.J., Shtasel, D.L., Gur, R.E., Kester, D.B., Mozley, L.H., Stafiniak, P. and Gur, R.C. (1994) Neuropsychological deficits in neuroleptic naive patients with first-episode schizophrenia. Arch. Gen. Psychiatry, 51: 124–131.

Selemon, L.D. and Goldman-Rakic, P.S. (1999) The reduced neuropil hypothesis: a circuit based model of schizophrenia. Biol. Psychiatry, 45: 17–25.

Silver, H., Feldman, P., Bilker, W. and Gur, R.C. (2003) Working memory deficit as a core neuropsychological dysfunction in schizophrenia. Am. J. Psychiatry, 160: 1809–1816.

Sitskoorn, M.M., Aleman, A., Ebisch, S.J., Appels, M.C. and Kahn, R.S. (2004) Cognitive deficits in relatives of patients with schizophrenia: a meta-analysis. Schizophr. Res., 71: 285–295.

Somogyi, P. (1977) A specific axo-axonal interneuron in the visual cortex of the rat. Brain Res., 136: 345–350.

Sugai, T., Kawamura, M., Iritani, S., Araki, K., Makifuchi, T., Imai, C., Nakamura, R., Kakita, A., Takahashi, H. and Nawa, H. (2004) Prefrontal abnormality of schizophrenia revealed by DNA microarray: impact on glial and neurotrophic gene expression. Ann N. Y. Acad. Sci, 1025: 84–91.

Talkowski, M.E., Seltman H., Bassett, A.S., Brzustowicz, L.M., Chen, X., Chowdari, K.V., Collier, D.A., Cordeiro, Q., Corvin, A.P., Deshpande, S., Egan, M.F., Ferrell, R.E., Gill, M., Kendler, K.S., Kirov, G., Levitt, P., Lewis, D.A., Li, T., Mirnics, K., Morris, D.W., O'Donnovan M.C., Owen, M.J., Sobell J.L., Thelma, B.K., Vallada, H., Weinberger, D.R., Williams, N.M., Wood, J. and Devlin, B. (2006) Meta-analysis of RGS4 polymorphisms with schizophrenia using genotypes of 13,807 individuals from 13 independent samples. Biol. Psychiatry, in press.

Thune, J.J., Uylings, H.B.M. and Pakkenberg, B. (2001) No deficit in total number of neurons in the prefrontal cortex in schizophrenics. J. Psychiatr. Res., 35: 15–21.

Tkachev, D., Mimmack, M.L., Ryan, M.M., Wayland, M., Freeman, T., Jones, P.B., Starkey, M., Webster, M.J., Yolken, R.H. and Bahn, S. (2003) Oligodendrocyte dysfunction in schizophrenia and bipolar disorder. Lancet, 362: 798–805.

Vawter, M.P., Crook, J.M., Hyde, T.M., Kleinman, J.E., Weinberger, D.R., Becker, K.G. and Freed, W.J. (2002) Microarray analysis of gene expression in the prefrontal cortex in schizophrenia: a preliminary study. Schizophr. Res., 58: 11–20.

Volk, D.W., Austin, M.C., Pierri, J.N., Sampson, A.R. and Lewis, D.A. (2000) Decreased GAD67 mRNA expression in a subset of prefrontal cortical GABA neurons in subjects with schizophrenia. Arch. Gen. Psychiatry, 57: 237–245.

Volk, D.W., Austin, M.C., Pierri, J.N., Sampson, A.R. and Lewis, D.A. (2001) GABA transporter-1 mRNA in the prefrontal cortex in schizophrenia: decreased expression in a subset of neurons. Am. J. Psychiatry, 158: 256–265.

Volk, D.W., Pierri, J.N., Fritschy, J.-M., Auh, S., Sampson, A.R. and Lewis, D.A. (2002) Reciprocal alterations in pre- and postsynaptic inhibitory markers at chandelier cell inputs to pyramidal neurons in schizophrenia. Cereb. Cortex, 12: 1063–1070.

Vreugdenhil, M., Jefferys, J.G., Celio, M.R. and Schwaller, B. (2003) Parvalbumin-deficiency facilitates repetitive IPSCs and gamma oscillations in the hippocampus. J. Neurophysiol., 89: 1414–1422.

Weinberger, D.R., Berman, K.F. and Zec, R.F. (1986) Physiologic dysfunction of dorsolateral prefrontal cortex in schizophrenia I. Regional cerebral blood flow evidence. Arch. Gen. Psychiatry, 43: 114–124.

Williams, N.M., Preece, A., Spurlock, G., Norton, N., Williams, H.J., McCreadie, R.G., Buckland, P., Sharkey, V., Chowdari, K.V., Zammit, S., Nimgaonkar, V., Kirov, G., Owen, M.J. and O'Donovan, M.C. (2004) Support for RGS4 as a susceptibility gene for schizophrenia. Biol. Psychiatry, 55: 192–195.

Wilson, F.A., O Scalaidhe, S.P. and Goldman-Rakic, P.S. (1994) Functional synergism between putative gamma-aminobutyrate-containing neurons and pyramidal neurons in prefrontal cortex. Proc. Natl. Acad. Sci. USA, 91: 4009–4013.

Woo, T.-U., Miller, J.L. and Lewis, D.A. (1997) Schizophrenia and the parvalbumin-containing class of cortical local circuit neurons. Am. J. Psychiatry, 154: 1013–1015.

Woo, T.-U., Whitehead, R.E., Melchitzky, D.S. and Lewis, D.A. (1998) A subclass of prefrontal gamma-aminobutyric acid axon terminals are selectively altered in schizophrenia. Proc. Natl. Acad. Sci. U.S.A, 95: 5341–5346.

Xu, B., Wratten, N., Charych, E.I., Buyske, S., Firestein, B.L. and Brzustowicz, L.M. (2005) Increased expression in dorsolateral prefrontal cortex of CAPON in schizophrenia and bipolar disorder. PLoS Med., 2: e263–e263.

Zaitsev, A.V., Gonzalez-Burgos, G., Povysheva, N.V., Kroner, S., Lewis, D.A. and Krimer, L.S. (2005) Localization of calcium-binding proteins in physiologically and morphologically characterized interneurons of monkey dorsolateral prefrontal cortex. Cereb. Cortex, 15: 1178–1186.

CHAPTER 8

Strategies for improving sensitivity of gene expression profiling: regulation of apoptosis in the limbic lobe of schizophrenics and bipolars

Francine M. Benes[1,2,*]

[1]*Program in Structural and Molecular Neuroscience, McLean Hospital, Mailman Research Center, 115 Mill Street, Belmont, MA 02178, USA*
[2]*Program in Neuroscience and Department of Psychiatry, Harvard Medical School, Boston, MA, USA*

Introduction

For over a century, schizophrenia (SZ) was believed to be a neurodegenerative disorder with an onset during late adolescence and early adulthood (Kraepelin, 1919). In the past two decades, this idea received support from brain imaging studies demonstrating volume loss in several different regions, including the hippocampus (HIPP) and the amygdala (AMYG) (Lawrie and Abukmeil, 1998; Harrison, 1999; Arnold, 2000). Using positron-emission tomography (PET), an increase of basal metabolism has been observed in HIPP of SZ subjects (Heckers et al., 1998) and this finding is consistent with a large number of postmortem studies suggesting that a decrease of inhibitory GABAergic activity may be present in the prefrontal cortex (Benes et al., 1991a, 1996b, 2000, 2006; Guidotti et al., 2000; Volk et al., 2000), anterior cingulate region (Benes et al., 1991a, 1992b; Woo et al., 2004), hippocampus (Benes et al., 1996a, 1997; Todtenkopf and Benes, 1998; Heckers et al., 2002) and amygdala (Simpson et al., 1989) (for a review see Benes and Berretta, 2001). Several other postmortem observations have pointed to the amygdala as a unifying factor that could account for this pattern (Benes et al., 1992a; Longson et al., 1996). The basolateral nucleus, which plays a pivotal role in the response to environmental stress (Antoniadis and McDonald, 2000; Davis and Shi, 2000; LeDoux, 2000), sends a significant innervation to sectors CA3 and CA2 of the HIPP (Pitkanen et al., 2000). Based on these observations, we have developed a rodent model for neural circuitry changes in postmortem studies of the limbic lobe in the psychotic disorders (Fig. 1). To study neural circuitry changes in SZ and bipolar disorder (BD), the GABAA antagonist, picrotoxin, is stereotaxically infused into the basolateral nucleus of the amygdala (Benes and Berretta, 2000). Within 2 h, a selective reduction of GABAergic terminals in CA3 and CA2, but not CA1, was observed (Berretta et al., 2001) and this suggested that the changes observed in the hippocampus of schizophrenic and bipolar subjects could potentially be related to excessive discharges of excitatory activity from the AMYG occurring in response to a defect of GABA neurotransmission in this region. The amygdala plays a central role in emotional responses and the encoding of context-dependent explicit memory by the hippocampus (Ledoux, 2000). There has been a growing interest in the possibility that the amygdala and hippocampus may contribute to the

---

*Corresponding author. Tel.: +1-617-855-2401; Fax: +1-617-855-3199; E-mail: fbenes@mclean.harvard.edu

DOI: 10.1016/S0079-6123(06)58008-2

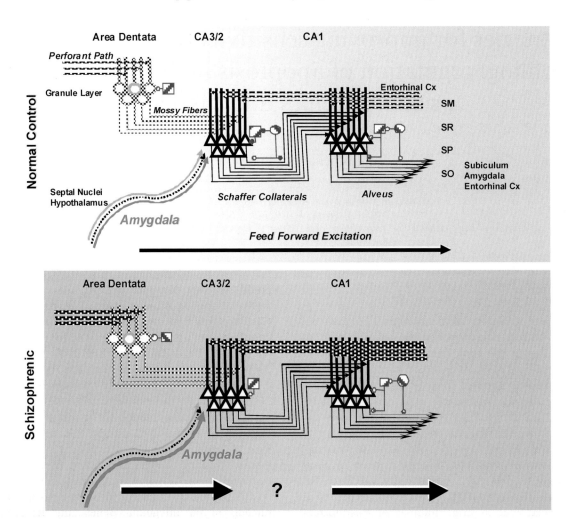

Fig. 1. A model for alterations in GABAergic integration in sectors CA3/CA2 of the hippocampus in subjects with schizophrenia, particularly in the stratum oriens and stratum radiatum where GABAergic interneurons are selective localized. A selective reduction of nonpyramidal neurons in sector CA2 and a robust upregulation of GABA-A receptor binding activity have been observed in post-mortem studies of normal controls and schizophrenics. When a high-resolution technique was employed, an upregulation of this receptor was found on interneurons in the pyamidal cell layer, suggesting there may be a reduction of both inhibitory and disinhibitory GABA cells and/or activity may be present in CA3/2 of schizophrenics.

pathophysiology of psychotic disorders (Tamminga et al., 1992; Benes, 2000).

In order to interpret these findings within the context of complex biochemical pathways, we have used gene expression profiling as a broad screening tool to evaluate the status of multiple transduction, signaling and metabolic pathways. In pursuing this strategy, it has become apparent that the use of such technology to study the central nervous system, particularly in relation to complex and subtle forms of neuropathology like those found in SZ and BD, it was clear that novel strategies would be required for analyzing several thousand genes showing changes in expression at any given time.

In the discussion that follows, a two-pronged approach to the study of the hippocampus in relation to SZ and BD is presented. This strategy includes parallel gene expression profiling (GEP) studies of this region from postmortem brains of SZ and BD subjects and from our rodent model of neural circuitry changes in these disorders observed in our postmortem findings. This approach makes it possible to control changes in gene expression in human hippocampus that may be related to medical illnesses, medications and the agonal state. It is assumed that when similar changes are observed in the rodent model, they are likely to be related to activation of the amygdala under conditions of stress. There are important potential limitations to such studies. These include: (1) issues related to quality control of the tissue and its RNA content, (2) the inherent lack of sensitivity that occurs in working with whole extracts of the hippocampus and (3) the need for validation of the findings. It is becoming broadly recognized that the hunt for disease genes requires methods for evaluating associations between multiple factors (Tsunoda et al., 2000) and assessments of reliability (Raffelsberger et al., 2002; Asyali et al., 2004). In the discussion that follows, these issues are addressed in relation to specific gene expression profiling studies of human postmortem hippocampus that have been conducted in our laboratory.

Quality controls

As shown in Table 1, there are several different controls that should be examined in gene expression profiling studies of human postmortem tissue. Among the most crucial is the use of cohorts of normal control, schizophrenic and bipolar subjects that are matched for basic demographic factors such as age and gender. Additionally, it is critical to include cases that are also matched with respect to the postmortem interval and tissue pH. These latter variables can have a profound influence on the quality of RNA contained within the samples. The pH of the samples, however, does not necessarily reflect the length of the postmortem interval and it appears that there may be other factors, such as medical illness, medications and the agonal phase that may contribute substantially to RNA degradation. Indeed, the 18S/28S ratio serves as an important index that may reflect negative influences on the integrity of RNA. As shown in Table 1, a ratio close to 1 is considered optimal and for the cohort shown this was well matched. In Fig. 2, 18S/28S ratios for a normal control and schizophrenic case are excellent, while that for a bipolar case shows a pattern indicative of RNA degradation. The latter case had to be replaced when it was determined that the degradation was not due to experimental artifact.

Other quality-control parameters include 3'–5' ratios for "house-keeping" genes that show no changes in regulation and serve as an internal control. Two such genes are β-actin and glycero-3-phosphate dehydrogenase (G3PDH), but it cannot be assumed that they are not showing differences in regulation. This makes it imperative that they are indeed "housekeeping" genes by comparing the expression values across the three groups. This is a critical control issue because they will be used to normalize the results for target genes when quantitative RT-PCR (qRT-PCR) is undertaken. Figure 3 shows the normalization of qRT-PCR data using G3PDH as a control. It can be seen that there are no differences among the three groups (upper right). This is an essential quality-control measure that ensures that the microarray results are valid by evaluating the presence of false positives and false negatives. Although a false-negative result raises questions about the validity of the microarray data, false negatives do occur to varying degree and indicate that the microarrays were unable to detect the changes that are inherently present among the groups.

Table 1. Comparison of demographics and RNA quality in HIFP of McLean 66 cohort

	AGE	PMI	L/R	Gender (M/F)	Brain pH	Total RNA	18/28S	% present	3'/5' G3PDH	3'/5' β-actin
Normal controls	58	20	12/15	19/8	6.43		1.07	45.9	1.44	2.33
Schizophrenics	56	21	8/12	13/7	6.44		1.09	45.9	1.56	2.43
Bipolar disorder	63	21	9/10	12/7	6.45		1.02	44.2	1.59	2.73

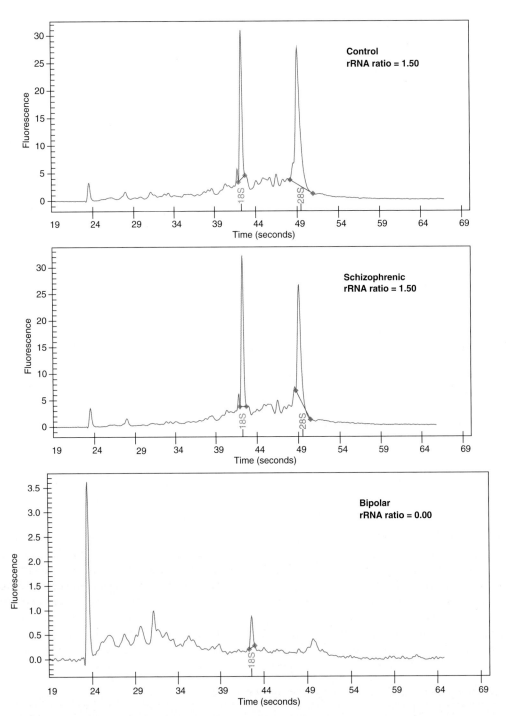

Fig. 2. Postmortem cohort of controls, schizophrenics and bipolars matched for age, postmortem interval and gender were arranged as triplets. The 18S/28S peaks for the control and schizophrenic suggest good-quality RNA is present. The peaks are essentially absent in the bipolar, making it necessary to reject this case from the cohort.

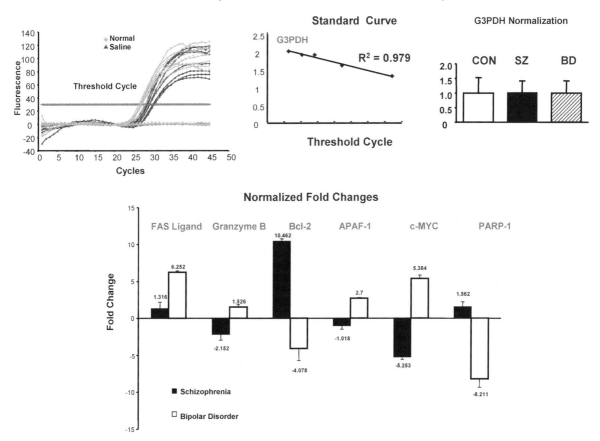

Fig. 3. FRET-based quantitative q RT-PCR showing melt curves (upper left), a standard curve for G3PDH (upper middle) and mRNA levels for G3PDH in the normal control (CON), schizophrenic (SZ) and bipolar (BD) groups (upper right). The lower panel shows the results for FAS ligand, granzyme B, Bcl-2, APAF-1, c-MYC and PARP-1 for schizophrenics versus bipolars. The normalized results show fold changes in the expected direction based on the microarray data (Benes et al., 2005).

Table 2. Comparison of RNA quality control parameters

	28S/18S	3'/5' G3PDH	3'/5' β-actin	% present calls
ACCx	1.09	2.53	3.86	36.6
PFCx	1.10	1.43	2.28	45.9
Hippocampus		2.28	2.86	44.2

Note: NBD, National Brain Databank; ACCx, anterior cingulate cortex; PFCx, dorsolateral prefrontal cortex.

The 3'–5' ratios can show significant variations among different regions of the brain. For example, in Table 2, the ratios for the dorsolateral prefrontal cortex are generally quite good. In the hippocampus, on the other hand, the β-actin ratio is similar to that shown for the prefrontal area, while the G3PDH ratio is much higher. For both of these regions, however, the percent of present calls are equivalently high, suggesting that the overall quality of the RNA is very good. For the anterior cingulate region, the ratios for G3PDH and β-actin are higher than those for the other two regions and the percent of present calls are considerably lower. Overall, these data suggest that the quality of the RNA in the cingulate region is inferior to that of the prefrontal area and hippocampus. What factors accounts for the differences in the 3′–5′ ratios is not clear; however, in practice, particularly when the 18S/28S ratio is also considered, the percent of present calls provide the most reliable index of RNA quality (Table 2).

There may be other differences among brain regions that must be considered when dealing with gene expression profiling data. For example, when RNA from identical cases is hybridized to microarrays and analyzed using the same platform (e.g. Affymetrix), it becomes apparent that it is not appropriate to extrapolate from one region to another. As shown in Fig. 4, when the GEP data from normal controls are plotted against similar data from bipolars and schizophrenics, the resulting scatterplots show a remarkably tight distribution. In contrast, when the same data are plotted on a regional basis (Fig. 5, the distribution of the findings for the anterior cingulate region and prefrontal cortex is relatively linear ($R^2 = 0.76$–0.89). On the other hand, those for either of these cortical areas relative to the hippocampus show a marked degree of spread that is distinctly curvilinear in nature ($R^2 = 0.20$–0.22). It is well established that the

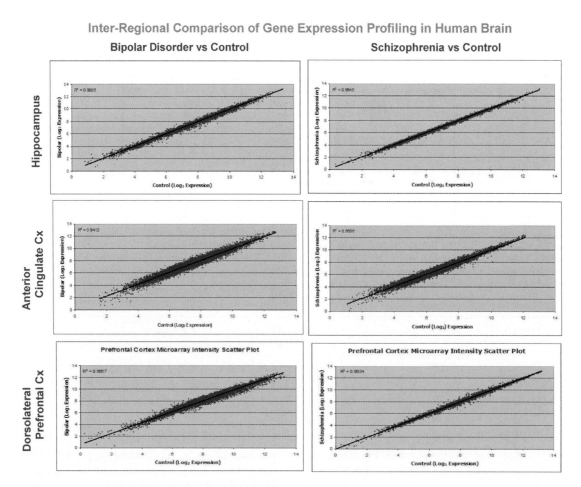

Fig. 4. Scatterplots comparing expression data for bipolar disorder versus normal controls (left) and schizophrenics versus normal controls (right) in the hippocampus, anterior cingulate region and dorsolateral prefrontal cortex. In all cases, the R^2 is equal to 0.94–0.99, suggesting a relatively tight distribution of expression data.

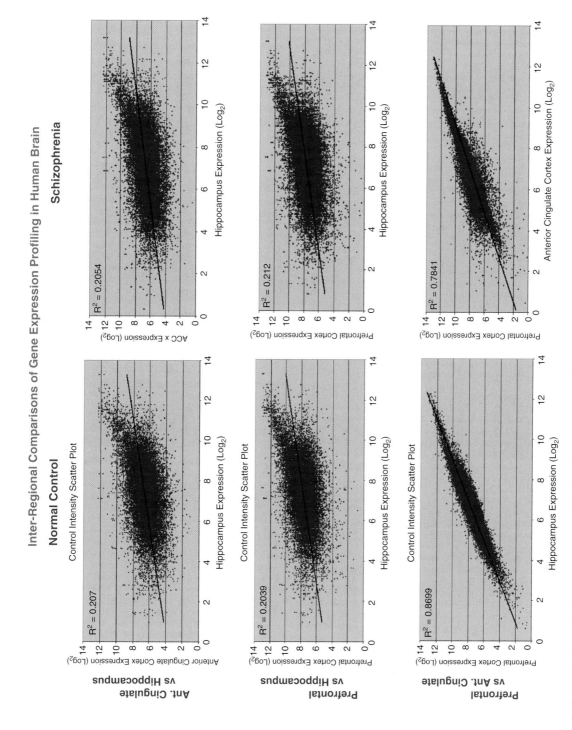

Fig. 5. A set of scatterplots of gene expression profiling data from normal controls (left) and schizophrenics (right) expressed on an inter-regional basis. The data for the prefrontal cortex versus the anterior cingulate region are relatively linear, while those for either of the latter two regions with respect to the hippocampus show a marked degree of scatter and a departure from linearity.

cytoarchitectural organization of the prefrontal and anterior cingulate cortices is quite similar, while that of the hippocampus is markedly different from these latter two regions. These differences reflect the fact that the hippocampus is the first cortical area to appear in late reptilian forms during vertebrate evolution. Overall, these findings underscore the importance of extrapolating the gene expression profiles of one brain region to that of another.

Analyzing gene expression profiling data in brain systems

Recent postmortem evidence has suggested that apoptotic cell death might be involved in the pathophysiology of SZ and/or BD (Margolis et al., 1994; Jarskog et al., 2000). For example, several cell counting studies from this laboratory had demonstrated a reduction of interneurons in the anterior cingulate cortex of schizophrenics, although the magnitude of this change is much stronger in affective disorder (Benes et al., 1991a, 2001; Todtenkopf et al., 2005). More recently, study using *in situ* end-labeling reported that there is a marked reduction of single-stranded DNA breaks in the anterior cingulate cortex of schizophrenics (Benes et al., 2003). This finding suggested that in schizophrenia there may either be a failure of apoptotic signaling pathways to proceed to DNA damage or alternatively, there might be an activation of a DNA "repair" enzyme (Benes et al., 2003). We postulated that genes associated with the apoptotic cascade might be down-regulated in SZ and up-regulated in BD. To explore the question of whether there is an abnormal regulation of apoptosis in SZ and BD, GEP has presented itself as a powerful screening tool, as it permits a broad evaluation at the level of cellular and molecular function. In an initial study in which this technology was employed and validated using qRT-PCR, genes associated with the electron transport chain were found to be down-regulated in BDs, but not in SZs (Konradi et al., 2004). Contrary to our working hypothesis, the apoptosis pathways did not show appreciable changes in the hippocampus of either group. As shown in Table 1, the quality-control parameters, including the 3′–5′ ratios for G3PDH and β-Actin, percent present calls, and the 28S/18S ratios, were all equivalent for the three groups and it seems unlikely that the changes in the electron transport genes in bipolars are attributable to differences in the quality of RNA among the groups.

It is important to emphasize that the human hippocampus is composed of many different subregions, sublaminae and cellular subtypes that make it relatively difficult to detect subtle changes in the expression of genes in whole extracts of this region. This is particularly true when very stringent approaches to microarray analysis are employed. To improve the sensitivity of GEP in these studies, we have employed a low stringency approach to the analysis (Benes et al., 2004) of a microarray database that was previously reported (Konradi et al., 2004). As shown in Fig. 6, there were striking differences in the distribution of apoptosis genes in SZs versus BDs. There are some genes that showed changes in the same direction in both groups. For example, an up-regulation of the pro-apoptotic factors RIP (receptor-interfering kinase), caspase 2 and the anti-apoptotic factors, MDM-2 and Bcl-x, were observed. In contrast, the pro-apoptotic factor BAX showed decreased expression in both the SZ and BD groups. The remainder of the apoptosis genes showed fundamental differences in regulation in the two disorders.

Assessing the significance of differences in functional biopathways/clusters of genes

It is becoming well recognized that low signal represented in the form of small fold changes (e.g. 1.05–1.25) represents a problem for GEP studies of the brain (Tsunoda et al., 2000; Asyali et al., 2004). Replication and validations of results using qRT-PCR are accepted strategies that can help to establish the reliability of gene expression changes (Raffelsberger et al., 2002). The presence of small fold changes becomes much more acute when complex biological systems, such as brain tissue, where any region-of-interest under study can contain several different neuronal and non-neuronal cell types defined by their anatomical detail, functional specialization, locations within complex

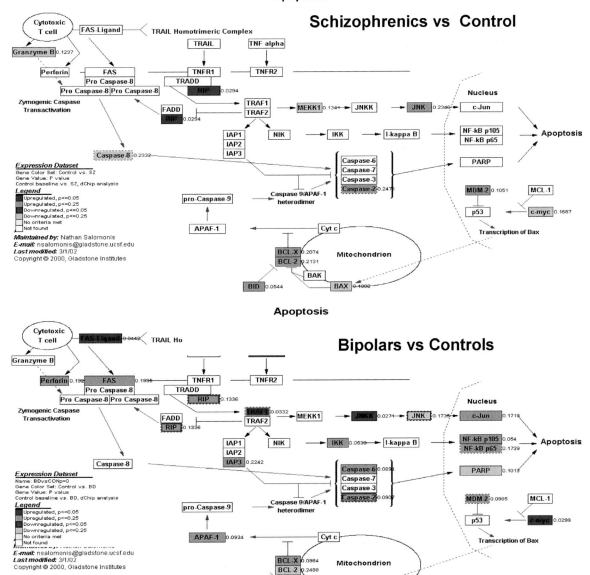

Fig. 6. A set of diagrams showing the apoptosis pathways schizophrenics versus normal controls (upper) and bipolar disorder versus normal controls (lower). The genes show either an up-regulation (red) or down-regulation (blue) and the intensity of the color varies according to whether the inclusion criterion for a particular gene was $p = 0.05$ ("high stringency") or $p = 0.25$ ("low stringency"). The p-values are shown to the right for each gene. Using low stringency criteria, the apoptosis pathways for the bipolar subjects show 24 out of 44 genes with changes in expression. Of these, 19 are up-regulated and only 5 are down-regulated. For the schizophrenics, eight genes were up-regulated and five were down-regulated. In both cases, pivotal genes at the beginning of the apoptosis casdade show changes in regulation. Overall, the bipolars show an increase of pro-apoptotic changes not seen in the schizophrenics.

circuits and preferential distributions in several different subregions. In this setting, small fold changes may belie the presence of large and functionally meaningful changes in gene expression in specific cell populations.

We have encountered this problem in our studies of the hippocampus in relation to psychotic disorders. This region in human brain typically has five different subregions and a total of nine different layers across these subregions (Rosene and Van Hoesen, 1987). In addition to these subregional and laminar specializations, there are many different types of cells present in the hippocampus (e.g. see Fig. 7). In addition to the principle projections neurons (i.e. pyramidal cells), there are at least four different types of local circuit neurons (interneurons), two different types of non-neuronal glial cell types and other non-neuronal cells associated with connective tissue and the vascular supply. This arrangement results in more than 45 different neuronal subtypes and countless other non-neuronal cell types dispersed across the complex hippocampal circuitry. This situation is made even more complex by the fact that these various cell types have many different size and numerical density distributions. For example, interneurons in the human hippocampus account for only 10% of the total number of neurons (Benes et al., 1998). In addition, interneurons are approximately half the size of these pyramidal cells (Benes et al., 1991b), making it likely that the overall representation of mRNA for interneurons in whole hippocampal extracts may be less than 5%, particularly if non-neuronal cells are also taken into account. Given this complexity, microarray-based GEP presents unique challenges when used as a broad screening tool, because it is likely that most of the genes associated with various subpopulations of neurons and non-neuronal cells will relatively small fold changes as a result of the dilution effect inherent in a complex neural system. This situation contrasts sharply with that seen in other areas of research where relatively homogeneous populations of cells, e.g. those in culture or those removed from the liver, are under investigation. In these latter cases, the fold-changes that may be detected may range from 2- to 10-fold (Raffelsberger et al., 2002).

Using a low stringency approach for analyzing GEP data in postmortem studies

For studies of the brain where small fold changes are inevitable, an important issue is how to test the significance of differences between two groups. The first method is to use an alpha level of significance $p = 0.05$. With this traditional statistical approach, it is likely that a relatively small number of genes will show significant changes in their level of expression when the fold changes are small. In fact, it is likely that the vast majority of genes will

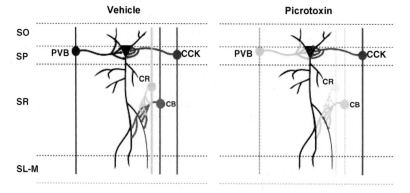

Fig. 7. Model showing changes in subtypes of GABA cells in rat HIPP following amygdalar activation. Note the presence of four different types of GABAergic interneuron identified by the presence of parvalbumin (PVB), calbindin (CB), calretinin (CR) and cholecystokin (CCK) in their cytoplasm. Each type forms synapses on different portions of the projection neuron and are found in different layers. Adapted from Berretta et al. (2004).

not be significant and that there may potentially be hundreds or even thousands of Type II statistical errors when an ANOVA is used as the statistical test. This traditional statistical approach also limits the amount of information that can be derived from a GEP study and hampers the ability of investigators to generate novel hypotheses. A second approach is one in which the p-value for significance is increased beyond $p = 0.05$. By doing so, increasing numbers of genes will be erroneously found to be significant when a simple ANOVA is used. Unlike the first example, the higher alpha levels will tend to cause unacceptably large numbers of Type I statistical errors.

An alternative approach, however, also employs a higher p-value, but in this case, it is not used as an estimate of significance, but rather as a criterion for inclusion of genes into a much larger post hoc analysis based on biological function. The cut-off for the inclusionary p-value is established empirically by examining a large number of functional pathways or clusters (like those found on the GenMapp website at www.genmapp.org). Within each functional pathway or cluster, there will be progressively larger numbers of genes "lighting up" as the inclusionary p-value is increased. An appropriate inclusionary p-value can be set at a point where genes begin to appear in many different pathways and/or clusters that originally showed only one or perhaps none. By increasing or decreasing the inclusionary p-value, non-specific background noise for post hoc analyses can also be proportionately increased or decreased. If the inclusionary p-value is set too low, the integrity of the analysis will become compromised; in much the same way that turning up the gain for detecting light rays will eventually compromise the signal-to-noise ratio for a spectrophotometric assay.

A critical issue that emerges from the above discussion is what statistic to use in order to test the significance of differences in GEP between the two groups. Obviously, Type I and II statistical errors are both undesirable and to be avoided at all costs. There are two statistics that can be potentially used for post hoc analyses when a low stringency inclusionary criterion is used. The first is the Fischer Exact Test, which essentially examines all possible permutations for associations between genes in pathways or clusters and generates a p-value of significance. We have found that when this statistic is used with large numbers of genes and pathways, the p-values of significance for any given pathway generally fall in the range of 10^{-2}–10^{-6}. This statistic is incorporated into the Ingenuity Pathways Analysis and it seems to be best suited for analyses that are directed toward establishing more generic functional associations.

One strategy is to individually consider the number of genes in each pathway or cluster that show changes in regulation by satisfying the inclusionary criterion (Benes et al., 2004a). If the pathway or cluster contains only a small number of genes, but a relatively large proportion of these "light up" (i.e. showing changes in increases or decreases of expression), then it may be likely that this set of changes may of biological significance. On the other hand, if the number of genes "lighting up" is small, but the pathway contains a large number of potential candidates, then it is less likely that this pathway or cluster may show changes that are functionally significant. This type of analysis can be operationalized by employing a variation of probability theory that is adapted to the unique problems presented by small fold changes in GEP studies of the brain. By multiplying all the p-values for each gene within a particular pathway included in the post hoc analysis, a cumulative probability can be obtained and used to assess the significance of the changes in the biopathway or cluster as a whole. The equation is as follows:

$$P_C = [(P_1 P_2 P_3 ... P_i)/(n_i/n_t)] N_p$$

where P_c is the composite probability, $P_1 ... P_i$ are the individual p-values for each gene in a given functional pathway or cluster, n_i the number of genes in each pathway meeting the inclusionary criterion for entry into the post hoc analysis, n_t the total number of genes in the respective pathways or clusters and N_p the total number of pathways included in the post hoc analysis.

This equation provides two separate corrections for multiple comparisons by multiplying by both N_p and n_t; this is comparable to the standard Bonferroni correction used in parametric statistics. The α level of significance for each GenMapp biopathway or cluster is established by examining the P_c for the various pathways or clusters included in the analysis and noting those in which there are only two or less genes that meet the criterion for inclusion in the analysis. In our experience, most GenMapp pathways have shown one or no genes meeting the stringency criteria and the P_c tended to be greater than 5×10^{-6} or lower. In our rat study (Benes et al., 2004a), some pathways showed three or more genes meeting the stringency criteria for inclusion and the resulting P_c tended to be less than 5×10^{-4}. In our human studies, many more genes showed changes in regulation and this caused the post hoc alpha-level to be much lower (i.e. 5×10^{-6}). In visually inspecting each of the pathways or clusters in the human hippocampus GEP data, it was noted that there were several in the SZs and BDs that showed only scattered numbers of included genes and, in these cases, the P_c generally exceeded the alpha-level of significance and were not considered to be statistically significant with respect to the normal control group.

It is worthwhile to compare the metric with the Fischer Exact Test. The Fischer Exact Test considers the entire database and begins with the null assumption that there are no associations among the genes in the database with any of the functional pathways represented in the Ingenuity software. It considers each and every possible interaction in relation to known functional pathways and eventually generates a p-value of significance. When performed in this way the Fischer Exact Test assigned a p-value of significance = 10^{-1} for the apoptosis signaling pathway and was not considered to show significance, even though 32 genes showing changes in regulation were found to be associated with this pathway. With the post hoc probability approach discussed above, the P_c was equal to 2.9×10^{-27} for apoptosis in the BDs. Very importantly, this analysis also makes it straightforward to directly examine the GenMapp apoptosis pathway to see which genes show changes, how robust those changes are and whether they are increased or decreased. This approach has great strength for investigators who are attempting to understand that gene regulation is changing within specific molecular pathways.

Taken together, this post hoc metric approach affords a method for analyzing gene expression profiling data with the low signal intensity that is inherent to complex neural systems. This analytic strategy is based on probability theory, one of the most fundamental mathematical constructs, and it views changes in gene regulation in relation to other functional parameters embodied within the functional pathways and clusters.

GEP studies in hippocampus of bipolar and schizophrenic subjects

Postmortem evidence, gathered over the past 15 years, has suggested that BD and, to a lesser extent, SZ may involve apoptotic cell death (Margolis et al., 1994; Jarskog et al., 2000). Several cell counting studies demonstrated a reduction of interneurons in the anterior cingulate cortex of schizophrenics, although this change has consistently shown a stronger covariation with affective disorder (Benes et al., 1991a; Benes et al., 2001; Todtenkopf et al., 2005). A more recent study using in situ end-labeling reported that there is a marked reduction of single-stranded DNA breaks in the anterior cingulate cortex of schizophrenics (Benes et al., 2003). This finding suggested that in schizophrenia there may either be a failure of apoptotic signaling pathways to proceed to DNA damage or alternatively, there might be an activation of a DNA "repair" enzyme (Benes et al., 2003). We postulated that genes associated with the apoptotic cascade might be down-regulated in SZ and up-regulated in BD. To explore the question of whether there is an abnormal regulation of apoptosis in SZ and BD, GEP has presented itself as a powerful screening tool, as it permits a broad evaluation at the level of cellular and molecular function. In an initial study in which this technology was employed and validated using qRT-PCR, genes associated with the electron transport chain were found to be down-regulated in BDs, but not in SZs (Konradi et al., 2004). On contrary to our working hypothesis, the apoptosis pathways did not show appreciable changes in the hippocampus of either group.

We have used GenMapp (www.genmapp.org) to evaluate changes in expression at the level of functional clusters and biopathways, rather than at the level of isolated genes. Using these strategies, we report here for the first time changes in the expression of genes associated with the apoptotic cascade in the hippocampus of BD and SZ subjects. For the BDs, 24 out of a total of 44 genes in the apoptosis pathways satisfied the low stringency criterion for inclusion in the analysis (post hoc composite $P_c = 2.9 \times 10^{-27}$). As depicted in Fig. 6, there were several up-regulated pro-apoptotic genes, including FAS ligand, FAS receptor (Shin et al., 1998; Hu et al., 2000) perforin (Ohara et al., 2003), TNFa (Thome et al., 1998; Hu et al., 2000), c-Jun (Mielke et al., 1999), c-myc (Alarcon-Vargas et al., 2002), BAK (Viktorsson et al., 2003), APAF-1 and caspases 2 (Ferrer et al., 2000) and 8 (Hu et al., 2000; Northington et al., 2001). Other genes that are thought to inhibit apoptosis, such as TRAF1, IKK, IAP3, NF-κβ (Micheau et al., 2001) and Bcl-2 (Adams and Cory, 1998) also showed increased expression in the BD group, but these changes would tend to counteract the influence of the 10 upregulated pro-apoptotic genes, particularly when other key pro-apoptotic factors, such as JNKK and JNK (Yang et al., 1997) were found to be down-regulated. The DNA repair enzyme, polyadenosine diphosphate-ribosyl polymerase PARP-1 (Bouchard et al., 2003), also showed a decrease in regulation.

When the BDs were broken down according to neuroleptic exposure (see Supplementary Materials), mRNA expression for the pro-apoptotic factors, FAS ligand, RIP, BID, TRAF1, FADD, MDM-2, caspase 2, p53 and c-myc, as well as the anti-apoptotic factors, NIK, IKK, IAP3, were all increased in the drug naïve BDs, while the pro-apoptotic factors JNKK, and JNK, as well as the anti-apoptotic factors, IAP2, NF-κβ-p105, and PARP all showed decreased expression. The neuroleptic-free BD subjects also showed a decreased expression of PARP. For the neuroleptic-treated BDs, pro-apoptotic factors, such as perforin, TNF-α, caspase 6, c-Jun, BAX, APAF-1 and caspase 2, all showed increased expression. Conversely, anti-apoptotic factors, such as IKK, NF-κβ-p105, NF-κβ-p65, MCL-1, Bcl-2 and Bcl-x, were all upregulated in the BDs receiving neuroleptic. As shown in Table 2, there was an overall increase of anti-apoptotic changes in gene expression in the BDs, suggesting that neuroleptics may suppress apoptotic cell death in this disorder. The potential effect of mood-stabilizers on changes in apoptosis gene expression was also considered; however, all of the BD subjects were actively treated with these agents at the time of death.

Numerous previous experiments suggest that increases in oxidative stress promote an environment in which the accumulation of free radicals potentiates apoptosis (Warner et al., 2004). To assess whether this potential mechanism may have contributed to cellular dysfunction, we identified a panel of 21 genes involved in the primary or secondary detoxification of reactive oxygen species (ROS) and are shown in Table 3. Genes showing changes in expression in BDs included glutathione peroxidase 4, glyoxylase, esterase D-formylglutathione hydrolase, glutathione synthetase, glutathione S-transferase (the 3, A2, M5 and omega isoforms), catalase and superoxide dismutase (SOD). Neuronal nitric oxide synthase (NOS1) was up-regulated.

In the SZ group, significant changes in the expression of genes associated with apoptosis were also observed (post hoc $P_c = 4.3 \times 10^{-9}$), but there were differences with respect to the specific genes affected and the direction of the changes (refer to Fig. 3). Four pro-apoptotic genes (RIP, BID, JNK and caspase 2), and two pro-survival genes (Bcl-x and Bcl-2) showed an overall increase of expression. Other pro-apoptotic genes, such as granzyme B (Ohara et al., 2003; Sun et al., 2004), caspase 8 (Wang et al., 1999; Northington et al., 2001), MEKK1 (Alarcon-Vargas et al., 2002; Boldt et al., 2003) and c-myc (Alarcon-Vargas et al., 2002), showed decreased expression in the SZ group). These latter changes would tend to suppress the apoptotic potential of hippocampal cells, as these four factors are believed to play a critical role in facilitating the progression of apoptosis. When the SZ group was broken down according to "low" (CPZ-equivalent dose<500 mg per day; average $= 189 \pm 173$ mg per day) and "high" (CPZ-equivalent dose>500 mg per day; average $= 816 \pm 290$ mg per day) dose neuroleptic exposure (Table 2 and Supplementary Materials), the pro-apoptotic factors, FAS ligand, perforin,

Table 3. Comparison of expression profiling results for genes associated with anti-oxidant reactions in normal controls versus bipolars and schizophrenics

Bipolar disorder			
Lactoperoxidase	U39573	1.09	0.064
Heme oxygenease 1	Z82244	1.14	0.131
Superoxide dismutase 1	X02317	−1.21	0.081
Catalase	AL035079	−1.17	0.124
Glutathione peroxidase 2 (gastrointestinal)	X53463	1.11	0.041
Glutathione peroxidase 4 (phospholipid hydroperoxidase)	X71973	−1.23	0.014
Glutathione peroxidase 1	X13710	−1.14	0.134
Glyoxalase 1	NM006708	−1.32	0.014
Microsomal glutathione S-transferase 3	AF026977	−1.39	0.002
Hydroxyacyl glutathione hydrolase	X90999	−1.36	0.028
Glutathione S-transferase A4	AF025887	−1.24	0.046
Glutathione S-transferase M3 (brain)	AF043105	−1.19	0.198
Esterase D/formylglutathione hydrolase	AF112219	−1.31	0.044
Glutathione synthetase	U34683	−1.23	0.015
Glutathione S-transferase A2	M16594	1.16	0.045
Glutathione S-transferase M5	L02321	1.14	0.032
Glutathione S-transferase like glutathione transferase omega	U90313	−1.25	0.021
C-terminal PDZ domain ligand of neuronal nitric oxide synthase	AB007933	−1.12	0.090
Human neuronal nitric oxide synthase (NOS1) gene, exon 29	U17326	1.17	0.030
Human inducible nitric oxide synthase gene, promoter and exon 1	D29675	1.13	0.083
Nitric oxide synthase 3 (endothelial cell)	M93718	1.23	0.059
Schizophrenia			
Glutathione synthetase	U34683	−1.18	0.050

TRAIL, caspase 8, MEKK1, p53, c-myc, BAX and BAK, showed decreased expression in SZs receiving low-dose neuroleptic. Some pro-survival genes, such as TRAF1 and MDM-2, showed decreased expression, whereas others, such as BCL-2, showed increased expression. In the "high" dose sub-group, the pattern observed was quite different. Several pro-apoptotic genes, including perforin, TRAIL, RIP, TNFa, caspase 2 and BAK, were up-regulated, whereas anti-apoptotic genes, such as IAP3, MCL-1 and PARP, were down-regulated. Unlike the subjects with BD, the subjects with SZ showed no difference in the number of genes showing pro-apoptotic changes in expression. On the other hand, the number of genes showing anti-apoptotic changes in expression was markedly reduced, suggesting that neuroleptics might have some ability to increase apoptotic potential in hippocampal cells of SZs. The DNA repair enzyme, PARP, showed a decrease of expression in the SZs treated with high-dose neuroleptic and this change could potentially increase the amount of DNA fragmentation present in these subjects (Benes et al., 2003). This observation further supports the view that the decrease of DNA damage in SZ is probably not due to a neuroleptic effect (Benes et al., 2003).

As shown in Fig. 3, the microarray data for several apoptotic genes, including FAS ligand, granzyme B, c-myc, PARP, BAK, Bcl-2 and APAF-1, were validated using qRT-PCR. In each case, the direction of change in expression for the CON, SZ and BD groups were consistent with that seen using the microarray approach, except that the magnitude of these differences were generally much greater. For example, in the case of c-myc, the fold changes were − 5.2 and + 5.4, respectively, in the SZs and BDs. Although there was no change in PARP expression in the SZs when neuroleptic effects were not considered, the BDs showed a − 8.2-fold decrease when compared to either the SZ or CON groups. Overall, the magnitude of the differences between the groups that were detected with qRT-PCR were much larger than those obtained with the microarrays and provided an important validation of the microarray results.

This study reports the results of a novel post hoc analysis of an extant microarray database, together with GenMapp biopathways and clusters, to obtain a more inclusive understanding of how complex aspects of transduction, signaling and metabolism may be altered in SZ and BD. Indeed, with this new methodology, it has been possible to detect marked changes in the regulation of genes associated with the apoptosis cascade and would not have been detected, given the relative insensitivity of the standard approaches that employ an alpha level of $p = 0.05$ (Benes et al., 2004). The more sensitive analysis described above has revealed robust changes in the expression of apoptosis genes in hippocampal cells in both SZ and BD subjects. Although there is some overlap in the genes showing differences in expression when compared to normal controls, the preponderance of such changes has been found to be remarkably different in SZs when compared to BDs. The hypothesis that apoptosis may play a role in the pathophysiology of SZ and BD can be traced to earlier microscopic studies in the anterior cingulate cortex (Benes et al., 1991b, 2001) and hippocampus (Benes et al., 1998), which suggested a loss of interneurons occurs in both disorders. These changes were found to be far more striking in BDs (30–35% reduction) when compared to SZs (12–15% reduction) and it was postulated that there may be a more marked activation of apoptosis in affective disorder than in schizophrenia. A subsequent study demonstrated that there was a paradoxical reduction in the amount of DNA damage present in the anterior cingulate cortex of schizophrenic subjects (Benes et al., 2003). Indeed, the findings reported here are consistent with this latter observation, as there was a down-regulation of several pro-apoptotic genes, such as granzyme B, caspase 8, c-myc and BAX. Based on the results reported here, it seems more likely that GABA cell pathology in SZ (Benes and Berretta, 2001) may by related to an disturbance of intracellular signaling pathways, rather than to overt cell loss, as appears to be the case in BD (Woo et al., 2004).

A noteworthy aspect of these findings is the observation that neuroleptic exposure was associated with a decrease of apoptotic potential in BDs, but an increase in SZs. Both typical (Achour et al., 2003) and atypical (Wei et al., 2003) antipsychotic drugs have been found to promote cell survival, although the atypical agents may be more effective in this regard (Qing et al., 2003). Indeed, several atypical antipsychotic drugs appear to protect against DNA damage (Qing et al., 2003). To date, only one neuroleptic, the typical agent perphenazine, has been associated with increased DNA fragmentation (Gil-ad et al., 2001), although one study has reported that clozapine may act as a hapten and increase inflammatory potential (Haack et al., 2003). Contrary to the latter report, clozapine has also been found to activate Akt (Kang et al., 2004), a pro-survival factor that, in turn, inhibits glycogen synthase kinase-3b (GSK-3b), a protein that drives intracellular signaling toward cell death via the Wnt pathway. Lithium carbonate, a standard mood-stabilizing agent, is also believed to inhibit GSK-3b (Li et al., 2002). Both lithium and valproate have also been found to be associated with increased expression of Bcl-2 (Manji et al., 2000a, b) and ultimately influence both signal transduction (Manji et al., 1993; Chen et al., 2000; Li et al., 2002) and intracellular signaling cascades that are fundamental to cell survival. Accordingly, the up-regulation of anti-apoptotic genes in bipolar subjects is not explainable by exposure of the subjects to neuroleptic drugs and/or mood-stabilizing agents.

Overall, the data reported in this study are consistent with the working hypothesis that apoptosis may contribute to cell dysfunction in BD (Benes et al., 2003). As depicted in Fig. 8, additional clusters of genes that play a central role in the clearance of free radicals generated by mitochondrial oxidation reactions, such as glutathione synthase, catalase and SOD (Warner et al., 2004), also showed substantial decreases in expression in bipolars, but not schizophrenics. An up-regulation of NOS1 could potentially contribute to apoptosis through increased excitotoxicity and or the generation of peroxynitrite radicals, which would increase due to diminished clearance by SOD as suggested by the microarray data. This suggests that the accumulation of ROS associated with the oxidative stress would tend to potentiate damage to DNA, proteins and lipids (Pollack and Leeuwenburgh,

Schizophrenia

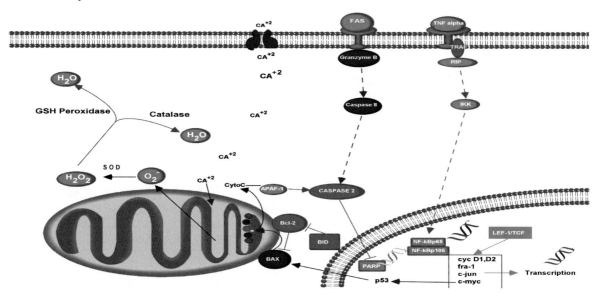

Bipolar Disorder

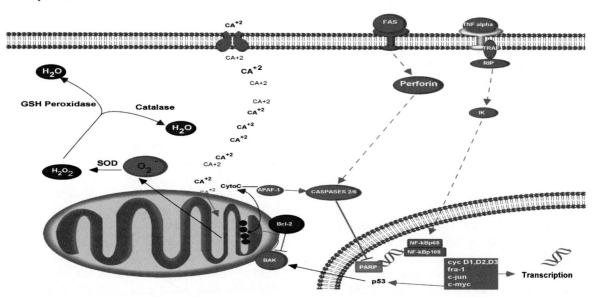

Fig. 8. Schematic diagrams depicting genes associated with the apoptosis cascade that show changes in expression and schizophrenia (upper) and bipolar disorder (lower). In schizophrenia, as postulated in an earlier study, (Benes et al., 2003) an overall down-regulation of the apoptosis cascade, including granzyme B and caspase 8, in the cell periphery and the mitochondrial-associated anti-apoptotic factors, BAX and c-myc, was observed. In bipolars, pro-apoptotic factors, including two death ligands, FAS and TNF—, as well as the FAS receptor, perforin, c-myc and BAK were up-regulated. NF-κβp65 and p105 were also up-regulated, but the latter may play a role in stimulating the activity of the DNA repair enzyme, PARP-1, that was found to be down-regulated. Anti-oxidant genes, such as SOD, catalase and GSH peroxidase, were significantly down-regulated in BDs suggesting that reactive oxygen, such as O_2^-, that could be generated by the decreased mitochondrial electron transport activity previously reported (Konradi et al., 2004) might also contribute to DNA damage in BDs. Preliminary evidence is also suggesting that there is increased expression of L-type voltage-gated calcium channel activity (LVGCC) in BDs and this would exacerbate this situation. In SZs, anti-oxidant genes showed little change in regulation and LVGCC activity is probably also down-regulated. Overall, these changes in bipolars would promote apoptotic injury or death, while those in schizophrenics would promote cell survival.

2001). In schizophrenia, however, a down-regulation of the apoptotic cascade may represent either an adaptive compensation to oxidative stress or a non-adaptive dysregulation of this pathway (Benes et al., 2003). The absence of regulatory changes in the anti-oxidant pathways is consistent with this pattern. Additionally, we also have preliminary evidence suggesting that an L-type voltage-gated calcium channel is up-regulated in bipolars and down-regulated in schizophrenics. As shown in Fig. 8, these latter changes would tend to increase and decrease, respectively, the apoptotic response of hippocampal cells (Tanaka et al., 2004; Yagami et al., 2004) in BDs versus SZs. It is important to emphasize that the interpretation of simultaneous changes in the expression of many genes that comprise complex signaling and metabolic pathway, such as apoptosis, is probably not straightforward, as it is likely that such changes may not be additive in nature. For example, it is theoretically possible that the activity of a single anti-apoptotic gene could exert a stronger inhibitory effect than the summed influence of multiple pro-apoptotic factors. Given this caveat, however, the detection of differences in the regulation of apoptosis-related genes in BD and SZ may potentially lead to the development of specific and rational therapies directed at unique aspects of neuronal cell dysfunction in each disorder.

Conclusions

The above discussion illustrates the power of GEP for the study of cellular and molecular mechanisms in neuropsychiatric diseases, such as SZ and BD. It is important to emphasize, however, that the need for quality control at every stage of this work is imperative. The development of methods for evaluating RNA quality and validating the transcriptome is critical for studies of this type. Additionally, it is critical to take into account the inherent complexity of neural systems, such as the hippocampus, and the limitations that this places on the sensitivity of GEP for detecting changes when they may occur in particular subtypes of neuron defined by their respective subregional and laminar distributions. One strategy that has been recently applied employs a post hoc metric that provides a means for interrogating GEP databases on the basis account of functionally defined clusters of genes. While this strategy does make it possible to detect differences in a psychiatric disorder, the use of RNA extracts from homogenates of the whole hippocampus makes it impossible to know which neuronal or non-neuronal types might be responsible for the changes found. This limitation will be overcome by the use of laser capture microdissection (LCM) which makes it possible to remove distinct subpopulations of neurons from particular subregions and layers. While this technology is very time-consuming and expensive to use, it will improve the inherent sensitivity of GEP studies. Indeed, preliminary findings from this laboratory are already demonstrating this, as the fold changes and p-values for individual genes are 1–2 orders of magnitude greater than those obtained from whole extracts.

The importance of microarray-based GEP lies with its ability to provide information from 20,000 or more genes simultaneously. When the data analysis is combined with other strategies that discriminate genes that show unique changes in expression in SZ, but not BD and vice versa, it is possible to begin the arduous process of identifying genes associated with the endophenotype for each of these disorders. Such information will eventually lead to the unraveling of why some individuals show a susceptibility to SZ and others for BD.

References

Achour, A., Lu, W., Arlie, M., Cao, L. and Andrieu, J.M. (2003) T cell survival/proliferation reconstitution by trifluoperazine in human immunodeficiency virus-1 infection. Virology, 315: 245–258.

Adams, J.M. and Cory, S. (1998) The Bcl-2 protein family: arbiters of cell survival. Science, 281: 1322–1326.

Alarcon-Vargas, D., Tansey, W.P. and Ronai, Z. (2002) Regulation of c-myc stability by selective stress conditions and by MEKK1 requires AA 127–189 of c-myc. Oncogene, 21: 4384–4391.

Antoniadis, E.A. and McDonald, R.J. (2000) Amygdala, hippocampus and discriminative fear conditioning to context. Behav. Brain Res., 108: 1–19.

Arnold, S.E., 2000. Hippocampal Pathology. Oxford University Press, Oxford, UK, pp. 57–80.

Asyali, M.H., Shoukri, M.M., Demirkaya, O. and Khabar, K.S.A. (2004) Assessment of reliability of microarray data and estimation of signal thresholds using mixture modeling. Nucleic Acids Res., 32: 2323–2335.

Benes, F.M. (2000) Emerging principles of altered neural circuitry in schizophrenia. Brain Res. Brain Res. Rev., 31: 251–269.

Benes, F.M. and Berretta, S. (2000) Amygdalo-entorhinal inputs to the hippocampal formation in relation to schizophrenia. Ann. NY Acad. Sci., 911: 293–304.

Benes, F.M. and Berretta, S. (2001) GABAergic interneurons: implications for understanding schizophrenia and bipolar disorder. PG — 1–27. Neuropsychopharmacology, 25: 1–27.

Benes, F.M., Burke, R.E., Walsh, J., Matzilevich, D., Berretta, S., Minns, M. and Konradi, C. (2004a) Acute amygdalar activation induces an upregulation of multiple monoamine G protein coupled pathways in rat hippocampus. Mol. Psychiatry, 9: 932–945.

Benes, F.M., Burke, R.E., Walsh, J., Berretta, S., Matzilevich, D., Minns, M. and Konradi, C. (2004) Acute amygdalar activation induces an upregulation of multiple monoamine G protein coupled pathways in rat hippocampus. Mol. Psychiatry, 9: 932–945.

Benes, F.M., Khan, Y., Vincent, S.L. and Wickramasinghe, R. (1996a) Differences in the subregional and cellular distribution of GABAA receptor binding in the hippocampal formation of schizophrenic brain. Synapse, 22: 338–349.

Benes, F.M., Kwok, E.W., Vincent, S.L. and Todtenkopf, M.S. (1998) A reduction of nonpyramidal cells in sector CA2 of schizophrenics and manic depressives [see comments]. Biol. Psychiatry, 44: 88–97.

Benes, F.M., McSparren, J., Bird, E.D., SanGiovanni, J.P. and Vincent, S.L. (1991a) Deficits in small interneurons in prefrontal and cingulate cortices of schizophrenic and schizoaffective patients. Arch. Gen. Psychiatry, 48: 996–1001.

Benes, F.M., Sorensen, I. and Bird, E.D. (1991b) Reduced neuronal size in posterior hippocampus of schizophrenic patients. Schizophr. Bull., 17: 597–608.

Benes, F.M., Sorensen, I., Vincent, S.L., Bird, E.D. and Sathi, M. (1992a) Increased density of glutamate-immunoreactive vertical processes in superficial laminae in cingulate cortex of schizophrenic brain. Cereb. Cortex, 2: 503–512.

Benes, F.M., Todtenkopf, M.S., Logiotatos, P. and Williams, M. (2000) Glutamate decarboxylase(65)-immunoreactive terminals in cingulate and prefrontal cortices of schizophrenic and bipolar brain. J. Chem. Neuroanat., 20: 259–269.

Benes, F.M., Vincent, S.L., Alsterberg, G., Bird, E.D. and SanGiovanni, J.P. (1992b) Increased GABAA receptor binding in superficial layers of cingulate cortex in schizophrenics. J. Neurosci., 12: 924–929.

Benes, F.M., Vincent, S.L., Marie, A. and Khan, Y. (1996b) Up-regulation of GABAA receptor binding on neurons of the prefrontal cortex in schizophrenic subjects. Neuroscience, 75: 1021–1031.

Benes, F.M., Vincent, S.L. and Todtenkopf, M. (2001) The density of pyramidal and nonpyramidal neurons in anterior cingulate cortex of schizophrenic and bipolar subjects. Biol. Psychiatry, 50: 395–406.

Benes, F.M., Walsh, J., Bhattacharyya, S., Sheth, A. and Berretta, S. (2003) DNA fragmentation decreased in schizophrenia but not bipolar disorder. Arch. Gen. Psychiatry, 60: 359–364.

Benes, F.M., Wickramasinghe, R., Vincent, S.L., Khan, Y. and Todtenkopf, M. (1997) Uncoupling of GABA-A and benzodiazepine receptor binding activity in the hippocampal formation of schizophrenic brain. Brain Res., 755: 121–129.

Benes, F.M., Matzilevich, D., Burke, R.E. and Walsh, J. (2006) The expression of proapoptosis genes is increased in bipolar disorder, but not in schizophrenia. Mol. Psychiatry, 11: 241–251.

Berretta, S., Lange, N., Bhattacharyya, S., Sebro, R., Garces, J. and Benes, F.M. (2004) Long-term effects of amygdala GABA receptor blockade on specific subpopulations of hippocampal interneurons. Hippocampus, 14: 876.

Berretta, S., Munno, D.W. and Benes, F.M. (2001) Amygdalar activation alters the hippocampal GABA system: "partial" modelling for postmortem changes in schizophrenia. J. Comp. Neurol., 431: 129–138.

Boldt, S., Weidle, U.H. and Kolch, W. (2003) The kinase domain of MEKK1 induces apoptosis by dysregulation of MAP kinase pathways. Exp. Cell Res., 283: 80–90.

Bouchard, V.J., Rouleau, M. and Poirier, G.G. (2003) PARP-1, a determinant of cell survival in response to DNA damage. Exp. Hematol., 31: 446–454.

Chen, G., Masana, M.I. and Manji, H.K. (2000) Lithium regulates PKC-mediated intracellular cross-talk and gene expression in the CNS in vivo. Bipolar Disord., 2: 217–236.

Davis, M. and Shi, C. (2000) The amygdala. Curr. Biol., 10: R131.

Ferrer, I., Lopez, E., Blanco, R., Rivera, R., Krupinski, J. and Marti, E. (2000) Differential c-Fos and caspase expression following kainic acid excitotoxicity. Acta Neuropathol. (Berl.), 99: 245–256.

Gil-ad, I., Shtaif, B., Shiloh, R. and Weizman, A. (2001) Evaluation of the neurotoxic activity of typical and atypical neuroleptics: relevance to iatrogenic extrapyramidal symptoms. Cell Mol. Neurobiol., 21: 705–716.

Guidotti, A., Auta, J., Davis, J.M., Gerevini, V.D., Dwivedi, Y., Grayson, D.R., Impagnatiello, F., Pandey, G., Pesold, C., Sharma, R., Uzunov, D. and Costa, E. (2000) Decrease in reelin and glutamic acid decarboxylase67 (GAD67) expression in schizophrenia and bipolar disorder: a postmortem brain study. Arch. Gen. Psychiatry, 57: 1061–1069.

Haack, M.J., Bak, M.L., Beurskens, R., Maes, M., Stolk, L.M. and Delespaul, P.A. (2003) Toxic rise of clozapine plasma concentrations in relation to inflammation. Eur. Neuropsychopharmacol., 13: 381–385.

Harrison, P.J. (1999) The neuropathology of schizophrenia. A critical review of the data and their interpretation. Brain, 122: 593–624.

Heckers, S., Rausch, S.L., Goff, D., Savage, C.R., Schacter, D.L., Fischman, A.J. and Alpert, N.M. (1998) Impaired recruitment of the hippocampus during conscious recollection in schizophrenia. Nat. Neurosci., 1: 318–323.

Heckers, S., Stone, D., Walsh, J., Shick, J., Koul, P. and Benes, F.M. (2002) Differential hippocampal expression of glutamic acid decarboxylase 65 and 67 messenger RNA in bipolar disorder and schizophrenia. Arch. Gen. Psychiatry, 59: 521–529.

Hu, W.H., Johnson, H. and Shu, H.B. (2000) Activation of NF-kappaB by FADD, Casper, and caspase-8. J. Biol. Chem., 275: 10838–10844.

Jarskog, L.F., Gilmore, J.H., Selinger, E.S. and Lieberman, J.A. (2000) Cortical bcl-2 protein expression and apoptotic regulation in schizophrenia. Biol. Psychiatry, 48: 641–650.

Kang, U.G., Seo, M.S., Roh, M.S., Kim, Y., Yoon, S.C. and Kim, Y.S. (2004) The effects of clozapine on the GSK-3-mediated signaling pathway. FEBS Lett., 560: 115–119.

Konradi, C., Eaton, M., Walsh, J., Benes, F.M. and Heckers, S. (2004) Molecular evidence for mitochondrial dysfunction in bipolar disorder. Arch. Gen. Psychiatry, 61: 300–308.

Kraepelin, E., 1919. Dementia Praecox and Paraphrenia. E. and S. Livingstone, Edinburg.

Lawrie, S.M. and Abukmeil, S.S. (1998) Brain abnormality in schizophrenia. A systematic and quantitative review of volumetric magnetic resonance imaging studies. Br. J. Psychiatry, 172: 110–120.

Ledoux, J. (2000) The amygdala and emotion. In: Aggleton, J.P. (Ed.), The Amygdala. Oxford University Press, Oxford, pp. 289–310.

LeDoux, J.E. (2000) Emotion circuits in the brain [In Process Citation]. Annu. Rev. Neurosci., 23: 155–184.

Li, X., Bijur, G.N. and Jope, R.S. (2002) Glycogen synthase kinase-3beta, mood stabilizers, and neuroprotection. Bipolar Disord., 4: 137–144.

Longson, D., Deakin, J.F. and Benes, F.M. (1996) Increased density of entorhinal glutamate-immunoreactive vertical fibers in schizophrenia. J. Neural. Transm., 103: 503–507.

Manji, H.K., Etcheberrigaray, R., Chen, G. and Olds, J.L. (1993) Lithium decreases membrane-associated protein kinase C in hippocampus: selectivity for the alpha isozyme. J. Neurochem., 61: 2303–2310.

Manji, H.K., Moore, G.J. and Chen, G. (2000a) Clinical and preclinical evidence for the neurotrophic effects of mood stabilizers: implications for the pathophysiology and treatment of manic-depressive illness. Biol. Psychiatry, 48: 740–754.

Manji, H.K., Moore, G.J., Rajkowska, G. and Chen, G. (2000b) Neuroplasticity and cellular resilience in mood disorders. Mol. Psychiatry, 5: 578–593.

Margolis, R.L., Chuang, D.M. and Post, R.M. (1994) Programmed cell death: implications for neuropsychiatric disorders. Biol. Psychiatry, 35: 946–956.

Micheau, O., Lens, S., Gaide, O., Alevizopoulos, K. and Tschopp, J. (2001) NF-kappaB signals induce the expression of c-FLIP. Mol. Cell Biol., 21: 5299–5305.

Mielke, K., Brecht, S., Dorst, A. and Herdegen, T. (1999) Activity and expression of JNK1, p38 and ERK kinases, c-Jun N-terminal phosphorylation, and c-jun promoter binding in the adult rat brain following kainate-induced seizures. Neuroscience, 91: 471–483.

Northington, F.J., Ferriero, D.M. and Martin, L.J. (2001) Neurodegeneration in the thalamus following neonatal hypoxia-ischemia is programmed cell death. Dev. Neurosci., 23: 186–191.

Ohara, T., Morishita, T., Suzuki, H., Masaoka, T. and Ishii, H. (2003) Perforin and granzyme B of cytotoxic T lymphocyte mediate apoptosis irrespective of *Helicobacter pylori* infection: possible act as a trigger of peptic ulcer formation. Hepatogastroenterology, 50: 1774–1779.

Pitkanen, A., Pikkarainen, M., Nurminen, N. and Ylinen, A. (2000) Reciprocal connections between the amygdala and the hippocampal formation, perirhinal cortex, and postrhinal cortex in rat. A review [In Process Citation]. Ann. NY Acad. Sci., 911: 369–391.

Pollack, M. and Leeuwenburgh, C. (2001) Apoptosis and aging: role of the mitochondria. J. Gerontol. A Biol. Sci. Med. Sci., 56: B475–B482.

Qing, H., Xu, H., Wei, Z., Gibson, K. and Li, X.M. (2003) The ability of atypical antipsychotic drugs vs. haloperidol to protect PC12 cells against MPP+–induced apoptosis. Eur. J. Neurosci., 17: 1563–1570.

Raffelsberger, W., Dembele, D., Neubauer, M.G., Gottardis, M.M. and Gronemeyer, H. (2002) Quality indicators increase the reliability of microarray data. Genomics, 80: 385–394.

Rosene, D.L. and Van Hoesen, G.W. (1987) The hippocompal formation of the primate brain. In: Peters, A. and Jones, E.G. (Eds.), Cerebral Cortex. Plenum Press, New York, pp. 345–456.

Shin, S.W., Park, J.W., Suh, M.H., Suh, S.I. and Choe, B.K. (1998) Persistent expression of Fas/FasL mRNA in the mouse hippocampus after a single NMDA injection. J. Neurochem., 71: 1773–1776.

Simpson, M.D., Slater, P., Deakin, J.F., Royston, M.C. and Skan, W.J. (1989) Reduced GABA uptake sites in the temporal lobe in schizophrenia. Neurosci. Lett., 107: 211–215.

Sun, J., Bird, C.H., Thia, K.Y., Matthews, A.Y., Trapani, J.A. and Bird, P.I. (2004) Granzyme B encoded by the commonly-occurring human RAH allele retains pro-apoptotic activity. J. Biol. Chem., 279: 16907–16911.

Tamminga, C.A., Thaker, G.K., Buchanan, R., Kirkpatrick, B., Alphs, L.D., Chase, T.N. and Carpenter, W.T. (1992) Limbic system abnormalities identified in schizophrenia using positron emission tomography with fluorodeoxyglucose and neocortical alterations with deficit syndrome. Arch. Gen. Psychiatry, 49: 522–530.

Tanaka, T., Nangaku, M., Miyata, T., Inagi, R., Ohse, T., Ingelfinger, J.R. and Fujita, T. (2004) Blockade of calcium influx through L-type calcium channels attenuates mitochondrial injury and apoptosis in hypoxic renal tubular cells. J. Am. Soc. Nephrol., 15: 2320–2333.

Thome, M., Hofmann, K., Burns, K., Martinon, F., Bodmer, J.L., Mattmann, C. and Tschopp, J. (1998) Identification of CARDIAK, a RIP-like kinase that associates with caspase-1. Curr. Biol., 8: 885–888.

Todtenkopf, M.S. and Benes, F.M. (1998) Distribution of glutamate decarboxylase65 immunoreactive puncta on pyramidal and nonpyramidal neurons in hippocampus of schizophrenic brain. Synapse, 29: 323–332.

Todtenkopf, M.S., Vincent, S.L. and Benes, F.M. (2005) A cross-study meta-analysis and three-dimensional comparison of cell counting in the anterior cingulate cortex of schizophrenic and bipolar brain. Schizophr. Res., 73: 79–89.

Tsunoda, T., Yamada, R., Tanaka, T., Ohnishi, Y. and Kamatani, N. (2000) Environmental factor dependent maximum likelihood method for association study targeted to personalized medicine. Genome Informatics, 11: 96–105.

Viktorsson, K., Ekedahl, J., Lindebro, M.C., Lewensohn, R., Zhivotovsky, B., Linder, S. and Shoshan, M.C. (2003) Defective stress kinase and Bak activation in response to ionizing radiation but not cisplatin in a non-small cell lung carcinoma cell line. Exp. Cell Res., 289: 256–264.

Volk, D.W., Austin, M.C., Pierri, J.N., Sampson, A.R. and Lewis, D.A. (2000) Decreased glutamic acid decarboxylase67 messenger RNA expression in a subset of prefrontal cortical gamma-aminobutyric acid neurons in subjects with schizophrenia. Arch. Gen. Psychiatry, 57: 237–245.

Wang, C.Y., Guttridge, D.C., Mayo, M.W. and Baldwin Jr., A.S. (1999) NF-kappaB induces expression of the Bcl-2 homologue A1/Bfl-1 to preferentially suppress chemotherapy-induced apoptosis. Mol. Cell. Biol., 19: 5923–5929.

Warner, D.S., Sheng, H. and Batinic-Haberle, I. (2004) Oxidants, antioxidants and the ischemic brain. J. Exp. Biol., 207: 3221–3231.

Wei, Z., Bai, O., Richardson, J.S., Mousseau, D.D. and Li, X.M. (2003) Olanzapine protects PC12 cells from oxidative stress induced by hydrogen peroxide. J. Neurosci. Res., 73: 364–368.

Woo, T.U., Walsh, J.P. and Benes, F.M. (2004) Density of glutamic acid decarboxylase 67 messenger RNA-containing neurons that express the N-methyl-D-aspartate receptor subunit NR2A in the anterior cingulate cortex in schizophrenia and bipolar disorder. Arch. Gen. Psychiatry, 61: 649–657.

Yagami, T., Ueda, K., Sakaeda, T., Itoh, N., Sakaguchi, G., Okamura, N., Hori, Y. and Fujimoto, M. (2004) Protective effects of a selective L-type voltage-sensitive calcium channel blocker, S-312-d, on neuronal cell death. Biochem. Pharmacol., 67: 1153–1165.

Yang, D.D., Kuan, C.Y., Whitmarsh, A.J., Rincon, M., Zheng, T.S., Davis, R.J., Rakic, P. and Flavell, R.A. (1997) Absence of excitotoxicity-induced apoptosis in the hippocampus of mice lacking the Jnk3 gene. Nature, 389: 865–870.

CHAPTER 9

Assessment of genome and proteome profiles in cocaine abuse

Scott E. Hemby*

Department of Physiology and Pharmacology, Wake Forest University School of Medicine, Winston-Salem, NC 27157, USA

Abstract: Until recently, knowledge of the impact of abuse drugs on gene and protein expression in the brain was limited to less than 100 targets. With the advent of high-throughput genomic and proteomic techniques investigators are now able to evaluate changes across the entire genome and across thousands of proteins in defined brain regions and generate expression profiles of vulnerable neuroanatomical substrates in rodent and non-human primate drug abuse models and in human post-mortem brain tissue from drug abuse victims. The availability of gene and protein expression profiles will continue to expand our understanding of the short- and long-term consequences of drug addiction and other addictive disorders and may provide new approaches or new targets for pharmacotherapeutic intervention. This chapter will review gene expression data from rodent, non-human primate and human post-mortem studies of cocaine abuse and will provide a preliminary proteomic profile of human cocaine abuse and explore how these studies have advanced our understanding of addiction.

Keywords: microarray; RNA amplification; gene expression; molecular fingerprint; qPCR; transcriptome; proteome; brain; post-mortem; monkey

Introduction

The efforts to complete sequencing of the human genome have enabled new endeavors into the function of these genes in human disorders and have provided a wealth of knowledge about the molecular underpinnings of behavior. The next challenge in addiction biology is the utilization of this information to determine the function of the genes and proteins in the context of human disease. The advent of high-throughput screening technologies has produced a paradigm shift in the manner in which scientists are able to detect and identify molecular mechanisms related to disease. Microarray and proteomic analysis strategies allow the simultaneous assessment of thousands of genes and proteins of known and unknown function — thereby enabling a global biological view of addictive disorders. Broad-scale evaluations of gene and protein expression are well suited to the study of drug abuse, particularly in light of the complexity of the brain compared with other tissues, the multigenic nature of drug addiction, the vast representation of expressed genes in the brain, and our relatively limited knowledge of the molecular pathology of this illness.

The content of this chapter will include recent studies employing genomic and proteomic strategies to develop a comprehensive understanding of the changes induced by cocaine, a commonly abused stimulant. Furthermore, the chapter will focus on studies employing rodent and non-human primate models as well as studies examining the

*Corresponding author. Tel.:336-716-8620; Fax: 336-716-8501;
E-mail: shemby@wfubmc.edu

neuropathology identified in post-mortem human tissue of individuals with chronic histories of illicit substance abuse. The chapter is limited to studies on cocaine due to the fact that this is the most-studied abused drug with respect to genomic and proteomic strategies and thus may provide an investigative template for studying other abuse substances.

The use and abuse of illicit drugs has continued to increase and poses one of the most significant public health care concerns in American society. A recent report indicates that approximately 13.6 million Americans are current users of illicit drugs (e.g. marijuana, 11 million; cocaine, 1.8 million; heroin 130,000) and over 4 million Americans meet the diagnostic criteria for dependence on illicit drugs (SAMHSA, 2002). Despite intense behavioral and biological research, few effective pharmacotherapeutic strategies exist, with the arguable exception of methadone and LAAM treatment programs for opiate dependence. In order to devise effective treatment strategies, it is necessary to understand the interactions of behavioral, pharmacological and biochemical factors that underlie use and abuse. Substance abuse is the culmination of a number of contributing factors spanning scientific disciplines from behavior to molecular biology. As such, to understand the biology of addiction requires a multidisciplinary approach to identify the contributing factors, synthesize the information in the appropriate biological context and eventually relate this context to the behavioral abnormality. The development of new and innovative medications for drug addiction requires multidisciplinary research approaches examining the spectrum of drug-induced effects from behavior to the biological and biochemical effects in discrete neuronal populations.

A generally accepted tenet in drug abuse research is that drugs can function as reinforcing stimuli. Hence, with respect to drug abuse, the reinforcing effects of certain drugs contribute largely to their abuse liability. A significant amount of research investigating the neurobiology of drug abuse is conducted in animal models which closely resemble characteristics of human drug intake l. Criteria should include, but not be limited to the following: (1) behaviors are contingent upon drug delivery, (2) behaviors are engendered and maintained by drug delivery, and (3) drug delivery increases the frequency of those behaviors. The self-administration paradigm meets these criteria, unlike the other procedures, and is widely accepted as an appropriate model for studying the reinforcing effects of drugs. Generally, the self-administration paradigm involves the emission of specific behavior(s) (e.g. lever-press; nose-poke) that is maintained by drug administration (e.g. intravenous, oral, or intracranial). Advantages of self-administration include the following: (1) substances abused by humans can function as positive reinforcing stimuli under laboratory conditions, (2) general concordance between substances abused by humans and those self-administered by laboratory animals, (3) a variety of species readily acquire and maintain self-administration under a number of operant schedules and (4) the ability to generate clear dose-effect curves using this procedure (Hemby et al., 1997b; Hemby, 1999). Procedures such as place conditioning are hindered by the lack of objectively quantifiable behaviors, lack of dose dependency and most importantly by the fact that drug administration is not contingent on the behavior of the animal.

The concept of the contingency is critical for researchers attempting to draw conclusions regarding the involvement of specific neural substrates in drug reinforcement. The majority of studies investigating the neurobiological basis of drug administration have used experimenter-controlled drug administration and extrapolated the relevance of those findings to reinforcement mechanisms (Di Chiara and Imperato, 1988). However, a growing body of literature has demonstrated pronounced neurochemical differences resulting from the context and contingency of drug administration (self-administered versus experimenter delivered) (Wilson et al., 1994; Hemby et al., 1995, 1997a,b). Neurobiological differences between rats self-administering drugs and rats receiving experimenter-administered infusions are based on the context of drug presentation and suggest inferences of reinforcement mechanisms drawn from studies using experimenter-drug administration protocols may be misleading. These studies clearly indicate a need for reliance on accepted behavioral models when asserting relevance of biological findings to behavioral phenomena such as reinforcement. While reinforcement does not solely explain drug abuse, it allows for the quantification of the initiation and maintenance of drug self-administration.

Neuroanatomy of cocaine addiction

Similar to other psychiatric illnesses, drug abuse is a heterogeneous disorder with multiple causes all of which can lead to the same functional endpoint — namely addiction. While the regulation of individual transcripts and proteins have been suggested as mediators of the addictive process, a more probable scenario is that the coordinate regulation of multiple genes and proteins in defined neuroanatomical loci are either the mediators of addictive behaviors or are modulated by chronic drug use. Over the past 20 years, the driving theoretical construct in drug abuse research has been the psychomotor-stimulant theory of addiction which attempts to provide a unifying theory for the neurobiological basis of all abused drugs (Wise and Bozarth, 1987). The theory indicates that both the stimulant and the reinforcing effects of all abused drugs are mediated by a common neural mechanism, the mesolimbic dopamine system. The pathway originates in the mesencephalon, ventral tegmental area (VTA) and projects to several basal forebrain regions including the nucleus accumbens (NAc), ventral caudate-putamen, bed nucleus of the stria terminalis, diagonal band of Broca, olfactory tubercles, prefrontal and anterior cingulate cortices. Administration of drugs that are abused by humans lead to activation of this pathway in humans, non-human primates and rodents (Porrino, 1993; Lyons et al., 1996; Volkow et al., 1997). Activation of this circuit has been correlated with subjective reports of craving and euphoria in cocaine addicts (Volkow et al., 1997; Childress et al., 1999).

Dopaminergic projections from the VTA to the NAc have been implicated in the reinforcing effects of psychomotor stimulants (cocaine and amphetamine) and alcohol, whereas the role of this pathway in opiate reinforcement remains controversial (Hemby et al., 1997b). Previous studies have shown that rats will self-administer cocaine, amphetamine, opiates, and alcohol directly into regions of this pathway. Altering the functional integrity of the mesolimbic pathway by dopamine-selective neurotoxic lesions and dopamine D1 and D2 receptor blockade attenuate psychomotor stimulant self-administration. Similar manipulations of the other monoamines serotonin and norepinephrine fail to significantly influence drug intake. Thirdly, microdialysis studies indicate that extracellular dopamine concentrations are elevated during cocaine and amphetamine self-administration sessions (Hemby et al., 1997b). Taken together, the most recent research indicates that the neurobiological substrates of drug abuse are not the same across all dug classes and probably involve a myriad of neurotransmitter and receptor systems.

Functional genomics

Over the past 10 years, approximately 20 studies have employed various high-throughput gene expression strategies to examine stimulant-induced changes in various brain regions of animal models and humans. Several obstacles prevent the assimilation of the results from these studies into an overarching understanding of stimulant-induced transcriptional regulation such as species, brain regions, route and contingency of administration, dose and duration of drug administration, length of time since the final drug administration, experimental variables in microarray analysis, validation of findings with alternative techniques, etc. Although several studies have examined the effects of stimulants on gene expression, there is minimal literature on stimulant-induced proteomic analysis on a broad scale; however, preliminary data will be presented on proteomic analysis of human cocaine overdose victims.

Rodent studies: non-contingent administration

Several studies have examined the effects of cocaine administration on the coordinate expression of genes in rodent brain regions associated with the mesocorticolimbic pathway, including the NAc (Toda et al., 2002), prefrontal cortex (PFC) (Freeman et al., 2002; Toda et al., 2002), hippocampus (Freeman et al., 2001a), lateral hypothalamus (Ahmed et al., 2005) and VTA (Backes and Hemby, 2003). In the one study, rats were administered cocaine three times per day (15 mg/kg; intraperitoneal) for 14 days (Freeman et al., 2002) as an analogous "binge" paradigm, and gene expression was evaluated in the hippocampus using RNA pools. Using stringent inclusion criteria of 50% induction or 33% reduction, the authors noted only five transcripts were differentially regulated — all were upregulated in the cocaine-treated rats: protein kinase A alpha (PKAcα), metabotropic glutamate receptor 5 (mGluR5) and voltage-gated potassium channel 1.1 (Kv1.1), survival of motor neuron (SMN) and protein phosphatase 2A alpha subunit (PP2Aα). From this set, only mGluR5, PKCα, and Kv1.1 showed analogous changes in protein levels in this region. Interestingly, the authors note that protein tyrosine kinase 2 (PYK2), protein kinase C epsilon (PKCε) and β catenin, proteins found to be elevated in the NAc of cynomolgus monkeys, were also elevated in the hippocampus of cocaine-treated rats suggesting these changes are not region or treatment-specific regimen.

In a separate study, changes in gene expression in the PFC of the same subjects (Freeman et al., 2002) were examined by screening 588 rat genes (BD Bioscience Clontech Atlas cNDA Expression Array). Cocaine administration induced the expression of activity-regulated cytoskeletal protein (ARC), NGFI-B and HMG-CoA synthase I and decreased the expression of casein kinase II alpha (CKIIa), glycogen synthase 3 alpha (GSK3α), and fos-related antigen (FRA1). The upregulation of NGFI-B was confirmed by quantitative PCR; however the remaining encoded proteins of the differentially expressed transcripts were assessed by Western blot analysis. Interestingly, only ARC protein levels were increased in the PFC similar to the mRNA levels — which may be due in part to the somatodendritic localization of ARC in neurons. The authors also examined proteins that had been shown to be upregulated in the hippocampus of rats and NAc of monkeys administered cocaine including PYK2, mitogen-activated kinase I (MEK), β-catenin, PKCα, PKCε, – of which only PYK2 was found to be upregulated in the frontal cortex of cocaine-reated rats. The study provides confirmatory data from previous studies showing increased ARC mRNA expression following cocaine administration (Fosnaugh et al., 1995; Tan et al., 2000; Ujike et al., 2002) as well as extending current knowledge on the ability of cocaine to induce genes and protein involved in neuroplasticity.

Additional insight into prefrontal and striatal synaptic dysfunction came from a cDNA microarray study which screened 1176 rat genes (BD Bioscience Clontech Atlas cNDA Expression Array) in samples of NAc core, NAc shell, striatum and dorsal PFC of rats following 3 weeks of withdrawal from 7 days of cocaine administration (intraperitoneal; 15 mg/kg on days 1 and 7, 30 mg/kg on days 2–6) (Toda et al., 2002). Nine genes were identified with at least 40% increase or 29% decrease relative to controls in one of the four brain regions studied. In the PFC, the authors noted a significant downregulation of the neurotrophic tyrosine kinase receptor type 2 (Ntrk2) in the PFC of cocaine-treated rats. Ntrk2 is the receptor for brain-derived neurotrophic factor (BDNF) previously shown to be involved in the behavioral effects of cocaine in the VTA and NAc (Berhow et al., 1996; Horger et al., 1999; Pierce and Bari, 2001; Freeman and Pierce, 2002). Though not significantly different at the protein level in the PFC, protein levels of the Ntrk2 truncated isoforms p95 and p145 were upregulated in the core of the NAc — a region receiving inputs from the distal regions such as the VTA, hippocampus, etc. Interestingly, the NAc core region exhibited changes in the expression of five transcripts: mitochondrial ATP synthase subunit D (ATP5H), adenosine receptor 1 (ADORA1/A1), leukocyte common antigen-related tyrosine phosphatase (LAR), RET ligand 2 (Retl2) (also known as glial cell line-derived neurotrophic factor family receptor alpha 2; Gfra2). The authors also identified a cocaine-induced downregulation of gastric inhibitory peptide (GIP) mRNA (also known as glucose-dependent insulinotropic polypeptide) — recently shown to be upregulated by chronic clozapine administration in the striatum (Sondhi et al.,

2006) suggesting mediation of this transcript by dopamine given the reciprocal regulation by cocaine and clozapine. More recently, Gip was shown to be expressed in rat hippocampus and involved in a regulatory function in progenitor cell proliferation in the dentate gyrus (Nyberg et al., 2005). Examination of transcript-encoded transcripts showed significantly elevated levels of adenosine 1 receptor protein in the NAc core which may represent a compensatory response to the cocaine-induced upregulation of the D1/Gs signaling cascade documented previously (Nestler, 2001; Scheggi et al., 2004; Zhang et al., 2005), a decreased Gi/Go function (Nestler et al., 1990), elevated adenosine levels (Manzoni et al., 1998), or some combination thereof.

Kreek and colleagues further examined cocaine-induced gene expression in the striatum following acute (3 hourly injection of 15 mg/kg for 1 day) and chronic (3 hourly injections of 15 mg/kg for 3 days) "binge" administration using the Affymetrix rat genome U34A containing approximately 8000 gene/EST clusters (Yuferov et al., 2003). The authors noted 117 upregulated and 22 downregulated transcripts as a result of cocaine administration. Upregulated transcripts included immediate-early genes, "effector" and scaffolding proteins and receptors and signal transduction proteins, while downregulated transcripts was comprised primarily of transcripts related to mitochondrial function along with transcripts encoding signal transduction proteins. RNAse protection assays were used to confirm differential expression as noted by array analysis. In addition to expanding our understanding of cocaine-induced regulation of several gene families and pathways, the authors revealed upregulation of the Per2 clock gene and the somatostatin receptor 2 following "binge" cocaine administration. Previously, disruption of Per genes have been shown to block cocaine-induced sensitization in *Drosophila* (Andretic et al., 1999) and mice (Abarca et al., 2002); however, the localization to the striatum is interesting in that previous studies have found expression limited to the suprachiasmatic nucleus (Masubuchi et al., 2000). The elevated expression of SSTR2 may possibly reflect a less-studied mechanism of cocaine-regulated dopamine release in the striatum as noted by the authors. Additional studies that examine the cellular origin and localization of the Per 2 transcript and protein and the role of SSTR2 in the behavioral effects of cocaine are warranted.

Rodent studies: self-administration

The previous studies have expanded the knowledge base of the cocaine's effects in the brain and provided novel insights into the pharmacological effects of cocaine in various brain regions; however, all used the non-contingent administration of cocaine and thus may have limited applicability to understanding the abuse liability/reinforcing effects of cocaine. As discussed in the Introduction, inferences of reinforcement mechanisms drawn from studies using experimenter drug administration protocols may be misleading as several studies have shown significant differences between experimenter- and self-administered drugs of abuse (Wilson et al., 1994; Hemby et al., 1995, 1997a, b; Hemby, 1999). To date, two studies have combined rodent intravenous self-administration procedures with functional genomics procedures. Ahmed and colleagues examined gene expression profiles in samples of NAc, lateral hypothalamus, septum, VTA, medial PFC and amygdala from rats self-administering cocaine or serving as controls using pooled samples of RNA on the Affymetrix Neurobiology RNU434 chips (Ahmed et al., 2005). The cocaine self-administration group was divided into two subgroups: short access (ShA; 1 h/day; 250 mg/infusion) and long access (LhA; 6 h/day; 250 mg/infusion access) in which one press of a level resulted in the delivery of the dose of cocaine through the intravenous catheter. This procedure results in a marked escalation of cocaine intake within the first hour of access and has been proposed as a model of compulsive drug intake (Ahmed and Koob, 1998, 1999; Ahmed et al., 2002). Interestingly, the lateral hypothalamus exhibited the greatest number of genes that were regulated by cocaine self-administration access (ShA and LhA) and by the escalation paradigm (LhA versus ShA) when compared to the other brain regions studied and differential expression of select transcripts were confirmed by qPCR. Transcripts altered by the escalation paradigm were members of several functional classes including functional and structural plasticity, receptors, synthetic and metabolic enzymes, neurotransmitter release, and proteins coding for neuronal growth and survival.

The aforementioned studies utilized dissected brain regions from rats to generate molecular profiles of cocaine administration. As noted in the previous section on the neuroanatomical basis of reinforcement, the circuitry that mediates the reinforcing effects of cocaine and others drugs of abuse is well-defined and includes dopaminergic cell bodies in the VTA that projects to several forebrain and cortical regions. The advent of discrete cell microdissection and laser capture microdissection (LCM) combined with RNA amplification strategies makes it possible to evaluate expression patterns in defined cell populations in the brain (Ginsberg et al., 1999, 2000, 2004; Hemby et al., 2002; Fasulo and Hemby, 2003). Whereas previous studies have examined regional gene expression profiles in the VTA as a function of cocaine administration, the effects of cocaine self-administration on VTA dopamine neurons remain largely unknown even though these cells are a critical substrate of drug reinforcement. To this end, the expression profile of 95 transcripts following 1 or 20 days of intravenous cocaine self-administration was assessed in dopamine neurons of the VTA in rats (Backes and Hemby, 2003). Tyrosine hydroxylase immunopositive cells were microdissected from the VTA using LCM microdissection and aRNA amplification was used to provide a linear amplification of the mRNA from each rat (Van Gelder et al., 1990; Eberwine et al., 1992; Eberwine, 2001; Hemby et al., 2002). Five GABA-A receptor subunit mRNAs ($\alpha 4$, $\alpha 6$, $\beta 2$, $\gamma 2$, and δ) were downregulated at both 1 and 20 days of cocaine self-administration. In contrast, the catalytic subunit of protein phosphatase 2A (PP2α), GABA-A $\alpha 1$ and Gα_{i2} were significantly increased at both time points. Additionally, calcium/calmodulin-dependent protein kinase IIα (CaMKIIα) mRNA levels were increased initially followed by a slight decrease after 20 days, whereas neuronal nitric oxide synthase (nNOS) mRNA levels were initially decreased but returned to near control levels by day 20. These results indicate that alterations of specific GABA-A receptor subtypes and other signal transduction transcripts appear to be specific neuroadaptations associated with cocaine self-administration. Moreover, as subunit composition determines the functional properties of GABA-A receptors, the observed changes may indicate alterations in the excitability of dopamine transmission underlying long-term biochemical and behavioral effects of cocaine.

Transgenic mouse studies

In an elegant series of experiments, Nestler and colleagues utilized ΔFosB and CREB-inducible transgenic mice with targets know to be involved in the behavioral effects of cocaine to ascertain their effects on the down-stream regulation of gene expression. Previous studies have shown that repeated cocaine administration leads to sustained elevation of ΔFosB levels in brain regions associated with the behavioral effects of cocaine (Hope et al., 1994; Moratalla et al., 1996; Nestler, 2001; Nestler et al., 2001; McClung and Nestler, 2003; Perrotti et al., 2005; Brenhouse and Stellar, 2006). Using the ΔFosB-inducible transgenic mouse model, the investigators were able to demonstrate increased levels of cyclin-dependent kinase 5 (cdk5) mRNA following induction and similarly increased following chronic cocaine administration (Bibb et al., 2001) using a 588 cDNA mouse array (BD Bioscience Clontech Atlas cNDA Expression Array). More importantly, a functional role of cdk5 in cocaine-mediated behaviors was shown by antagonism of cdk5 in the striatum and attenuation of kainate peak currents in the striatum following cocaine administration (Bibb et al., 2001). In a separate study using the ΔFosB-inducible transgenic mouse model, the authors employed the higher density Affymetrix DNA mouse array and found significantly higher levels of NFκB mRNA and protein in the transgenic mice and similar elevations in NFκB protein levels in wild-type mice administered cocaine (20 mg/kg; 14 days) (Ang et al., 2001).

Comparison of the effects of ΔFosB- and CREB-inducible transgenic mouse models on transcription in the NAc revealed that the majority of transcripts induced by CREB occurred after 2 weeks of expression and were sustained at 8 weeks of expression (McClung and Nestler, 2003). Conversely, ΔFosB expression generated dichotomous patterns of gene expression at 2 and 8 weeks with the 2-week expression pattern for ΔFosB similar to CREB expression. The longer ΔFosB expression was similar to effects observed following expression of the dominant-negative CREB. Interestingly, acute cocaine administration (5 days; 10 mg/kg) induced 21% of the genes induced by CREB expression alone whereas chronic cocaine administration (15 mg/g; 20 days) induced 27% of the genes induce by ΔFosB expression alone, leading the authors to conclude that the effects of short-term cocaine administration are more dependent on CREB, whereas chronic administration is dependent on ΔFosB. The list of genes attributable to the induction of CREB and ΔFosB is lengthy and will not be reviewed in here entirely for the sake of brevity; however it is important to note that these studies have significantly expanded the knowledge of transcriptional regulation by these transcription factors and the understanding of the neuroadaptive effects of cocaine administration.

Using a similar approach, Caron and colleagues examined the striatal transcriptomes of three transgenic mouse models, dopamine, norepinephrine, and vesicular monoamine 2 transporter knockouts and a cocaine-treated mouse model using the Affymetrix mouse Genechips (MG U74v2 Set) containing approximately 36,000 gene clusters (Yao et al., 2004). Twenty-six transcripts were altered in all three knockouts and six genes were also found to be altered following chronic cocaine administration (20 mg/kg per day for 5 days followed by 14 days of withdrawal) — adenylate cyclase 1 (signal transduction and plasticity), Pin/Dic-2 (involved in NOS activity and signaling) and post-synaptic density protein 95 kDa (PSD-95; involved in scaffolding of NMDA receptors and plasticity). In situ hybridization indicated a significant decrease in PSD-95 levels in the NAc and striatum of all knockdowns and the cocaine-treated groups, and qPCR confirmed similar decreases in the whole striatum — separate qPCR assessments in NAc and caudate-putamen were not performed. Similarly PSD-95 protein levels were decreased in the NAc, caudate-putamen and in whole striatum of all three knock-outs and the cocaine-treated mice. In addition, all four groups exhibited altered synaptic plasticity of cortical accumbal plasticity.

Non-human primates

One of the first published studies to utilize array technology examined the effects of chronic intramuscular injections of cocaine in cynomolgus monkeys on gene expression in the NAc using a low-density human macroarray from Clonetech consisting of 588 probes (Freeman et al., 2001b). Pools of mRNA from each group were hybridized to two separate arrays leading to the identification of 18 transcripts designated as differentially expressed and included. Unfortunately, the complete list of differentially expressed transcripts is not provided in the manuscript and the website containing the complete dataset is no longer functional. Of the 18 differentially expressed transcripts, eight were selected for post-hoc analysis using Western blot procedures. Four of the eight selected encoded proteins exhibited significant increases in abundance (as hypothesized from the array data) and included PKAα subunit (catalytic; PKAα), the beta subunit of cell adhesion tyrosine kinase, MEK1 and β-catenin. Differences in the protein expression of the remaining four targets did not agree with the array data, which could be due to several factors including post-transcriptional degradation, differences in spatial trafficking of mRNA and protein in neurons, or more practical factors such as the extrapolation of data from pooled RNA samples. An additional limitation of this study is the cross-species hybridization of monkey cDNA (generated using human PCR primers) with human extended oligo probes. The generation of targets for the Clontech assay is a PCR-based method in which primers are used which correspond to the human cDNA sequence. In this case, the overriding assumption is that the *Macaca fascicularis* cDNA is identical to the human cDNA sequence for the transcripts of interest such that the primers would readily anneal to the monkey cDNA and prime the PCR reaction. The lack of specificity of the human primers for cynomolgus cDNA may lead to an underestimation of the abundance of target transcripts and/or may represent the amplification of multiple transcripts in the cynomolgus monkeys.

Nonetheless, the authors aptly point out that the confirmed targets are members of a common biochemical pathway that interact with CREB and AP-1 proteins shown previously to be regulated in rodent models following cocaine administration.

More recently, Hemby and colleagues have used a non-human primate cocaine self-administration model to validate protein and mRNA changes observed in human post-mortem tissue of cocaine-overdose victims (Hemby et al., 2005b). Unfortunately, attempts to recapitulate changes observed in cocaine overdose victims and non-human primate models in rodent self-administration models have not succeeded (Tang et al., 2004; Hemby et al., 2005a). Additional studies are needed to specifically address the ability of the rodent model to recapitulate biochemical changes observed in the primate brain. Whereas rodent models have provided significant information on drug-induced alterations, non-human primate models more closely approximate the anatomy and biochemical milieu of the human brain. For instance, differences between rodents and primates in frontal lobe anatomy (Preuss, 1995) are likely to be reflected in prefrontal–accumbal glutamatergic neurotransmission. In addition, mid-brain dopamine projections in rodents have been ascribed to different midbrain nuclei; however, studies in primates suggest a more complex pattern (Lynd-Balta and Haber, 1994; Williams and Goldman-Rakic, 1998). The use of non-human primates may allow the development of a more clear and clinically relevant characterization of the biochemical changes associated with cocaine use.

Human post-mortem studies

Understanding the consequences of long-term cocaine abuse on post-mortem brain tissues requires vigorous investigation with the benefit of revealing whether the adaptations observed in rodent and non-human primates are applicable to human brain, and which changes are state or trait markers in human drug abusers. Findings in post-mortem brains often provide the first leads that can be investigated in living brain, for example the loss of dopamine in Parkinson's disease (Kish et al., 1988), changes in the levels of the dopamine transporter (Little et al., 1993a, b; Staley et al., 1994a, b;) or opiate system (Hurd and Herkenham, 1993; Staley et al., 1997) with chronic cocaine exposure, and the downregulation of the nicotinic ACh receptor after chronic nicotine (Breese et al., 1997). Although there are many difficulties with post-mortem brain studies, this approach is one of the most promising ways to view biochemical changes relevant to human drug abusers and to educate the public about the consequences of cocaine abuse. Whereas animal studies have advanced our understanding of the neurobiological basis of drug addiction, the evaluation of similar questions in human tissue are few, yet are essential. By assessing changes in defined biochemical pathways in human post-mortem tissue, the fundamental molecular and biochemical processes associated with long-term cocaine use can be ascertained.

Bannon and colleagues examined gene expression in the NAc of post-mortem brain tissue of human cocaine abusers and controls using Affymetrix Human U133A and U133B arrays with represent over 39,000 transcripts (Albertson et al., 2004). Forty-nine transcripts were present in all pairs ($n = 10$) of cocaine and control cases and were differentially expressed in the NAc of cocaine abusers. Transcripts were members of several functional classes including signal transduction, transcriptional and translational processing, neurotransmission and synaptic function, glia, structural and cell adhesion, receptors/transporters/ion channels, cell cycle and growth, and lipid and protein processing. The authors noted a significant upregulation of cocaine and amphetamine-related transcript (CART), a transcript previously discovered following cocaine administration in rats (Douglass et al., 1995; Douglass and Daoud, 1996). In addition, several myelin-associated transcripts were significantly decreased in the NAc of cocaine abusers including myelin basic protein (MBP), proteolipid protein 1 (PLP) and myelin-associated oligodendrocyte basic protein (MOBP) and a significant increase in T-cell differentiation protein (MAL2) — which were confirmed by qPCR. Immunohistochemistry revealed a similar decrease in MBP immunoreactivity in the NAc of these subjects as well. These data provide molecular basis of previous studies which suggested altered white-matter density and myelin expression in cocaine abusers (Volkow et al., 1988; Wiggins and Ruiz, 1990; Lim et al., 2002).

In a separate cohort, Hemby and colleagues used targeted macroarrays consisting of 96 cDNAs to compare gene and protein expression patterns between cocaine overdose victims and age-matched controls in the VTA and lateral substantia nigra (l-SN) (Tang et al., 2003). Evaluated transcripts included ionotropic glutamate receptor (iGluR) subunits, GABAA receptor subunits, dopamine receptors, G-protein subunits, regulators of G-protein signaling and other GTPases, transcriptional regulation, cell growth and death, and others (CART, cannabinoid receptor 1, and serotonin receptors 2A, 2C, and 3). Array analysis revealed significant upregulation of numerous transcripts in the VTA, but not l-SN, of cocaine overdose victims including NMDAR1, GluR2, GluR5, and KA2 receptor mRNAs. Corresponding Western blot analysis revealed VTA-selective upregulation of CREB, NR1, GluR2, GluR5, and KA2 protein levels in cocaine overdose victims. These results indicate that selective alterations of CREB and certain iGluR subunits appear to be associated with chronic cocaine use in humans in a region-specific manner. Extending these studies, we recently examined the extent of altered iGluR subunit expression in the NAc and putamen in cocaine overdose victims (Hemby et al., 2005b). Results revealed statistically significant increases in the NAc, but not in the putamen, of NR1 and GluR2/3 with trends in GluR1 and GluR5 in cocaine-overdose victims (COD). In order to determine that changes were related to cocaine intake and not to other factors in the COD victims, the effects of cocaine intravenous self-administration in rhesus monkeys for 18 months (unit dose of 0.1 mg/kg/injection and daily drug intake of 0.5 mg/kg/session) were examined. Statistically significant elevations were observed for NR1, GluR1, GluR2/3, and GluR5 ($P<0.05$) and a trend toward increased NR1 phosphorylated at Serine 896 ($p = 0.07$) in the NAc but not putamen of monkeys self-administering cocaine compared to controls (Hemby et al., 2005b). These results extend previous results by demonstrating an upregulation of NR1, GluR2/3, and GluR5 in the NAc and suggest these alterations are pathway specific and likely mediate in part the persistent drug intake and craving in the human cocaine abuser.

Proteomics

Whereas several studies have assessed gene and subsequent protein expression as a function of cocaine administration in humans and animal models, to date there are few studies using high-throughput proteomic technologies to examine drug-induced global protein expression patterns in brain regions (Freeman and Hemby, 2004; Freeman et al., 2005; Kim et al., 2005). In order to begin to fill this void in the field of the neurobiology of cocaine addiction, our lab has embarked on several studies in rhesus monkey cocaine self-administration models and in human post-mortem tissue from COD victims. Initial efforts have focused on changes in the NAc given the role of this brain region in the addictive processes of cocaine and the growing gene expression databases. In a preliminary study, cytosolic fractions of NAc proteins from human COD and controls ($n = 5$/group) were separated and quantified by two-dimensional difference gel electrophoresis (2-DIGE) and identified by matrix-assisted laser desorption/ionization time-of-flight (MALDI ToF/ToF) mass spectroscopy (see Chapter 4 for detailed explanation of procedures). Greater than 1000 spots were detected across the five pairs (COD and controls) of which 340 spots were excised, digested in-gel with trypsin, and subsequently analyzed by MALDI ToF/ToF (see Supplemental Table I). Fifty-two percent of the spots were positively identified including 11 upregulated proteins including DJ1 (Parkinson's disease 7 (PARK7; autosomal recessive, early onset)) , ubiquitin carboxyl-terminal esterase L1 (UCHL1; PARK5), lamda crystallin, endothelial monocyte-activating polypeptide 2

Table 1. Identified and matched proteins

Spot #	Protein GI #	Protein name	Theoretical MW	Theoretical pI	Peptide count	Mascot score	Confidence interval	t-test value	Average ratio	Protein homolgues/other protein names
526	28595	Aldolase A; fructose-bisphosphate aldolase	39851.5	8.3	15	196	100		-1.1286	ALDOA
769	136066	Triosephosphate isomerase	26909.8	6.45	4	74	99.944	0.05221	-1.3549	TIM
772	136066	Triosephosphate isomerase	26909.8	6.45	4	74	99.944	0.1059	1.3564	TIM
775	136066	Triosephosphate isomerase	26909.8	6.45	4	74	99.944	0.111	-1.1886	TIM
459	180570	Creatine kinase	42876.4	5.3	17	571	100	0.5089	-1.0726	CKB
401	285975	Rab GDI	51088	5.94	13	200	100	0.654	-1.0426	GDP dissociation inhibitor 2, GDI2
259	334284	GP120	58060.7	5.35	7	87	99.997	0.2132	-1.3476	
660	387016	Phosphoglycerate mutase	28867.8	8.77	2	62	99.176	0.09345	-1.1645	PGAM2, phosphoglycerate mutase 2 (muscle)
450	423123	Tpr protein	238769.7	5.05	20	63	99.315	0.1194	-1.2695	
951	494781	Fatty acid-binding protein	14774.7	6.34	6	236	100	0.4423	1.0726	
162	763431	Albumin	52047.8	5.69	7	80	99.986	0.06642	-1.8004	
639	999892	Chain A, triosephosphate isomerase	26806.8	6.51	5	274	100		2.9063	
221	1465733	Cytosolic NADP(+)-dependent malic enzyme	63858.9	5.88	4	59	98.318	0.4025	-1.3355	ME1
531	2118269	Zebrin II	39797.4	6.67	8	198	100		1.4454	Similar to human Aldolase C
322	2183299	Aldehyde dehydrogenase 1	55427.2	6.3	12	219	100	0.6272	1.0482	ALDH1A1
741	2737906	Plasminogen-related protein A	7983.9	8.44	5	61	99.01	0.2901	1.8309	LOC285189
705	2914390	Chain B, hemoglobin mutant	15834.2	6.76	4	84	99.995	0.4753	-1.1197	
953	2981643	Chain B, hemoglobin	15980.2	6.75	4	93	99.999	0.6047	-1.0872	
864	2982080	Familial als mutant G37r, chain A	16122	5.87	2	133	100	0.158	1.7454	
861	2982080	Familial als mutant G37r, chain A	16122	5.87	2	133	100	0.9453	-1.0013	
797	3205211	Non-muscle myosin heavy chain	72555.1	5.18	9	62	99.117	0.05802	1.4379	
413	3766197	ATP-specific succinyl-CoA synthetase beta subunit	46732.3	5.84	4	62	99.117	0.3475	1.5502	Succinate-CoA ligase, ADP-forming, beta subunit; SUCLA2
112	3811317	Tryptophan hydroxylase isoform 1	6476.5	9.7	4	56	96.486	0.8722	1.3864	TPH
150	4389275	Serum albumin	67988.5	5.69	25	679	100	0.1761	2.0119	
161	4389275	Serum albumin	68424.7	5.67	35	975	100	0.4513	1.597	
523	4502561	Capping protein (actin filament), gelsolin-like	38778.6	5.88	3	67	99.727	0.2784	-1.2687	CAPG
280	4503377	Hydropyrimidinase-like 2; collapsin response mediator	62710.7	5.95	14	325	100	0.3267	-1.3266	CRMP2; DRP2; DPYSL2
282	4503377	Hydropyrimidinase-like 2; collapsin response mediator	62710.7	5.95	14	380	100	0.6647	-1.0535	CRMP2; DRP2; DPYSL2

272		Rab GDI-alpha	51177.4	5	8	131	100	0.2966	1.5102	GDP Dissociation Inhibitor 1; GDI1; oligophrenin 2; OPHN2; RHOGDI
261	4503971	GDP Dissociation inhibitor 1	51177.4	5	9	112	100	0.3715	1.3108	GDI1
58	4504165	Gelsolin	86043.3	5.9	15	258	100	0.2562	-1.6123	GSN
385	4504169	Glutathione synthetase	52523.3	5.67	17	346	100	0.06022	-2.0733	GSS; GSHS; MGC14098
391	4504169	Glutathione synthetase	52523.3	5.67	13	227	100	0.4307	-1.1292	GSS; GSHS; MGC14098
667	4505585	Platelet-activating factor acetylhydrolase	25724.2	5.57	2	60	98.724	0.8345	1.0289	PAFAH1B2; platelet-activating factor acetylhydrolase, isoform Ib, beta subunit 30 kDa
366	4506019	Protein phosphatase 3, catalytic subunit, alpha isoform	52172.7	5.82	7	72	99.912	0.9398	-1.0099	Calcineurin A alpha, PPP3CA, PP2BCA
275	4506089	Mitogen-activated protein kinase 4	63039.9	6.05	8	64	99.417	0.06896	-1.9117	MAPK4, p63MAPK
314	4506089	Mitogen-activated protein kinase 4	63039.9	6.05	11	72	99.916	0.3501	-1.2684	MAPK4, p63MAPK
920	4507793	Ubiquitin-conjugating enzyme E2N	17184	6.13	2	66	99.694	0.5976	-1.0522	Uniquitin-conjugating enzyme 13, UBC13; bendless, ubchen
633	4557032	Lactate dehydrogenase B	36900.2	5.71	13	559	100	0.1127	-1.4306	LDHB
783	4557032	Lactate dehydrogenase B	36900.2	5.71	13	559	100	0.5403	-1.1423	LDHB
636	4557032	Lactate dehydrogenase B	36900.2	5.71	8	219	100	0.8655	1.0184	LDHB
862	4557797	Nucleoside-diphosphate kinase 1 isoform b	17308.7	5.83	8	274	100		1.1544	Non-metastatic cells 1; NME1; NM23A
95	4557871	Transferrin	79280.5	6.81	9	105	100	0.1594	2.0797	TF
97	4557871	Transferrin	79280.5	6.81	9	105	100	0.8981	1.443	TF
418	4758426	Guanine deaminase	51483.8	5.44	11	323	100	0.2742	-1.1109	GDA
684	4758484	Glutathione S-transferase omega 1	27833.1	6.23	7	132	100	0.434	-1.0991	GSTO1
708	4758484	Glutathione S-transferase omega 1	27833.1	6.23	7	132	100	0.4908	1.1479	
647	4758638	Peroxiredoxin 6	25133.2	6	9	326	100	0.1888	1.2268	PRDX6
764	4758638	Peroxiredoxin 6	25133.2	6	9	326	100	0.6421	1.0663	
567	4759036	Regucalcin;senescence marker protein-30	33801.7	5.89	5	64	99.48	0.3902	1.2342	RGN, SMP30
168	4827056	WD repeat-containing protein 1 isoform 2	58593.2	6.41	3	59	98.279	0.09829	-1.8217	WDR1
514	4885063	Aldolase C, fructose-bisphosphate;	39830.4	6.41	15	422	100	0.3166	-1.1063	ALDOC
557	5031777	Isocitrate dehydrogenase 3 (NAD+) alpha precursor	40022.2	6.47	6	95	100	0.7043	-1.0422	IDH3A
873	5031851	Stathmin 1; metablastin;	17291.9	5.76	5	147	100	0.547	1.0666	Leukemia-associated phosphoprotein p18; LAP18
549	5174391	Aldo-keto reductase family 1, member A1	36892	6.32	12	263	100	0.08951	-1.492	ALDR1

Table 1 (continued)

Spot #	Protein GI #	Protein name	Theoretical MW	Theoretical pI	Peptide count	Mascot score	Confidence interval	t-test value	Average ratio	Protein homologues/other protein names
644	5174539	Cytosolic malate dehydrogenase	36631.1	6.91	8	206	100	0.06213	−1.639	MDH1
768	5174539	Cytosolic malate dehydrogenase	36631.1	6.91	5	159	100	0.6754	1.1273	MDH1
762	5174539	Cytosolic malate dehydrogenase	36631.1	6.91	5	159	100	0.8077	1.0359	MDH1
170	5729877	HSP70 protein 8 isoform 1	71082.3	5.37	23	437	100	0.1846	−1.8623	HSPA8
561	5803187	Transaldolase 1; dihydroxyacetone transferase; glycerone transferase	37687.5	6.36	11	178	100	0.4273	1.3777	TALDO1
189	6005938	Utrophin, dystrophin-related protein	396472.1	5.21	22	60	98.809	0.07498	−1.7499	Dystrophin-like protein, DMDL, DRP1
354	6137677	Mitochondrial aldehyde dehydrogenase	54394.4	5.7	7	111	100	0.2883	−1.1245	
121	6470150	BiP protein	71001.6	5.23	15	90	99.999	0.06848	−1.4204	HSPA5; heat shock 70 kDa protein 5 (glucose-regulated protein, 78 kDa)
576	6688197	PAP-inositol-1,4-phosphatase	33743.3	5.46	9	243	100	0.1324	1.1991	3(2′), 5′-bisphosphate nucleotidase 1; BPNT1
892	6806898	Synuclein, alpha	11365	7.88	3	67	99.727	0.9546	1.0552	SNCA
99	6912526	Nasopharyngeal epithelium-specific protein 1	46224	9.99	9	64	99.526	0.1088	1.6362	NESG1
441	7670399	Unnamed protein product	43689.4	6.1	10	149	100	0.09942	1.6662	MEK1
637	7677074	Lamda crystallin	33793.2	5.68	6	171	100	0.7142	−1.0275	CRYL1
755	8393948	Phosphoglycerate mutase 2	28907.9	8.85	2	70	99.857	0.1175	1.3015	Pgam2; Pgmut; PGAM-M; D14Mgh1
402	9966913	Actin-related protein 3-beta	40185.1	5	6	126	100	0.5354	1.0727	ARP11
638	10092677	Hypothetical protein dJ37E16.5	32077.4	6.12	6	81	99.99	0.06068	1.5844	
724	10092677	Hypothetical protein dJ37E16.5	32077.4	6.12	8	261	100	0.897	1.1513	
725	10092677	Hypothetical protein dJ37E16.5	32077.4	6.12	6	81	99.99		1.7072	
556	10241724	Hypothetical protein	31816.9	5.84	7	14	100	0.09672	1.3355	
116	10433666	Unnamed protein product	88418.5	6.68	9	56	96.868	0.3911	1.2484	Ring finger protein 20, RNF20
337	10434221	Unnamed protein product	63177.7	8.73	9	57	97.144	0.1442	−1.4787	Hypothetical protein FLJ10498, FLJ10498
439	11374664	Isocitrate dehydrogenase (NADP) (EC 1.1.1.42), cytosolic	46596.5	6.19	6	67	99.757	0.7066	−1.1504	Isocitrate/isopropylmalate dehydrogenase
242	12804225	CCT5, chaperonin-containing TCP1, subunit 5 (epsilon)	59886.9	5.45	11	159	100	0.1577	−1.611	
680	12860410	Unnamed protein product	15612.5	10.08	5	51	88.887	0.6782	1.1754	AU RNA binding protein/enoyl-coenzyme A hydratase

516	13279173	Similar to COP9	46524.8	5.5	15	346	100	0.1907	1.1899	COP9 constitutive photomorphogenic homolog subunit 4, COPS4
728	13435960	Similar to hypothetical protein FLJ23571	41024.6	9.4	10	60	98.664	0.3842	−1.1525	Hypothetical protein DKFZp434B227
534	13435960	Similar to hypothetical protein FLJ23571	41024.6	9.4	7	50	87.531	0.6963	−1.0929	Hypothetical protein DKFZp434B227
317	13623415	Fascin 1	55151.3	6.84	18	311	100	0.1424	−1.3277	FSCN1
149	13676857	HSP70 protein 2	69977.9	5.56	20	467	100	0.885	1.0331	HSPA2
332	13938355	Unknown	55708.4	5.4	15	345	100	0.07185	−1.6655	ATPase, H+ transporting, lysosomal 56/58 kDa, V1 subunit B, isoform 2, ATP6V1B2
359	15099973	Thrombospondin immunoglobulin heavy chain variable region	12778.3	8.67	4	54	94.557	0.3817	1.2396	
888	15680064	Similar to stathmin 1/oncoprotein 18	17325.9	5.76	4	58	98.069	0.3664	1.0782	STMN1
880	15824412	Neuronal protein 22	22629.2	6.84	6	143	100	0.3978	1.2247	NP22, NP25
669	15930083	Calbindin 2	31663.6	5.06	7	83	99.993		1.1627	Calretinin, calbindin 29 kDa
760	16198390	Unknown (protein for MGC:27286)	33535.7	5.4	6	238	100	0.1244	1.2993	CGI-150 protein
659	16198390	Unknown (protein for MGC:27286)	33535.7	5.4	4	78	99.98	0.1916	1.1935	CGI-150 protein
581	16307182	Similar to transaldolase 1	35534.5	9.07	9	124	100	0.8523	−1.0085	TALDO1
473	16924319	Unknown (protein for IMAGE:3538275)	40819.4	5.78	16	644	100	0.913	−1.0861	Actin
643	17389815	Triosephosphate isomerase 1	26909.8	6.45	4	74	99.944	0.09687	1.7767	TPI
348	18202063	Endothelial-monocyte-activating polypeptide II (EMAP-II)	39975.2	9.37	7	60	98.809	0.1527	−1.3207	Small inducible cytokine subfamily E, memeber 1
950	18202063	Endothelial-monocyte-activating polypeptide II (EMAP-II)	39975.2	9.37	5	53	93.893	0.499	1.1811	Small inducible cytokine subfamily E, memeber 1
117	18256043	Glycyl-tRNA synthetase	81798.7	6.24	8	98	100	0.1313	−1.7071	Gars
911	18307562	Unnamed protein product	69825.2	9.55	7	62	99.249	0.1397	7.2612	
295	18307562	Unnamed protein product	69825.2	9.55	5	49	85.009	0.3296	−1.1903	
224	18307562	Unnamed protein product	69825.2	9.55	8	58	97.678	0.4037	−1.1735	
301	19705447	CDC-ike kinase 3	59262.5	9.53	8	61	99.01	0.3224	−1.1692	Clk3
896	19716076	Myeloid cell nuclear-differentiation factor	46244.3	9.72	8	64	99.492	0.1551	1.41	
325	19913428	ATPase, H+ transporting, lysosomal 56/58 kD, V1 subunit B, isoform 2	56807	5.57	11	216	100	0.1551	−1.3285	ATP6V1B2
443	19923206	Glutamate-ammonia ligase	42664.5	6.43	9	173	100	0.1807	−1.4274	GLUL
71	20072188	Aconitase 2	86252.3	7.62	21	472	100	0.1309	−1.8516	
934	20385874	Beta-tropomyosin	17808.9	4.6	6	63	99.315	0.4222	1.5841	
541	20563689	Mannose phosphate isomerase isoform	29908.3	5.99	4	80	99.988	0.8909	1.135	MPI

Table 1 (continued)

Spot #	Protein GI #	Protein name	Theoretical MW	Theoretical pI	Peptide count	Mascot score	Confidence interval	t-test value	Average ratio	Protein homolgues/other protein names
585	20862467	Hypothetical protein XP_164064	14450.5	9.57	5	56	96.566	0.5297	1.0918	
400	20864657	Similar to Retrovirus-related POL polyprotein	21374.9	9.35	6	65	99.547	0.6174	−1.1668	Similar to Cas-Br-M ectropic retroviral-transforming sequence b
712	20865698	Similar to protein phosphatase 1, regulatory (inhibitor) subunit 12A	22293.5	9.19	6	51	90.541	0.07508	1.1249	PPP1R12A
186	20868874	Hypothetical protein XP_160082	25606.8	6.41	6	54	94.43	0.09973	−1.5525	
800	20887601	Hypothetical protein XP_157898	21094.5	8.58	6	58	98.069	0.1626	1.3765	
981	20892463	RIKEN cDNA 1300010H20	13156.8	9.79	4	54	95.036	0.9819	1.0705	Similar to NADH:ubiquinone oxidoreductase B15 subunit
455	20892491	Similar to creatine kinase, brain	19243.9	7.82	4	59	98.279	0.9706	1.0978	
55	20901108	Hypothetical protein XP_157013	13206.1	4.7	4	53	93.893	0.4558	−1.169	
616	20978314	GTP-ase ran	24606.6	6.6	2	52	90.757	0.1235	−1.5667	RAN, member RAS oncogene family
795	20978314	GTP-ase ran	24606.6	6.6	2	52	90.757	0.9104	1.1077	RAN, member RAS oncogene family
48	20984919	Similar to interferon-inducible protein 10 (IP-10) receptor	89422.8	5.14	24	429	100	0.09089	−1.9122	
311	21313234	RIKEN cDNA 1300006M19	57494.2	8.87	7	58	98.024	0.05004	−1.6899	
570	22041696	Similar to ribosomal protein L7a, cytosolic	13035.2	10.46	5	55	95.677	0.1715	−1.1965	
658	22748619	Tropomyosin 3	28262.3	4.72	7	58	97.833	0.2513	−1.1211	TPM3, alpha-tropomyosin 3
615	22748619	Tropomyosin 3	28262.3	4.72	8	69	99.839	0.6958	−1.0386	TPM3, alpha-tropomyosin 3
361	23208520	DNA polymerase kappa	11938	9.04	5	56	97.009	0.5065	−1.0928	
284	23308577	PHGDH, phosphoglycerate dehydrogenase	57355.7	6.29	5	124	100	0.06459	−1.6773	3-Phosphoglycerate dehydrogenase
796	23395758	TPA: aflatoxin B1-aldehyde reductase	40019.9	6.7	7	63	99.403	0.7007	−1.0226	Aldo-keto reductase family 7, member A2 (aflatoxin aldehyde reductase), AKR7A2
255	24987750	Protein phosphatase 3, catalytic subunit, alpha isoform	43482.6	5.9	7	78	99.98	0.07135	−1.3094	Calcineurin A alpha, PPP3CA, PP2BCA
252	24987750	Protein phosphatase 3, catalytic subunit, alpha isoform	42696	5.26	6	61	98.836	0.1063	−1.4861	Calcineurin A alpha, PPP3CA, PP2BCA

399	24987750	Protein phosphatase 3, catalytic subunit, alpha isoform	3380.3	5.26	8	152	100	0.1853	−1.1264	Calcineurin A alpha, PPP3CA, PP2BCA
723	24987750	Protein phosphatase 3, catalytic subunit, alpha isoform	43380.3	5.26	5	89	99.998	0.8823	1.3475	Calcineurin A alpha, PPP3CA, PP2BCA
106	25020592	Hypothetical protein XP_206488	11818.2	10.44	4	49	84.66	0.7306	1.6247	
375	25777739	Aldehyde dehydrogenase 9A1	54679.3	5.69	13	299	100	0.07027	−1.6495	ALDH9A1
376	25777739	Aldehyde dehydrogenase 9A1	54679.3	5.69	13	299	100	0.453	−1.1089	ALDH9A1
378	26330804	Unnamed protein product	14753.8	10.76	4	52	91.949	0.1878	1.2013	RIKEN cDNA 5730406M06 gene, 5730406M06Rik
687	26336324	Unnamed protein product	47225	8.58	6	56	96.32		3.1461	RIKEN cDNA 1500032A09 gene, 1500032A09Rik
409	27480797	Similar to hypothetical protein DKFZp434D0917.1	26958.3	8.97	4	54	94.43	0.949	1.0125	
453	27503783	Similar to mitochondrial translational release factor 1	52843	8.75	5	54	94.168	0.4426	1.4356	RF1; MTTRF1; MGC47721
964	27574235	Chain B deoxyhemoglobin	16090.3	6.75	3	108	100	0.2269	1.6911	
949	27574235	Chain B deoxyhemoglobin	16090.3	6.75	3	108	100	0.2754	1.3964	
642	27658930	Similar to ATP-dependent chromatin remodeling protein SNF2H	42705.1	9.1	6	55	95.775	0.2464	−1.1293	
789	27658930	Similar to ATP-dependent chromatin remodeling protein SNF2H	42705.1	9.15	6	51	90.095	0.265	−1.1694	
624	27677648	Similar to 60S ribosomal protein L7	17571.1	9.43	5	50	87.241	0.5553	−1.0843	
278	27707686	Similar to ribosomal protein L19	14078	9.87	4	52	91.949	0.2165	−1.3871	
296	27714549	Similar to ribosomal protein L24	12066.3	11.2	4	50	86.943	0.1397	−1.5458	
351	27717139	Similar to 60S ribosomal protein L29	13101.2	10.94	4	52	90.757	0.8939	−1.0125	
411	27960434	Colon cancer autoantigen protein	83725.6	6.11	10	61	98.986	0.5941	1.153	Serologically defined colon cancer antigen 8; Sdccag8
702	28376635	Rab37	24268.2	5.97	7	64	99.417	0.9899	1.0827	
335	28422545	UDP glucose pyrophosphorylase 2	57075.8	8.16	11	101	100	0.1491	−1.3007	UGPP2, UDPG
265	28552838	Hypothetical protein XP_289117	12857.5	9.25	4	57	97.569	0.06322	−1.6146	

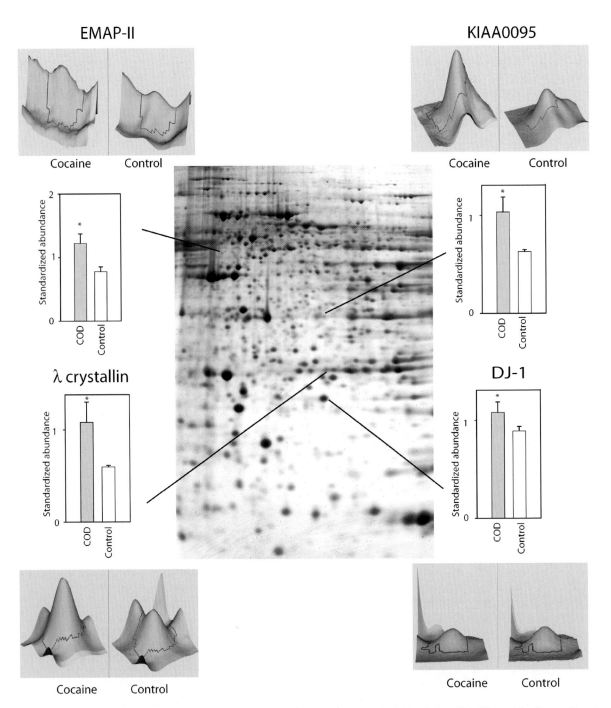

Fig. 1. Preliminary data of representative proteins exhibiting increased abundance in COD victims. Signal intensities for specific gel spots from COD victims and control subjects were compared. Included in the figure are the proteins quantified by the 2DIGE technique using the normalization by Cy2-labeled pool sample and have statistical significance difference in expression profiles between the two groups (*$p<0.05$, t-test). Examples of proteins are provided with representative 3-D plots of individual COD and control spots.

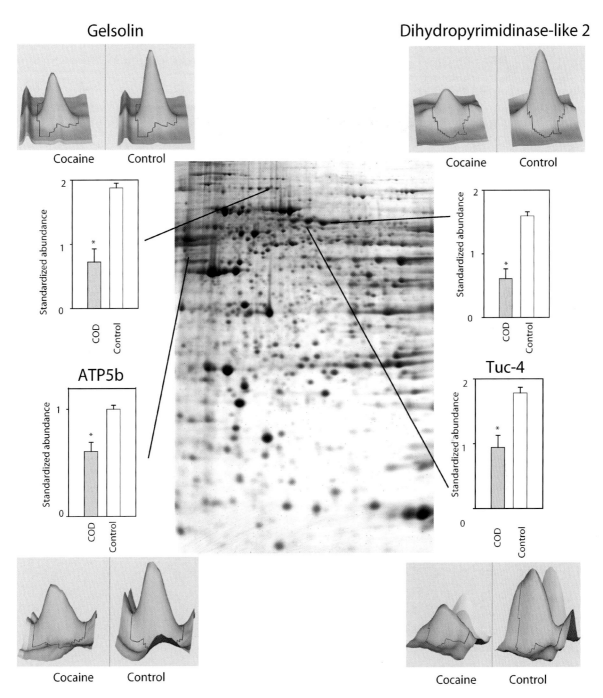

Fig. 2. Preliminary data of representative proteins exhibiting decreased abundance in COD victims. Signal intensities for specific gel spots from COD victims and control subjects were compared. Included in the figure are the proteins quantified by the 2DIGE technique using the normalization by Cy2-labeled pool sample and have statistical significant difference in expression profiles between the two groups (*$p<0.05$, t-test). Examples of proteins are provided with representative 3-D plots of individual COD and control spots.

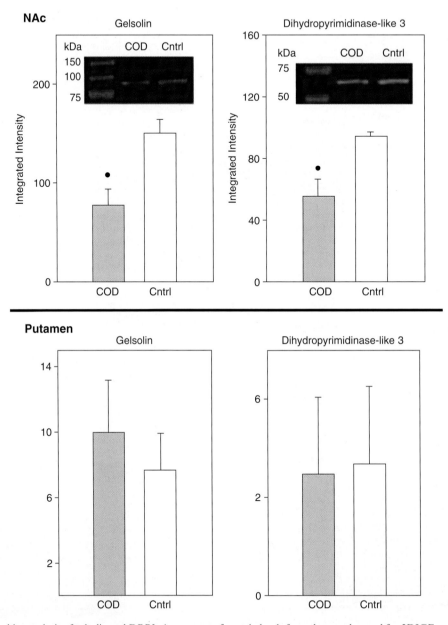

Fig. 3. Western blot analysis of gelsolin and DRP3. Assessment of protein levels from the samples used for 2DIGE revealed significant decreases in gelsolin and DRP3 in agreement with the 2DIGE analyis. Moreover, these changes were specific to the NAc and not observed in the putamen. (*$p<0.05$, t-test). β tubulin was used as a loading control and no differences were observed for this protein.

(EMAP-II) and others (Fig. 1). DJ1, a redox-sensitive chaperone that protects neurons against oxidative stress and cell death, and UCHL1, a neuronal de-ubiquitinating enzyme, are both associated with Parkinson Disease (Abou-Sleiman et al., 2006). Combined with elevated a-synuclein levels in human COD victims (Qin et al., 2005), these data support the suggestion by Deborah Mash and colleagues that chronic cocaine use induce Parkinson-like pathology in striatal regions. Eighteen-positively identified proteins were found to be downregulated in the NAc of COD victims including gelsolin, ATP5b, dihydropyrimidinase-like 3 (DRP3/TUC-4) and dihydropyrimidinase-like 2 (DRP2) (Fig. 2). Decreased expression of gelsolin and DRP3/TUC-4 and gelsolin was confirmed by immunoblotting (Fig. 3) Gelsolin has been reported to exhibit antiapoptotic properties in neurons (Harms et al., 2004) and fibroblasts (Ahn et al., 2003a, b) such that decreased gelsolin expression may render NAc cells more susceptible to apopotosis and oxidative stress due to cocaine exposure. DRP2 and 3 are generally associated with nerve terminal activity, more specifically axonal restructuring and decreased expression may imply decreased plasticity of NAc cells with chronic cocaine exposure. Efforts are underway to assimilate genomic and proteomic databases in a more systematic manner. The application of proteomics holds great promise to understanding the biology of psychiatric diseases, including substance abuse disorders. Further investigation of the changes found and a more comprehensive examination of the human proteome, which may provide the biological understanding and identification of novel therapeutic targets for treatment of cocaine dependence.

Conclusion

In conclusion, relevant gene expression profiles for cocaine abuse and other substance abuse disorders are being generated expanding our knowledge of drug-induced changes in the brain that may underlie persistent drug taking and relapse. Results from rodent, non-human primate and human post-mortem studies indicate significant impairments in neuronal function and plasticity in several brain regions. To date the majority of studies have utilized rodents to model human cocaine intake, however growing evidence indicates the need to refine rodent and non-human primate models to better recapitulate human drug intake and associated neuropathologies. As in other psychiatric and neurological illnesses, researchers should identify the molecular pathologies associated with cocaine addiction in humans and attempt to recapitulate such biological alterations in animal models.

The neurobiological and molecular characteristics of cocaine addiction, although specific to cocaine, may generalize to other drug dependencies. Understanding the coordinated involvement of multiple proteins with chronic cocaine abuse provides insight into the molecular basis of drug dependence and may offer novel targets for pharmacotherapeutic intervention. Although significant advances have been made in the identification of neurochemical and neurobiological substrates involved in the behavioral effects of abused drugs, the relationship between these effects and resultant alterations in gene and protein expression remains in its infancy. The relationship between altered gene and protein expression and the addictive effects of specific drugs remains understudied. The application of this information to the development of treatment strategies has not been fruitful for several reasons. One explanation is that research in the areas of neurobehavioral pharmacology and molecular biology has proceeded in relative isolation of each other. To date, there have been few published studies combining models of self-administration with genomic and proteomic approaches. Other possible explanations include (1) the inappropriate use of experimental models, (2) reliance on non-neuronal systems or neuronal tissue not directly involved in the reinforcing effects of the drug, and (3) the lack of definable neural substrates at the cellular or biochemical level. The combination of appropriate behavioral models of drug reinforcement, specific neurobiological systems and state-of-the-art molecular techniques will provide the most pertinent data for understanding the molecular basis of drug reinforcement and for potentially establishing novel targets for pharmacotherapeutic intervention.

A more detailed understanding of the molecular and biochemical cascades in specific neuronal populations and the interactions between well-defined neuronal populations within discrete brain regions could lead to a greater knowledge of the basic neurobiological processes involved in drug reinforcement. Future efforts investigating the biological basis of drug reinforcement should be directed at specific cellular targets in brain regions considered to be involved in drug reinforcement. The integration of basic neuroscience and behavior offers the most productive avenue for delineating the complexity of the neurobiological underpinnings of drug reinforcement and the subsequent development of effective pharmacotherapies to treat addiction.

Acknowledgments

Support for this project comes from NIH grants DA013234 and DA DA013772. Special appreciation to Nilesh Yannu, Kaitlin Duschene, Wenxue Tang, and Willard Freeman for technical assistance. I would like to express my appreciation for the altruism and support of families of the patients studied.

References

Abarca, C., Albrecht, U. and Spanagel, R. (2002) Cocaine sensitization and reward are under the influence of circadian genes and rhythm. Proc. Natl. Acad. Sci. USA, 99: 9026–9030.

Abou-Sleiman, P.M., Muqit, M.M.K. and Wood, N.W. (2006). Expanding insights of mitochondrial dysfunction in Parkinson's disease, 7: 207–219.

Ahmed, S.H., Kenny, P.J., Koob, G.F. and Markou, A. (2002) Neurobiological evidence for hedonic allostasis associated with escalating cocaine use. Nat. Neurosci., 5: 625–626.

Ahmed, S.H. and Koob, G.F. (1998) Transition from moderate to excessive drug intake: change in hedonic set point. Science, 282: 298–300.

Ahmed, S.H. and Koob, G.F. (1999) Long-lasting increase in the set point for cocaine self-administration after escalation in rats. Psychopharmacology, 146: 303–312.

Ahmed, S.H., Lutjens, R., van der Stap, L.D., Lekic, D., Romano-Spica, V., Morales, M., Koob, G.F., Repunte-Canonigo, V. and Sanna, P.P. (2005) Gene expression evidence for remodeling of lateral hypothalamic circuitry in cocaine addiction. Proc. Natl. Acad. Sci. USA, 102: 11533–11538.

Ahn, J.S., Jang, I.S., Kim, D.I., Cho, K.A., Park, Y.H., Kim, K., Kwak, C.S. and Chul Park, S. (2003a) Aging-associated increase of gelsolin for apoptosis resistance. Biochem. Biophys. Res. Commun., 312: 1335–1341.

Ahn, J.S., Jang, I.S., Rhim, J.H., Kim, K., Yeo, E.J. and Park, S.C. (2003b) Gelsolin for senescence-associated resistance to apoptosis. Ann. N. Y. Acad. Sci., 1010: 493–495.

Albertson, D.N., Pruetz, B., Schmidt, C.J., Kuhn, D.M., Kapatos, G. and Bannon, M.J. (2004) Gene expression profile of the nucleus accumbens of human cocaine abusers: evidence for dysregulation of myelin. J. Neurochem., 88: 1211–1219.

Andretic, R., Chaney, S. and Hirsh, J. (1999) Requirement of circadian genes for cocaine sensitization in Drosophila. Science, 285: 1066–1068.

Ang, E., Chen, J., Zagouras, P., Magna, H., Holland, J., Schaeffer, E. and Nestler, E.J. (2001) Induction of nuclear factor-kappaB in nucleus accumbens by chronic cocaine administration. J. Neurochem., 79: 221–224.

Backes, E. and Hemby, S.E. (2003) Discrete cell gene profiling of ventral tegmental dopamine neurons after acute and chronic cocaine self-administration. J. Pharmacol. Exp. Ther., 307: 450–459.

Berhow, M.T., Hiroi, N. and Nestler, E.J. (1996) Regulation of ERK (extracellular signal-regulated kinase), part of the neurotrophin signal transduction cascade, in the rat mesolimbic dopamine system by chronic exposure to morphine or cocaine. J. Neurosci., 16: 4707–4715.

Bibb, J.A., Chen, J., Taylor, J.R., Svenningsson, P., Nishi, A., Snyder, G.L., Yan, Z., Sagawa, Z.K., Ouimet, C.C., Nairn, A.C., Nestler, E.J. and Greengard, P. (2001) Effects of chronic exposure to cocaine are regulated by the neuronal protein Cdk5. Nature, 410: 376–380.

Breese, C.R., Marks, M.J., Logel, J., Adams, C.E., Sullivan, B., Collins, A.C. and Leonard, S. (1997) Effect of smoking history on [3 H]nicotine binding in human postmortem brain. J. Pharmacol. Exp. Ther., 282: 7–13.

Brenhouse, H.C. and Stellar, J.R. (2006) c-Fos and deltaFosB expression are differentially altered in distinct subregions of the nucleus accumbens shell in cocaine-sensitized rats. Neuroscience, 137: 773–780.

Childress, A.R., Mozley, P.D., McElgin, W., Fitzgerald, J., Reivich, M. and O'Brien, C.P. (1999) Limbic activation during cue-induced cocaine craving. Am. J. Psychiatry, 156: 11–18.

Di Chiara, G. and Imperato, A. (1988) Drugs abused by humans preferentially increase synaptic dopamine concentration in the mesolimbic system of freely moving rats. Proc. Natl. Acad. Sci. USA, 85: 5274–5278.

Douglass, J. and Daoud, S. (1996) Characterization of the human cDNA and genomic DNA encoding CART: a cocaine- and amphetamine-regulated transcript. Gene, 169: 241–245.

Douglass, J., McKinzie, A.A. and Couceyro, P. (1995) PCR differential display identifies a rat brain mRNA that is transcriptionally regulated by cocaine and amphetamine. J. Neurosci., 15: 2471–2481.

Eberwine, J. (2001) Single-cell molecular biology. Nat. Neurosci., 4(Suppl): 1155–1156.

Eberwine, J., Yeh, H., Miyashiro, K., Cao, Y., Nair, S., Finnell, R., Zettel, M. and Coleman, P. (1992) Analysis of gene expression in single live neurons. Proc. Natl. Acad. Sci. USA, 89: 3010–3014.

Fasulo, W.H. and Hemby, S.E. (2003) Time-dependent changes in gene expression profiles of midbrain dopamine neurons following haloperidol administration. J. Neurochem., 87: 205–219.

Fosnaugh, J.S., Bhat, R.V., Yamagata, K., Worley, P.F. and Baraban, J.M. (1995) Activation of arc, a putative "effector" immediate early gene, by cocaine in rat brain. J. Neurochem., 64: 2377–2380.

Freeman, A.Y. and Pierce, R.C. (2002) Neutralization of neutrophin-3 in the ventral tegmental area or nucleus accumbens differentially modulates cocaine-induced behavioral plasticity in rats. Synapse, 46: 57–65.

Freeman, W.M., Brebner, K., Amara, S.G., Reed, M.S., Pohl, J. and Phillips, A.G. (2005) Distinct proteomic profiles of amphetamine self-administration transitional states. Pharmacogenom. J., 5: 203–214.

Freeman, W.M., Brebner, K., Lynch, W.J., Robertson, D.J., Roberts, D.C. and Vrana, K.E. (2001a) Cocaine-responsive gene expression changes in rat hippocampus. Neuroscience, 108: 371–380.

Freeman, W.M., Brebner, K., Patel, K.M., Lynch, W.J., Roberts, D.C. and Vrana, K.E. (2002) Repeated cocaine self-administration causes multiple changes in rat frontal cortex gene expression. Neurochem. Res., 27: 1181–1192.

Freeman, W.M. and Hemby, S.E. (2004) Proteomics for protein expression profiling in neuroscience. Neurochem. Res., 29: 1065–1081.

Freeman, W.M., Nader, M.A., Nader, S.H., Robertson, D.J., Gioia, L., Mitchell, S.M., Daunais, J.B., Porrino, L.J., Friedman, D.P. and Vrana, K.E. (2001b) Chronic cocaine-mediated changes in non-human primate nucleus accumbens gene expression. J. Neurochem., 77: 542–549.

Ginsberg, S.D., Crino, P.B., Hemby, S.E., Weingarten, J.A., Lee, V.M., Eberwine, J.H. and Trojanowski, J.Q. (1999) Predominance of neuronal mRNAs in individual Alzheimer's disease senile plaques. Ann. Neurol., 45: 174–181.

Ginsberg, S.D., Elarova, I., Ruben, M., Tan, F., Counts, S.E., Eberwine, J.H., Trojanowski, J.Q., Hemby, S.E., Mufson, E.J. and Che, S. (2004) Single-cell gene expression analysis: implications for neurodegenerative and neuropsychiatric disorders. Neurochem. Res., 29: 1053–1064.

Ginsberg, S.D., Hemby, S.E., Lee, V.M., Eberwine, J.H. and Trojanowski, J.Q. (2000) Expression profile of transcripts in Alzheimer's disease tangle-bearing CA1 neurons. Ann. Neurol., 48: 77–87.

Harms, C., Bosel, J., Lautenschlager, M., Harms, U., Braun, J.S., Hortnagl, H., Dirnagl, U., Kwiatkowski, D.J., Fink, K. and Endres, M. (2004) Neuronal gelsolin prevents apoptosis by enhancing actin depolymerization. Mol. Cell Neurosci., 25: 69–82.

Hemby, S.E. (1999) Recent advances in the biology of addiction. Curr. Psychiatry Rep., 1: 159–165.

Hemby, S.E., Co, C., Koves, T.R., Smith, J.E. and Dworkin, S.I. (1997a) Differences in extracellular dopamine concentrations in the nucleus accumbens during response-dependent and response-independent cocaine administration in the rat. Psychopharmacology (Berl.), 133: 7–16.

Hemby, S.E., Ginsberg, S.D., Brunk, B., Trojanowski, J.Q. and Eberwine, J.H. (2002) Gene expression profile for schizophrenia: discrete neuron transcription patterns in the entorhinal cortex. Arch. Gen. Psychiatry, 59: 631–640.

Hemby, S.E., Horman, B. and Tang, W. (2005a) Differential regulation of ionotropic glutamate receptor subunits following cocaine self-administration. Brain Res., 1064: 75–82.

Hemby, S.E., Johnson, B.A. and Dworkin, S.I. (1997b) Neurobiological basis of drug reinforcement. In: Johnson, B.A. and Roache, J.D. (Eds.), Drug Addiction and Its Treatment: Nexus of Neuroscience and Behavior. Lippincott-Raven Publishers, Philadelphia, pp. 137–169.

Hemby, S.E., Martin, T.J., Co, C., Dworkin, S.I. and Smith, J.E. (1995) The effects of intravenous heroin administration on extracellular nucleus accumbens dopamine concentrations as determined by in vivo microdialysis. J. Pharmacol. Exp. Ther., 273: 591–598.

Hemby, S.E., Tang, W., Muly, E.C., Kuhar, M.J., Howell, L. and Mash, D.C. (2005b) Cocaine-induced alterations in nucleus accumbens ionotropic glutamate receptor subunits in human and non-human primates. J. Neurochem., 95: 1785–1793.

Hope, B.T., Nye, H.E., Kelz, M.B., Self, D.W., Iadarola, M.J., Nakabeppu, Y., Duman, R.S. and Nestler, E.J. (1994) Induction of a long-lasting AP-1 complex composed of altered Fos-like proteins in brain by chronic cocaine and other chronic treatments. Neuron, 13: 1235–1244.

Horger, B., Iyasere, C., Berhow, M., Messer, C.J., Nestler, E.J. and Taylor, J.R. (1999) Enhancement of locomotor activity and conditioned reward to cocaine by brain-derived neurotrophic factor. J. Neurosci., 19: 4110–4122.

Hurd, Y.L. and Herkenham, M. (1993) Molecular alterations in the neostriatum of human cocaine addicts. Synapse, 13: 357–369.

Kim, S.Y., Chudapongse, N., Lee, S.M., Levin, M.C., Oh, J.T., Park, H.J. and Ho, I.K. (2005) Proteomic analysis of phosphotyrosyl proteins in morphine-dependent rat brains. Brain Res. Mol. Brain Res., 133: 58–70.

Kish, S.J., Shannak, K. and Hornykiewicz, O. (1988) Uneven pattern of dopamine loss in the striatum of patients with idiopathic Parkinson's disease. Pathophysiologic and clinical implications. N. Engl. J. Med., 318: 876–880.

Lim, K.O., Choi, S.J., Pomara, N., Wolkin, A. and Rotrosen, J.P. (2002) Reduced frontal white matter integrity in cocaine dependence: a controlled diffusion tensor imaging study. Biol. Psychiatry, 51: 890–895.

Little, K.Y., Kirkman, J.A., Carroll, F.I., Breese, G.R. and Duncan, G.E. (1993a) [125I]RTI-55 binding to cocaine-sensitive dopaminergic and serotonergic uptake sites in the human brain. J. Neurochem., 61: 1996–2006.

Little, K.Y., Kirkman, J.A., Carroll, F.I., Clark, T.B. and Duncan, G.E. (1993b) Cocaine use increases [3 H]WIN 35428 binding sites in human striatum. Brain Res., 628: 17–25.

Lynd-Balta, E. and Haber, S.N. (1994) The organization of midbrain projections to the ventral striatum in the primate. Neuroscience, 59: 609–623.

Lyons, D., Friedman, D.P., Nader, M.A. and Porrino, L.J. (1996) Cocaine alters cerebral metabolism within the ventral striatum and limbic cortex of monkeys. J. Neurosci., 16: 1230–1238.

Manzoni, O., Pujalte, D., Williams, J. and Bockaert, J. (1998) Decreased presynaptic sensitivity to adenosine after cocaine withdrawal. J. Neurosci., 18: 7996–8002.

Masubuchi, S., Honma, S., Abe, H., Ishizaki, K., Namihira, M., Ikeda, M. and Honma, K. (2000) Clock genes outside the suprachiasmatic nucleus involved in manifestation of locomotor activity rhythm in rats. Eur. J. Neurosci., 12: 4206–4214.

McClung, C.A. and Nestler, E.J. (2003) Regulation of gene expression and cocaine reward by CREB and DeltaFosB. Nat. Neurosci., 6: 1208–1215.

Moratalla, R., Elibol, B., Vallejo, M. and Graybiel, A.M. (1996) Network level changes in expression of inducible Fos-Jun proteins in the striatum during chronic cocaine treatment and withdrawal. Neuron, 17: 147–156.

Nestler, E.J. (2001) Molecular basis of long-term plasticity underlying addiction. Nat. Neurosci. Rev., 2: 119–128.

Nestler, E.J., Barrot, M. and Self, D.W. (2001) DeltaFosB: a sustained molecular switch for addiction. Proc. Natl. Acad. Sci. USA, 98: 11042–11046.

Nestler, E.J., Terwilliger, R.Z., Walker, J.R., Sevarino, K.A. and Duman, R.S. (1990) Chronic cocaine treatment decreases levels of the G protein subunits Gi alpha and Go alpha in discrete regions of rat brain. J. Neurochem., 55: 1079–1082.

Nyberg, J., Anderson, M.F., Meister, B., Alborn, A.M., Strom, A.K., Brederlau, A., Illerskog, A.C., Nilsson, O., Kieffer, T.J., Hietala, M.A., Ricksten, A. and Eriksson, P.S. (2005) Glucose-dependent insulinotropic polypeptide is expressed in adult hippocampus and induces progenitor cell proliferation. J. Neurosci., 25: 1816–1825.

Perrotti, L.I., Bolanos, C.A., Choi, K.H., Russo, S.J., Edwards, S., Ulery, P.G., Wallace, D.L., Self, D.W., Nestler, E.J. and Barrot, M. (2005) DeltaFosB accumulates in a GABAergic cell population in the posterior tail of the ventral tegmental area after psychostimulant treatment. Eur. J. Neurosci., 21: 2817–2824.

Pierce, R.C. and Bari, A.A. (2001) The role of neurotrophic factors in psychostimulant-induced behavioral and neuronal plasticity. Rev. Neurosci., 12: 95–110.

Porrino, L.J. (1993) Functional consequences of acute cocaine treatment depend on route of administration. Psychopharmacology (Berl.), 112: 343–351.

Preuss, T.M. (1995) Do rats have prefrontal cortex? The Rose–Woolsey–Akert Program reconsidered. J. Cogn. Neurosci., 7: 1–24.

Qin, Y., Ouyang, Q., Pablo, J. and Mash, D.C. (2005) Cocaine abuse elevates alpha-synuclein and dopamine transporter levels in the human striatum. Neuroreport, 16: 1489–1493.

SAMHSA (2002) Results from the 2001 national household survey on drug abuse: Volume I. Summary of National Findings. Substance Abuse and Mental Health Services Administration, Rockville, MD.

Scheggi, S., Rauggi, R., Gambarana, C., Tagliamonte, A. and De Montis, M.G. (2004) Dopamine and cyclic AMP-regulated phosphoprotein-32 phosphorylation pattern in cocaine and morphine-sensitized rats. J. Neurochem., 90: 792–799.

Sondhi, S., Castellano, J.M., Chong, V.Z., Rogoza, R.M., Skoblenick, K.J., Dyck, B.A., Gabriele, J., Thomas, N., Ki, K., Pristupa, Z.B., Singh, A.N., MacCrimmon, D., Vorug-anti, P., Foster, J. and Mishra, R.K. (2006) cDNA array reveals increased expression of glucose-dependent insulinotropic polypeptide following chronic clozapine treatment: role in atypical antipsychotic drug-induced adverse metabolic effects. Pharmacogenom. J., 6: 131–140.

Staley, J.K., Basile, M., Flynn, D.D. and Mash, D.C. (1994a) Visualizing dopamine and serotonin transporters in the human brain with the potent cocaine analogue [125I]RTI-55: in vitro binding and autoradiographic characterization. J. Neurochem., 62: 549–556.

Staley, J.K., Hearn, W.L., Ruttenber, A.J., Wetli, C.V. and Mash, D.C. (1994b) High affinity cocaine recognition sites on the dopamine transporter are elevated in fatal cocaine overdose victims. J. Pharmacol. Exp. Ther., 271: 1678–1685.

Staley, J.K., Rothman, R.B., Rice, K.C., Partilla, J. and Mash, D.C. (1997) Kappa2 opioid receptors in limbic areas of the human brain are upregulated by cocaine in fatal overdose victims. J. Neurosci., 17: 8225–8233.

Tan, A., Moratalla, R., Lyford, G.L., Worley, P. and Graybiel, A.M. (2000) The activity-regulated cytoskeletal-associated protein arc is expressed in different striosome-matrix patterns following exposure to amphetamine and cocaine. J. Neurochem., 74: 2074–2078.

Tang, W., Wesley, M., Freeman, W.M., Liang, B. and Hemby, S.E. (2004) Alterations in ionotropic glutamate receptor subunits during binge cocaine self-administration and withdrawal in rats. J. Neurochem., 89: 1021–1033.

Tang, W.-X., Fasulo, W.H., Mash, D.C. and Hemby, S.E. (2003) Molecular profiling of midbrain dopamine regions in cocaine overdose victims. J. Neurochem., 85: 911–924.

Toda, S., McGinty, J.F. and Kalivas, P.W. (2002) Repeated cocaine administration alters the expression of genes in corticolimbic circuitry after a 3-week withdrawal: a DNA macroarray study. J. Neurochem., 82: 1290–1299.

Ujike, H., Takaki, M., Kodama, M. and Kuroda, S. (2002) Gene expression related to synaptogenesis, neuritogenesis, and MAP kinase in behavioral sensitization to psychostimulants. Ann. N. Y. Acad. Sci., 965: 55–67.

Van Gelder, R.N., von Zastrow, M.E., Yool, A., Dement, W.C., Barchas, J.D. and Eberwine, J.H. (1990) Amplified RNA synthesized from limited quantities of heterogeneous cDNA. Proc. Natl. Acad. Sci. USA, 87: 1663–1667.

Volkow, N.D., Valentine, A. and Kulkarni, M. (1988) Radiological and neurological changes in the drug abuse patient: a study with MRI. J. Neuroradiol., 15: 288–293.

Volkow, N.D., Wang, G.J., Fischman, M.W., Foltin, R.W., Fowler, J.S., Abumrad, N.N., Vitkun, S., Logan, J., Gatley, S.J., Pappas, N., Hitzemann, R. and Shea, C.E. (1997) Relationship between subjective effects of cocaine and dopamine transporter occupancy. Nature, 386: 827–830.

Wiggins, R.C. and Ruiz, B. (1990) Development under the influence of cocaine. I. A comparison of the effects of daily cocaine treatment and resultant undernutrition on pregnancy and early growth in a large population of rats. Metab. Brain Dis., 5: 85–99.

Williams, S.M. and Goldman-Rakic, P.S. (1998) Widespread origin of the primate mesofrontal dopamine system. Cereb. Cortex, 8: 321–345.

Wilson, J.M., Nobrega, J.N., Corrigall, W.A., Coen, K.M., Shannak, K. and Kish, S.J. (1994) Amygdala dopamine levels are markedly elevated after self- but not passive-administration of cocaine. Brain Res., 668: 39–45.

Wise, R.A. and Bozarth, M.A. (1987) A psychomotor stimulant theory of addiction. Psychol. Rev., 94: 469–492.

Yao, W.D., Gainetdinov, R.R., Arbuckle, M.I., Sotnikova, T.D., Cyr, M., Beaulieu, J.M., Torres, G.E., Grant, S.G. and Caron, M.G. (2004) Identification of PSD-95 as a regulator of dopamine-mediated synaptic and behavioral plasticity. Neuron, 41: 625–638.

Yuferov, V., Kroslak, T., Laforge, K.S., Zhou, Y., Ho, A. and Kreek, M.J. (2003) Differential gene expression in the rat caudate putamen after "binge" cocaine administration: advantage of triplicate microarray analysis. Synapse, 48: 157–169.

Zhang, D., Zhang, L., Tang, Y., Zhang, Q., Lou, D., Sharp, F.R., Zhang, J. and Xu, M. (2005) Repeated cocaine administration induces gene expression changes through the dopamine D1 receptors. Neuropsychopharmacology, 30: 1443–1454.

Hemby & Bahn (Eds.)
Progress in Brain Research, Vol. 158
ISSN 0079-6123
Copyright © 2006 Elsevier B.V. All rights reserved

CHAPTER 10

Neuronal gene expression profiling: uncovering the molecular biology of neurodegenerative disease

Elliott J. Mufson[1,*], Scott E. Counts[1], Shaoli Che[2,3] and Stephen D. Ginsberg[2,3,4]

[1]*Department of Neurological Sciences, Rush University Medical Center, Chicago, IL 60612, Orangeburg, NY 10962, USA*
[2]*Center for Dementia Research, Nathan Kline Institute, Orangeburg, NY 10962, USA*
[3]*Department of Psychiatry, New York University School of Medicine, Orangeburg, NY 10962, USA*
[4]*Department of Physiology and Neuroscience, New York University School of Medicine, Orangeburg, NY 10962, USA*

Abstract: The development of gene array techniques to quantify expression levels of dozens to thousands of genes simultaneously within selected tissue samples from control and diseased brain has enabled researchers to generate expression profiles of vulnerable neuronal populations in several neurodegenerative diseases, including Alzheimer's disease, Parkinson's disease, schizophrenia, multiple sclerosis, and Creutzfeld–Jakob disease. Intriguingly, gene expression analysis reveals that vulnerable brain regions in many of these diseases share putative pathogenetic alterations in common classes of genes, including decrements in synaptic transcript levels and increments in immune response transcripts. Thus, gene expression profiles of diseased neuronal populations may reveal mechanistic clues to the molecular pathogenesis underlying various neurological diseases and aid in identifying potential therapeutic targets. This chapter will review how regional and single cell gene array technologies have advanced our understanding of the genetics of human neurological disease.

Keywords: microarray; Alzheimer's disease; Parkinson's disease; schizophrenia; RNA amplification; aging

Introduction

The onset of human neurological brain disease arises when a threshold number of neurons stops performing their normal functions, lose their ability to respond to changes in the external and internal milieu, are disconnected from their projection sites and eventually degenerate. Since the brain is a complex structure with heterogeneous neuronal (e.g., pyramidal neurons and interneurons) and non-neuronal (e.g., glial cells, epithelial cells, and vascular elements) cell populations, determining the factor(s) that differentiate vulnerable neurons from those that are not affected during the course of a disorder is a major challenge facing researchers in the field of the neurobiology of disease. Understanding neuronal structure–function interactions has been an ongoing area of basic research since the time of Aristotle. Although for centuries it has been possible to visualize select cytoarchitectonic elements of the brain, it was not until the mid-1800s that the classic work of Emilo Golgi and Ramón y Cajal initiated a debate about the structural and connectional organization of the brain (Berchtold and Cotman, 1998). Over the course of the past 150 years, great strides have been made in defining the chemical phenotype, afferent and efferent connections, and ultrastructure of the neuronal circuitry that orchestrates brain structure–function responses. Many of these basic biologic techniques have been applied to the study of neurological

*Corresponding author. Tel.: +1-312-563-3558;
Fax +1-312-563-3571; E-mail: emufson@rush.edu

DOI: 10.1016/S0079-6123(06)58010-0

disorders including amyotrophic lateral sclerosis (ALS), Alzheimer's disease (AD), Parkinson's disease (PD), schizoprehenia, and other progressive, late-onset brain abnormalities. For example, chemical anatomic studies have revealed that the entorhinal cortex glutamatergic projection neurons to the hippocampus (Hyman et al., 1984, 1990) and cholinergic basal forebrain projection neurons to the cortex and hippocampus (Davies and Maloney, 1976; Whitehouse et al., 1982) are selectively vulnerable in AD, whereas the dopaminergic nigrostriatal projection neurons are predominately affected in PD (Fearnley and Lees, 1991) and spinal motor neurons are affected in ALS (Martin et al., 2000). The variation in morphology, connectivity, physiology, chemistry and, ultimately, function of these cell types suggest that distinct and dynamic gene expression programs orchestrate the phenotypic differences important for their normative function and survival. As such, dysregulation of the homeostatic balance of gene expression profiles within these brain cell systems may be a cause or a consequence of an extrinsic or intrinsic insult that results from human neurologic disease. Therefore, a precise determination of neuronal transcriptional changes of diseased neurons will provide new insights into the cellular vulnerability underlying the pathogenesis of human neurological disorders.

The development of high-throughput gene expression profiling methods, including microarray technology, has enabled researchers to quantitatively assess the expression levels of multiple (e.g., dozens to thousands) genes simultaneously within the selected tissue samples from control and diseased brain to elucidate the molecular fingerprint of vulnerable neuronal populations in neurodegenerative disease (Baranzini, 2004; Ginsberg et al., 2004). Expression profiling is performed by extracting RNAs from sample tissues, amplifying RNA when necessary, labeling RNA/amplified RNA probes, and hybridizing labeled RNA to array platforms (Ginsberg and Che, 2002; Ginsberg, 2005). Quantification of hybridization signal intensity is performed to evaluate the relative level of signal intensity of each feature on the array platform. Gene expression is evaluated with the aid of univariate and multivariate statistics and informatics software, which enables high-throughput coordinate analyses.

Regional analysis of gene expression is a widely used paradigm due to the relatively large amounts of RNA that can be extracted from carefully dissected frozen postmortem human brain tissues, as evidenced by reports on several neurodegenerative and neuropsychiatric disorders including AD (Ho et al., 2001; Loring et al., 2001; Colangelo et al., 2002; Blalock et al., 2004; Lukiw, 2004), ALS (Malaspina et al., 2001; Jiang et al., 2005), PD (Grunblatt et al., 2004; Hauser et al., 2005), schizophrenia (Hakak et al., 2001; Middleton et al., 2002; Mimmack et al., 2002; Pongrac et al., 2002), and multiple sclerosis (Becker et al., 1997; Whitney et al., 1999; Chabas et al., 2001; Ramanathan et al., 2001). One advantage of regional gene expression analysis is that, in most cases, extracted RNA is sufficient to generate significant hybridization signal intensity for large-scale microarrays, thus allowing for the analysis of thousands of target genes without additional RNA amplification protocols. However, regional expression profiles cannot discern molecular signatures in discrete neuronal populations, nor can regional assessments evaluate differences in adjacent neuronal and non-neuronal populations. More recently, techniques were developed to analyze gene expression profiles of single neurons harvested from human postmortem brain tissue. One of the key features of single cell or single population gene profiling is that different cell types can be discriminated based upon their distinct neurochemical phenotype. For example, cholinergic basal forebrain (CBF) neurons (Mufson et al., 2002) and midbrain dopaminergic neurons (Fasulo and Hemby, 2003; Tang et al., 2003) can be identified by selective expression of phenotypic markers and isolated for single cell RNA amplification and subsequent microarray analysis. Moreover, cells that lack a distinct or selective phenotype can be analyzed using a variety of cell stains for Nissl bodies including cresyl violet and thionin (Luo and Jackson, 1999; Kamme et al., 2003; Ginsberg and Che, 2004) for downstream genetic applications. Discrimination of adjacent cell types enables the differentiation of neuronal cells from glia, vascular epithelia, and other non-neuronal cell types within the brain and is optimal for understanding pathogenetic mechanisms underlying the selective vulnerability of distinct neuronal populations in neurodegenerative disorders.

The present chapter will review studies directed at understanding the genetics of human neurological disease using regional and single cell gene array technologies. The findings described will concentrate on investigations using tissue derived from people with AD, PD, schizophrenia, multiple sclerosis (MS), and Creutzfeld–Jakob disease (CJD). We will also review expression-profiling data associated with normal aging, as age is a significant risk factor for many human neurodegenerative diseases. We will not discuss data obtained from animal models of human neurological disease since there are many pathobiological differences between rodent and non-human primate correlates of human neurodegenerative diseases.

Alzheimer's disease

AD is the most common cause of dementia and affects approximately 50% of people over 85 years of age (Evans et al., 1989) and 5% of those over the age of 60 years worldwide (Prentice, 2001). The pathologic hallmarks of this disease are the formation of amyloid-containing senile plaques (SPs) and hyperphosphorylated tau-containing neurofibrillary tangles (NFTs), which are distributed throughout association neocortex and limbic cortex including the hippocampus and entorhinal cortex. Although there is a widespread decline in various neurotransmitter-containing cell bodies as the disease progresses, the most consistent losses are seen in long projection neurons such as the cholinergic neurons of the basal forebrain (Davies and Maloney, 1976; Whitehouse et al., 1981; Mufson et al., 1989) and glutamatergic neurons within the entorhinal cortex and neocortex (Hyman et al., 1984, 1986). During the last several years, several investigations have been performed to determine the molecular profile of these and other pathological aspects of AD.

Determination of RNA within senile plaques and neurofibrillary tangles in AD

The polypeptide composition of SPs and NFTs has been characterized in the AD brain (Selkoe, 1997; Ginsberg et al., 1999b). However, little data exists on the non-proteinaceous components of these lesions. To determine the presence of RNA species in SPs and NFTs within brain, tissue containing these lesions were stained using acridine orange (AO) histofluorescence, alone or in combination with thioflavine-S (TS) and immunohistochemistry to identify SPs and NFTs in AD and related neurodegenerative diseases (Ginsberg et al., 1997, 1998). A quantitative analysis of double-labeled sections revealed that approximately 55% of TS-positive SPs are AO positive, whereas approximately 80% of TS-stained NFTs are AO positive (Ginsberg et al., 1997, 1998). The observed sequestration of RNA in SPs and NFTs in AD and related disorders formed the foundation for single cell gene profiling analysis of these lesions.

Single cell gene array analysis of hippocampal senile plaques in AD

Single cell gene profiling analysis was applied to SPs immunostained for amyloid protein in sections of AD hippocampus. The expression profile of SPs was compared to individual CA1 neurons and surrounding neuropil of control-aged brains using single cell RNA amplification coupled with custom-designed cDNA arrays (Ginsberg et al., 1999a). SPs harbor two distinct populations of genes: high-abundance genes including those encoding the amyloid precursor protein (APP), tau, bcl-2, bax, cyclic AMP response element-binding protein (CREB), protein phosphatase (PP) subunits, and several ionotropic glutamate receptor subunits (GluRs), and low-abundance genes such as neurofilament subunits and glial-enriched mRNAs (Ginsberg et al., 1999a). The presence of mRNA species in extracellular SPs was validated by combined TS and in situ hybridization (ISH) using a probe directed against CREB as well as PCR amplification of APP isoforms from single SPs (Ginsberg et al., 1999a). Data from these studies confirmed the existence of multiple mRNA species within individual, extracellular hippocampal SPs in AD. Moreover, the expression profile of mRNAs amplified from SPs is predominantly neuronal. These observations suggest that SPs sequester the remnants of degenerating and/or dying neurons and their processes. Although microglial cells have been detected within SPs, and astrocytosis occurs around SPs (Itagaki et al., 1989; Eikelenboom and Veerhuis, 1996; Ginsberg et al., 1997), relatively low levels of glial-derived mRNAs have been reported in SPs, which may reflect a lower abundance of RNAs in glial cells (Sarnat et al., 1987).

Single cell gene analysis of hippocampal NFTs in AD

The factor(s) initiating the formation of NFTs in AD remain unknown, despite being investigated with a variety of histopathological and biochemical methodologies. We have hypothesized that alterations in the expression of specific mRNAs may reflect mechanisms underlying the formation of NFTs in affected neurons (Ginsberg et al., 2000). In this investigation, the relative abundance of multiple mRNAs was assessed in NFT-bearing versus normal hippocampal CA1 neurons microaspirated from AD and neuropathologically confirmed normal aged control brains (Fig. 1A and B). Each brain was confirmed to have abundant cytoplasmic RNAs by AO histofluorescence prior to use in the expression profiling studies (Ginsberg et al., 1997, 1998).

Single cell RNA amplification was performed in combination with high-density microarrays [>18,000 expressed-sequence tagged cDNAs (ESTs)] and custom-designed (approximately 120 cDNAs/ESTs relevant to neuroscience) array platforms (Ginsberg et al., 2000). In this investigation, total hybridization signal intensity on both platforms was down-regulated in single AD NFT-bearing compared to normal CA1 neurons by approximately 28–36%, consistent with other semi-quantitative studies of polyadenylated mRNA expression in AD (Griffin et al., 1990; Harrison et al., 1991). Notably, a correspondence of 85% between the two array platforms was found (73/86 cDNAs/ESTs) in terms of the direction (e.g., down-regulation, up-regulation, or no change) of select genes with dual representation (Ginsberg et al., 2000). For example, all five ESTs linked to the PP2A-Aβ subunit (Unigene-NCBI annotation PPP2R1B) on the high-density microarray were down-regulated in NFT-bearing neurons, consistent with a similar level of relative

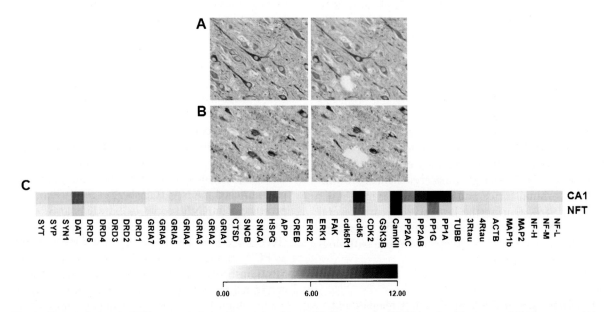

Fig. 1. Custom-designed cDNA array analysis of normal and neurofibrillary tangle (NFT)-bearing hippocampal CA1 neurons. (A) Photomicrographs illustrating a neurofilament-immunoreactive normal CA1 neuron before (left) and after aspiration (right) from an aged control brain. (B) Phosphorylated tau immunoreactive CA1 NFT bearing neurons in Alzheimer's disease (AD) brain before (left) and after aspiration (right). Scale bar (A and B) = 25 μm. (C) Dendrogram illustrating relative expression levels of transcripts from normal control CA1 pyramidal neurons and NFT-bearing CA1 neurons in AD.

down-regulation observed on the custom-designed cDNA array platform (Ginsberg et al., 2000). Relative to normal CA1 neurons, those harboring NFTs in AD revealed significant reductions in several classes of mRNAs that are known to encode proteins implicated in AD neuropathology including protein phosphatases/kinases, cytoskeletal elements, and GluR subunits. In addition, the synaptic-related markers synaptophysin (2.2-fold), synaptotagmin (2.0-fold), synapsin I (2.5-fold), α-synuclein (2.5-fold), and β-synuclein (2.3-fold) were all down-regulated in NFT-containing cells (Ginsberg et al., 2000), suggesting a role for NFT formation in synaptic dysfunction. APP mRNA was also decreased approximately 2-fold on both high-density and custom-designed cDNA arrays. Together these findings suggest that neurons harboring NFTs express less APP mRNA in affected regions of the AD brain. Moreover, a 2 to 4-fold decrease in the expression of the mRNAs for dopamine (DA) receptors DRD1–DRD5 and the DA transporter was seen in NFTs in AD versus non-tangle-bearing neurons in control brains (Ginsberg et al., 1999b, 2000). These findings are consistent with data showing reduced DRD2 receptor binding in the AD hippocampus (Joyce et al., 1993; Ryoo and Joyce, 1994). These observations underscore the advantages of single cell mRNA analyses since antibodies and ligand-based studies have not been able to discriminate unequivocally between DA receptor subtypes. A summary of representative mRNA changes in NFT-containing CA1 neurons is presented in Fig. 1C.

Evaluation of the high-density cDNAs microarrays demonstrated that several ESTs are up-regulated in CA1 neurons displaying NFTs in AD. However, of the 50 most up-regulated ESTs, just one was linked to a protein of known cellular function, cathepsin D (Ginsberg et al., 2000). Furthermore, an approximate 2-fold up-regulation of cathepsin D was observed on the custom-designed array platform, and paired helical filament tau (PHFtau) positive NFTs co-localize with increased cathepsin D immunoreactivity in CA1 neurons (Ginsberg et al., 1999b, 2000). Up-regulation of the acid hydrolase cathepsin D was consistent with a growing literature indicating that activation of the endosomal-autophagic-lysosomal system is an early alteration in AD, suggesting that it may act as a novel biomarker of this disorder (Cataldo et al., 1995, 1996; Nixon et al., 2000). These data illustrate an experimental strategy to employ both high-density and custom-designed cDNA microarrays and single cell RNA amplification to identify changes in the expression levels of various transcripts in NFT-bearing AD as compared to normal control CA1 neurons.

Regional gene expression profiling in the hippocampus in AD

Some of the first expression profiling studies in the hippocampus in AD were performed by Paul Coleman's group at University of Rochester School of Medicine (Cheetham et al., 1997; Chow et al., 1998). RNA was amplified from single CA1/subiculum pyramidal neurons from early- and late-stage AD brains using the amplified antisense (aRNA) method of Eberwine (Eberwine et al., 1992). Radiolabeled-RNA probes were hybridized to custom-designed membranes containing 20 selected cDNAs. Multivariate analysis revealed that 5 of the 20 genes were differentially regulated between the early- and late-stage AD groups. Specifically, cyclin D1, HSP27 chaperone, and glutamate decarboxylase were down-regulated in late compared to early AD, whereas wee1 and α1-antichympotrypsin were up-regulated in late compared to early AD (Chow et al., 1998). These results indicated that dynamic changes in gene expression related to neurotransmitter metabolism, stress responses, and potentially cell-cycle regulation could be detected in the hippocampus during the progression of AD. This small seminal study set the stage for large-scale regional analysis of hippocampal gene expression changes in the aged control and AD brain (Loring et al., 2001; Colangelo et al., 2002; Blalock et al., 2004; Lukiw, 2004).

Recently, a large-scale microarray analysis was performed on frozen hippocampus obtained from aged controls and incipient, moderate or severe AD cases using the Affimetrix HG-U133A GeneChip (Blalock et al., 2004). To control for multiple comparison and false-positive errors, target array genes that produced no signal or were unidentified (i.e., ESTs not linked to genes of known function) were excluded from the analysis. Resulting target hybridization signal intensities were correlated with the Mini Mental State Examination (MMSE) score for cognitive assessment and NFT pathology scores across all groups (Blalock et al., 2004). This investigation revealed incipient AD-related changes in genes encoding transcription factors and regulators of cellular proliferation and differentiation, including tumor suppressors (e.g., retinoblastoma-binding proteins), oligodendrocyte growth factors (e.g., sterol regulatory element-binding factor), and protein kinase A (PKA) modulators (e.g., A kinase anchoring molecules) (Blalock et al., 2004). These findings led the authors to suggest a novel hypothesis for AD pathogenesis, whereby a genomically orchestrated up-regulation of tumor suppressor-mediated differentiation and oligodendrocyte stimulation induces the spread of the disease processes along myelinated axons. In addition, genes involved in cell adhesion, apoptosis, lipid metabolism, and inflammatory processes were up-regulated in moderate/severe AD, whereas genes involved in energy metabolism, protein folding and transport and synaptic transmission were down-regulated (Blalock et al., 2004). Interestingly, 89 genes within these functional categories were shown to correlate with both antemortem MMSE and postmortem NFT index (Blalock et al., 2004), suggesting a contribution of NFTs to transcriptional dysregulation during the progression of AD.

The results of this study are consistent with and elaborate upon an earlier report of hippocampal gene expression changes using frozen CA1 samples from control and mild/moderate AD subjects analyzed on the Affimetrix U95Av2 GeneChip (Colangelo et al., 2002). Statistical strength was controlled by including only those target genes that differed by a factor of 3 or more between control and AD cases. The investigators found a general depression in gene transcription in the CA1 region in AD brain, with significant down-regulation of functional classes of genes encoding transcription and chromatin modifying factors (Colangelo et al., 2002). In addition, genes encoding synaptic regulatory elements (e.g., synapsin I) and neurotrophic factors (e.g., brain-derived neurotrophic factor) were also significantly down-regulated. Intriguingly, choline acetyltransferase (ChAT, the synthetic enzyme for acetylcholine) was down-regulated 3-fold in the CA1 of the mild/moderate AD cases (Colangelo et al., 2002), perhaps foreshadowing the loss of ChAT activity observed in the hippocampus of end-stage AD subjects (Perry et al., 1977; Davies, 1979; Henke and Lang, 1983). Similar to Blalock et al. (2004), the authors also report an up-regulation of genes encoding apoptotic and pro-inflammatory factors in AD (Colangelo et al., 2002). Taken together, these studies support the notion that the hippocampus, while undergoing a general reduction in transcription that may compromise metabolic function and synaptic transmission, is actively responding to cytotoxic stressors during the progression of AD. These specific alterations in gene expression may represent physiological causes or consequences related to hippocampal degeneration and provide a platform for downstream proteomic and mechanistic studies to understand the pathophysiology underlying memory deficits in AD.

Regional gene expression profiling in frontal and temporal neocortex in AD

The frontal cortex exhibits synaptic loss in AD (DeKosky and Scheff, 1990; Scheff and Price, 2003) that correlates with cognitive function. Hence, synaptic gene expression changes in the AD frontal cortex may provide mechanistic clues to perturbations in cortical synaptic efficacy that contribute to AD symptoms. Gene expression studies in the frontal cortex in AD include the down-regulation of several genes encoding synaptic regulatory elements (Yao et al., 2003). In one study, RNA from frozen samples of the superior frontal gyrus of aged control and AD subjects was amplified using biotinylated nucleotides and hybridized to Affimetrix HuFL oligonucleotide arrays. Quantitative analysis focused only on targets related to synaptic function (12 genes out of 6800 cDNAs/ESTs on the array). All 12 genes were statistically "present" using an average difference algorithm that compares the signal intensities of perfect to mismatched target oligonucleotides. Of these 12 genes, there was a specific down-regulation in those encoding synaptic trafficking proteins related to vesicle biogenesis, including dynamin I and amphiphysin (Yao et al., 2003). These changes were corroborated by real-time quantitative PCR (qPCR), and the down-regulation of dynamin I was further corroborated by immunoblotting.

An additional insight into neocortical synaptic dysfunction came from a high-throughput cDNA microarray study which screened 6794 human genes (Incyte VI human microchips) in samples of superior temporal cortex from cases characterized as moderate dementia (clinical dementia rating CDR = 2) (Ho et al., 2001). Thirty-two genes (25 known and 7 ESTs not linked to known genes) were identified with at least 1.8-fold different expression relative to cases characterized by normal cognitive status (CDR = 0). Specifically, the authors noted a down-regulation of the synaptic vesicle protein synapsin IIa, which plays an important role in vesicle docking and fusion with the presynaptic plasma membrane (Greengard et al., 1993). Taken together, these studies indicate that putative deficits in neocortical synaptic function in AD may be related to selective decreases in synaptic transcripts regulating presynaptic vesicle formation and trafficking. Vesicle turnover and trafficking deficits may ultimately result in reduced membrane fusion probability for neurotransmitter release. The pathogenic mechanisms underlying the observed reductions in synaptic transcripts in the neocortex are unclear. However, evidence from the hippocampus and neocortex presented above and elsewhere (Callahan and Coleman, 1995) suggests that reduced synaptic mRNA expression may be related to NFT deposition, which target corticocortical projection neurons in AD (Hof et al., 1995; Morrison and Hof, 2002).

Regional gene expression profiling in other AD-related brain regions

Loring and colleagues examined gene expression profiles in samples of striatum, cerebellum, amygdala, and cingulate cortex from control and AD subjects using the Incyte Unigene Lifearray microarray (Loring et al., 2001). After removing all genes that varied between control and AD striatum and cerebellum, two regions relatively spared in AD, 118 genes were differentially expressed at least 2-fold between control and AD amygdala and cingulate cortex, two regions involved in higher cognitive function that accumulate significant AD-related pathology (Brady and Mufson, 1990; Vogt et al., 1997). Of these genes, 87 were down-regulated and 31 were up-regulated in AD relative to controls. Down-regulated genes were members of several functional classes including signal transduction (e.g., glycogen synthase kinase, mitogen-activated protein kinase), energy metabolism (e.g., mitochondrial enzymes), and protein degradation, whereas up-regulated genes were related to inflammation (e.g., complement C3), stress responses (e.g., heat-shock proteins), and protein synthesis (Loring et al., 2001).

Single cell analysis of cholinergic basal forebrain (CBF) neurons in AD

CBF neurons supply the majority of cholinergic innervation to the cerebral cortex and hippocampal formation, and are key anatomic substrates of memory and attention systems in the brain (Mesulam et al., 1983; Baxter and Chiba, 1999; Bartus, 2000). CBF neurons of the nucleus basalis (NB) are selectively vulnerable in AD (Davies and Maloney, 1976; Whitehouse et al., 1981) and NB neuron degeneration correlates with disease duration and cognitive decline (Mufson et al., 1989; Bierer et al., 1995). However, the molecular mechanism(s) associated with CBF cytopathology and cellular dysfunction is unknown. Single cell RNA amplification and custom-designed cDNA array technology was employed to examine the expression of several functional classes of mRNAs found in NB neurons from normal aged and AD subjects (Mufson et al., 2002; Counts et al., 2003). Synaptic transcripts were selectively down-regulated in these neurons in AD, with significant reductions in synaptophysin and synaptotagmin but not synaptobrevin or SNAP29 mRNA (Fig. 2A) (Mufson et al., 2002). Intriguingly, synaptotagmin function is related to vesicle–presynaptic membrane fusion and neurotransmitter release, consistent with the molecular evidence presented above suggesting that perturbations in presynaptic vesicle trafficking comprise a common event in vulnerable neuronal populations in AD. In

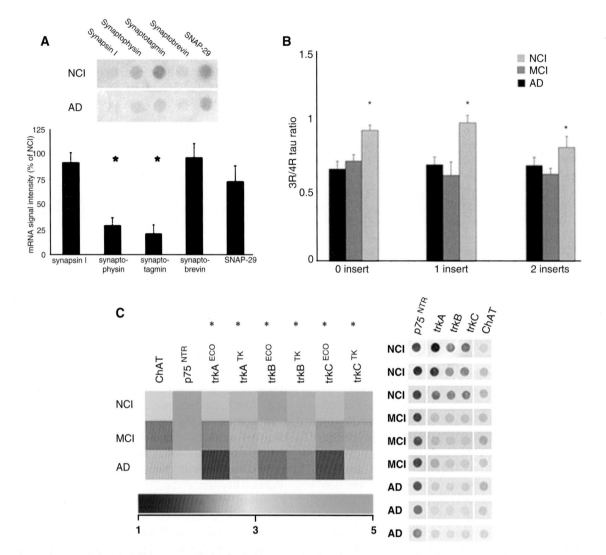

Fig. 2. Custom-designed cDNA array analysis of mRNA expression levels in cholinergic basal forebrain (CBF) neurons during the progression of AD. (A) Representative gene array (top) and histogram (bottom) illustrating down-regulation of synaptophysin ($25 \pm 7\%$ of control) and synaptotagmin ($23 \pm 6\%$ of control) expression in AD CBF neurons as compared to aged controls. $*p = 0.01$ via ANOVA with Neuman–Keuls post hoc test for multiple comparisons. (B) Histogram showing a significant reduction in the ratio of 3-repeat tau to 4-repeat tau (3R/4R) transcripts in CBF neurons in subjects with mild cognitive impairment (MCI) or AD relative to subjects with no cognitive impairment (NCI). The decrease in 3R/4R tau was significant regardless of the number of inserts (0, 1 or 2) in the 5′ region of the tau transcripts (see text). $*p < 0.001$ via ANOVA with Neuman–Keuls post hoc test for multiple comparisons. (C) Dendrogram (left) and representative gene array (right) showing relative expression levels of $p75^{NTR}$, trkA, trkB, trkC, and ChAT transcripts derived from individual CBF neurons from NCI, MCI, and AD subjects. No significant differences are found for ChAT and $p75^{NTR}$ gene expression. In contrast, statistically significant down-regulation ($*p < 0.001$) of trkA, trkB, and trkC are observed in MCI and AD. ESTs identifying ECD and TK domains display down-regulation. The decrement of trk gene expression in MCI is intermediate relative to AD, indicating a step-down effect in expression levels from NCI to MCI to AD.

contrast to synaptic transcripts, mRNAs encoding APP and Notch were unchanged between the control and AD groups, whereas cathepsin D mRNAs were up-regulated in AD. In addition, subunits of protein phosphatase PP1 (Unigene-NCBI annotation PPP1CA and PPP1CC) but not subunits of protein phosphatase PP2 (PPP2R1A and PPP2R1B) mRNAs were down-regulated in CBF neurons in AD (Mufson et al., 2002). This observation is interesting in light of observations that experimental down-regulation of PP1 activity leads to increased tau hyperphosphorylation (Merrick et al., 1997; Liu et al., 2005) and that CBF neurons are an early target for NFT formation (Braak and Braak, 1991; Sassin et al., 2000).

To investigate further the role of tau in CBF neuronal dysfunction, the expression of tau transcripts was analyzed in single NB neurons aspirated from postmortem tissue obtained from subjects with an antemortem clinical diagnosis of no cognitive impairment (NCI), mild cognitive impairment (MCI), or early-stage AD (Ginsberg et al., 2006a). The adult human brain contains six tau isoforms ranging from 48 to 67 kDa, which are expressed through alternative splicing of a single tau gene on chromosome 17 (Goedert et al., 1989a, b). Three tau isoforms contain three tandem repeats in the carboxy-terminus of the molecule (3Rtau), while three isoforms contain four tandem repeats (4Rtau) in this region. The expression levels of the six tau transcripts within individual CBF neurons did not vary among the NCI, MCI, and AD cases. However, when a 3Rtau/4Rtau ratio was calculated, a significant shift in the ratio was observed with a decrement in 3Rtau in relation to 4Rtau levels for all tau transcripts examined in CBF neurons from the MCI and AD groups (Fig. 2B) (Ginsberg et al., 2006a). These data suggest a subtle, yet pervasive shift in the dosage of 3Rtau and 4Rtau within vulnerable CBF neurons in MCI and AD (Ginsberg et al., 2006a). Interestingly, a similar shift does not occur during normal aging (Ginsberg et al., 2006a). Shifts in the ratio of tau transcripts may be a fundamental mechanism whereby normal tau function is dysregulated contributing to NFT formation.

To investigate additional molecular mechanisms underlying CBF neurodegeneration in AD, custom-designed cDNA array analysis was also used to test whether levels of neurotrophin receptor transcripts are altered in CBF neurons during the progression of AD (Ginsberg et al., 2006b). CBF neuron survival requires appropriate binding, internalization and retrograde transport of the prototypic neurotrophin, nerve growth factor (NGF), which is synthesized and secreted by cells in the cortex and hippocampus (Sofroniew et al., 2001; Lad et al., 2003; Mufson et al., 2003). NGF exerts functional consequences for cholinergic NB neuronal survival by interacting with at least two neurotrophin receptors, the low-affinity pan-neurotrophin receptor $p75^{NTR}$ and the high-affinity NGF-specific receptor tyrosine kinase trkA (Kaplan and Miller, 2000; Sofroniew et al., 2001; Teng and Hempstead, 2004). trkB and trkC are also localized to CBF neurons, albeit at lower levels than trkA (Salehi et al., 1996; Mufson et al., 2002). Trk receptors, along with $p75^{NTR}$, are produced within CBF neurons and transported anterogradely to the cortex where they bind NGF and other members of this family of neurotrophins (Kaplan and Miller, 2000; Sofroniew et al., 2001; Teng and Hempstead, 2004). Custom cDNA array analysis revealed that individual CBF neurons displayed a significant down-regulation of trkA, trkB, and trkC mRNA expression during the progression of AD (Fig. 2C) (Ginsberg et al., 2006b). An intermediate reduction was observed in MCI, with the greatest decrement in mild-to-moderate AD as compared to controls. Importantly, trk down-regulation is associated with cognitive decline measured by a Global Cognitive Score (GCS) comprised of a battery of 19 different neuropsychological tests (Bennett et al., 2002) and the MMSE (Ginsberg et al., 2006b). In contrast, there was a lack of regulation of $p75^{NTR}$ expression (Ginsberg et al., 2006b). A 'step down' dysregulation of trk expression may, in part, underlie cholinergic NB neuron demise associated with cognitive impairment in AD. Trk defects may also represent a molecular marker for the transition from NCI to MCI, and from MCI to frank AD.

Single cell profiling of galanin hyperinnervated CBF neurons in AD

In late-stage AD, fibers within the CBF containing the neuropeptide galanin (GAL) thicken and hyperinnervate NB neurons (Chan-Palay, 1988; Mufson et al., 1993; Counts et al., 2003). Functionally, GAL has been suggested to play a crucial role in the regulation of cholinergic neuronal tone and in the regulation of amyloid in AD (Counts et al., 2003; Ding et al., 2005). However, the molecular consequences of this unique plasticity response in AD are unclear. To investigate this phenomenon, single non-hyperinnervated NB cholinergic neurons from control cases and NB cholinergic neurons displaying GAL fiber overexpression (GAL+) or those lacking GAL innervation (GAL−) from AD cases were microaspirated and processed for RNA amplification and custom designed cDNA array analysis (Counts et al., 2003, 2005). Results indicate a significant (~90%) down-regulation of PP1 subunit mRNAs in GAL− NB neurons from AD brains compared to NB neurons from control brains; however, there was no down-regulation in PP1 subunit transcripts in the GAL+ NB neurons from AD cases (Counts et al., 2003). This apparent maintenance of PP1 expression by GAL hyperinnervation may be neuroprotective for remaining CBF neurons by preventing tau hyperphosphorylation (see above). Additional preliminary data show that ChAT transcripts are up-regulated in GAL+ but not GAL− NB neurons in AD compared to controls and that a down-regulation of the neuronal activation marker c-fos in GAL−NB neurons is not found in GAL+ NB neurons in AD (Counts et al., 2005). These observations suggest that GAL hyperinnervation is neuroprotective and regulates cholinergic tone in the CBF neurons. Interestingly, examination of the three known GAL receptors, GALR1, GALR2, and GALR3, revealed no change in the expression of these genes with CBF neurons between control, MCI, and AD cases (Counts et al., 2006).

Summary of gene expression profiling in AD

Studies profiling gene expression changes in different brain areas vulnerable to AD pathology have utilized a wide variety of array platforms and statistical criteria to analyze the genetic profiles at both the regional level and with single cell resolution. As such, it is not surprising that the studies outlined above have identified many unique gene expression changes related to AD. However, the results from these studies indicate that several functional classes of genes are similarly altered in AD independent of brain region or array methodology (Fig. 3). For instance, select transcripts encoding synaptic proteins, particularly those regulating vesicle trafficking and neurotransmitter release, appear to be down-regulated in vulnerable neuron populations in AD, supporting the notion that impaired synaptic efficacy in brain regions mediating cognitive function (e.g., hippocampus, CBF, polymodal association neocortex) contributes to the clinical presentation of AD. Moreover, putative functional deficits resulting from select alterations in synaptic transcript stoichiometry may provide a molecular underpinning to the observed structural synaptic degeneration described in these same areas in MCI and AD (Scheff et al., 2006). The consistent observation that reductions in specific synaptic-related markers are associated with the deposition of NFTs suggests that therapies aimed at preventing NFT formation may ameliorate the disease process in part by restoring the expression of synaptic components vital to normative function. Other functional classes of genes that are down-regulated in AD brain include those regulating transcription, energy metabolism, and protein turnover, suggesting a general

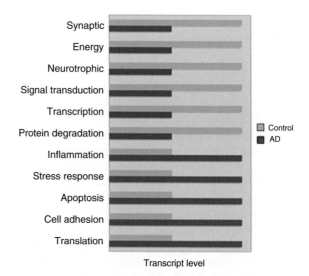

Fig. 3. Schematic diagram summarizing general increases or decreases in functional classes of mRNAs in the AD brain relative to the aged control brain, based on regional and single cell gene expression analysis described in the text.

depression of homeostatic cellular activity. Whether these changes in gene expression reflect causes or consequences of pathogenic events in AD remains to be determined. For instance, it is tempting to speculate that defective signaling along pathways regulating energy metabolism may comprise a seminal event in AD pathogenesis, as reduced cell charge would presumably compromise overall cellular function. On the other hand, these studies also show that classes of genes regulating pro-inflammatory responses, stress responses, and pro- and anti-apoptotic pathways are up-regulated in AD brain. These increases in gene expression may be more reflective of a veritable siege state within vulnerable neuronal populations that underlie neurodegenerative processes.

Several approaches have been recently employed to establish gene expression profiles that hold greater promise for detecting pathogenetic events in AD. For instance, the analysis of expression profiles in vulnerable neuronal populations from subjects with MCI, a putative preclinical stage of AD, may point to the earliest changes in gene expression related to the onset of dementia. Along these lines, the observation discussed above that the 3Rtau/4Rtau transcript ratio is reduced in CBF neurons in subjects with MCI suggests that alterations in tau isoform mRNAs may contribute to the abundant NFT deposition that occurs in these cells during the progression of AD (Ginsberg et al., 2006a). Another study detected very early AD-related gene expression changes by comparing the genetic signature of entorhinal cortex layer II (ECII) neurons in control brains with non-NFT and NFT-bearing ECII neurons in AD brains (Dunckley et al., 2006). For example, PAK7, which stabilizes microtubules and promotes neuritogenesis, was down-regulated ~2.7-fold in AD non-NFT neurons compared to controls (Dunckley et al., 2006). Thus, genes that are most likely early contributors to NFT formation may show dysregulation in AD non-NFT neurons compared to control neurons.

Recently, an intriguing strategy using skin fibroblasts, rather than brain tissue, has been used to generate gene expression profiles from subjects with familial, early-onset forms of AD associated with presenilin and APP mutations (Hardy, 1997) as compared with fibroblasts from non-demented siblings (Nagasaka et al., 2005). This approach involved comparing gene expression profiles of skin fibroblasts derived from three families with APPswe, APParc, and PS1 H163Y mutations, respectively. Although the genes analyzed were not reported in the text (but can be found online at the Alzheimer Research Forum web page (www.alzforum.org)), the authors identified a common gene expression profile among the three carrier groups independent of gene mutation that distinguished them from their siblings (Nagasaka et al., 2005). While the relevance of these peripheral gene expression changes to familial AD-related events in the brain remain unclear, results may point to systemic pathogenetic alterations associated with the onset of aggressive, early-onset forms of AD that have important implications for the more common, sporadic forms of the disorder.

Parkinson's disease

PD is the second most frequent chronic neurodegenerative disease after AD. It is characterized by a series of cardinal clinical symptoms including tremor, bradykinesia, hypokinesia, and postural disturbance. Key pathological features are the loss of striatal DA, a progressive degeneration of dopaminergic perikarya within the substantia nigra pars compacta (SN_{pc}) and the presence of hyaline eosinophilic cytoplasmic inclusions known as Lewy bodies (LBs) (Fearnley and Lees, 1991). While the physiological relevance and clinical significance of dopaminergic neurons within the SN_{pc} are well recognized, the pathogenetic mechanisms underlying their degeneration and the formation of LBs are the subject of intense investigation. The application of microarray technology to study gene expression profiles in affected tissues derived postmortem from PD subjects may reveal alterations in mRNA levels related to the selective vulnerability of the nigrostriatal DA system in this disorder.

Regional gene profiling of the substantia nigra in PD

In a recent study, high-density microarray analysis performed on RNA extracted from the SN_{pc} of PD patients revealed alterations in the expression of 137 genes, with 68 down-regulated and 69 up-regulated (Grunblatt et al., 2004) (Fig. 4). The down-regulated genes encoded functional classes of proteins involved in signal transduction, dopaminergic transmission, ion transport, and energy metabolism. Of particular interest was the down-regulation of SKP1, a single copy gene product involved in the formation of E3 ubiquitin ligase and several proteasome subunits. Decreased ubiquitin–proteasome pathway activity in SN_{pc} neurons may account for the abnormal accumulation of proteins such as α-synuclein, the main component of LBs (Spillantini et al., 1997; Trojanowski and Lee, 1998). This study also revealed an up-regulation of genes involved in a number of biological processes including cell adhesion and inflammation/stress (Grunblatt et al., 2004). A notable up-regulated gene product in the latter category was EGLN1, an iron and oxidative stress sensor that negatively regulates cell survival/proliferation, glucose and iron metabolism genes (Lee et al., 2004), potentially exacerbating antioxidant and survival mechanisms within SN_{pc} neurons.

A separate microarray analysis compared the genetic signature of postmortem SN_{pc} tissue

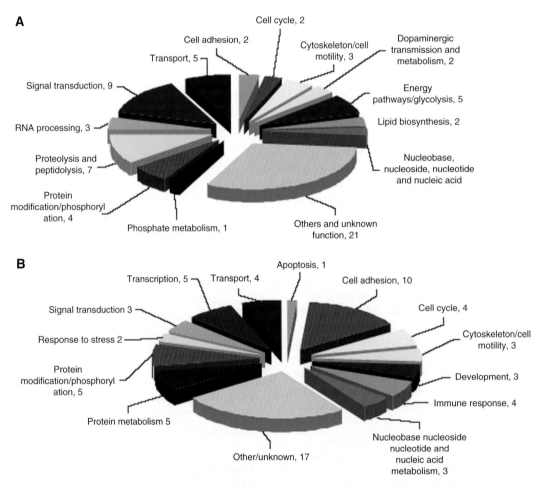

Fig. 4. Functional cluster analysis of genes involved in biological processes categorized according to gene ontology in aged control and Parkinson's disease (PD) subjects. Pie chart showing the distribution of down-regulated (A) and up-regulated (B) genes in PD samples compared to control brains. The number of altered genes in the different functional groups is indicated. Reprinted with permission from Grunblatt et al. (2004).

obtained from patients diagnosed with PD, progressive supranuclear palsy (PSP), frontotemporal dementia with parkinsonism (FTDP), and aged control subjects (Hauser et al., 2005). This investigation used Affymetrix U133A oligonucleotide arrays and identified 142 genes that were differentially expressed between PD and controls. In addition, 96 genes were differentially expressed between the combined PSP and FTD cases as compared to aged controls. Increases in the heat-shock proteins HSPA1A and HSPA1B in PD, PSP, and FTDP were observed compared to control SN_{pc} (Hauser et al., 2005). These data suggest a common response in progressive, late-onset neurological disorders to mitigate the toxic effects of misfolded protein. In addition, there was a 2-fold down-regulation in the PARK5 familial PD-linked ubiquitin carboxyl-terminal–hydrolase L1 (UCH-L1) gene in PD (Liu et al., 2002) but not PSP or FTDP, lending further support to the notion that deficits in the ubiquitin–proteasome pathway contribute to PD pathogenesis. Down-regulation of UCH-L1 is also observed in the adult male zebra finch within specific neurons responsible for song generation that undergo spontaneous replacement, providing a provocative comparator between neurodegeneration and neuronal replacement (Lombardino et al., 2005). Interestingly, PD, but not PSP or FTDP, cases expressed decreased levels of transcripts associated with synaptic vesicle trafficking and neurotransmitter secretion. For example, there was a 2-fold reduction in the gene encoding synaptobrevin, a trafficking protein required for docking of multiple membrane proteins and for neuronal exocytosis (Sorensen, 2005). The mRNA encoding STXBP1, which binds to syntaxin on the target membrane, and synaptotagmin were also reduced significantly in the SN_{pc} in PD. Finally, there was a PD-selective reduction in nuclear-encoded mitochondrial transcripts (e.g., cytochrome *c* oxidase subunits, mitochondrial ATPases) in the SN_{pc} compared to controls (Hauser et al., 2005). Taken together, these profiling studies reveal alterations in the transcription of multiple genes either independently or cooperatively during the progression of PD, leading potentially to reduced energy metabolism, protein turnover, synaptic transmission, and increased stress-related responses. These physiological changes may contribute to the demise of dopaminergic neurons in the SNpc resulting in the PD clinical phenotype.

Gene expression profiling of Lewy body-containing SN_{pc} neurons in PD

LBs are a pathological hallmark of PD and are found in the cell body, axon, and dendrites of neurons (Trojanowski and Lee, 1998; Lu et al., 2005). LBs contain multiple proteins, of which α-synuclein is the chief component (Spillantini et al., 1997). The exact function of these cellular inclusions remains a mystery. LBs have been proposed to be neuroprotective by sequestering the toxic form of α-synuclein (Conway et al., 2000, 2001). Alternatively, LBs may occur as an epiphenomenon of the primary pathology and have little or no effect on neuronal survival. For instance, LBs are also composed of abnormal neurofilaments, which may reflect a general aberration of the cytoskeleton (Hill, 1996). Therefore, it is of great importance that the functional significance of this inclusion be resolved. In a recent investigation, Lu et al. (2005) used laser capture microdissection (LCM) combined with arbitrarily primed PCR of isolated RNA to generate a molecular fingerprint of SN_{pc} dopaminergic neurons with and without LBs in PD. An antibody directed against human α-synuclein was used to identify LB-harboring SN_{pc} neurons and non-LB-containing neurons were identified by neuromelanin (Lu et al., 2005). Differentially expressed EST fragments of interest were then validated by qPCR and sequenced. In general, DA neurons without LBs preferentially expressed genes beneficial to cell survival (e.g., KLHL1, Stch, and BPAG1), whereas genes preferentially expressed in DA cells harboring LBs appear to support a cytotoxic role of LBs. For example, the up-regulation of ubiquitin specific protease 8 (USP8) in LB-containing neurons suggests that the high state of ubiquitination and related protein aggregation induces a compensatory up-regulation of de-ubiquitination enzymes (Lu et al., 2005). These findings are consistent with immunohistochemical investigations showing that LB-positive neurons contained pro-apoptotic molecules (i.e., caspase-3, Bax), whereas anti-apoptotic molecules (i.e., Bclx-x$_1$) were mainly found in LB-negative cells in the SN$_{pc}$ in PD (Hartmann et al., 2000, 2001, 2002). A caveat of this study is that α-synuclein is not a select marker for LBs, as many normal cells express this protein. Therefore, neuron selection may be confounded by false positives.

Summary of gene expression profiling in PD

It has been hypothesized that the accumulation of proteins, particularly α-synuclein, into LB inclusions in SN$_{pc}$ neurons in PD is the result of pathogenic alterations in the ubiquitin–proteasome pathway resulting in reduced protein catabolism. Gene expression profiles of vulnerable brain regions in PD described above support this view, as several genes encoding proteasomal subunits and enzymes related to protein ubiquitination are reportedly down-regulated in the SN$_{pc}$ of PD patients (Liu et al., 2002; Choi et al., 2004; Grunblatt et al., 2004). Thus, down-regulation in gene expression may result in reduced proteasomal pathway function and subsequent misfolded protein accumulation.

Expression profiling studies also revealed down-regulation of genes critical to mitochondrial function and energy production, supporting several lines of evidence from animal models that experimental perturbations (e.g., complex I inhibitors (Betarbet et al., 2000; Sherer et al., 2002)) of the respiratory chain results in selective PD-like pathology in the SN$_{pc}$. The mechanisms underlying these reductions in mitochondrial gene expression are unclear, but accumulations of oxidized DNA have been reported in DA neurons in the SN$_{pc}$ in PD (Alam et al., 1997; Zhang et al., 1999), potentially confounding transcription of respiratory chain components. Aberrant mitochondrial function results in subthreshold production of ATP and increased oxidative stress via excess production of reactive oxygen species (ROS), potentially creating a vicious cycle that further damages mitochondria and elevates ROS levels. Intriguingly, a recent proteomic study identified that the PARK5 gene product, UCH-L1, is down-regulated in the SN$_{pc}$ in PD and is a specific target for protein oxidation (Choi et al., 2004), suggesting a link between mitochondrial dysfunction, oxidative stress and reduced proteasomal activity.

Schizophrenia

Schizophrenia is a chronic psychiatric illness affecting approximately 1% of the general population (Johns and van Os, 2001). Clinical features appear in late-adolescence to early-adulthood, including a mixture of positive and negative symptoms relating to the integration of sensory and cognitive information in the brain. The etiology of schizophrenia remains unknown; however, several investigations implicate a multigenic predisposition, neurodevelopmental dysfunction, and environmental events as potential causative agents (Johns and van Os, 2001; Lencz et al., 2001; Lewis, 2002). Neuroimaging and neuropsychological testing of schizophrenics indicate that the temporal lobe, including the hippocampus and entorhinal cortex, is an affected area of the brain in this disorder (Kurtz et al., 2001; Gur and Gur, 2002). However, pathologic examination has revealed relatively little neurodegeneration in the schizophrenic brain (Trojanowski and Arnold, 1995; Arnold et al., 1998). The absence of frank neuropathology is a common caveat in determining molecular substrates for neuropsychiatric disorders due to the absence of hallmark pathologic lesions such as those observed in other neurological disorders (e.g., AD, PD). Since cognitive information organized within the frontal cortex as well as sensory cortical information is processed through the temporal neocortex and funneled to the hippocampus via the stellate cells of the entorhinal cortex, these areas make intriguing candidates for probing disease-related differences in gene expression related to schizophrenia.

Regional gene expression profiling in frontal cortex in schizophrenia

Postmortem regional studies of prefrontal cortex from schizophrenics using microarrays reveal consistent alterations in several classes of relevant transcripts including synaptic-related markers (e.g., synapsin II (Pongrac et al., 2002)), G protein-coupled signaling molecules (e.g., RGS4, 14-3-3 (Pongrac et al., 2002), PKA-regulatory subunit II, PKCβ (Hakak et al., 2001), oligodendrocyte-related markers (e.g. MAG myelin-associated glycoprotein and the Her3 neuregulin receptor (Middleton et al., 2002)), metabolic pathways (e.g., tricarboxylic acid cycle components (Middleton et al., 2002)) and apolipoprotein L isoforms (Mimmack et al., 2002). Although these regional studies provide an exciting molecular fingerprint of an important affected brain region in schizophrenia, these expression profiles cannot delineate alterations that occur selectively within affected neuronal populations from adjacent neuronal, non-neuronal, epithelial cells, and vascular elements.

Single cell gene profiling in the entorhinal cortex in schizophrenia

Single cell gene expression profiling technology was applied to the study of layer II/III entorhinal cortex stellate cells obtained from the brains of eight schizophrenics and nine age-matched neurologically normal controls (Hemby et al., 2002). High-density microarray platforms containing approximately 18,240 and 9710 genes were utilized in addition to custom-designed cDNA arrays containing approximately 48 ESTs that were regulated on the high-density platforms (Hemby et al., 2002). The large-scale arrays revealed significant differences between schizophrenics and control layer II/III entorhinal cortex stellate cells including various G protein-coupled receptor signaling transcripts, GluR subunits, and synaptic-related markers. Corroborative expression profiling using custom-designed arrays revealed similar significant decreases in G protein-signaling mRNAs including the stimulatory GTP-binding factor GNAS1, GluRs including the AMPA receptor GRIA3, kainate receptor GRIK1, and the NMDA receptor GRIN1, as well as synaptic-related markers such as synaptophysin, SNAP-23, and SNAP-25 (Hemby et al., 2002), suggesting that deficits in pre- and postsynaptic glutamatergic signaling within the entorhinal cortex may contribute to this complex neuropsychiatric disease. These findings may be influenced by the age of the sample population, environmental factors associated with neuropsychiatric disorders, medication history, and the constellation of clinical symptoms of schizophrenia. Thus, understanding the coordinated changes in hundreds of genes simultaneously in a human disease with multigenic etiologies such as schizophrenia will hopefully provide insight into common molecular pathogenic pathways underlying the disease process and may ultimately provide new targets for pharmacotherapy.

Multiple sclerosis

MS is an inflammatory disease of the central nervous system (CNS) characterized by myelin loss, gliosis, varying degrees of axonal pathology, and progressive neurological dysfunction. The etiology of MS is multifactorial, with underlying genetic susceptibility factors, which likely interact with as yet unknown environmental agents (Compston, 1994; Baranzini, 2004). Although the pathogenesis of MS is unclear, it has been hypothesized that a microbial mimic activates lymphocytes in the periphery and invades the CNS. This mimic becomes attached to endothelial cell receptors, which cross the blood–brain barrier (BBB) to enter the interstitial matrix of T cells. These cells are then reactivated in situ by fragments of myelin antigens exposed in the context of major histocompatibility complex (MHC) molecules located on the surface of antigen-presenting cells (e.g., macrophages, microglia, and, perhaps, astrocytes) resulting in the release of pro-inflammatory cytokines which further open the BBB and stimulate chemotaxis (Baranzini, 2004). This process activates a second wave of inflammatory cell recruitment as well as leakage of pathogenic antibodies and other plasma proteins into the CNS. Pathological hallmarks of the MS brain include predominantly white matter plaques.

Gene profiling in multiple sclerosis

An initial approach to analyze the transcriptional signature of the MS plaque examined a cDNA library made from mRNA isolated from postmortem CNS lesions in a patient with the primary progressive form of the disease (Becker et al., 1997). Despite a lack of statistical power, the authors identified numerous inflammatory genes and putative autoantigens in MS but not in normal control libraries. Another study used spotted custom cDNA microarrays to investigate levels of transcriptional expression of more than 5000 genes in normal white matter and CNS lesions from the same MS case (Whitney et al., 1999). Sixty-two differential gene expression patterns were found, including the Duffy chemokine receptor, interferon regulatory factor-2, and tumor necrosis factor alpha (TNFα) (Whitney et al., 1999). In a study using large-scale EST sequencing of cDNA libraries derived from active MS lesions, αB-crystallin mRNA was overexpressed in MS plaques as well as prostaglandin D synthase, prostatic-binding protein, ribosomal protein L17, and osteopontin (OPN, or early T-cell activation gene-1), which exhibits pleiotropic functions including roles in tissue remodeling, cell survival and cell immunity (Chabas et al., 2001). Another study evaluated relapsing-remitting MS cases and controls by probing cDNA arrays with radiolabeled cDNA (Ramanathan et al., 2001). This investigation revealed that 34 of 4000 genes examined differed significantly between MS and control subjects (Ramanathan et al., 2001). Recently, several microarray-based studies aimed at describing the molecular signature of the MS lesion have been published (Lock et al., 2002; Graumann et al., 2003; Mycko et al., 2003; Lindberg et al., 2004). Despite differences in plaque activity, tissue harvesting, processing procedures and data analysis, these studies reported an up-regulation of inflammation-related genes (mainly HLA class II and immunoglobulin genes), especially at the periphery of active plaques, which may reflect an active immune response in the lesions. The expression of immunoglobulin genes displayed a differential regulation based upon plaque stage (i.e., a greater expression in acute as compared to chronic plaques) (Lock et al., 2002; Graumann et al., 2003; Mycko et al., 2003; Lindberg et al., 2004). Similarly, the gene encoding granulocyte colony-stimulating factor (G-CSF), which plays a key role in T cell and dendritic cell function, was also up-regulated in many active plaques (Lock et al., 2002; Graumann et al., 2003; Mycko et al., 2003; Lindberg et al., 2004).

Creutzfeld–Jakob disease

CJD is a fatal neurodegenerative disease characterized by gliosis, neuronal loss, and spongiform degeneration. The majority of CJD cases are sporadic in nature with no known etiology, but ~15% of CJD cases are inherited due to coding mutations in the prion protein gene (PrP) or are acquired by infection (Prusiner, 1998). In CJD, prion proteins are converted into abnormal insoluble aggregates with a β-pleated sheet structure. CJD usually has a rapid disease course with death occurring within 6 months of diagnosis, urging the need for comprehensive screening of potential molecular targets. Using RNA from frontal cortex of control and CJD subjects to probe Affimetrix HGU133A microarrays, it has been demonstrated that the expression of 287 genes were altered more than 2-fold in CJD, including a prominent up-regulation of genes encoding immune and stress-response factors (e.g., CD58, S100β), cell cycle regulatory proteins (cyclin D1, CDK5 subunit p35) and cell-death genes (p53), whereas genes involved in G protein-signaling and synaptic regulation (SNAP25, synaptotagmin, synaptophysin) were decreased in CJD (Xiang et al., 2005). Intriguingly, four of the up-regulated genes in CJD (e.g., inositol 4,5 bisphosphate 3-kinase B) are also increased in the aging brain (see below), suggesting a link between CJD pathogenesis and age-related neurodegenerative processes.

Gene profiling in the aged brain

Aging is the major risk factor for many progressive, late-onset neurological disorders including AD, ALS, and PD. However, it is still controversial whether there is significant neuronal degeneration during the normal human aging process. A region of the brain that has received extensive investigation in normal aging is the hippocampal complex. The hippocampus proper, entorhinal cortex, and subicular complex compose the hippocampal formation and represent an important conduit between anatomical structure and physiological function, notably the transmission of information related to learning and memory under normal conditions and deficits in cognitive function associated with the degeneration of these structures in age-related neurodegenerative disease (Amaral, 1987; Rosene and Van Hoesen, 1987; Hyman et al., 1990). The hippocampus proper consists of anatomically defined subfields with distinct cell types: Ammon's horn (CA), which is partitioned into CA1 pyramidal and CA3 pyramidal neurons, with an intermediate CA2 zone. CA4 is considered part of the hilar region that abuts the dentate gyrus. The dentate gyrus consists principally of granule cells. The entorhinal cortex receives diverse multimodal cortical input from the neocortex and projects into the hippocampal formation within the dentate gyrus. In addition, there are differences in connectivity and physiological function between CA1, CA3, and dentate gyrus granule cells (Malinow et al., 2000; McCormick and Contreras, 2001; Dragoi et al., 2003). Therefore, cellular and molecular events involved in hippocampal complex function can be segregated based upon cytoarchitechtonics and connectional anatomy to identify unique region-specific processes associated with normal senescence and putative pathogenetic events associated with age-related neurodegenerative disease.

Single cell profiling of aged CA1 and CA3 hippocampal neurons

A recent report focused upon the genetic signature of individual neuronal populations within the aged human hippocampus (Ginsberg and Che, 2005). In this study, tissue was stained for Nissl substance using cresyl violet histochemistry in conjunction with population cell cDNA array analysis to compare and contrast expression profiles of CA1 and CA3 pyramidal neurons along with the surrounding hippocampal neuropil in postmortem brain tissues accrued from cognitively normal human subjects.

Expression profile analysis revealed interesting molecular signatures of the two cell types. Approximately 16% of the transcripts analyzed were expressed differentially between the two cell types (Ginsberg and Che, 2005). A summary of these region-specific differences in gene expression is presented in Fig. 5. In particular, 12 genes displayed significantly higher relative expression levels in CA1 as compared to CA3 pyramidal neurons, including several excitatory GluRs such as the kainate receptor KA1 (GRIK3) and NMDA receptor subunits NR1 (GRIN1) and NR2B (GRIN2B), GABA Aα1 (GABRA1), and GABA Aα2 (GABRA2) neurotransmitter receptors, and DA receptors DRD1, DRD2, and DRD5. Significantly, higher CA1 mRNA expression levels were also observed for activity regulated cytoskeletal-associated protein (arc), bcl-2, cyclooxygenase 2 (COX-2), and EEA1, an early endosome component (Ginsberg and Che, 2005).

In contrast, eight genes were found to have significantly higher relative expression levels in CA3 pyramidal neurons as compared to CA1 pyramidal neurons accessed from the same tissue sections including the excitatory AMPA receptors GluR1 (GRIA1) and GluR2 (GRIA2), cytoskeletal elements such as β- and γ-actin, AD-related APP and PD-related α-synuclein, the trkA high-affinity NGF receptor, and immediate-early gene junD (Ginsberg and Che, 2005). These differential levels of gene expression illustrate the importance of evaluating individual populations of neurons within a functional circuit.

The aforementioned findings were derived using tissue from aged normal subjects and reveal individual differences in gene expression, which presumably contribute to the unique 'molecular fingerprint' of CA1 and CA3 pyramidal neurons. These differences may represent molecular substrates for the selective vulnerability of these two cell types during aging, which may play a role in their degeneration during the progression of a disease. For instance, differences in GluR subunit stoichiometry between CA1 and CA3 neurons (e.g., higher levels of NMDA receptor subunits in CA1) may underlie the relative vulnerability of CA1 cells in AD (Hyman et al., 1984; Ginsberg et al., 1999b). Continued analysis of both cell types is warranted along with assessments of other principal neuronal cells of the hippocampal formation, including dentate gyrus granule cells, entorhinal stellate cells, subicular neurons as well as the constellation of interneuronal populations that reside throughout the hippocampus. Understanding

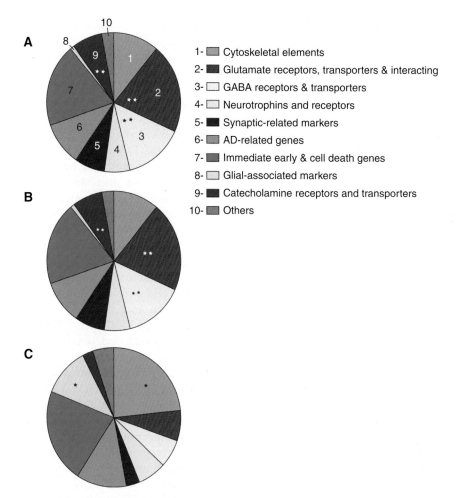

Fig. 5. Venn diagrams illustrating the percentage of representation of specific classes of transcripts to the overall hybridization signal intensity of CA1 pyramidal neurons (A), CA3 pyramidal neurons (B), and regional hippocampal dissections (C). No significant differences exist between the classes of transcripts for CA1 and CA3 pyramidal neurons. Note the significantly increased representation of cytoskeletal elements and glial-associated messages (single asterisk) in the regional hippocampal dissection as compared to CA1 and CA3 pyramidal neurons. Double asterisks indicate a significant increased representation of several classes of transcripts related to receptors and synaptic markers in CA1 and CA3 pyramidal neurons as compared to regional hippocampal dissections. Reprinted with permission from Ginsberg and Che (2005).

coordinated involvement of multiple genes within the human hippocampus provides insight into the molecular basis of structure–function integration of this important neuronal circuit. Moreover, experimental paradigms such as microarray analysis combined with regional and single cell RNA molecular fingerprinting may help develop novel agents that target systems and/or circuits that are affected selectively and/or specifically during aging as well as disease pathogenesis, thus reducing the potential problems of drug interactions and unwanted side effects.

Gene regulation during the course of normal aging within the frontal cortex

Alteration in frontal cortex function is also associated with the aging process and most likely plays a role in the cognitive decline seen in the elderly.

During a person's normal life span the time at which age-related brain changes occur is unclear. Lu et al. (2004) used Affymetrix HG-U95Av2 oligonucleotide arrays (~12,000 probe sets) for transcriptional profiling of human frontal cortex obtained from 30 subjects ranging in age from 26 to 106 years (Fig. 6). This study revealed that genes related to synaptic function and plasticity were among those most significantly affected in the aging human frontal cortex (Lu et al., 2004). For example, various neurotransmitter receptors that are centrally involved in synaptic plasticity showed significant down-regulation after the age of 40, including GRIA, GRIN2A, and GABRA subunits (Fig. 6). Several of these genes are also altered in aged human hippocampal neurons (Ginsberg and Che, 2005). In addition, the expression of genes that mediate synaptic vesicle release and recycling were also significantly reduced, notably synaptobrevin, synapsin II, RAB3A, and SNAPs. Classes of genes related to signal transduction that mediate long-term potentiation (LTP) and memory storage, including calmodulin 1 and CAM kinase IIα, were also age down-regulated (Lu et al., 2004). Furthermore, members of the protein kinase C (PKC) and mitogen-activated protein kinase (MAPK) signaling pathways displayed decreased expression. The activation state of PKC was also reduced, as indicated by decreased levels of activated phosphorylated forms. Together these observations suggest that calcium homeostasis and pre- and postsynaptic neuronal signaling may be affected in the aged frontal cortex.

Genes that were up-regulated in the aging human frontal cortex were associated with those mediating stress responses (e.g., heat-shock protein 70 and α-crystallin), antioxidant defense (non-selenium glutathione peroxidase, paraoxonase, and selenoprotein P), metal ion homeostasis (metallothioneins), inflammation (TNFα), and DNA repair (8-oxoguanine DNA glycosylase and uracil DNA glycosylase) (Fig. 6) (Lu et al., 2004). With respect to the observed increases in DNA repair gene expression, qPCR evaluation of promoters of 30 different genes showed increased DNA damage (Lu et al., 2004). Thus, DNA damage appears to be pervasive in the

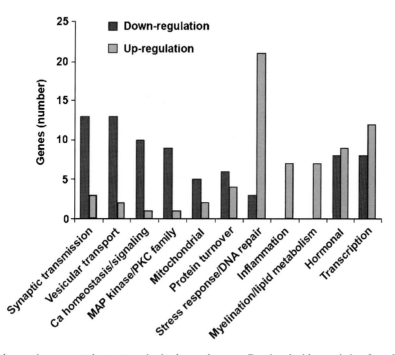

Fig. 6. Relative changes in gene ontology categories in the aged cortex. Reprinted with permission from Lu et al. (2004).

aging human frontal cortex. Taken together, these observations suggest that accelerated DNA damage, particularly in promoter regions, may contribute to reduced gene expression in the human brain after 40 years of age. Interestingly, middle-aged people (40–70 years of age) show much greater cortical genetic heterogeneity than people under 40 or over 70, suggesting that individuals approach old age at different rates (Lu et al., 2004). It is intriguing to note that many of the same functional classes of genes altered during the aging process are also changed in neurodegenerative disease, indicating that many aspects of brain dysfunction in neurodegenerative disease may be related to mechanisms that accelerate brain aging.

Conclusions

This review focused upon the power and potential of regional or single cell/population cell analysis for genetic fingerprinting of normal and diseased postmortem human brain using RNA amplification coupled with cDNA microarrays and qPCR-based analyses. Over time, this paradigm will accrue a database of expression profiles that will enable a coordinated comparison of molecular fingerprints from a myriad of specific cell types that are vulnerable to aging and neurodegeneration in several neurodegenerative and neuropsychiatric disorders. The application of microarray technology has generated significant interest across disciplines and spans a multiplicity of biological systems. However, the brain remains a difficult organ to study, in part due to the nuclear, laminar, and cellular heterogeneity of brain regions and cell types (Serafini, 1999; Colantuoni et al., 2000; Ginsberg et al., 2004). Thus, a combination of single cell analysis with microarrays is a highly desirable paradigm, whereby expression profiles of single populations of neuronal and non-neuronal subtypes can be evaluated independently and compared under normal and pathological conditions. The pattern of mRNA expression in a subpopulation of single neurons may provide more informative data than results derived from whole brain or regional tissue homogenates, as each neuronal subtype most likely will have a unique molecular signature in normal and disease states (Ginsberg, 2005). Moreover, simultaneous comparison of multiple genes from relevant classes of transcripts will aid in the elucidation of molecular mechanisms underlying the aging process as well as the pathogenesis of neurodegenerative and neuropsychiatric disorders. Finally, single cell mRNA analysis has the potential for development of novel pharmacotherapeutic agents that target vulnerable gene(s) and gene products within specific cell types.

Abbreviations

ACT	antichymotrypsin
AD	Alzheimer's disease
ALS	amyotrophic lateral sclerosis
AMPA	α-amino-3-hydroxy-5-methyl-isoxanole-4-proprionic acid
AO	acridine orange
APP	amyloid precursor protein
arc	activity regulated cytoskeletal-associated protein
aRNA	antisense ribonucleic acid
BBB	blood–brain barrier
CA	Ammon's horn hippocampal region
CamKII	calcium/calmodulin dependent protein kinase II
cAMP	cyclic adenosine monophosphate
CBF	cholinergic basal forebrain
cdk	cyclin dependent kinase
cDNA	complementary deoxyribonucleic acid
CJD	Creutzfeld-Jakob disease
ChAT	choline acetyltransferase
CNS	central nervous system
COX-2	cyclooxygenase 2
CREB	cAMP response element binding protein
CTSD	cathepsin D
DA	dopamine
DAT	DA transporter
DRD	dopamine receptor subtype
ECII	entorhinal cortex layer II
ERK	extracellular-regulated kinase, also MAPK
EST	expressed-seuqence tagged cDNA
FAK	focal adhesion kinase

FTDP	frontotemporal dementia with parkinsonism
GABA	γ-aminobutyric acid
GABARA	GABA$_A$ receptor
GAL	galanin
GALR	GAL receptor
GCS	global cognitive score
G-CSF	granulocyte colony stimulating factor
GluR	glutamate receptor
GRIA	AMPA GluR subunit
GRIK	kainate GluR subunit
GRIN	NMDA GluR subunit
GSK	glycogen synthase kinase
HSP	heat-shock protein
ISH	in situ hybridization
LB	Lewy body
LTP	long-term potentiation
MAP	microtubule-associated protein
MAPK	mitogen-activated protein kinase
MCI	mild cognitive impairment
MHC	major histocompatibility complex
MMSE	Mini Mental State Examination
mRNA	messenger RNA
MS	multiple sclerosis
NB	nucleus basalis
NCI	no cognitive impairment
NFL-H	neurofilament heavy chain
NFL-L	neurofilament light chain
NFL-M	neurofilament medium chain
NFT	neurofibrillary tangle
NGF	nerve growth factor
NMDA	N-methyl-D-aspartate
OPN	osteopontin
qPCR	quantitative real-time polymerase chain reaction
p75^{NTR}	low-affinity neurotrophin receptor
PD	Parkinson's disease
PHF	paired helical filament
PKA	protein kinase A
PKC	protein kinase C
PP	protein phosphatase subunit
PS1	presenilin-1
PSP	progressive supranuclear palsy
ROS	reactive oxygen species
3Rtau	3 repeat tau
4Rtau	4 repeat tau
SCN	synuclein
SN$_{pc}$	substantia nigra pars compacta
SP	senile plaque
SYN	synapsin
SYP	synaptophysin
SYT	synaptotagmin
TNF	tumor necrosis factor
trk	high-affinity neurotrophin receptor tyrosine kinase
TS	thioflavine-S
TUBB	β-tubulin
UCH	ubiquitin carboxyl-terminal-hydrolase
USP	ubiquitin-specific protease

Acknowledgments

We thank I. Elarova, S. Fang, M. D. Ruben, and N. Mohammad for expert technical assistance. Support for this project comes from NIH grants NS43939, NS48447 AG14449, AG10161, AG10688, AG09466, AG17617, AG26032, and the Alzheimer's Association. We express our appreciation to the families of the patients studied and to the altruism and support of the hundreds of nuns, priests, and brothers participating in the Religious Orders Study. A list of participating groups can be found in the website: http://www.rush.edu/rumc/page-R12394.html.

References

Alam, Z.I., Jenner, A., Daniel, S.E., Lees, A.J., Cairns, N., Marsden, C.D., Jenner, P. and Halliwell, B. (1997) Oxidative DNA damage in the parkinsonian brain: an apparent selective increase in 8-hydroxyguanine levels in substantia nigra. J. Neurochem., 69: 1196–1203.

Amaral, D.G. (1987) Memory: anatomical organization of candidate brain regions. In: Mountcastle, V.B., Plum, F. and Geiger, S.R. (Eds.), Handbook of Physiology. Section 1: The Nervous System. V. Higher Functions of the Brain, Part 1. American Physiological Society, Bethesda, MD, pp. 211–294.

Arnold, S.E., Trojanowski, J.Q., Gur, R.E., Blackwell, P., Han, L.Y. and Choi, C. (1998) Absence of neurodegeneration and neural injury in the cerebral cortex in a sample of elderly patients with schizophrenia. Arch. Gen. Psychiatry, 55: 225–232.

Baranzini, S.E. (2004) Gene expression profiling in neurological disorders: toward a systems-level understanding of the brain. Neuromol. Med., 6: 31–51.

Bartus, R.T. (2000) On neurodegenerative diseases, models, and treatment strategies: lessons learned and lessons forgotten a generation following the cholinergic hypothesis. Exp. Neurol., 163: 495–529.

Baxter, M.G. and Chiba, A.A. (1999) Cognitive functions of the basal forebrain. Curr. Opin. Neurobiol., 9: 178–183.

Becker, K.G., Mattson, D.H., Powers, J.M., Gado, A.M. and Biddison, W.E. (1997) Analysis of a sequenced cDNA library from multiple sclerosis lesions. J. Neuroimmunol., 77: 27–38.

Bennett, D.A., Wilson, R.S., Schneider, J.A., Evans, D.A., Beckett, L.A., Aggarwal, N.T., Barnes, L.L., Fox, J.H. and Bach, J. (2002) Natural history of mild cognitive impairment in older persons. Neurology, 59: 198–205.

Berchtold, N.C. and Cotman, C.W. (1998) Evolution in the conceptualization of dementia and Alzheimer's disease: Greco-Roman period to the 1960s. Neurobiol. Aging, 19: 173–189.

Betarbet, R., Sherer, T.B., MacKenzie, G., Garcia-Osuna, M., Panov, A.V. and Greenamyre, J.T. (2000) Chronic systemic pesticide exposure reproduces features of Parkinson's disease. Nat. Neurosci., 3: 1301–1306.

Bierer, L.M., Haroutunian, V., Gabriel, S., Knott, P.J., Carlin, L.S., Purohit, D.P., Perl, D.P., Schmeidler, J., Kanof, P. and Davis, K.L. (1995) Neurochemical correlates of dementia severity in Alzheimer's disease: relative importance of the cholinergic deficits. J. Neurochem., 64: 749–760.

Blalock, E.M., Geddes, J.W., Chen, K.C., Porter, N.M., Markesbery, W.R. and Landfield, P.W. (2004) Incipient Alzheimer's disease: microarray correlation analyses reveal major transcriptional and tumor suppressor responses. Proc. Natl. Acad. Sci. USA, 101: 2173–2178.

Braak, H. and Braak, E. (1991) Neuropathological stageing of Alzheimer-related changes. Acta Neuropathol., 82: 239–259.

Brady, D.R. and Mufson, E.J. (1990) Amygdaloid pathology in Alzheimer's disease: a qualitative and quantitative analysis. Dementia, 1: 5–17.

Callahan, L.M. and Coleman, P.D. (1995) Neurons bearing neurofibrillary tangles are responsible for selected synaptic deficits in Alzheimer's disease. Neurobiol. Aging, 16: 311–314.

Cataldo, A.M., Barnett, J.L., Berman, S.A., Li, J., Quarless, S., Bursztajn, S., Lippa, C. and Nixon, R.A. (1995) Gene expression and cellular content of cathepsin D in Alzheimer's disease brain: evidence for early up-regulation of the endosomal–lysosomal system. Neuron, 14: 671–680.

Cataldo, A.M., Hamilton, D.J., Barnett, J.L., Paskevich, P.A. and Nixon, R.A. (1996) Properties of the endosomal–lysosomal system in the human central nervous system: disturbances mark most neurons in populations at risk to degenerate in Alzheimer's disease. J. Neurosci., 16: 186–199.

Chabas, D., Baranzini, S.E., Mitchell, D., Bernard, C.C., Rittling, S.R., Denhardt, D.T., Sobel, R.A., Lock, C., Karpuj, M., Pedotti, R., Heller, R., Oksenberg, J.R. and Steinman, L. (2001) The influence of the proinflammatory cytokine, osteopontin, on autoimmune demyelinating disease. Science, 294: 1731–1735.

Chan-Palay, V. (1988) Galanin hyperinnervates surviving neurons of the human basal nucleus of Meynert in dementias of Alzheimer's and Parkinson's disease: a hypothesis for the role of galanin in accentuating cholinergic dysfunction in dementia. J. Comp. Neurol., 273: 543–557.

Cheetham, J.E., Coleman, P.D. and Chow, N. (1997) Isolation of single immunohistochemically identified whole neuronal cell bodies from post-mortem human brain for simultaneous analysis of multiple gene expression. J. Neurosci. Methods, 77: 43–48.

Choi, J., Levey, A.I., Weintraub, S.T., Rees, H.D., Gearing, M., Chin, L.S. and Li, L. (2004) Oxidative modifications and down-regulation of ubiquitin carboxyl-terminal hydrolase L1 associated with idiopathic Parkinson's and Alzheimer's diseases. J. Biol. Chem., 279: 13256–13264.

Chow, N., Cox, C., Callahan, L.M., Weimer, J.M., Guo, L. and Coleman, P.D. (1998) Expression profiles of multiple genes in single neurons of Alzheimer's disease. Proc. Natl. Acad. Sci. USA, 95: 9620–9625.

Colangelo, V., Schurr, J., Ball, M.J., Pelaez, R.P., Bazan, N.G. and Lukiw, W.J. (2002) Gene expression profiling of 12633 genes in Alzheimer hippocampal CA1: transcription and neurotrophic factor down-regulation and up-regulation of apoptotic and pro-inflammatory signaling. J. Neurosci. Res., 70: 462–473.

Colantuoni, C., Purcell, A.E., Bouton, C.M. and Pevsner, J. (2000) High throughput analysis of gene expression in the human brain. J. Neurosci. Res., 59: 1–10.

Compston, A. (1994) The epidemiology of multiple sclerosis: principles, achievements, and recommendations. Ann. Neurol., 36(2): S211–S217.

Conway, K.A., Lee, S.J., Rochet, J.C., Ding, T.T., Williamson, R.E. and Lansbury Jr., P.T. (2000) Acceleration of oligomerization, not fibrillization, is a shared property of both alpha-synuclein mutations linked to early-onset Parkinson's disease: implications for pathogenesis and therapy. Proc. Natl. Acad. Sci. USA, 97: 571–576.

Conway, K.A., Rochet, J.C., Bieganski, R.M. and Lansbury Jr., P.T. (2001) Kinetic stabilization of the alpha-synuclein protofibril by a dopamine-alpha-synuclein adduct. Science, 294: 1346–1349.

Counts, S.E., Chen, E.Y., Che, S., Ikonomovic, M.D., Wuu, J., Ginsberg, S.D., DeKosky, S.T. and Mufson, E.J. (2006) Galanin fiber hypertrophy within the cholinergic nucleus basalis during the progression of Alzheimer's disease. Dement. Geriatr. Cogn. Disord., 21: 205–214.

Counts, S.E., He, B., Che, S., Ginsberg, S.D. and Mufson, E.J. (2005) cDNA array analysis of galanin-hyperinnervated cholinergic basal forebrain neurons in Alzheimer's disease. Soc. Neurosci., 80: 12 Washington, DC.

Counts, S.E., Perez, S.E., Ginsberg, S.D., De Lacalle, S. and Mufson, E.J. (2003) Galanin in Alzheimer disease. Mol. Interv., 3: 137–156.

Davies, P. (1979) Neurotransmitter-related enzymes in senile dementia of the Alzheimer type. Brain Res., 171: 319–327.

Davies, P. and Maloney, A.J. (1976) Selective loss of central cholinergic neurons in Alzheimer's disease. Lancet, 2: 1403.

DeKosky, S.T. and Scheff, S.W. (1990) Synapse loss in frontal cortex biopsies in Alzheimer's disease: correlation with cognitive severity. Ann. Neurol., 27: 457–464.

Ding, X., Mactavish, D., Kar, S. and Jhamandas, J.H. (2005) Galanin attenuates beta-amyloid (Abeta) toxicity in rat cholinergic basal forebrain neurons. Neurobiol. Dis, 21: 413–420.

Dragoi, G., Harris, K.D. and Buzsaki, G. (2003) Place representation within hippocampal networks is modified by long-term potentiation. Neuron, 39: 843–853.

Dunckley, T., Beach, T.G., Ramsey, K.E., Grover, A., Mastroeni, D., Walker, D.G., LaFleur, B.J., Coon, K.D., Brown, K.M., Caselli, R., Kukull, W.A., Higdon, R., McKeel, D., Morris, J.C., Hulette, C., Schmechel, D., Reiman, E.M., Rogers, J. and Stephan, D.A. (2006). Gene expression correlates of neurofibrillary tangles in Alzheimer's disease. Neurobiol. Aging, in press.

Eberwine, J., Yeh, H., Miyashiro, K., Cao, Y., Nair, S., Finnell, R., Zettel, M. and Coleman, P. (1992) Analysis of gene expression in single live neurons. Proc. Natl. Acad. Sci. USA, 89: 3010–3014.

Eikelenboom, P. and Veerhuis, R. (1996) The role of complement and activated microglia in the pathogenesis of Alzheimer's disease. Neurobiol. Aging, 17: 673–680.

Evans, D.A., Funkenstein, H.H., Albert, M.S., Scherr, P.A., Cook, N.R., Chown, M.J., Hebert, L.E., Hennekens, C.H. and Taylor, J.O. (1989) Prevalence of Alzheimer's disease in a community population of older persons higher than previously reported. JAMA, 262: 2551–2556.

Fasulo, W.H. and Hemby, S.E. (2003) Time-dependent changes in gene expression profiles of midbrain dopamine neurons following haloperidol administration. J. Neurochem., 87: 205–219.

Fearnley, J.M. and Lees, A.J. (1991) Ageing and Parkinson's disease: substantia nigra regional selectivity. Brain, 114: 2283–2301.

Ginsberg, S.D. (2005) RNA amplification strategies for small sample populations. Methods, 37: 229–237.

Ginsberg, S.D. and Che, S. (2002) RNA amplification in brain tissues. Neurochem. Res., 27: 981–992.

Ginsberg, S.D. and Che, S. (2004) Combined histochemical staining, RNA amplification, regional, and single cell cDNA analysis within the hippocampus. Lab. Invest., 84: 952–962.

Ginsberg, S.D. and Che, S. (2005) Expression profile analysis within the human hippocampus: comparison of CA1 and CA3 pyramidal neurons. J. Comp. Neurol., 487: 107–118.

Ginsberg, S.D., Che, S., Counts, S.E. and Mufson, E.J. (2006a) Shift in the ratio of 3-repeat tau and 4-repeat tau mRNAs in individual cholinergic basal forebrain neurons in mild cognitive impairment and Alzheimer's disease. J. Neurochem., 96: 1401–1408.

Ginsberg, S.D., Che, S., Wuu, J., Counts, S.E. and Mufson, E.J. (2006b) Down-regulation of trk but not p75NTR gene expression in single cholinergic basal forebrain neurons mark the progression of Alzheimer's disease. J. Neurochem., 97: 475–487.

Ginsberg, S.D., Crino, P.B., Hemby, S.E., Weingarten, J.A., Lee, V.M.-Y., Eberwine, J.H. and Trojanowski, J.Q. (1999a) Predominance of neuronal mRNAs in individual Alzheimer's disease senile plaques. Ann. Neurol., 45: 174–181.

Ginsberg, S.D., Crino, P.B., Lee, V.M.-Y., Eberwine, J.H. and Trojanowski, J.Q. (1997) Sequestration of RNA in Alzheimer's disease neurofibrillary tangles and senile plaques. Ann. Neurol., 41: 200–209.

Ginsberg, S.D., Elarova, I., Ruben, M., Tan, F., Counts, S.E., Eberwine, J.H., Trojanowski, J.Q., Hemby, S.E., Mufson, E.J. and Che, S. (2004) Single-cell gene expression analysis: implications for neurodegenerative and neuropsychiatric disorders. Neurochem. Res., 29: 1053–1064.

Ginsberg, S.D., Galvin, J.E., Chiu, T.S., Lee, V.M.-Y., Masliah, E. and Trojanowski, J.Q. (1998) RNA sequestration to pathological lesions of neurodegenerative diseases. Acta Neuropathol. (Berl.), 96: 487–494.

Ginsberg, S.D., Hemby, S.E., Lee, V.M.-Y., Eberwine, J.H. and Trojanowski, J.Q. (2000) Expression profile of transcripts in Alzheimer's disease tangle-bearing CA1 neurons. Ann. Neurol., 48: 77–87.

Ginsberg, S.D., Schmidt, M.L., Crino, P.B., Eberwine, J.H., Lee, V.M.-Y. and Trojanowski, J.Q. (1999b) Molecular pathology of Alzheimer's disease and related disorders. In: Peters, A. and Morrison, J.H. (Eds.), Neurodegenerative and Age-related Changes in Structure and Function of Cerebral Cortex. Kluwer, New York, pp. 603–653.

Goedert, M., Spillantini, M.G., Jakes, R., Rutherford, D. and Crowther, R.A. (1989a) Multiple isoforms of human microtubule-associated protein tau: sequences and localization in neurofibrillary tangles of Alzheimer's disease. Neuron, 3: 519–526.

Goedert, M., Spillantini, M.G., Potier, M.C., Ulrich, J. and Crowther, R.A. (1989b) Cloning and sequencing of the cDNA encoding an isoform of microtubule-associated protein tau containing four tandem repeats: differential expression of tau protein mRNAs in human brain. EMBO J., 8: 393–399.

Graumann, U., Reynolds, R., Steck, A.J. and Schaeren-Wiemers, N. (2003) Molecular changes in normal appearing white matter in multiple sclerosis are characteristic of neuroprotective mechanisms against hypoxic insult. Brain Pathol., 13: 554–573.

Greengard, P., Valtorta, F., Czernik, A.J. and Benfenati, F. (1993) Synaptic vesicle phosphoproteins and regulation of synaptic function. Science, 259: 780–785.

Griffin, W.S., Ling, C., White 3rd, C.L. and Morrison-Bogorad, M. (1990) Polyadenylated messenger RNA in paired helical filament-immunoreactive neurons in Alzheimer disease. Alzheimer Dis. Assoc. Disord., 4: 69–78.

Grunblatt, E., Mandel, S., Jacob-Hirsch, J., Zeligson, S., Amariglo, N., Rechavi, G., Li, J., Ravid, R., Roggendorf, W., Riederer, P. and Youdim, M.B. (2004) Gene expression profiling of parkinsonian substantia nigra pars compacta; alterations in ubiquitin-proteasome, heat shock protein, iron and oxidative stress regulated proteins, cell adhesion/cellular matrix and vesicle trafficking genes. J. Neural Transm., 111: 1543–1573.

Gur, R.C. and Gur, R.E. (2002) Neuroimaging applications in elderly patients. Am. J. Geriatr. Psychiatry, 10: 5–11.

Hakak, Y., Walker, J.R., Li, C., Wong, W.H., Davis, K.L., Buxbaum, J.D., Haroutunian, V. and Fienberg, A.A. (2001)

Genome-wide expression analysis reveals dysregulation of myelination-related genes in chronic schizophrenia. Proc. Natl. Acad. Sci. USA, 98: 4746–4751.

Hardy, J. (1997) Amyloid, the presenilins and Alzheimer's disease. Trends Neurosci., 20: 154–159.

Harrison, P.J., Barton, A.J., Najlerahim, A., McDonald, B. and Pearson, R.C. (1991) Regional and neuronal reductions of polyadenylated messenger RNA in Alzheimer's disease. Psychol. Med., 21: 855–866.

Hartmann, A., Hunot, S., Michel, P.P., Muriel, M.P., Vyas, S., Faucheux, B.A., Mouatt-Prigent, A., Turmel, H., Srinivasan, A., Ruberg, M., Evan, G.I., Agid, Y. and Hirsch, E.C. (2000) Caspase-3: a vulnerability factor and final effector in apoptotic death of dopaminergic neurons in Parkinson's disease. Proc. Natl. Acad. Sci. USA, 97: 2875–2880.

Hartmann, A., Michel, P.P., Troadec, J.D., Mouatt-Prigent, A., Faucheux, B.A., Ruberg, M., Agid, Y. and Hirsch, E.C. (2001) Is Bax a mitochondrial mediator in apoptotic death of dopaminergic neurons in Parkinson's disease? J. Neurochem., 76: 1785–1793.

Hartmann, A., Mouatt-Prigent, A., Vila, M., Abbas, N., Perier, C., Faucheux, B.A., Vyas, S. and Hirsch, E.C. (2002) Increased expression and redistribution of the antiapoptotic molecule Bcl-xL in Parkinson's disease. Neurobiol. Dis., 10: 28–32.

Hauser, M.A., Li, Y.J., Xu, H., Noureddine, M.A., Shao, Y.S., Gullans, S.R., Scherzer, C.R., Jensen, R.V., McLaurin, A.C., Gibson, J.R., Scott, B.L., Jewett, R.M., Stenger, J.E., Schmechel, D.E., Hulette, C.M. and Vance, J.M. (2005) Expression profiling of substantia nigra in Parkinson disease, progressive supranuclear palsy, and frontotemporal dementia with parkinsonism. Arch. Neurol., 62: 917–921.

Hemby, S.E., Ginsberg, S.D., Brunk, B., Arnold, S.E., Trojanowski, J.Q. and Eberwine, J.H. (2002) Gene expression profile for schizophrenia: discrete neuron transcription patterns in the entorhinal cortex. Arch. Gen. Psychiatry, 59: 631–640.

Henke, H. and Lang, W. (1983) Cholinergic enzymes in neocortex, hippocampus and basal forebrain of non-neurological and senile dementia of Alzheimer-type patients. Brain Res., 267: 281–291.

Hill, W.D. (1996) Altered neurofilament expression does not contribute to Lewy body formation. Am. J. Pathol., 149: 728–729.

Ho, L., Guo, Y., Spielman, L., Petrescu, O., Haroutunian, V., Purohit, D., Czernik, A., Yemul, S., Aisen, P.S., Mohs, R. and Pasinetti, G.M. (2001) Altered expression of a-type but not b-type synapsin isoform in the brain of patients at high risk for Alzheimer's disease assessed by DNA microarray technique. Neurosci. Lett., 298: 191–194.

Hof, P.R., Giannakopoulos, P., Vickers, J.C., Bouras, C. and Morrison, J.H. (1995) The morphologic and neurochemical basis of dementia: aging, hierarchical patterns of lesion distribution and vulnerable neuronal phenotype. Rev. Neurosci., 6: 97–124.

Hyman, B.T., Van Hoesen, G.W. and Damasio, A.R. (1990) Memory-related neural systems in Alzheimer's disease: an anatomic study. Neurology, 40: 1721–1730.

Hyman, B.T., Van Hoesen, G.W., Damasio, A.R. and Barnes, C.L. (1984) Alzheimer's disease: cell-specific pathology isolates the hippocampal formation. Science, 225: 1168–1170.

Hyman, B.T., Van Hoesen, G.W., Kromer, L.J. and Damasio, A.R. (1986) Perforant pathway changes and the memory impairment of Alzheimer's disease. Ann. Neurol., 20: 472–481.

Itagaki, S., McGeer, P.L., Akiyama, H., Zhu, S. and Selkoe, D. (1989) Relationship of microglia and astrocytes to amyloid deposits of Alzheimer disease. J. Neuroimmunol., 24: 173–182.

Jiang, Y.M., Yamamoto, M., Kobayashi, Y., Yoshihara, T., Liang, Y., Terao, S., Takeuchi, H., Ishigaki, S., Katsuno, M., Adachi, H., Niwa, J., Tanaka, F., Doyu, M., Yoshida, M., Hashizume, Y. and Sobue, G. (2005) Gene expression profile of spinal motor neurons in sporadic amyotrophic lateral sclerosis. Ann. Neurol., 57: 236–251.

Johns, L.C. and van Os, J. (2001) The continuity of psychotic experiences in the general population. Clin. Psychol. Rev., 21: 1125–1141.

Joyce, J.N., Kaeger, C., Ryoo, H. and Goldsmith, S. (1993) Dopamine D2 receptors in the hippocampus and amygdala in Alzheimer's disease. Neurosci. Lett., 154: 171–174.

Kamme, F., Salunga, R., Yu, J., Tran, D.T., Zhu, J., Luo, L., Bittner, A., Guo, H.Q., Miller, N., Wan, J. and Erlander, M. (2003) Single-cell microarray analysis in hippocampus CA1: demonstration and validation of cellular heterogeneity. J. Neurosci., 23: 3607–3615.

Kaplan, D.R. and Miller, F.D. (2000) Neurotrophin signal transduction in the nervous system. Curr. Opin. Neurobiol., 10: 381–391.

Kurtz, M.M., Moberg, P.J., Gur, R.C. and Gur, R.E. (2001) Approaches to cognitive remediation of neuropsychological deficits in schizophrenia: a review and meta-analysis. Neuropsychol. Rev., 11: 197–210.

Lad, S.P., Neet, K.E. and Mufson, E.J. (2003) Nerve growth factor: structure, function and therapeutic implications for Alzheimer's disease. Curr. Drug Target CNS Neurol. Disord., 2: 315–334.

Lee, J.W., Bae, S.H., Jeong, J.W., Kim, S.H. and Kim, K.W. (2004) Hypoxia-inducible factor (HIF-1)alpha: its protein stability and biological functions. Exp. Mol. Med., 36: 1–12.

Lencz, T., Cornblatt, B. and Bilder, R.M. (2001) Neurodevelopmental models of schizophrenia: pathophysiologic synthesis and directions for intervention research. Psychopharmacol. Bull., 35: 95–125.

Lewis, D.A. (2002) In pursuit of the pathogenesis and pathophysiology of schizophrenia: where do we stand? Am. J. Psychiatry, 159: 1467–1469.

Lindberg, R.L., De Groot, C.J., Certa, U., Ravid, R., Hoffmann, F., Kappos, L. and Leppert, D. (2004) Multiple sclerosis as a generalized CNS disease — comparative microarray analysis of normal appearing white matter and lesions in secondary progressive MS. J. Neuroimmunol., 152: 154–167.

Liu, F., Grundke-Iqbal, I., Iqbal, K. and Gong, C.X. (2005) Contributions of protein phosphatases PP1, PP2A, PP2B and PP5 to the regulation of tau phosphorylation. Eur. J. Neurosci., 22: 1942–1950.

Liu, Y., Fallon, L., Lashuel, H.A., Liu, Z. and Lansbury Jr., P.T. (2002) The UCH-L1 gene encodes two opposing enzymatic activities that affect alpha-synuclein degradation and Parkinson's disease susceptibility. Cell, 111: 209–218.

Lock, C., Hermans, G., Pedotti, R., Brendolan, A., Schadt, E., Garren, H., Langer-Gould, A., Strober, S., Cannella, B., Allard, J., Klonowski, P., Austin, A., Lad, N., Kaminski, N., Galli, S.J., Oksenberg, J.R., Raine, C.S., Heller, R. and Steinman, L. (2002) Gene-microarray analysis of multiple sclerosis lesions yields new targets validated in autoimmune encephalomyelitis. Nat. Med., 8: 500–508.

Lombardino, A.J., Li, X.C., Hertel, M. and Nottebohm, F. (2005) Replaceable neurons and neurodegenerative disease share depressed UCHL1 levels. Proc. Natl. Acad. Sci. USA, 102: 8036–8041.

Loring, J.F., Wen, X., Lee, J.M., Seilhamer, J. and Somogyi, R. (2001) A gene expression profile of Alzheimer's disease. DNA Cell Biol., 20: 683–695.

Lu, L., Neff, F., Alvarez-Fischer, D., Henze, C., Xie, Y., Oertel, W.H., Schlegel, J. and Hartmann, A. (2005) Gene expression profiling of Lewy body-bearing neurons in Parkinson's disease. Exp. Neurol., 195: 27–39.

Lu, T., Pan, Y., Kao, S.Y., Li, C., Kohane, I., Chan, J. and Yankner, B.A. (2004) Gene regulation and DNA damage in the ageing human brain. Nature, 429: 883–891.

Lukiw, W.J. (2004) Gene expression profiling in fetal, aged, and Alzheimer hippocampus: a continuum of stress-related signaling. Neurochem. Res., 29: 1287–1297.

Luo, L.G. and Jackson, I.M. (1999) Advantage of double labeled in situ hybridization for detecting the effects of glucocorticoids on the mRNAs of protooncogenes and neural peptides (TRH) in cultured hypothalamic neurons. Brain Res. Brain Res. Protoc., 4: 201–208.

Malaspina, A., Kaushik, N. and de Belleroche, J. (2001) Differential expression of 14 genes in amyotrophic lateral sclerosis spinal cord detected using gridded cDNA arrays. J. Neurochem., 77: 132–145.

Malinow, R., Mainen, Z.F. and Hayashi, Y. (2000) LTP mechanisms: from silence to four-lane traffic. Curr. Opin. Neurobiol., 10: 352–357.

Martin, L.J., Price, A.C., Kaiser, A., Shaikh, A.Y. and Liu, Z. (2000) Mechanisms for neuronal degeneration in amyotrophic lateral sclerosis and in models of motor neuron death. Int. J. Mol. Med., 5: 3–13.

McCormick, D.A. and Contreras, D. (2001) On the cellular and network bases of epileptic seizures. Annu. Rev. Physiol., 63: 815–846.

Merrick, S.E., Trojanowski, J.Q. and Lee, V.M.-Y. (1997) Selective destruction of stable microtubules and axons by inhibitors of protein serine/threonine phosphatases in cultured human neurons. J. Neurosci., 17: 5726–5737.

Mesulam, M.M., Mufson, E.J., Levey, A.I. and Wainer, B.H. (1983) Cholinergic innervation of cortex by the basal forebrain: cytochemistry and cortical connections of the septal area, diagonal band nuclei, nucleus basalis (substantia innominata), and hypothalamus in the rhesus monkey. J. Comp. Neurol., 214: 170–197.

Middleton, F.A., Mirnics, K., Pierri, J.N., Lewis, D.A. and Levitt, P. (2002) Gene expression profiling reveals alterations of specific metabolic pathways in schizophrenia. J. Neurosci., 22: 2718–2729.

Mimmack, M.L., Ryan, M., Baba, H., Navarro-Ruiz, J., Iritani, S., Faull, R.L., McKenna, P.J., Jones, P.B., Arai, H., Starkey, M., Emson, P.C. and Bahn, S. (2002) Gene expression analysis in schizophrenia: reproducible up-regulation of several members of the apolipoprotein L family located in a high-susceptibility locus for schizophrenia on chromosome 22. Proc. Natl. Acad. Sci. USA, 99: 4680–4685.

Morrison, J.H. and Hof, P.R. (2002) Selective vulnerability of corticocortical and hippocampal circuits in aging and Alzheimer's disease. Prog. Brain Res., 136: 467–486.

Mufson, E.J., Bothwell, M. and Kordower, J.H. (1989) Loss of nerve growth factor receptor-containing neurons in Alzheimer's disease: a quantitative analysis across subregions of the basal forebrain. Exp. Neurol., 105: 221–232.

Mufson, E.J., Cochran, E., Benzing, W. and Kordower, J.H. (1993) Galaninergic innervation of the cholinergic vertical limb of the diagonal band (Ch2) and bed nucleus of the stria terminalis in aging, Alzheimer's disease and Down's syndrome. Dementia, 4: 237–250.

Mufson, E.J., Counts, S.E. and Ginsberg, S.D. (2002) Gene expression profiles of cholinergic nucleus basalis neurons in Alzheimer's disease. Neurochem. Res., 27: 1035–1048.

Mufson, E.J., Ikonomovic, M.D., Styren, S.D., Counts, S.E., Wuu, J., Leurgans, S., Bennett, D.A., Cochran, E.J. and DeKosky, S.T. (2003) Preservation of brain nerve growth factor in mild cognitive impairment and Alzheimer disease. Arch. Neurol., 60: 1143–1148.

Mycko, M.P., Papoian, R., Boschert, U., Raine, C.S. and Selmaj, K.W. (2003) cDNA microarray analysis in multiple sclerosis lesions: detection of genes associated with disease activity. Brain, 126: 1048–1057.

Nagasaka, Y., Dillner, K., Ebise, H., Teramoto, R., Nakagawa, H., Lilius, L., Axelman, K., Forsell, C., Ito, A., Winblad, B., Kimura, T. and Graff, C. (2005) A unique gene expression signature discriminates familial Alzheimer's disease mutation carriers from their wild-type siblings. Proc. Natl. Acad. Sci. USA, 102: 14854–14859.

Nixon, R.A., Cataldo, A.M. and Mathews, P.M. (2000) The endosomal–lysosomal system of neurons in Alzheimer's disease pathogenesis: a review. Neurochem. Res., 25: 1161–1172.

Perry, E.K., Gibson, P.H., Blessed, G., Perry, R.H. and Tomlinson, B.E. (1977) Neurotransmitter enzyme abnormalities in senile dementia. Choline acetyltransferase and glutamic acid decarboxylase activities in necropsy brain tissue. J. Neurol. Sci., 34: 247–265.

Pongrac, J., Middleton, F.A., Lewis, D.A., Levitt, P. and Mirnics, K. (2002) Gene expression profiling with DNA microarrays: advancing our understanding of psychiatric disorders. Neurochem. Res., 27: 1049–1063.

Prentice, T. (2001). The World Health Reoprt 2001. Mental health: new understanding, new hope. World Health Organization, Geneva.

Prusiner, S.B. (1998) The prion diseases. Brain Pathol., 8: 499–513.

Ramanathan, M., Weinstock-Guttman, B., Nguyen, L.T., Badgett, D., Miller, C., Patrick, K., Brownscheidle, C. and Jacobs, L. (2001) In vivo gene expression revealed by cDNA arrays: the pattern in relapsing-remitting multiple sclerosis patients compared with normal subjects. J. Neuroimmunol., 116: 213–219.

Rosene, D.L. and Van Hoesen, G.W. (1987) The hippocampal formation in the primate brain. A review of some comparative aspects of cytoarchitecture and connections. In: Jones, E.G. and Peters, A. (Eds.) Cerebral Cortex, Vol. 6. Plenum, New York, pp. 345–455.

Ryoo, H.L. and Joyce, J.N. (1994) Loss of dopamine D2 receptors varies along the rostrocaudal axis of the hippocampal complex in Alzheimer's disease. J. Comp. Neurol., 348: 94–110.

Salehi, A., Verhaagen, J., Dijkhuizen, P.A. and Swaab, D.F. (1996) Co-localization of high-affinity neurotrophin receptors in nucleus basalis of Meynert neurons and their differential reduction in Alzheimer's disease. Neuroscience, 75: 373–387.

Sarnat, H.B., Curry, B., Rewcastle, N.B. and Trevenen, C.L. (1987) Gliosis and glioma distinguished by acridine orange. Can. J. Neurol. Sci., 14: 31–35.

Sassin, I., Schultz, C., Thal, D.R., Rub, U., Arai, K., Braak, E. and Braak, H. (2000) Evolution of Alzheimer's disease-related cytoskeletal changes in the basal nucleus of Meynert. Acta Neuropathol. (Berl.), 100: 259–269.

Scheff, S.W. and Price, D.A. (2003) Synaptic pathology in Alzheimer's disease: a review of ultrastructural studies. Neurobiol. Aging, 24: 1029–1046.

Scheff, S.W., Price, D.A., Schmitt, F.A. and Mufson, E.J. (2006). Hippocampal synaptic loss in early Alzheimer's disease and mild cognitive impairment. Neurobiol. Aging, in press.

Selkoe, D.J. (1997) Alzheimer's disease: genotypes, phenotypes, and treatments. Science, 275: 630–631.

Serafini, T. (1999) Of neurons and gene chips. Curr. Opin. Neurobiol., 9: 641–644.

Sherer, T.B., Betarbet, R., Stout, A.K., Lund, S., Baptista, M., Panov, A.V., Cookson, M.R. and Greenamyre, J.T. (2002) An in vitro model of Parkinson's disease: linking mitochondrial impairment to altered alpha-synuclein metabolism and oxidative damage. J. Neurosci., 22: 7006–7015.

Sofroniew, M.V., Howe, C.L. and Mobley, W.C. (2001) Nerve growth factor signaling, neuroprotection, and neural repair. Annu. Rev. Neurosci., 24: 1217–1281.

Sorensen, J.B. (2005) SNARE complexes prepare for membrane fusion. Trends Neurosci., 28: 453–455.

Spillantini, M.G., Schmidt, M.L., Lee, V.M.-Y., Trojanowski, J.Q., Jakes, R. and Goedert, M. (1997) Alpha-synuclein in Lewy bodies. Nature, 388: 839–840.

Tang, W.X., Fasulo, W.H., Mash, D.C. and Hemby, S.E. (2003) Molecular profiling of midbrain dopamine regions in cocaine overdose victims. J. Neurochem., 85: 911–924.

Teng, K.K. and Hempstead, B.L. (2004) Neurotrophins and their receptors: signaling trios in complex biological systems. Cell Mol. Life Sci., 61: 35–48.

Trojanowski, J.Q. and Arnold, S.E. (1995) In pursuit of the molecular neuropathology of schizophrenia. Arch. Gen. Psychiatry, 52: 274–276.

Trojanowski, J.Q. and Lee, V.M.-Y. (1998) Aggregation of neurofilament and alpha-synuclein proteins in Lewy bodies: implications for the pathogenesis of Parkinson disease and Lewy body dementia. Arch. Neurol., 55: 151–152.

Vogt, L.J., Nimchinsky, E.A. and Hof, P.R. (1997) Primate cingulate cortex chemoarchitecture and it's disruption in Alzheimer's disease. In: Bloom, F., Bjorklund, A. and Hokfelt, T. (Eds.) The Primate Nervous System, Part I, Vol. 13. Elsevier, Amsterdam, pp. 455–528.

Whitehouse, P.J., Price, D.L., Clark, A.W., Coyle, J.T. and DeLong, M.R. (1981) Alzheimer disease: evidence for selective loss of cholinergic neurons in the nucleus basalis. Ann. Neurol., 10: 122–126.

Whitehouse, P.J., Price, D.L., Struble, R.G., Clark, A.W., Coyle, J.T. and DeLong, M.R. (1982) Alzheimer's disease and senile dementia: loss of neurons in the basal forebrain. Science, 215: 1237–1239.

Whitney, L.W., Becker, K.G., Tresser, N.J., Caballero-Ramos, C.I., Munson, P.J., Prabhu, V.V., Trent, J.M., McFarland, H.F. and Biddison, W.E. (1999) Analysis of gene expression in mutiple sclerosis lesions using cDNA microarrays. Ann. Neurol., 46: 425–428.

Xiang, W., Windl, O., Westner, I.M., Neumann, M., Zerr, I., Lederer, R.M. and Kretzschmar, H.A. (2005) Cerebral gene expression profiles in sporadic Creutzfeldt–Jakob disease. Ann. Neurol., 58: 242–257.

Yao, P.J., Zhu, M., Pyun, E.I., Brooks, A.I., Therianos, S., Meyers, V.E. and Coleman, P.D. (2003) Defects in expression of genes related to synaptic vesicle trafficking in frontal cortex of Alzheimer's disease. Neurobiol. Dis., 12: 97–109.

Zhang, J., Perry, G., Smith, M.A., Robertson, D., Olson, S.J., Graham, D.G. and Montine, T.J. (1999) Parkinson's disease is associated with oxidative damage to cytoplasmic DNA and RNA in substantia nigra neurons. Am. J. Pathol., 154: 1423–1429.

CHAPTER 11

Epileptogenesis-related genes revisited

Katarzyna Lukasiuk[1,*], Michal Dabrowski[1], Alicja Adach[1] and Asla Pitkänen[2,3]

[1] The Nencki Institute of Experimental Biology, Polish Academy of Sciences, Pasteur 3, 02-093 Warsaw, Poland
[2] A.I.Virtanen Institute for Molecular Sciences, University of Kuopio, PO Box 1627, FIN-70 211 Kuopio, Finland
[3] Department of Neurology, Kuopio University Hospital, PO Box 1777, FIN-70 211 Kuopio, Finland

Abstract: The main goal of this study was to identify common features in the molecular response to epileptogenic stimuli across different animal models of epileptogenesis. Therefore, we compared the currently available literature on the global analysis of gene expression following epileptogenic insult to search for (i) highly represented functional gene classes (GO terms) within data sets, and (ii) individual genes that appear in several data sets, and therefore, might be of particular importance for the development of epilepsy due to different etiologies. We focused on two well-described models of brain insult that induce the development of spontaneous seizures in experimental animals: status epilepticus and traumatic brain injury. Additionally, a few papers describing gene expression in rat and human epileptic tissue were included for comparison. Our analysis revealed that epileptogenic insults induce significant changes in gene expression within a subset of pre-defined GO terms, that is, in groups of functionally linked genes. We also found individual genes for which expression changed across different models of epileptogenesis. Alterations in gene expression appear time-specific and underlie a number of processes that are linked with epileptogenesis, such as cell death and survival, neuronal plasticity, or immune response. Particularly, our analysis highlighted alterations in gene expression in glial cells as well as in genes involved in the immune response, which suggests the importance of gliosis and immune reaction in epileptogenesis.

Keywords: brain; epilepsy; gene expression; microarrays; status epilepticus; traumatic brain injury; inflammation

Introduction

Epilepsies are the second most-common neurologic disorder after stroke (Porter, 1993). It is estimated that approximately 0.8% of the population is affected by some form of epilepsy. In approximately 30% of cases, epilepsy is a result of an insult to the brain, such as traumatic brain injury (TBI), stroke, brain infection, prolonged complex febrile seizures, or status epilepticus (SE) (Hauser, 1997). In such cases, the initial insult is commonly followed by a latency period (epileptogenesis) that can last for months or years before the appearance of spontaneous seizures and epilepsy diagnosis (Pitkanen and Sutula, 2002).

During the latency period, several phenomena can occur in parallel, including neuronal loss, dendritic and axonal plasticity, neurogenesis, gliosis, remodeling of the extracellular matrix, and alterations in gene expression (Parent et al., 1997; Bazan and Serou, 1999; Clark and Wilson, 1999; Coulter and DeLorenzo, 1999; Endo et al., 1999; Covolan et al., 2000; Wu et al., 2000). Such phenomena might reflect a response of the brain to the insult. At least some of the molecular alterations are also involved in the chronic remodeling of neuronal circuits, which eventually leads to the development

*Corresponding author. Tel.: +48-22-58-92-238; Fax: +48-22-822-53-42; E-mail: k.lukasiuk@nencki.gov.pl

of epilepsy. Hypothetically, identification of key molecular changes will provide a better understanding of epileptogenesis and point to targets that can be used to modify the epileptogenic process and, in the most optimistic scenario, develop antiepileptogenic treatments.

Studies of gene expression following potentially epileptogenic events like SE, ischemia, or TBI have a long history, and a vast amount of data has been gathered using traditional molecular biology methods (Nedivi et al., 1993; Koistinaho and Hokfelt, 1997; Zagulska-Szymczak et al., 2001). Nevertheless, these studies have focused on a limited number of pre-selected genes at a time. Recent technologic developments allow for the analysis of gene expression at the level of the whole transcriptome (microarrays and serial analysis of gene expression), and thus, provide an unbiased insight into the ensemble of molecular events that occur in the brain following various types of injuries. Brain trauma can trigger alterations in gene expression that partly represent a normal physiologic response to the injury. On the other hand, each of the components of the pathologic circuitry reorganization requires an expression of a particular set of genes.

Just a few years ago, there were less than a handful of large-scale molecular profiling studies of epileptogenic brain (Hendriksen et al., 2001; Elliott et al., 2003; Lukasiuk et al., 2003). A conspicuous feature of these data is that there is a very little overlap between the data sets of genes with altered expression (Lukasiuk and Pitkanen, 2004). This might be related to technical issues, including the brain area or cell population selected for the analysis, the use of different experimental platforms with a limited selection of gene probes, the limited sensitivity of methods for global analysis of gene expression in brain tissue, and the use of different kinds of animal models at different stages of disease development (Lukasiuk and Pitkanen, 2004). There are also specific problems in data interpretation related to analysis of the whole transcriptome. For example, there is only a limited amount of information or no information available on the biologic function of most of the altered genes.

The recent explosion in the application of large-scale molecular profiling in studies investigating the consequences of epileptogenic brain insults as well as advances in bioinformatics has provided a large amount of new data that can be analyzed with novel tools to extract the meaningful information from the noise. For example, databases allowing functional annotations of genes are updated regularly (e.g., http://www.geneontology.org/), and innovative tools that allow for the analysis of gene expression data have been developed.

The present study was fueled by the idea that an unbiased analysis of gene expression will highlight the most-prominent metabolic pathways or other phenomena that underlie reorganization of the epileptogenic circuitry in the brain, and eventually, guide our efforts to identify candidate targets for antiepileptogenic treatments. Here, we reanalysed and compared lists of genes that are regulated by potentially epileptogenic stimuli and are available from publications. We particularly searched for (i) highly represented functional gene classes (GO terms) within the data sets, and (ii) individual genes that appeared in several data sets, and therefore, could be of particular importance for epileptogenesis.

Methods

Selection of papers for analysis and creation of gene lists

The main goal of this project was to identify common features in the molecular responses to epileptogenic stimuli across different animal models. Therefore, papers that describe alterations in the transcriptome following SE or TBI, which are known to trigger epileptogenesis in experimental animals, were selected for analysis, as summarized in Table 1. A few papers describing gene expression in epileptic tissue were also included for comparison. The analysis required extensive database searches, and therefore, only papers providing gene identifiers (e.g., GenBank Accession Numbers, UniGene Cluster ID, or Clone ID) were enrolled (Table 1). Data sets presented in selected papers were used to construct a database that was used for further analysis. The database was constructed in such a way to enable independent

Table 1. Summary of references from which the gene lists were extracted for analysis of GO terms and search of common genes

Reference	Model	Early time period							Middle time period				Late time period	
		30 min	2 h	3 h	4 h	6 h	8 h	9 h	1 day	3 days	4 days	8 days	14 days	Chronic epilepsy
(1) Becker et al. (2003)	SE (pilocarpine) epileptic rats with low[a] or high[b] seizure frequency													•
(2) Elliott et al. (2003)	SE (pilocarpine), rat												•	
(3) Tang et al. (2002)	SE (kainic acid), rat													
(4) Hendriksen et al. (2001)	SE (angular bundle stimulation), rat											•		
(5) Lukasiuk et al. (2003)	SE (amygdala stimulation), rat												• •	•
(6) Kobori et al. (2002)	TBI (CCI), mouse		•											
(7) Long et al. (2003)	TBI (CCI), rat				•				•	•				
(8) Li et al. (2004)	TBI (severe[a] and mild[b] FPI), rat			•	• •				•					
(9) Matzilevich et al. (2002)	TBI (CCI), rat	•			•	•			•		•			
(10) Natale et al. (2003)	TBI (lateral FPI), rat[a]; TBI (CCI), mouse[b]			•			•		•	•				
(11) Raghavendra Rao et al. (2003)	TBI (CCI), rat							•	•					
(12) Arai et al. (2003)	Genetically epileptic rats													•
(13) Bo et al. (2002)	Genetically epileptic rats													•
(14) Becker et al. (2002)	Human temporal-lobe epilepsy													•
(15) Kim et al. (2003)	Human cortical dysplasia													•

Abbreviations: CCI, controlled cortical impact; FPI, fluid percussion injury; SE, status epilepticus; TBI, traumatic brain injury; •, time point investigated in a given manuscript.
[a]Low vs. high seizure frequency in Becker et al. (2003). [b]Severe vs. mild FPI in Li et al. (2004) or rat vs. mouse in Natale et al. (2003).

analysis of each experimental time point from each dataset.

Description of animal models and gene array analysis

Status epilepticus

Three of the five papers that described gene expression after SE used convulsive drugs (kainic acid or pilocarpine) and the other two used electric stimulation to induce SE. Becker et al. (2003) induced SE in rats using intraperitoneal injection of pilocarpine. SE was terminated 2 h later with an administration of diazepam. Gene expression in the dentate gyrus (DG) and the CA1 region of the hippocampus was studied at 3 days and 14 days after SE or in chronically epileptic animals with the Affymetrix U34A microarrays. Genes with a greater than 2-fold expression change compared to the control value and with a P value of less than 0.05 from three replicates for each gene were considered to have altered expression. For the present analysis, only the data derived from epileptic animals were used, as gene accession numbers were provided only for this data set (http://www.meb.uni-bonn.de/neuropath/neurogenomics.html).

Epileptic animals were sacrificed 46–58 days after the induction of SE and divided into two groups: animals exhibiting a low seizure frequency (< 2 seizures/day) and those with a high seizure frequency (> 10 seizures/day). For the purpose of the present analysis, we created two data sets; one containing genes altered in animals with low seizure frequency (in this manuscript referred to as "Becker — low sz SE rat") and one containing genes altered in animals with high seizure frequency (referred to as "Becker — high sz SE rat"). Genes from both the DG and CA1 were combined in the data set (see Tables 2 and 3).

Elliott et al. (2003) compared gene expression between epileptogenesis and development. SE was triggered in adult rats with intraperitoneal injection of pilocarpine and discontinued with diazepam at 2 h. Animals were sacrificed 14 days later and the DG was dissected and processed for analysis with Affymetrix U34A microarrays. For the present analysis, we selected genes with altered expression at 14 days after induction of SE from the database provided by the authors (http://homecaregroup.org/templatesnew/departments/BID/neurology-bpe/uploaded_documents/RCE-JNeurosciMarch03).

Tang et al. (2002) investigated the brain transcriptome response to several stimuli, including ischemic stroke, intracerebral hemorrhage, hypoglycemia, hypoxia, and kainate-induced seizures (i.e., SE). For the present analysis, we selected only the data on gene expression alterations following kainate-induced SE. In this experiment, rats were injected with kainate subcutaneously and sacrificed 24 h later. Affymetrix U34A arrays were used to measure gene expression in the parietal cortex. The authors provided lists of genes that were up- or downregulated to the greatest extent by ischemia, hemorrhage, kainate, or hypoglycemia (Table 2 in Tang et al., 2002). From these data sets, only alterations in gene expression that occurred after kainate-induced SE were selected for analysis.

Hendriksen et al. (2001) searched for changes in gene expression during the early phase of epileptogenesis using Serial Analysis of Gene Expression, which is an alternative to the microarray strategy for global analysis of gene expression. Epileptogenesis was induced by SE that was triggered by electrical stimulation of the angular bundle. SE was terminated at 4 h with pentobarbital. Animals were sacrificed 8 days later and the hippocampi were used for gene expression analysis. Analysis of tag frequency in control and experimental groups identified 92 differentially expressed genes in the hippocampus of rats 8 days after SE, which were used in the present analysis.

Finally, we studied gene expression in an amygdala stimulation model of temporal lobe epilepsy (Lukasiuk et al., 2003) and sacrificed rats at 1, 4, or 14 days after stimulation. The 14-days group was further divided into two subgroups: 14-day epileptogenesis and 14-day epilepsy groups, on the basis of the absence or presence of spontaneous seizures, respectively. Alterations in gene expression were monitored in the hippocampus and temporal lobe. For this analysis we created four gene lists based on the timing of sacrifice after stimulation and the presence of spontaneous seizures: 1, 4, and 14-day epileptogenesis, and 14-day

Table 2. Results of analysis of biologic function GO terms

Biological function GO terms	Early time period													Middle time period														Late time period									
	30 min–4 h										6–9 h			1 day									3–8 days				14 days			Chronic epilepsy						
	Li - severe FPI	Kobori –CCI mouse	Matzilevich – CCI rat	Raghavendra Rao CCI rat	Li - moderate FPI	Li - severe FPI	Long – CCI rat	Natale - FPI rat	Natale - CCI mouse	Kobori –CCI mouse	Natale - FPI rat	Natale - CCI mouse	Raghavendra Rao CCI rat	Lukasiuk – SE rat	Tang - SE rat	Kobori –CCI mouse	Li - moderate FPI	Li - severe FPI	Long – CCI rat	Matzilevich – CCI rat	Natale - FPI rat	Natale - CCI mouse	Raghavendra Rao CCI rat	Kobori –CCI mouse	Natale - FPI rat	Natale - CCI mouse	Lukasiuk – SE rat	Hendriksen - SE	Lukasiuk – SE rat	Elliott - SE	Becker - low sz, rat	Becker - high sz., rat	Lukasiuk – SE rat	Becker human TLE	Bo – epileptic rats	Kim – human epilepsy
Behavior	•		•	•	•																															
Regulation of transcription	•		•		•	•		•			•																									
Ion homeostasis		•			•					•						•							•								•					
sRegulation of protein kinase activity		•														•																				
Intracellular signaling cascade		•			•					•						•	•						•													
Apoptosis		•																									•									
Taxis			•	•		•		•			•		•							•				•									•			
Cell cycle			•			•																														
Cell motility				•				•			•		•				•			•		•														
Response to stress				•									•		•							•			•							•				
Immune response				•								•	•		•				•		•	•									•	•				
Protein folding				•											•																					
Protein transport					•												•	•																		
Response to wounding					•																				•						•	•				
Cell surface receptor-linked signal transduction						○						•						•	•	•																
Ion transport						•																									•					
Cellular protein metabolism							•				•																									
Neurophysiological process									•			•												•												
Protein biosynthesis												•									•						•								•	
Organogenesis												•									•				•						•	•				
Bone remodeling												•			•									•												
Neurogenesis														•				•									•		•							
Extracellular matrix organization and biogenesis			•											•													•				•					
Response to biotic stimulus															•																•	•				
Cell death															•												•									
Protein localization																	•	•																		
intracellular, signaling cascade																	•						•													
Proton transport																	•																			
RNA metabolism																		•											•							
Cell adhesion																			•		•			•												
Nitric oxide biosynthesis																			•			•														
Response to abiotic stimulus																								•									•			
Cytoskeleton organization and biogenesis																										•			•							
Development cell differentiation																										•			•							
Defense response													•																		•	•				
Cell organization and biogenesis																																	•		•	•
Regulation of development														•	•																					

Note: The symbol • indicates a GO term that was over-represented and ○ indicates GO terms under-represented at $p<0.05$ in a given gene list. GO terms over-represented or under-represented in only one gene list were omitted.

Table 3. Results of analysis of molecular function GO terms

Molecular function GO terms	Early time period			Middle time period			Late time period	
	30 min–4 h	8–9 h		1 day		3–8 days	14 days	Chronic epilepsy

Column studies (left to right):
- Early 30 min–4 h: Li - moderate FPI; Li - severe FPI; Matzilevich – CCI rat; Raghavendra Rao CCI rat; Li - moderate FPI; Li - severe FPI; Long – CCI rat; Natale - FPI rat; Natale - CCI mouse
- Early 8–9 h: Natale - FPI rat; Natale - CCI mouse; Raghavendra Rao CCI rat; Lukasiuk – SE rat; Tang - SE rat
- Middle 1 day: Li - moderate FPI; Li - severe FPI; Long – CCI rat; Matzilevich – CCI rat; Natale - FPI rat; Natale - CCI mouse; Raghavendra Rao CCI rat; Kobori – CCI mouse
- Middle 3–8 days: Natale - FPI rat; Natale - CCI mouse; Lukasiuk – SE rat; Hendriksen - SE
- Late 14 days: Lukasiuk – SE rat; Elliott - SE; Becker - low sz, rat; Becker - high sz., rat
- Late Chronic epilepsy: Lukasiuk – SE rat; Bo - epileptic rats; Kim – human epilepsy

Row labels (molecular function GO terms):
- Diacylglycerol binding
- Cation-binding
- transcriptional activator activity
- Purine nucleotide binding
- Hydrolase activity
- Transmembrane receptor activity
- Receptor-binding cytokine activity
- Carbohydrate binding
- Protein kinase activity
- Receptor-signaling protein activity
- Phospholipid binding
- Phospholipase inhibitor activity
- Protein dimerization activity
- Cytokine binding
- Receptor binding
- Receptor-binding G-protein-coupled receptor binding
- Calmodulin binding
- Polysaccharide binding
- Receptor-binding signal transducer activity
- Cytoskeletal protein binding
- Immunoglobulin binding
- Kinase-binding
- transferase activity
- Nucleic acid-binding
- Oxidoreductase activity
- Protein binding
- receptor binding–cytokine binding
- Peptidase activity
- GTPase regulator activity
- Structural constituent of ribosome

Note: The symbol ● indicates a GO term that was over-represented and ○ indicates GO terms under-represented at $p < 0.05$ in a given gene list. GO terms over-represented or under-represented in only one gene list were omitted

epilepsy. Each list contained genes with altered expression in either the hippocampus, temporal lobe, or both.

Traumatic brain injury

Another epileptogenic etiology included in the analysis was TBI. TBI induced by lateral fluid percussion (FPI) induces the development of epilepsy in a substantial proportion of rats (D'Ambrosio et al., 2004; Pitkanen et al., 2006). For comparison, we also included data from another commonly used TBI model, that is, controlled cortical impact (CCI) injury, in which the occurrence of spontaneous seizures remains to be studied (see Pitkanen and McIntosh, in press).

Kobori et al. (2002) investigated changes in gene expression in the cerebral cortex following CCI injury in mice. Animals were sacrificed 2 h, 6 h, 24 h, 3-day, or 14-day post injury. A block of cortex ipsilateral to the injury, containing the prefrontal, motor, and somatosensory cortices, was used for gene expression analysis. Corresponding tissue from naive animals was used as a control. Gene expression was analyzed with Mouse Unigen1 arrays (Incyte Genomics). A change in the expression of 1.5-fold or more was considered to be "altered". For the present analysis, we created separate gene lists for each time point from the data provided in the article (see Table 1 in Kobori et al., 2002).

Long et al. (2003) analyzed changes in gene expression in the ipsilateral hippocampus at 4 and 24 h after CCI injury in mice using an in-house-produced microarray containing 6400 mouse cDNA probes. A gene was considered to be differentially expressed if the expression differed by a factor of 1.5 at any post-injury time point. For our analysis, we created gene lists for the 4- and 24-h time points by extracting data from Table 2 using the above criteria (1.5-fold change) for each respective time point.

Differential gene expression in the rat hippocampus following moderate versus severe lateral FPI was studied by Li et al. (2004). Tissue was dissected at 0.5, 4, or 24 h after trauma and used for hybridization to the Affymetrix U34A gene chip. Changes in the expression level greater than 2-fold were considered meaningful. We decided to divide the data from each time point into two separate gene lists: one containing genes altered by moderate injury and the other containing genes altered by severe injury. Data were extracted from the table provided on the authors' web page (http://www.birc.ucla.edu/pdf/hippocampus.pdf). We refer to these lists as: "Li-moderate FPI" and "Li-severe FPI", respectively (see Tables 1–3).

The gene expression after CCI injury in rat was investigated by Matzilevich et al. (2002). Ipsilateral hippocampi were isolated at 3 or 24 h after injury and used for hybridization to the Affymetrix U34A microarray. Data provided in Table 1 of the original paper showing the changes in gene expression at the two time points were used for the present analysis.

Natale et al. (2003) studied gene expression following lateral FPI in rat and CCI injury in mice. A block of the parieto-occipital cortex containing the lesion area was isolated at 4 h, 8 h, 1 day, or 3 days after brain trauma. The Affymetrix U34A and Affymetrix Mu74Av2 gene chips were used to assess gene expression in rats and mice, respectively. Genes indicated in bold font (with level of expression changed by a factor >2) in Table 3 of Natale et al. (2003) were considered to be significantly altered. Moreover, for our analysis we generated separate lists for each time point from each model (Table 3 in Natale et al., 2003). We refer to these gene lists as "Natale – CCI mice" and "Natale – FPI rat" (see Tables 2 and 3).

Raghavendra Rao et al. (2003) provided another data set on gene expression following CCI injury in rat. They collected rat cerebral cortex at 3 h, 9 h, and 1 day after injury and analyzed it with the Affymetrix U34A gene chip. For our analysis, gene lists for each time point were created by combining data from Tables 1 and 2 presented by Raghavendra Rao et al. (2003).

Chronic temporal-lobe epilepsy humans and genetic epilepsy in rat

The data at different post-injury time points provide information about the time period during which the epileptogenesis occurs. Recent studies investigated gene expression in human epileptic tissue as well as in tissue from spontaneously seizing rats with genetic epilepsy. These data provide insight into gene expression in tissue beyond the epileptogenic period; that is, in neuronal networks that are matured enough to generate spontaneous seizures.

Arai et al. (2003) investigated gene expression in spontaneously epileptic Ihara rats. These animals have a genetically programmed hippocampal microdysgenesis that is supposed to give origin to spontaneous seizures. Hippocampal tissue was investigated with Serial Analysis of Gene Expression analysis and the data were compared to data from normal Wistar rats.

Differences in gene expression between genetic epilepsy-prone P77PMC rats and normal Wistar rats were analyzed using the Atlas™ Rat cDNA Expression Array (Bo et al., 2002). Fifteen differentially expressed genes were detected in the cerebral cortex.

Becker et al., (2002) studied gene expression in hippocampal specimens obtained from patients who were operated on for drug-refractory temporal-lobe epilepsy (TLE). Characteristically, the seizure focus was in the hippocampus, which showed remarkable neuronal damage (ammon horn sclerosis, AHS). An Atlas Human Neurobiology array was used for molecular profiling. Tissue from nonepileptic tumor patients served as the control. The list of differentially expressed genes is referred to as 'Becker – human TLE' (see Table 2).

Finally, the last paper used for the analysis describes the gene expression profile in human cortical dysplasia (Kim et al., 2003). Tissue was analyzed using the Atlas Human 1.2 array and data from human dysplastic cortex was compared to that in nondysplastic cortex obtained from a 2-year-old boy with intractable epilepsy and medial temporal lobe ganglioglioma.

Analysis of representation of GO terms

To detect functional gene classes with particular importance in the development of epilepsy, we analyzed the representation of GO terms in gene lists that were generated from the data provided by the papers described above. Each gene list contained genes for which the expression level changed at a given time point according to a given paper. Each gene list was analyzed separately. For this purpose, we used GOstat software available at http://gostat.wehi.edu.au (Beissbarth and Speed, 2004). GOstat allows one to search for functional annotations based on GO terms that are listed in the Gene Ontology database (http://www.geneontology.org) The GO consortium provides tree-structured and controlled vocabularies (ontologies) that describe gene products in terms of their associated biologic processes, cellular components, and molecular functions in a species-independent manner. GOstat provides statistics about GO terms, indicating the most representative GO terms for a given data set. The list of interest is compared to a list containing all genes present on the microarray (or to the complete set of annotated genes in the GO gene-associations database for the data derived from a non-microarray type of experiment). The statistical significance of differences was then calculated with Fisher's Exact Test. The correction for multiple testing was performed with a Benjamini false-discovery rate (Beissbarth and Speed, 2004). The output of the program was a list of p-values for the associated GO terms for the list of genes provided. In our analysis, a p value of less than 0.05 was considered significant. The analysis was performed using the GO database modified on March 7, 2005.

The results of the analysis of GO terms according to biologic function and molecular function are presented in Tables 2 and 3, respectively. The tables contain gene lists that are arranged according to the time points after treatment, and according to the corresponding GO terms with significantly changed representation in a given gene list. On our web pages (http://www.uku.fi/aivi/neuro/research_epilepsy.shtml and http://www.nencki.gov.pl/labs/trlab.htm), we provide extended versions of the tables that also contain gene names represented by each over-represented or under-represented GO term in each of the gene lists that we analyzed. We encourage readers to use these tables to address their specific interests.

Search for common genes

To find a particular gene with altered expression, we had to re-annotate all data. For this purpose, we took the GenBank/EMBL/DDBJ accession numbers (essentially as provided by the authors) of the sequences corresponding to the genes with altered expression. The second parameter collected for our data set was the information about the time point, at which the altered expression occurred. The primary identifiers were mapped to mRNA (UniGene) and gene (LocusLink) identifiers using the data downloaded from the National Center for Biotechnology Information (http://www.ncbi.nlm.nih.gov/entrez). The accession numbers were first mapped to corresponding UniGene ID(s) and LocusLink ID(s) (Entrez: Gene ID), using the data from the UniGene database (built on 17 March 2005), separately for rat, mouse, and human. Each LocusLink ID was mapped to its corresponding HomoloGene group id (HID), using the data from the HomoloGene database. The re-annotation resulted in mapping 1699 distinct accession numbers to 1327 distinct Unigene cluster IDs (mRNA), 1214 distinct LocusLink IDs (genes), and 1014 HIDs. The gene symbols and full names used in this paper are the current versions from the UniGene database. The re-annotation was performed using a purpose-built relational database on a MySQL server platform. To permit comparison between the lists of genes across species, we compared the lists of HIDs corresponding to the accession numbers reported in each of the compared papers. The comparisons were performed with SQL queries embedded in a scripting language.

Gene lists were created for each time window from every paper. The analysis was performed at different time windows following the insult: (1) early-time points: 30 min to 9 h, (2) middle-time points: 1–4 days, and (3) late-time points: >14 days. The 8-day time point was omitted from the analysis.

Results and discussion

Our analysis revealed that various epileptogenic insults induce statistically significant changes in gene expression in functionally linked genes that were predefined as GO terms. The number of over-represented GO terms associated with a particular gene list was, however, variable. In some cases, many biologic or molecular processes were indicated, whereas others had no significant changes. The lack of significant findings in some data sets could relate, for example, to the small number of genes available for evaluation. Furthermore, data analysis as well as the cut-off criterion for "alteration" in gene expression varies between the studies, which can bias the number of GO terms that were attributed to each gene list in the present analysis. Despite these caveats, comparison of over-represented and under-represented GO terms between different gene lists revealed interesting patterns. We particularly focused on GO terms that were represented in multiple gene lists. We hypothesize that a given biologic or molecular phenomena connected with these terms is important for the process of epileptogenesis.

GO terms referring to biologic process

The first 24 h after insult
The representation of statistically significant GO terms describing biologic process was time dependent. Within hours after epileptogenic brain insult, the GO term 'regulation of transcription' was overrepresented. Interestingly, this GO term did not become significant at any later time point. The notion that transcription factors that regulate transcriptional activity of other genes are involved in the early response of the brain to epileptogenic stimuli was described earlier based on various methodologies (Dutcher et al., 1999; Zagulska-Szymczak et al., 2001). The strong representation of transcription factors in the current analysis precedes the changes in the expression level of secondary genes that presumably encode events leading to the development of epilepsy.

Many of the transcription factors detected by our analysis can contribute to ongoing neurodegeneration, or more generally, to cell stress. For example, JUNB and ATF2 are associated with neuronal injury, and ETS1 has a role in the inhibition of apoptosis by activating the transcription of cyclin-dependent kinase inhibitor p21WAF1/Cip1 (Belluardo et al., 1995; Zhang et al., 2003; Lindwall and Kanje, 2005). On the other hand, changes in the expression of some transcription factors might induce succeeding events that are presumably crucial for epileptogenesis, such as neuronal plasticity. For example, EGR1 can be involved in synaptic plasticity, BHLHB2 regulates neurite outgrowth, HES1 inhibits cell adhesion molecule NCAM-dependent neurite outgrowth, and NR4A2 regulates the development and maintenance of neurons (Worley et al., 1991; Kabos et al., 2002; Jessen et al., 2003; Perlmann and Wallen-Mackenzie, 2004).

Another GO term that is represented early after brain insult is "immune response". Interestingly, this term is also overrepresented in chronically epileptic rats. In our datasets, the GO term "immune response" includes transcription factors that are involved in regulating the expression of inflammatory genes (CEBPB and IRF1), interleukins (IL6, IL1B and IL18), chemokines (CXCL2 and CCL3), cytokine-inducible NOS2, histocompatibility and surface antigens (CD14, CD74, CD1D1, and RT1-AW2), and complement proteins (C1QB, and C3). These data support the idea that inflammation is involved in post-traumatic brain pathology (Stoll et al., 2002).

The next significantly over-represented GO terms, "response to stress" and "response to wounding", overlap to some extent with "immune response" as these GO terms also host a number of genes that are involved in immune response, including interleukins, cytokines, and surface antigens, but also revealed other genes like heat-shock proteins (HSPB1, HSPA1A). Heat-shock proteins are induced by a variety of pathologic stimuli. For example, increased expression of HSPA1 proteins is linked to excitotoxicity (Ayala and Tapia, 2003). Further, increased HSPB expression is observed in human epileptic tissue (Bidmon et al., 2004).

Several genes under the GO term "cell mobility and taxis" also belong under the GO term "immune response". These include chemokines involved in immune cell trafficking, including CCL3, CXCL2, and CCL4, as well as CCR5, which is a chemokine receptor involved in lymphocyte migration. Another interesting gene revealed by our analysis belonging to "cell mobility and taxis" was S100A8, which is a calcium-binding protein with a role in inflammatory response.

Genes under the "ion homeostasis" term represent mainly zinc-binding metallothioneins MT1 and MT2. The involvement of metallothioneins in brain dysfunctions has been studied extensively. For example, there is an increased expression of these proteins following SE (Dalton et al., 1995; Montpied et al., 1998). Regulation of zinc release by metalliothionein III seems to be detrimental to neuronal survival (Lee et al., 2003). On the other hand, metallothionein I reduces inflammation, neurodegeneration, and cell death following kainic acid-induced SE (Penkowa et al., 2005).

Day 1 after brain insult
Genes under the GO term neurogenesis, become significant at 1 day after injury. Various types of brain injury, including SE and TBI can trigger hippocampal neurogenesis (Parent, 2003; Cha et al., 2004). Thus, overrepresentation of the GO term "neurogenesis" in our analysis is consistent with these observations. This group of genes includes SCRP1, a cysteine-rich protein with homology to the delta opioid receptor that has a role in neuronal development. Other genes revealed by our analysis were YWHAG, which is a 14-3-3-protein family member; INEXA, which is a neuronal intermediate filament protein with a role in neurogenesis that can be involved in neuronal regeneration; BHLHB3, which regulates neuronal differentiation (Namikawa et al., 1998); and others (see Table 2 in Supporting Information).[1] In addition the GO term "immune response", described above, was over-represented in many gene lists at the 1-day time point.

Days 3–8 after insult
GO terms such as "ion homeostasis", "taxis", "response to stress or wounding", "neurogenesis", and others were also overrepresented at 3–8 days. None of the GO terms, however, was significant on more than one list studied.

Chronic epilepsy
In epileptic tissue collected from humans or animals with spontaneous recurrent seizures (i.e., the end result of epileptogenic process), the presence of the GO terms related to "synaptic organization"

---

[1]Supporting information for the tables can be accessed at http://www.uku.fi/aivi/neuro/research_epilepsy.shtml or http://www.nencki.gov.pl/labs/trlab.htm.

(BDNF), "plasticity" (BDNF), and "transmission" (APOE, SNAP25) attracted our attention. Changes in the expression of these particular genes are also observed in epileptic tissue using other methods (Montpied et al., 1999; Binder, 2004; Zhang et al., 2004). This supports the idea that neuronal plasticity occurs for a longer period of time after insult, and can progress even after epilepsy diagnosis (Pitkanen and Sutula, 2002).

GO terms over-represented at various stages of the epileptic process
Several GO terms, including "immune response", "cell motility", "response to stress", "response to wounding", or "ion homeostasis", are represented across different time points after insult. In addition, many GO terms that group genes involved in signal transduction, including "regulation of protein kinase activity", "intracellular signaling cascade", and "cell surface receptor-linked signal transduction" were over- or under-represented at different time points after insult. Proteins coded by these genes include various kinases, phosphatases, and adaptor proteins that can participate in numerous and overlapping signaling pathways. (see Table 2, and Table 2 in Supporting Information).

GO terms referring to molecular function

At early time points, there is a striking overrepresentation of the GO term "transcriptional activator" (Table 3), which partially overlaps with the GO term "regulation of transcription" referring to biologic process discussed above. Further, genes grouped under the ubiquitous "receptor binding, cytokine activity" term are actually interleukins and chemokines already mentioned under the GO term "immune function".

One of the molecular function of GO terms that was overrepresented, particularly at earlier time points, is "purine nucleotide binding". This group includes genes encoding proteins that are involved in translation, and therefore, regulate protein synthesis like EEF1A1 and EEF2. Other genes in this category include GTPases that are involved in signal transduction (RAB and RAS proteins, RHOA, GNA12, and GNAZ), numerous kinases (MAPK, SGK, and JAK), a number of ADP-ribosylation factors of variable functions, and ATPases with variable function (ATP1A1 and PSMC2).

Our analysis also revealed a global alteration in genes encoding second messengers. In this context, the most frequently represented GO terms were related to signal transduction, including "diacylglycerol binding", "protein kinase activity", "protein kinase regulator activity", "transmembrane receptor activity", "receptor-binding G-protein-coupled receptor binding", "receptor-binding signal transducer activity", and "receptor-signaling protein activity". These data suggest widespread changes in cell metabolism during epileptogenesis, and are of particular importance because each signaling pathway can participate in a plethora of cellular processes. A detailed discussion of this issue, however, is far beyond the scope of this review.

Similarly altered genes in different post-injury time windows

Table 4 summarizes the genes that were altered in at least three gene lists at a given post-injury time window. That is, early post-injury between 30 min and 9 h (30 min–4 h and 6 h–9 h subwindows), middle period between 1 and 4 days (1 and 3–4 days subwindows), and late period ($\geqslant 14$ days and chronic epilepsy). We discuss only the genes with altered expression in at least three lists in a given time window. To provide more detailed information, the function of listed genes was annotated using DAVID and GeneCards (http://apps1.niaid.nih.gov/david/; http://www.genecards.org/index.shtml). This analysis is hampered by differences in animal models and microarray platforms. We did, however, detect some overlap in genes with altered expression between the gene lists, which supports that the analysis performed can provide valuable information. An extended table containing detailed data about each gene is available at http://www.uku.fi/aivi/neuro/research_epilepsy.shtml or http://www.nencki.gov. pl/labs/trlab.htm. Finally, the function of some of the commonly altered genes can be estimated based on available literature. Several diverse functions are attributed to some of these genes, suggesting their potential to control a myriad of alterations during epileptogenesis. The

Table 4. Summary of genes with altered expression in at least three different gene lists at a given time period

Symbol and gene name	Summary of function	Number of lists containing the gene						
		Early time period			Middle time period			Late time period	
		30 min–4 h (8)	6–9 h (4)	1 day (10)	3–4 days (5)	14 days (3)	Chronic epilepsy (6)
Tieg, TGFB inducible early growth response	Regulation of transcription; apoptosis; proliferation	$3^{6,8ab}$					
Nr4a1, nuclear receptor subfamily 4, group A, member 1	Regulation of transcription	$3^{6,8ab}$					
Mt1a, metallothionein	Nitric oxide-mediated signal transduction; zinc ion homeostasis	$4^{6,7,8ab}$		$3^{6,7,8a}$			
Ccl4, small inducible cytokine A4	Immune response	$5^{6,8ab,9,11}$					
Ccl3, chemokine (C-C motif) ligand 3	Immune response	$5^{6,8ab,9,11}$	$2^{6,11}$				
Hmox1, heme oxygenase (decycling) 1	Apoptosis	$3^{6,8ab}$		$2^{6,8a}$			
Lcn2, lipocalin 2	Immune response	$4^{6,8ab,10a}$	$3^{6,10ab}$	$5^{3,6,8ab,10a}$	$2^{6,10a}$		
Lgals3, lectin, galactose binding, soluble 3	Cell adhesion, immune response	$3^{6,8a,10b}$	$3^{6,10ab}$	$7^{3,6,8ab,9,10ab}$	$3^{6,10ab}$		
Ptgs2, prostaglandin-endoperoxide synthase 2	Cell proliferation; neuronal plasticity	$3^{8ab,9}$					
Il1b, interleukin 1 beta	Cell proliferation; immune response	$4^{8ab,9,11}$					
Fcgr3, Fc receptor, IgG, low affinity III	Immune response	$3^{8ab,9}$					
Gnai1, guanine-nucleotide binding protein, alpha inhibiting 1	Signal transduction, regulation of synaptic transmission	$3^{8ab,9}$		2^{8ab}			
Sgk, serum/glucocorticoid regulated kinase	Apoptosis, neuronal plasticity	$3^{8ab,9}$					
Mapk1, mitogen-activated protein kinase 1	Cell cycle; signal transduction	$3^{8ab,9}$					
Arc, activity regulated cytoskeletal-associated protein	Actin binding; neuronal plasticity	$3^{8ab,9}$					
Sdfr1, stromal cell-derived factor receptor 1	Synaptic membrane glycoprotein	$3^{8ab,9}$				1^{14}	
Atp1a1, ATPase, Na+/K+ transporting, alpha 1 polypeptide	Sodium/potassium-exchanging ATPase	$3^{8ab,9}$					
Irf1, interferon-regulatory factor 1	Cell proliferation, immune response; regulation of transcription	$5^{8ab,9,10a,11}$	$2^{10b,11}$	$3^{9,10a,11}$			
Egr2, early growth response 2	Regulation of transcription; neuronal plasticity	$3^{8ab,9}$					
Nfkb1, nuclear factor kappa B p105 subunit	Regulation of transcription; response to oxidative stress	$5^{8a,9,10ab,11}$	$3^{10ab,11}$	$4^{9,10ab,11}$			
Egr1, early growth response 1	Regulation of transcription, neuronal plasticity	$3^{8ab,9}$					
Jun, v-jun sarcoma virus 17 oncogene homolog	Regulation of cell cycle; regulation of transcription, neuronal plasticity	$4^{8ab,10ab}$		$3^{8a,10ab}$			
Zfp36, zinc-finger protein 36	Regulation of transcription	$3^{8ab,10a}$	$3^{6,10ab}$	$2^{6,10b}$			
F3, coagulation factor 3	Blood coagulation factor	$3^{8ab,10a}$		$3^{5,8b,9}$			
Hspb1, heat shock 27 kDa protein 1	Heat shock protein	$3^{8ab,11}$		2^{10ab}			
Col1a1, collagen, type 1, alpha 1	Cell adhesion	$3^{8b,9,10b}$		$4^{3,6,8b,9}$	2^{10ab}		
Cebpb, CCAAT/enhancer-binding protein (C/EBP), beta	Apoptosis; neuron differentiation; regulation of transcription	$5^{8a,9,10ab,11}$	$3^{10ab,11}$	$4^{9,10ab,11}$	2^{10ab}		
Cpg21, MAP-kinase phosphatase (cpg21)	Signal transduction	$3^{8a,9,11}$					
S100a8, S100 calcium-binding protein A8 (calgranulin A)	Calcium ion binding; immune response	$3^{8a,10ab}$	2^{10ab}	2^{10ab}	2^{10ab}		
Cd53, CD53 antigen	Cell survival	$2^{6,10b}$	$3^{6,10ab}$	$2^{6,10b}$	$3^{6,10ab}$		
Mrlcb, myosin-regulatory light chain	Cytoskeleton			$3^{5,8b,9}$			
Il1r2, interleukin 1 receptor, type II	Signal transduction	$2^{6,8b}$		2^{10ab}	1^5		
Spin2c, Serine protease inhibitor	Immune response	$2^{6,8a}$	$2^{6,10a}$	$5^{3,6,8a,9,10a}$	$2^{6,10a}$		
Lyz, lysozyme	Immune response						
Hspb1, heat shock 27 kDa protein 1	Heat shock protein, cell death			$4^{3,8ab,11}$			$2^{14,1b}$

Gene	Function			
Timp1, tissue inhibitor of metalloproteinase 1	Metalloendopeptidase inhibitor activity; vascular tissue remodeling	2[8ab]		4[3, 8ab, 10b]
Tuba, Tubulin, alpha 6	Cytoskeleton			3[3, 8ab]
Vim, vimentin	Cytoskeleton			5[3, 8ab, 9, 11]
Bzrp, benzodiazepin receptor	Extracellular space; integral to membrane; mitochondrion; receptor activity		2[10ab]	6[3, 8ab, 9, 10a, 11]
Hla-dmb, major histocompatibility complex, class II, DM beta	Immune response			3[3, 8ab]
Anxa2, annexin A2	Transformation	2[8a, 10a]		4[3, 8ab, 10a]
S100a10, S-100-related protein, clone 42C	Calcium ion binding, neuronal differentiation			5[3, 8ab, 10ab]
Arpc1b, actin-related protein 2/3 complex, subunit 1B	Cytoskeleton	2[8ab]	2[10ab]	6[3, 8ab, 9, 10ab]
Mgp, matrix Gla protein	Calcium ion binding		2[10ab]	3[3, 8a, 10b]
Cd74, CD74 antigen	Chaperone activity; immune response			2[3, 9]
Gadd45a, growth arrest and DNA damage-inducible 45 alpha	Cell cycle arrest;	2[8a, 9]	1[5]	3[3, 8a, 10a]
Spp1, secreted phosphoprotein 1	Immune response; apoptosis		2[10ab]	3[3, 10ab]
Akap12, A kinase (PRKA) anchor protein (gravin) 12	Signal transduction	2[6, 8a]		3[6, 8ab]
Spin2c, Serine protease inhibitor	Immune response			3[6, 8b, 9]
Pak1, p21 (CDKN1A)-activated kinase 1	Signal transduction			3[6, 8b, 11]
RT1Aw2, RT1 class 1b gene(Aw2)	Immune response			3[6, 8a, 9]
Scd2, stearoyl-Coenzyme A desaturase 2	Lipid metabolism in PNS			3[8a,9]
Klk6, kallikrein 6	Myelencephalon-specific protease			3[8a,9]
Tagln, transgelin	Muscle development			3[8a,9]
RT1-Da, similar to H-2 class II histocompatibility antigen	Immune response			4[8ab,9, 10b]
Cd63, CD63 antigen	Protein trafficking			4[8a,9, 10b]
Igfbp2, insulin-like growth factor-binding protein 2	Regulation of cell growth			3[8ab,9]	2[2, 6]
Mag, myelin-associated glycoprotein	Cell adhesion; caxonal–glial and glial–glial interactions		2[10ab]	
Ddx1, nuclear RNA helicase, DECD variant of DEAD box family	Transcription			3[8ab, 9]
Nfix, nuclear factor I/X	Regulation of transcription, olfactory gene expression			3[8ab, 9]
C1qb, complement component 1, q subcomponent, beta polypeptide	Immune response		2[10ab]	3[8ab, 10b]
Tpm4, tropomyosin 4	Cytoskeleton		2[0ab]	2[8a, 10b]
Tmp21, integral membrane protein Tmp21-I	Protein transport		2[0ab]	3[7, 8b, 11]
Rab3a, member RAS oncogene family	Protein transport; synaptic vesicle Endocytosis			
Nes, nestin	Neurogenesis; neuron differentiation			3[8a, 9, 11]
Cd9, CD9 antigen	Signal transduction; intercellular signaling in the nervous system		2[10ab]	2[8a, 10b]
Ctsl, cathepsin L	Proteolysis and peptidolysis		3[6, 10ab]	
Id3, Inhibitor of DNA binding 3, dominant-negative helix–loop-helix protein	Regulation of transcription		3[6, 10ab]	
Ctsd, cathepsin D	Proteolysis and peptidolysis				2[2, 6] 1[5]
Ctss, cathepsin S	Proteolysis and peptidolysis	1[6]			2[2, 6] 1[5]

Note: Superscript numbers refer to manuscripts containing given gene, numbered according to Table 1.
[a]Low vs. high seizure frequency in Becker et al. (2003). [b]Severe vs. mild FPI Li et al. (2004) or rat vs. mouse in Natale et al. (2003).

function of many other genes is barely defined. Further studies on such genes, however, can be of particular interest in guiding us to new avenues in epileptogenesis research. We will next highlight the candidate functions of some of the commonly altered genes.

Early post-injury time window
An overview of Table 4 reveals that most of the listed genes change their expression for only a limited time. This supports the idea that epileptogenesis is a dynamic process consisting of waves of parallel and serial events (Lukasiuk et al., 2003). Transcription factors Nr4A1, Irf1, Egr2, Nfkb1, Jun, Zfp36, Cebpb, and Tieg were highly represented at early time points; in particular, between 0.5 and 4 h. Nr4A1 is induced by seizures and ischemia, and participates in neuronal plasticity as well as in the initiation of apoptosis (Li et al., 2000; French et al., 2001; Lachance and Chaudhuri, 2004; Yakubov et al., 2004). Egr2 is involved in brain development, but might also have a role in adult neuronal plasticity (Williams et al., 1995). Nfkb is a transcription factor which is activated after different types of brain injuries, including SE (Rong and Baudry, 1996; Mattson and Camandola, 2001; Voutsinos-Porche et al., 2004). Importantly, this transcription factor participates in brain pathology on many levels. For example, by regulating the expression of genes involved in inflammation and the immune response (e.g., TNF-α, IL-6, and βAPP), by controlling proliferation and apoptosis (e.g., calbindin-28, IAP, Bcl-2, Mn-SOD) as well as synaptic plasticity (e.g., GluRs) or glial activation (GFAP) (Mattson and Camandola, 2001). Interestingly, the activation of Nfkb1 in neurons is neuroprotective, whereas in microglia it is detrimental (Mattson and Camandola, 2001). Therefore, the net effect of activation of Nfkb1 on the epileptogenic process is difficult to predict. The Cebpb transcription factor can also be engaged in multiple functions, such as immune response, neuronal injury, and neuronal plasticity (Taubenfeld et al., 2001; Cortes-Canteli et al., 2004).

In addition to Nfkb, Cebp, Tieg, and Nr4A1, some other commonly expressed genes can also contribute to cell death occurring at the early post-injury period in TBI and SE, including Hmoc1 or Sgk. The expression of Hmoc1, which is a marker of oxidative stress and has expression with antioxidative function, is increased in neurons and glia following various types of lesions (Munoz et al., 2005). Interestingly, Hmoc1 in astrocytes and microglia is associated with survival of these cell types, but can be detrimental for neurons (Lu and Ong, 2001). The expression of Sgk, a serum/glucocorticoid-regulated kinase, can also be induced by a variety of insults (Hollister et al., 1997). It has been suggested that Sgk reduces neurotoxicity in the brain (Bohmer et al., 2004). Taken together, analysis of the function of genes expressed at early post-injury period indicates that both pro-life and pro-death genes can be activated by epileptogenic insults.

Neuronal plasticity is one of the well-known neuronal alterations that occur during epileptogenesis, and has been proposed to contribute both to harmful epileptogenic as well as beneficial recovery processes. The present analysis indicates that genes involved in neuronal plasticity are already represented at the early post-injury period. Some of them are transcription factors that influence the expression of other genes with a role in neuronal plasticity (Egr1, Egr2, and Jun) (Abraham et al., 1991). Another interesting gene is Arc, an immediate-early gene that is activated by neuronal activity, including seizures. Its mRNA is transported to dendrites and translated in the synaptic region, resulting in protein synthesis, which is associated with an increase in synaptic strength (Steward and Worley, 2001).

Middle post-injury time window
Some of the immune response-related genes were activated at 1 day postinjury. These include several genes encoding histocompatibility antigens, like Hla-dmb, which is a component of a major histocompatibility complex II and is expressed in astrocytes (Soos et al., 1998), RT1Aw2 (a major histocompatibility complex I gene), RT1-Da (a major histocompatibility complex II gene), and CD74 (an invariant polypeptide of a major histocompatibility complex, class II antigen-associated gene). Spp1 encodes a cytokine that is involved in enhancing the production of interferon-gamma

and interleukins. It is upregulated in microglia in several pathologic conditions including SE, and has neuroprotective effects (Kim et al., 2002; Meller et al., 2005). Spin2c encodes a protease inhibitor and an acute phase protein. Extensive studies of Alzheimer's disease demonstrate that Spin2c is expressed in reactive astrocytes and participates in neuroinflammatory responses associated with amyloid deposition (Abraham, 2001). The complement component C1qb is induced by injury as well as in Alzheimer's disease (Pasinetti et al., 1992). In conclusion, the involvement of the immune system in the sequence of events following epileptogenic insult extends beyond the acute phase.

At the middle time points (1 or 3–4 days after epileptogenic insult), there are notable alterations in the expression of a number of genes that encode structural proteins. Most of the gene products are involved in organization of the cytoskeleton. Tuba, for example, is a scaffold protein linking dynamin with actin-regulating proteins. It is concentrated at the synapse and might have a role in membrane trafficking (Salazar et al., 2003). Vimentin, an intermediate filament is a marker of reactive astrocytes and is induced during epileptogenesis (Khurgel et al., 1995). Arpc1b also encodes a cytoskeletal protein. The gene product is implicated in the control of actin polymerization. Another actin-binding protein is Tpm4, which is involved in stabilizing the cytoskeletal actin filaments. Changes in the expression of these genes might relate to the structural remodeling of affected tissue.

Bzrp is the last of the commonly altered genes that changes its expression within a few days post-injury, and the function of which is worth highlighting. Bzrp is a peripheral benzodiazepine receptor that is expressed in the brain and is upregulated in many pathologic conditions. In the brain, the receptors are localized in the mitochondria of glial cells. They are implicated in epileptogenesis of EL mice, and their antagonists have anticonvulsant effects (Lang, 2002).

Early and middle post-injury time windows
Several commonly altered genes are expressed both within hours as well as within one or a few days postinjury. Interestingly, they appear to function in different aspects of the immune response. For example, expression of chemokines Ccl2 and Ccl5 in glia after neuronal injury probably guides leukocyte entry in the brain. IL-1b is induced in microglia following a variety of brain insults. This interleukin is proposed to have both antiepileptogenic, anticonvulsant, and proconvulsant actions (Sayyah et al., 2005; Vezzani et al., 2004). Other immune response genes expressed within hours to 1 day to few days after TBI or SE consist of Lcn2, Lgals3, transcription factor Irf1, S100a8, and Lyz. Lgals3 is involved in the regulation of immune cell homeostasis by amplifying the inflammatory cascade (Rabinovich et al., 2002), but it also seems to have a more specific function in neuronal survival and stabilization of neurites (Mahoney et al., 2000; Pesheva et al., 2000). Expression of 100a8 is induced in macrophages in chronic inflammation and in activated microglia (Postler et al., 1997). Lyz (lysozyme) is associated with the monocyte/macrophage system in tissues and body fluids. In the brain, it is also expressed in macrophages and possibly also in microglia (Ohmi et al., 2003).

Late post-injury time window
Unexpectedly, the expression of only two genes changed in at least three different gene lists at the late post-injury time period, that is, after 14 days or in chronic epilepsy. This might relate to the fact that we combined diverse models for the analysis of gene expression at late time points. Also, little overlap was observed in the alterations of gene expression between the early and late time points, which supports the idea that the neurobiology of established epilepsy differs remarkably from that of the dynamic period of early epileptogenesis. Interestingly, however, both genes with altered expression at the late time points encode cysteine proteases (Ctsd and Ctss), indicating the importance of proteolysis in the late post-injury period.

Concluding remarks

We used novel bioinformatics tools to compare the available literature on global analysis of gene expression following epileptogenic insults. We aimed at identifying (i) highly represented functional gene classes within the data sets and (ii) individual genes that appear in several data sets, and therefore, might be of particular importance for epileptogenesis. Analysis of their function indicated some trends in post-injury gene expression that have been somewhat underappreciated in the epilepsy literature. There were five major observations in the present analysis. First, numerous genes change their expression in more than two models of epileptogenesis. Second, alterations in gene expression are time specific, that is, some sets of genes are affected within hours after injury whereas other becomes altered one to several days after injury. Third, a closer look at altered individual genes as well as GO terms indicated an involvement of many processes previously linked with epileptogenesis; for example, cell death and survival, neuronal plasticity, and immune response. Interestingly, however, only few genes are involved in neuronal electrical properties. Fourth, surprisingly, a large majority of alterations in gene expression is localized in the glial cells. Fifth, the predominant function of genes with altered expression is immune response. Although the roles of inflammation and immune response in brain trauma and epilepsy are increasingly studied, they have hardly been considered main perpetrators of epileptogenic process. In the light of our study and recent reports on the neuroprotective effects of modulation of the inflammatory response, this subject appears to be a good candidate for further research.

There are, however, several caveats in the analysis. Interpretation of the functional consequences of an orchestrated expression of groups of genes is complicated by the fact that many genes have multiple functions, and it is difficult to predict which functions are important for epileptogenesis. In addition, the functions attributed to genes in databases that are used for the functional annotation of genes are less extensive than those known from the literature and listed in PubMed. Further, the function of many of the gene products in the brain is unknown. Finally, verification of the relevance of findings from molecular studies should be confirmed in in vivo models. Luckily, basic knowledge on the functions of genes, quality, and extent of databases, and sophistication of bioinformatic tools for data mining develop rapidly. These advances will undoubtedly reveal new information about existing and new databases to identify the molecular basis of epileptogenesis.

Abbreviations

AHS	ammon horn sclerosis
CCI	controlled cortical impact injury
FPI	lateral fluid percussion injury
GO	gene ontology
SE	status epilepticus
TBI	traumatic brain injury
TLE	temporal-lobe epilepsy

Acknowledgments

The work was supported by The Polish State Committee for Scientific Research grant No. 2 P04A 052 26 (to K.L.) and the Academy of Finland, the Sigrid Juselius Foundation, the Finnish Cultural Foundation, and the Paulo Foundation (to A.P.). We apologize to all distinguished colleagues whose original articles are not cited in this review. Due to the vast number of issues discussed and space constraints, we were forced to refer only to databases and review articles whenever possible.

References

Abraham, C.R. (2001) Reactive astrocytes and alpha1-antichymotrypsin in Alzheimer's disease. Neurobiol. Aging, 22: 931–936.

Abraham, W.C., Dragunow, M. and Tate, W.P. (1991) The role of immediate early genes in the stabilization of long-term potentiation. Mol. Neurobiol., 5: 297–314.

Arai, M., Amano, S., Ryo, A., Hada, A., Wakatsuki, T., Shuda, M., Kondoh, N. and Yamamoto, M. (2003) Identification of epilepsy-related genes by gene expression profiling in the hippocampus of genetically epileptic rat. Brain Res. Mol. Brain Res., 118: 147–511.

Ayala, G.X. and Tapia, R. (2003) Expression of heat shock protein 70 induced by 4-aminopyridine through glutamate-mediated excitotoxic stress in rat hippocampus in vivo. Neuropharmacology, 45: 649–660.

Bazan, N.G. and Serou, M.J. (1999) Second messengers, long-term potentiation, gene expression and epileptogenesis. Adv. Neurol., 79: 659–664.

Becker, A.J., Chen, J., Paus, S., Normann, S., Beck, H., Elger, C.E., Wiestler, O.D. and Blumcke, I. (2002) Transcriptional profiling in human epilepsy: expression array and single cell real-time qRT-PCR analysis reveal distinct cellular gene regulation. Neuroreport, 13: 1327–1333.

Becker, A.J., Chen, J., Zien, A., Sochivko, D., Normann, S., Schramm, J., Elger, C.E., Wiestler, O.D. and Blumcke, I. (2003) Correlated stage-and subfield-associated hippocampal gene expression patterns in experimental and human temporal lobe epilepsy. Eur. J. Neurosci., 18: 2792–2802.

Beissbarth, T. and Speed, T.P. (2004) GOstat: find statistically overrepresented Gene Ontologies within a group of genes. Bioinformatics, 20: 1464–1465.

Belluardo, N., Mudo, G., Dell'Albani, P., Jiang, X.H. and Condorelli, D.F. (1995) NMDA receptor-dependent and -independent immediate early gene expression induced by focal mechanical brain injury. Neurochem. Int., 26: 443–453.

Bidmon, H.J., Gorg, B., Palomero-Gallagher, N., Behne, F., Lahl, R., Pannek, H.W., Speckmann, E.J. and Zilles, K. (2004) Heat shock protein-27 is upregulated in the temporal cortex of patients with epilepsy. Epilepsia, 45: 1549–1559.

Binder, D.K. (2004) The role of BDNF in epilepsy and other diseases of the mature nervous system. Adv. Exp. Med. Biol., 548: 34–56.

Bo, X., Zhiguo, W., Xiaosu, Y., Guoliang, L. and Guangjie, X. (2002) Analysis of gene expression in genetic epilepsy-prone rat using a cDNA expression array. Seizure, 11: 418–422.

Bohmer, C., Philippin, M., Rajamanickam, J., Mack, A., Broer, S., Palmada, M. and Lang, F. (2004) Stimulation of the EAAT4 glutamate transporter by SGK protein kinase isoforms and PKB. Biochem. Biophys. Res. Commun., 324: 1242–1248.

Cha, B.H., Akman, C., Silveira, D.C., Liu, X. and Holmes, G.L. (2004) Spontaneous recurrent seizure following status epilepticus enhances dentate gyrus neurogenesis. Brain Dev., 26: 394–397.

Clark, S. and Wilson, W.A. (1999) Mechanisms of epileptogenesis. Adv. Neurol., 79: 607–630.

Cortes-Canteli, M., Wagner, M., Ansorge, W. and Perez-Castillo, A. (2004) Microarray analysis supports a role for ccaat/enhancer-binding protein-beta in brain injury. J. Biol. Chem., 279: 14409–14417.

Coulter, D.A. and DeLorenzo, R.J. (1999) Basic mechanisms of status epilepticus. Adv. Neurol., 79: 725–733.

Covolan, L., Ribeiro, L.T., Longo, B.M. and Mello, L.E. (2000) Cell damage and neurogenesis in the dentate granule cell layer of adult rats after pilocarpine- or kainate-induced status epilepticus. Hippocampus, 10: 169–180.

D'Ambrosio, R., Fairbanks, J.P., Fender, J.S., Born, D.E., Doyle, D.L. and Miller, J.W. (2004) Post-traumatic epilepsy following fluid percussion injury in the rat. Brain, 127: 304–314.

Dalton, T., Pazdernik, T.L., Wagner, J., Samson, F. and Andrews, G.K. (1995) Temporalspatial patterns of expression of metallothionein-I and -III and other stress-related genes in rat brain after kainic acid-induced seizures. Neurochem. Int., 27: 59–71.

Dutcher, S.A., Underwood, B.D., Walker, P.D., Diaz, F.G. and Michael, D.B. (1999) Patterns of immediate early gene mRNA expression following rodent and human traumatic brain injury. Neurol. Res., 21: 234–242.

Elliott, R.C., Miles, M.F. and Lowenstein, D.H. (2003) Overlapping microarray profiles of dentate gyrus gene expression during development- and epilepsy-associated neurogenesis and axon outgrowth. J. Neurosci., 23: 2218–2227.

Endo, A., Nagai, N., Urano, T., Takada, Y., Hashimoto, K. and Takada, A. (1999) Proteolysis of neuronal cell adhesion molecule by the tissue plasminogen activator-plasmin system after kainate injection in the mouse hippocampus. Neurosci. Res., 33: 1–8.

French, P.J., O'Connor, V., Voss, K., Stean, T., Hunt, S.P. and Bliss, T.V. (2001) Seizure-induced gene expression in area CA1 of the mouse hippocampus. Eur. J. Neurosci., 14: 2037–2041.

Hauser, W.A. (1997) Incidence and prevalence. In: Engel, J.J. and Pedley, T.A. (Eds.), Epilepsy: A Comprehensive Textbook. Philadelphia, Lippincott-Raven Publishers, pp. 47–57.

Hendriksen, H., Datson, N.A., Ghijsen, W.E., van Vliet, E.A., da Silva, F.H., Gorter, J.A. and Vreugdenhil, E. (2001) Altered hippocampal gene expression prior to the onset of spontaneous seizures in the rat post-status epilepticus model. Eur. J. Neurosci., 14: 1475–1484.

Hollister, R.D., Page, K.J. and Hyman, B.T. (1997) Distribution of the messenger RNA for the extracellularly regulated kinases 1, 2 and 3 in rat brain: effects of excitotoxic hippocampal lesions. Neuroscience, 79: 1111–1119.

Jessen, U., Novitskaya, V., Walmod, P.S., Berezin, V. and Bock, E. (2003) Neural cell adhesion molecule-mediated neurite outgrowth is repressed by overexpression of HES-1. J. Neurosci. Res., 71: 1–6.

Kabos, P., Kabosova, A. and Neuman, T. (2002) Neuronal injury affects expression of helix–loop–helix transcription factors. Neuroreport, 13: 2385–2388.

Khurgel, M., Switzer 3rd, R.C., Teskey, G.C., Spiller, A.E., Racine, R.J. and Ivy, G.O. (1995) Activation of astrocytes during epileptogenesis in the absence of neuronal degeneration. Neurobiol. Dis., 2: 23–35.

Kim, S.K., Wang, K.C., Hong, S.J., Chung, C.K., Lim, S.Y., Kim, Y.Y., Chi, J.G., Kim, C.J., Chung, Y.N., Kim, H.J. and Cho, B.K. (2003) Gene expression profile analyses of cortical dysplasia by cDNA arrays. Epilepsy Res., 56: 175–183.

Kim, S.Y., Choi, Y.S., Choi, J.S., Cha, J.H., Kim, O.N., Lee, S.B., Chung, J.W., Chun, M.H. and Lee, M.Y. (2002) Osteopontin in kainic acid-induced microglial reactions in the rat brain. Mol. Cells, 13: 429–435.

Kobori, N., Clifton, G.L. and Dash, P. (2002) Altered expression of novel genes in the cerebral cortex following experimental brain injury. Brain Res. Mol. Brain Res., 104: 148–158.

Koistinaho, J. and Hokfelt, T. (1997) Altered gene expression in brain ischemia. Neuroreport, 8: i–viii.

Lachance, P.E. and Chaudhuri, A. (2004) Microarray analysis of developmental plasticity in monkey primary visual cortex. J. Neurochem., 88: 1455–1469.

Lang, S. (2002) The role of peripheral benzodiazepine receptors (PBRs) in CNS pathophysiology. Curr. Med. Chem., 9: 1411–1415.

Lee, J.Y., Kim, J.H., Palmiter, R.D. and Koh, J.Y. (2003) Zinc released from metallothionein-iii may contribute to hippocampal CA1 and thalamic neuronal death following acute brain injury. Exp. Neurol., 184: 337–347.

Li, H., Kolluri, S.K., Gu, J., Dawson, M.I., Cao, X., Hobbs, P.D., Lin, B., Chen, G., Lu, J., Lin, F., Xie, Z., Fontana, J.A., Reed, J.C. and Zhang, X. (2000) Cytochrome c release and apoptosis induced by mitochondrial targeting of nuclear orphan receptor TR3. Science, 289: 1159–1164.

Li, H.H., Lee, S.M., Cai, Y., Sutton, R.L. and Hovda, D.A. (2004) Differential gene expression in hippocampus following experimental brain trauma reveals distinct features of moderate and severe injuries. J. Neurotrauma, 21: 1141–1153.

Lindwall, C. and Kanje, M. (2005) Retrograde axonal transport of JNK signaling molecules influence injury induced nuclear changes in p-c-Jun and ATF3 in adult rat sensory neurons. Mol. Cell Neurosci., 29: 269–282.

Long, Y., Zou, L., Liu, H., Lu, H., Yuan, X., Robertson, C.S. and Yang, K. (2003) Altered expression of randomly selected genes in mouse hippocampus after traumatic brain injury. J. Neurosci. Res., 71: 710–720.

Lu, X.R. and Ong, W.Y. (2001) Heme oxgenase-1 is expressed in viable astrocytes and microglia but in degenerating pyramidal neurons in the kainate-lesioned rat hippocampus. Exp. Brain Res., 137: 424–431.

Lukasiuk, K., Kontula, L. and Pitkanen, A. (2003) cDNA profiling of epileptogenesis in the rat brain. Eur. J. Neurosci., 17: 271–279.

Lukasiuk, K. and Pitkanen, A. (2004) Large-scale analysis of gene expression in epilepsy research: is synthesis already possible? Neurochem. Res., 29: 1169–1178.

Mahoney, S.A., Wilkinson, M., Smith, S. and Haynes, L.W. (2000) Stabilization of neurites in cerebellar granule cells by transglutaminase activity: identification of midkine and galectin-3 as substrates. Neuroscience, 101: 141–155.

Mattson, M.P. and Camandola, S. (2001) NF-κ B in neuronal plasticity and neurodegenerative disorders. J. Clin. Invest., 107: 247–254.

Matzilevich, D.A., Rall, J.M., Moore, A.N., Grill, R.J. and Dash, P.K. (2002) High-density microarray analysis of hippocampal gene expression following experimental brain injury. J. Neurosci. Res., 67: 646–663.

Meller, R., Stevens, S.L., Minami, M., Cameron, J.A., King, S., Rosenzweig, H., Doyle, K., Lessov, N.S., Simon, R.P. and Stenzel-Poore, M.P. (2005) Neuroprotection by osteopontin in stroke. J. Cereb. Blood Flow Metab., 25: 217–225.

Montpied, P., de Bock, F., Baldy-Moulinier, M. and Rondouin, G. (1998) Alterations of metallothionein II and apolipoprotein J mRNA levels in kainate-treated rats. Neuroreport, 9: 79–83.

Montpied, P., de Bock, F., Lerner-Natoli, M., Bockaert, J. and Rondouin, G. (1999) Hippocampal alterations of apolipoprotein E and D mRNA levels in vivo and in vitro following kainate excitotoxicity. Epilepsy Res., 35: 135–146.

Munoz, A.M., Rey, P., Parga, J., Guerra, M.J. and Labandeira-Garcia, J.L. (2005) Glial overexpression of heme oxygenase-1: a histochemical marker for early stages of striatal damage. J. Chem. Neuroanat., 29: 113–126.

Namikawa, K., Su, Q., Kiryu-Seo, S. and Kiyama, H. (1998) Enhanced expression of 14-3-3 family members in injured motoneurons. Brain Res. Mol. Brain Res., 55: 315–320.

Natale, J.E., Ahmed, F., Cernak, I., Stoica, B. and Faden, A.I. (2003) Gene expression profile changes are commonly modulated across models and species after traumatic brain injury. J. Neurotrauma, 20: 907–927.

Nedivi, E., Hevroni, D., Naot, D., Israeli, D. and Citri, Y. (1993) Numerous candidate plasticity-related genes revealed by differential cDNA cloning. Nature, 363: 718–722.

Ohmi, K., Greenberg, D.S., Rajavel, K.S., Ryazantsev, S., Li, H.H. and Neufeld, E.F. (2003) Activated microglia in cortex of mouse models of mucopolysaccharidoses I and IIIB. Proc. Natl. Acad. Sci. USA, 100: 1902–1907.

Parent, J.M. (2003) Injury-induced neurogenesis in the adult mammalian brain. Neuroscientist, 9: 261–272.

Parent, J.M., Yu, T.W., Leibowitz, R.T., Geschwind, D.H., Sloviter, R.S. and Lowenstein, D.H. (1997) Dentate granule cell neurogenesis is increased by seizures and contributes to aberrant network reorganization in the adult rat hippocampus. J. Neurosci., 17: 3727–3738.

Pasinetti, G.M., Johnson, S.A., Rozovsky, I., Lampert-Etchells, M., Morgan, D.G., Gordon, M.N., Morgan, T.E., Willoughby, D. and Finch, C.E. (1992) Complement C1qB and C4 mRNAs responses to lesioning in rat brain. Exp. Neurol., 118: 117–125.

Penkowa, M., Florit, S., Giralt, M., Quintana, A., Molinero, A., Carrasco, J. and Hidalgo, J. (2005) Metallothionein reduces central nervous system inflammation, neurodegeneration, and cell death following kainic acid-induced epileptic seizures. J. Neurosci. Res., 79: 522–534.

Perlmann, T. and Wallen-Mackenzie, A. (2004) Nurr1, an orphan nuclear receptor with essential functions in developing dopamine cells. Cell Tissue Res., 318: 45–52.

Pesheva, P., Kuklinski, S., Biersack, H.J. and Probstmeier, R. (2000) Nerve growth factor-mediated expression of galectin-3 in mouse dorsal root ganglion neurons. Neurosci. Lett., 293: 37–40.

Pitkanen, A., Kharatishvili, I., Nissinen, J. and McIntosh, T.K. (2006) Post-traumatic epilepsy induced by lateral fluid-percussion injury in rats. In: Pitkanen, A., Schwarzkroin, P.A. and Moshe, S. (Eds.), Animal Models of Epilepsy. Elsevier, London, pp. 465–476.

Pitkanen, A. and McIntosh, T.K. (2006) Animal models of post-traumatic epilepsy. J. Neurotrauma, 23: 241–261.

Pitkanen, A. and Sutula, T.P. (2002) Is epilepsy a progressive disorder? Prospects for new therapeutic approaches in temporal-lobe epilepsy. Lancet Neurol., 1: 173–181.

Porter, R.J. (1993) Classification of epileptic seizures and epileptic syndromes. In: Laidlaw, J., Richens, A. and Chadwick, D. (Eds.), A textbook of epilepsy. Edinburgh, Churchill Livingstone, 1–22.

Postler, E., Lehr, A., Schluesener, H. and Meyermann, R. (1997) Expression of the S-100 proteins MRP-8 and -14 in ischemic brain lesions. Glia, 19: 27–34.

Rabinovich, G.A., Baum, L.G., Tinari, N., Paganelli, R., Natoli, C., Liu, F.T. and Iacobelli, S. (2002) Galectins and their ligands: amplifiers, silencers or tuners of the inflammatory response? Trends Immunol., 23: 313–320.

Raghavendra Rao, V.L., Dhodda, V.K., Song, G., Bowen, K.K. and Dempsey, R.J. (2003) Traumatic brain injury-induced acute gene expression changes in rat cerebral cortex identified by GeneChip analysis. J. Neurosci. Res., 71: 208–219.

Rong, Y. and Baudry, M. (1996) Seizure activity results in a rapid induction of nuclear factor-kappa B in adult but not juvenile rat limbic structures. J. Neurochem., 67: 662–668.

Salazar, M.A., Kwiatkowski, A.V., Pellegrini, L., Cestra, G., Butler, M.H., Rossman, K.L., Serna, D.M., Sondek, J., Gertler, F.B. and De Camilli, P. (2003) Tuba, a novel protein containing bin/amphiphysin/Rvs and Dbl homology domains, links dynamin to regulation of the actin cytoskeleton. J. Biol. Chem., 278: 49031–49043.

Sayyah, M., Beheshti, S., Shokrgozar, M.A., Eslami-far, A., Deljoo, Z., Khabiri, A.R. and Haeri Rohani, A. (2005) Antiepileptogenic and anticonvulsant activity of interleukin-1 beta in amygdala-kindled rats. Exp. Neurol., 191: 145–153.

Soos, J.M., Morrow, J., Ashley, T.A., Szente, B.E., Bikoff, E.K. and Zamvil, S.S. (1998) Astrocytes express elements of the class II endocytic pathway and process central nervous system autoantigen for presentation to encephalitogenic T cells. J. Immunol., 161: 5959–5966.

Steward, O. and Worley, P.F. (2001) A cellular mechanism for targeting newly synthesized mRNAs to synaptic sites on dendrites. Proc. Natl. Acad. Sci. USA, 98: 7062–7068.

Stoll, G., Jander, S. and Schroeter, M. (2002) Detrimental and beneficial effects of injury-induced inflammation and cytokine expression in the nervous system. Adv. Exp. Med. Biol., 513: 87–113.

Tang, Y., Lu, A., Aronow, B.J., Wagner, K.R. and Sharp, F.R. (2002) Genomic responses of the brain to ischemic stroke, intracerebral haemorrhage, kainate seizures, hypoglycemia, and hypoxia. Eur. J. Neurosci., 15: 1937–1952.

Taubenfeld, S.M., Milekic, M.H., Monti, B. and Alberini, C.M. (2001) The consolidation of new but not reactivated memory requires hippocampal C/EBPbeta. Nat. Neurosci., 4: 813–818.

Vezzani, A., Moneta, D., Richichi, C., Perego, C. and De Simoni, M.G. (2004) Functional role of proinflammatory and anti-inflammatory cytokines in seizures. Adv. Exp. Med. Biol., 548: 123–133.

Voutsinos-Porche, B., Koning, E., Kaplan, H., Ferrandon, A., Guenounou, M., Nehlig, A. and Motte, J. (2004) Temporal patterns of the cerebral inflammatory response in the rat lithium-pilocarpine model of temporal lobe epilepsy. Neurobiol. Dis., 17: 385–402.

Williams, J., Dragunow, M., Lawlor, P., Mason, S., Abraham, W.C., Leah, J., Bravo, R., Demmer, J. and Tate, W. (1995) Krox20 may play a key role in the stabilization of long-term potentiation. Brain Res. Mol. Brain. Res., 28: 87–93.

Worley, P.F., Christy, B.A., Nakabeppu, Y., Bhat, R.V., Cole, A.J. and Baraban, J.M. (1991) Constitutive expression of zif268 in neocortex is regulated by synaptic activity. Proc. Natl. Acad. Sci. USA, 88: 5106–5110.

Wu, Y.P., Siao, C.J., Lu, W., Sung, T.C., Frohman, M.A., Milev, P., Bugge, T.H., Degen, J.L., Levine, J.M., Margolis, R.U. and Tsirka, S.E. (2000) The tissue plasminogen activator (tPA)/plasmin extracellular proteolytic system regulates seizure-induced hippocampal mossy fiber outgrowth through a proteoglycan substrate. J. Cell Biol., 148: 1295–1304.

Yakubov, E., Gottlieb, M., Gil, S., Dinerman, P., Fuchs, P. and Yavin, E. (2004) Overexpression of genes in the CA1 hippocampus region of adult rat following episodes of global ischemia. Brain Res. Mol. Brain Res., 127: 10–26.

Zagulska-Szymczak, S., Filipkowski, R.K. and Kaczmarek, L. (2001) Kainate-induced genes in the hippocampus: lessons from expression patterns. Neurochem. Int., 38: 485–501.

Zhang, C., Kavurma, M.M., Lai, A. and Khachigian, L.M. (2003) Ets-1 protects vascular smooth muscle cells from undergoing apoptosis by activating p21WAF1/Cip1: ETS-1 regulates basal and and inducible p21WAF1/Cip: ETS-1 regulates basal and inducible p21WAF1/Cip1 transcription via distinct cis-acting elements in the p21WAF/Cip1 promoter. J. Biol. Chem., 278: 27903–27909.

Zhang, Y., Vilaythong, A.P., Yoshor, D. and Noebels, J.L. (2004) Elevated thalamic low-voltage-activated currents precede the onset of absence epilepsy in the SNAP25-deficient mouse mutant coloboma. J. Neurosci., 24: 5239–5248.

CHAPTER 12

Functional genomics of sex hormone-dependent neuroendocrine systems: specific and generalized actions in the CNS

Anna W. Lee, Nino Devidze, Donald W. Pfaff* and Jin Zhou

Laboratory of Neurobiology and Behavior, Box 275, The Rockefeller University, New York, NY 10021, USA

Abstract: Sex hormone effects on hypothalamic neurons have been worked out to a point where receptor mechanisms are relatively well understood, a neural circuit for a sex steroid-dependent behavior has been determined, and several functional genomic regulations have been discovered and conceptualized. With that knowledge in hand, we approach deeper problems of explaining sexual arousal and generalized CNS arousal. After a brief summary of arousal mechanisms, we focus on three chemical systems which signal generalized arousal and impact hormone-dependent hypothalamic neurons of behavioral importance: histamine, norepinephrine and enkephalin.

Keywords: estrogen; hypothalamus; sex; estrogen receptor; arousal; histamine; norepinephrine; opioid; enkephalin; lordosis; courtship

The seven sections of this chapter progress from the explanation of sex steroid hormone effects on a specific, biologically important behavior, to considerations of sexual arousal, to a summary of mechanisms bearing on generalized CNS arousal. Especially telling are the actions of neurotransmitters signaling generalized arousal on neurons in the ventromedial hypothalamus that govern female sex behavior. Generalized arousal is fundamental to all cognitive and emotional functions, and therefore the systematic study of its mechanisms will be important for many aspects of medicine and public health.

Neural and genomic mechanisms for female mating behaviors

The first problem was to use hormone actions on neurons to establish that it was indeed possible to explain mechanisms for any mammalian behavior. The simplest behavior obtainable in its natural form in the laboratory and essential for reproduction is lordosis behavior. The primary sex behavior of female quadrupeds, lordosis, depends on defined physical signals: cutaneous stimuli and estrogens plus progestins (Pfaff, 1999). Estrogen-dependent transcription in ventromedial hypothalamic (VMH) cells allows permissive signals to the midbrain central gray, thus enabling the rest of the circuit. The entire neural circuit for the production of lordosis behavior (Fig. 1) has been worked out (Pfaff, 1980, 1999). In the absence of fear- or anxiety-provoking conditions, females under the influence of estrogens plus progestins will demonstrate courtship and then mating behaviors. During the normal female cycle, these behavioral components of reproduction are synchronized with ovulation. Thus, with the mediation of estrogens plus progestins, the neural, behavioral and endocrine preparations for reproduction are harmonized.

*Corresponding author. Tel.: 212-327-8666; Fax: 212-327-8664; E-mail: pfaff@mail.rockefeller.edu

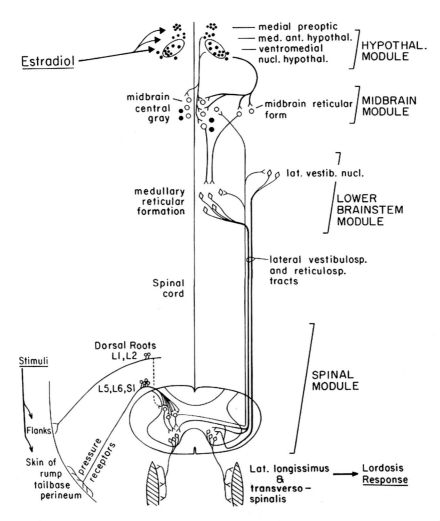

Fig. 1. Sketch of the basic neural circuit for producing the primary female reproductive behavior, lordosis. This was the first completed neural circuit for a vertebrate behavior. Cutaneous stimuli triggering pressure receptors on the flanks and then the hindquarters of the female rat evoke action potentials entering the spinal cord over dorsal roots L1, L2, L5, L6, and S1. Although the responses by neurons in appropriate parts of the dorsal horn are prompt and impressive, spinal tissue by itself cannot mediate lordosis behavior. The obligatory ascending fibers in the supraspinal loop travel to the brainstem in the anterolateral columns, terminating in considerable numbers in the medullary reticular formation and dorsocaudal lateral vestibular nucleus and, in less quantity, the midbrain (the peripeduncular region and the midbrain central gray). The main job of the hypothalamic module in this circuit is to add estrogen and progestin dependence. Operating through estrogen receptors expressed in ventromedial hypothalamic neurons, estradiol induces expression of the gene for the progesterone receptor, and circulating progesterone binds to that receptor. Their combined genomic and electrophysiological effects send an enabling signal back to the midbrain central gray over both axons that follow a sweeping lateral route and a smaller number of axons that follow a paraventricular route. Then, neurons in the dorsal lateral portion of the midbrain central gray, combined with a smaller number of neurons in the nearby mesencephalic reticular formation, send a descending signal that facilitates medullary reticulospinal neurons. Reticulospinal signals synergize with lateral vestibulospinal discharges to facilitate activity in the lumbar motor neurons that lie on the medial side of the ventral horn and are responsible for controlling the deep back muscles — the lateral longisimus and transversospinalis. On reception of adequate sensory input, these motor neurons allow muscle contraction that changes the posture of the back to 'concave up', which drives lordosis behavior, and thus permits fertilization. Modules in this circuit match embryological divisions of the developing mammalian central nervous system. From Pfaff et al. (1994).

Since (a) the hormone receptors discovered in neurons turned out to be transcription factors, (b) mating behaviors follow estrogen administration by more than 18 h, and (c) inhibitors of RNA and protein synthesis disrupt the estrogen effect, it was natural to look for genes whose induction comprised important mechanisms for the behavioral facilitation. Fig. 2 lists gene/neuronal and gene/glial modules found so far.

Throughout these studies, we compare two extremely similar transcription factors, estrogen receptors (ERs) alpha and beta. Likely gene duplication products, both have high affinity for estradiol and for genomic estrogen response elements (EREs) in vertebrate gene promoters (Jensen and Jacobson, 1962; Walter et al., 1985; Kuiper et al., 1996; Nilsson and Gustafsson, 2002). Yet, they have distinctly different neuroanatomical patterns of expression (Shughrue et al., 1997) and different functional consequences.

All of the genetic systems discussed below have the character that estrogen treatment turns them on in vivo and that they participate in facilitating estrogen-dependent female reproductive behaviors (Fig. 2). Mong and Pfaff (2004), quoted below, have tried systematically to conceive the biologically adaptive functions of these genes. Mong and Pfaff's thinking in that article, upon which this section depends, resulted in a theoretical approach called the 'GAPPS' model (discussed later).

Direct effects of estrogens, from gene induction to neural circuit to behavioral change

Hormone effects on neurotransmitter receptors in VMH neurons directly trigger the rest of the lordosis circuit to operate.

Noradrenergic alpha-1B receptors are induced (Etgen, 2002) in vivo in female rats by estrogen

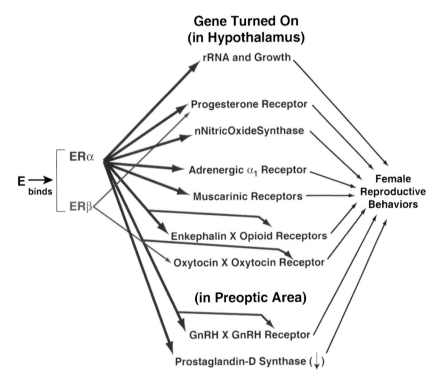

Fig. 2. Estrogen binds to the classical estrogen receptor, ERα and to ERβ in specific subsets of hypothalamic and preoptic neurons. Acting as transcription factors, ligand-activated ERs facilitate transcription of several genes whose products foster female reproductive behaviors in rats and mice. The routes of action of the 'downstream' genes listed are partially understood. Note that for neurochemical systems in which E activates transcription of genes for both ligand and the corresponding receptor, there is the possibility of a multiplicative effect on sex behavior. *Abbreviations*: E, estrogen; ER, estrogen receptor; VMH, ventromedial nucleus of the hypothalamus. From Pfaff (1999).

treatment in VMH cells which govern the rest of the lordosis behavior circuit (Kow et al., 1992; Petitti et al., 1992). Noradrenergic ascending afferents synapse on VMH neurons, coming in from the ventral noradrenergic bundle, which originates in arousal-related neuronal groups A1 and A2, and signals heightened arousal upon stimulation from the male. In biophysical studies, directly applied noradrenaline (NA) increases the electrical activity of VMH neurons (Kow et al., 1992). Beginning with G_q or G_{11} proteins activating phospholipase C (PLC), NA action will produce both diacylglycerol (DAG), a protein kinase C (PKC) activator, and inositol 1,4,5-trisphosphate (IP_3), which mobilizes intracellular calcium. This signal transduction route is predicted to lead to L-type Ca^{2+} channel opening as in the heart, but this needs to be established. The induction of alpha-1B receptors by estrogen treatment is consistent with the greater electrophysiological effectiveness of NA following estrogen, but the detailed step-by-step transduction route to the channel now provides a timely subject for analysis. Since these VMH neurons are at the top of the lordosis behavior circuit, the NA effect fosters reproductive behavior.

Muscarinic receptors responding to the neurotransmitter acetylcholine are also expressed in VMH neurons (Kow et al., 1995). Estrogen treatment increases their activities as determined electrophysiologically. Inputs to VMH neurons come from, among other places, the lateral dorsal nucleus of the tegmentum. Neurons there are part of the ascending arousal pathways, and would signal stimulation from the male upon mounting the female. We note that apparent redundancy between ascending noradrenergic and muscarinic cholinergic afferents to the hypothalamus helps to guarantee that the system will not fail, and so exemplifies a design characteristic prominent in brainstem arousal system neurobiology. In any case, inducing muscarinic receptors increases the VMH cellular electrophysiological response to acetylcholine. Whether the estrogen effect employs a membrane receptor, a signal transduction mechanism or a classical genomic facilitation is not yet known. However, it is known that the enhanced VMH neuronal output primes lower pathways in the circuit for lordosis behavior (Menard and Dohanich, 1990).

Indirect effects of estrogens, from gene induction to downstream genes to behavioral change

Some hormone effects occur early, long before the onset of reproductive behaviors, and set the stage for later developments.

Neuronal growth. In female rats, growth promotion by estrogens in VMH neurons follows from the stimulation of synthesis of ribosomal RNA, which precedes the elaboration of dendrites and synapses on VMH neurons observed after hormonal treatment. The earliest estrogen effect is the increase of transcription of ribosomal RNA (Jones et al., 1990), followed rapidly by morphological effects, including those in the nucleolus itself (Cohen et al., 1984) and a striking elaboration of rough endoplasmic reticulum in the cytoplasm (Cohen and Pfaff, 1981). Catherine Woolley and her colleagues (Woolley and Cohen, 2002) have shown, probably consequent to the phenomena above, a stimulatory effect of estrogen treatment on dendritic growth. In the female rat hypothalamus, Maya Frankfurt (1994) and Lori Flanagan (Flanagan-Cato, 2000; Flanagan-Cato et al., 2001) have reported that estrogens foster dendritic growth and an increased number of synapses (Carrer and Aoki, 1982).

Therefore, in VMH cells which control lordosis behavior circuitry, estrogens apparently provide the structural basis for increased synaptic activity and, therefore, greater sex-behavior-facilitating output. While the greater signaling capacity of these VMH cells thus proposed is consistent with the actual electrophysiological activity of such cells following estrogen treatment, the causal relation of structure to function still needs direct proof.

Amplification by progesterone. Administration of progesterone 24 or 48 h after estrogen priming greatly amplifies the effect of estrogen on mating behavior. This effect requires the nuclear progesterone receptor (PR), as it disappears after antisense DNA against PR mRNA has been administered onto VMH neurons (Pollio et al., 1993; Mani et al., 1994; Ogawa et al., 1994). It also disappears in PR knockout mice (Lydon et al., 1995). Since PR itself is a transcription factor, its induction by estrogen might imply that certain downstream genes would, consequently, be upregulated. With molecular probes directed to specific genes, studied primarily in female rats, several have been revealed as upregulated by progesterone: these include neuropeptide Y receptor (Xu et al., 2000), galanin (Brann et al., 1993; Rossmanith et al., 1996), oxytocin (Thomas et al., 1999), gonadotropin-releasing hormone (GnRH) (Cho et al., 1994; Petersen et al., 1995), mu (μ) opioid receptors (Petersen and LaFlamme, 1997), pro-opiomelanocortin (Petersen et al., 1993), glutamic acid decarboxylase (Unda et al., 1995), a glutamate receptor (Gu et al., 1999), and tyrosine hydroxylase (Arbogast and Voogt, 1993, 1994). The manners in which these particular downstream genes contribute to reproductive behaviors will be exciting to explore.

GnRH. The physiological importance of estrogenic elevation of gonadotropin-releasing hormone (GnRH, LHRH) mRNA levels under positive feedback conditions — as well as elevation of the receptor mRNA for GnRH — would be to synchronise reproductive behavior with the ovulatory surge of luteinizing hormone (LH). The same GnRH decapeptide which stimulates the ovulatory release of gonadotropins also facilitates mating behavior (Moss and McCann, 1973; Pfaff, 1973). In many small animals, synchrony of sex behavior with ovulation would be biologically adaptive because it eliminates unnecessary exposure to predation. In this respect, the behavioral effect of this neuropeptide is consonant with its peripheral physiological action.

GnRH also brings up the unusual case of an individual gene causally related to a human social behavior. During development in vertebrates ranging from fish to humans, GnRH neurons migrate from their birth place in the olfactory placode into the brain (Schwanzel-Fukada and Pfaff, 1989). A human with damage at the Kallmann's Syndrome (MacColl et al., 2002) locus on the X chromosome did not fail to express the GnRH gene in the appropriate neurons. Instead, the neurons failed to migrate out of the olfactory placode (Schwanzel-Fukada et al., 1989). A single gene for the Kall protein (Legouis et al., 1991; Ballabio and Camerino, 1992) accounts for the deficit. It is for an extracellular matrix protein which is necessary for the GnRH neuronal migration and which, in fact, decorates the migration route (Dellovade et al., 2003). A striking feature of the phenotype in men is important to note. They have no libido. Here is the causal route. The men have no sexual drive because they have little testosterone, because they have little LH and follicle-stimulation hormone circulating from the pituitary gland, because no GnRH is coming down the portal circulation to the pituitary from the hypothalamus, because there is no GnRH in the hypothalamus, because the GnRH neurons did not migrate during development into the brain, because of a mutation in the gene for the Kall protein. Therefore, we can causally connect, step-by-step, an individual gene to an important human social behavior, but at least six causal links are required. This causal route illustrates the complexity of gene/behavior relationships in humans.

Indirect effects of estrogens, from gene induction to intermediate behaviors

Some of the genes affected by estrogens work by altering other behaviors which then prepare the animal for the behavior in question, in this case mating.

Analgesia. In female mouse and rat hypothalamic neurons, the enkephalin gene is turned on rapidly by estrogens (Romano et al., 1988; Priest et al., 1995; Zhu et al., 2001), within about 30 min, and this is proven by in vitro transcription assays to represent a hormone-facilitated transcriptional facilitation (Vasudevan et al., 2001). The route of action of the enkephalin gene product, upon lordosis, would theoretically be indirect, through other behaviors. That is, we propose that, through the reduction of pain, enkephalins help to allow the female to engage in mating behavior despite the mauling she receives from the male. The strong somatosensory and interoceptive stimuli which ordinarily would be treated by the female as noxious, are now tolerable and allow successful mating to proceed (Bodnar et al., 2002). Hypothetically, the ability of estrogens to turn on genes for opioid receptors as well (Quinones-Jenab et al., 1997) has the potential of multiplying the hormone's effect on mating behavior sequences. Specificity among opioid receptors, as well as neuroanatomical site specificity, is observed in this course of action (Pfaus and Gorzalka, 1987a, b; Pfaus and Pfaff, 1992; Acosta-Martinez and Etgen, 2002). The indirect route of action of this multiplicative set of gene inductions is likely to allow the female to participate in sex behavior sequences.

Anxiety reduction. The oxytocin gene and the gene for its receptor are both expressed by hypothalamic neurons at higher levels in the presence of estrogens. The indirect route of action of this multiplicative set of gene inductions on mating behavior is likely through a behavioral link: anxiety reduction allows courtship and mating. This proposal is consistent with previous formulations: oxytocin has been conceived as protecting instinctive behaviors connected with reproduction, maternity and other social behaviors from the disruptive effects of stress (McCarthy et al., 1991). Indeed, oxytocin has an anxiolytic action in the presence of estrogens (which presumably elevate the oxytocin receptor gene product) (McCarthy et al., 1996).

Social recognition and aggression. The induction of the oxytocin gene by estrogens is an ER-β-dependent (Nomura et al., 2002b), behaviorally significant (Krezel et al., 2001) phenomenon. This makes sense, since only ER-β gene expression is found in oxytocinergic cells (Shughrue et al., 1997). In turn, oxytocinergic projections to the amygdala are thought to be important for social recognition in mice, which helps to prevent aggression (Insel and Young, 2000; Ferguson et al., 2001, 2002). Thus, the lack of social recognition by ER-β knockout mice (Choleris et al., 2003) could explain the hyperaggressiveness displayed by ER-β knockout male mice (Nomura et al., 2002a). Altogether, these data invoke the idea of a four-gene micronet important for social behaviors.

Newer findings. In order to lengthen the list of hormone-facilitated transcript levels in hypothalamic and preoptic area neurons and glial cells we used microarrays (Mong et al., 2003a) comparing estrogen treated with control ovariectomized female mice.

One of the findings so far has yielded significant implications: prostaglandin D synthase (PGDS) transcripts were dramatically suppressed by estradiol in the preoptic area (Mong et al., 2003b). Microarray results have been confirmed by Northern blot analysis and by in situ hybridization histochemistry. PGDS is a non-neuronal enzyme (Beuckmann et al., 2000), significantly involved in sleep (Ueno et al., 1982). Its suppression in a preoptic area cell group that fosters sleep leads to heightened arousal. Female mice microinjected in the preoptic area with a novel antisense moiety, 'locked nucleic acids' (Wahlestedt et al., 2000), showed heightened arousal following controlled application of sensory stimuli (Mong et al., 2003a). This result is consistent with heightened estrogen-dependent locomotor behaviors, which are components of estrogen-facilitated courtship responses.

Glutamine synthase (GS), also identified by microarray screens, is induced by estradiol treatment in basomedial hypothalamic tissue (Kia et al., 2002). Again, the microarray results have been confirmed by Northern blot analysis and in situ hybridization methods. This mRNA is intriguing because glutamine synthase is a glial enzyme involved in glutamate metabolism. In turn, a receptor for the neurotransmitter glutamate is the NMDA-2D receptor. This receptor is expressed in hypothalamic neurons which also express ER-α mRNA (Kia et al., 2002) and in fact is induced by estrogen at the transcriptional level (Watanabe et al., 1999). Together with the PGDS results, these GS data suggest hormone-facilitated glia/neuronal functional units.

Theory of orchestrated genomic responses to hormones by functional modules (GAPPS)

From this series of individual gene inductions by estrogens acting in the basal forebrain, and the recounting (above) of downstream genes and their physiological routes of action, there emerges a theoretical molecular 'formula', as quoted here from Mong and Pfaff (2004) which appears to account for some of the causal relations between sex hormones and female sex behaviors (Fig. 3). First, there is a hormone-dependent *growth* response, which permits hormone-facilitated, behavior-directing hypothalamic neurons a greater range of input/output connections and, thus, physiological power. Second, progesterone can *amplify* the estrogen effect, in part through the downstream genes listed above. Then, through indirect behavioral means — the reduction of anxiety and a partial analgesia — the female as an organism is *prepared* for engaging in reproductive behavior sequences. Here the genes for oxytocin (and its receptor) as well as the genes for the opioid peptide enkephalin (and its receptors) are important. Next, neurotransmitter receptor induction by estradiol *permits* the neural circuit for lordosis behavior to be activated. The noradrenergic α-1 receptor and the muscarinic acetylcholine receptors are key here, in the ventromedial nucleus of the hypothalamus. Finally, induction of the decapeptide which triggers ovulation, GnRH as well as its cognate receptor acts to *Synchronise* mating behavior with ovulation in a biologically adaptive fashion.

This theoretical formulation is intended to tie together disparate results from several transcriptional systems into one set of modules. Even so, the genomic mechanisms uncovered so far probably represent only a subset of the full range of neurochemical steps underlying sex behaviors.

MODULAR SYSTEMS DOWNSTREAM FROM HORMONE-FACILITATED TRANSCRIPTION RESPONSIBLE FOR A MAMMALIAN SOCIAL BEHAVIOR: "GAPPS".

- *Growth* (rRNA, cell body, synapses).
- *Amplify* (pgst/PR → → downstream genes).
- *Prepare* (indirect behavioral means; analgesia (ENK gene) and anxiolysis (OT gene).
- *Permit* (NE alpha-1b; muscarinic receptors).
- *Synchronize* (GnRH gene, GnRH Rcptr gene ------synchronizes with ovulation).

Fig. 3. Modular systems downstream from hormone-facilitated transcription responsible for a mammalian social behavior: 'GAPPS'. The 'GAPPS' model emphasizes the ability of sex hormones to make nerve cells *Grow*, to set up their own *Amplification*, to initiate *Preparative* behavioral steps, to have *Permissive* actions on the rest of the lordosis circuit, and to *Synchronize* sex behavior with ovulation. From Mong and Pfaff, (2004)

From lordosis to sexual arousal to generalized CNS arousal

Success in analyzing mechanisms underlying a simple sex behavior, lordosis, emboldened us to look ahead to attack the deeper concepts of sexual motivation, sexual arousal, and generalized CNS arousal. The ability of sex hormones to turn on sexual behaviors when all other aspects of the experiment are held constant gives proof-positive of the ability of sex hormones to raise sexual motivation (argued in Bodnar et al., 2002). In turn, motivational forces are divided into two types: specific motivational states (such as sex, hunger, thirst, fear, etc.) and generalized states that account for the activation of behavior (generalized arousal). With this train of reasoning, it was exciting to approach mechanisms for generalized arousal because it underlies the activation of all behaviors.

Generalized arousal can be assayed in the laboratory in its natural form (Frohlich et al., 2001), and has a permanence across the life span of the animal or human being. Because it occurs first in any chain of behavioral responses, its alterations can be causal to later behavioral alterations, while the reverse is not true. Assisted by a precise and complete operational definition (below), we have found generalized arousal easy to study in a genetically tractable organism like the mouse. It appears to be universal among higher animals in that it is triggered by giant medullary reticular neurons like Mauthner cells (Faber et al., 1991; Lee and Eaton, 1991; Lee et al., 1993; Zottoli and

Faber, 2000; Canfield, 2003), and can even be tied to stimulus salience and dopamine neurotransmission in Drosophila (van Swinderen and Greenspan, 2003; Kume et al., 2005). Generalized arousal is important in that its disorders in humans are disastrous, ranging from comatose vegetative and fatigue states, through attention deficit disorders, through problems with vigilance and mood (Pfaff, 2005). CNS arousal decline probably contributes to mental problems during aging. Although we usually think of problems related to arousal states that are too low, the opposite disorders can also cause trouble. Sleep disorders are frequent among modern adults, and difficulties with surgical anesthesia deserve attention.

Given that CNS arousal is crucial, its conceptualization has sometimes been vague. Therefore, we need to achieve an operational definition which is precise, complete and which yields quantitative, physical measures. The following operational definition has been proposed (Pfaff, 2005): a more aroused animal or human being (i) is more alert to sensory stimuli in all sensory modalities; (ii) emits more voluntary motor activity; and (iii) is more reactive emotionally. In addition to reviewing hormonal, neural and genetic mechanisms underlying generalized arousal (below), we conceptualize it in mathematical terms as follows.

During the last several years we have been seeking to formulate a mathematical description of arousal-related processes in the mammalian CNS. First, a meta-analysis of experimental data from five studies with mice, using principal components analysis, yielded the estimate that among arousal-related measures there is a generalized arousal component that accounts for about one-third of the variance (Garey et al., 2003). In that same paper, we presented the simplest form of an equation portraying the state of arousal in the mammalian brain as an increasing function not only of generalized arousal (accounting for about one-third of the data), but also of several specific forms of arousal (sexual, hunger, thirst, salt hunger, fear, pain, etc., accounting for the rest of the data related to arousal). That formulation was incomplete. At least three equations may be needed to describe three different levels of the molecular and biophysical forces leading to the activation of behavior. Working with Professor Martin Braun (Department of Mathematics at Queens College, CUNY) we can think of three families of equations which may serve to 'state the set of problems' ripe for molecular neurobiological discovery. Eqs. (1), (4), and (7) are likely to be the most important, for a reason stated below. The mathematics of arousal is open to investigation.

First, inputs. Small changes in the state of arousal (A) of the mammalian CNS can be described as a compound increasing function (F) of a generalized arousal force supplemented by many specific forms of arousal (sexual, hunger, thirst, salt hunger, fear, pain, temperature, etc.). The manner in which these various forces augment each other is not known. Below are equations that hypothetically picture their relations to each other as additive (Eq. (1)), multiplicative (Eq. (2)) or exponential (Eq. (3)).

$$F_1(As_1) + F_2(As_2) \cdots + \cdots F_n(As_n) + F_g(A_g) \quad (1)$$

$$F_1(As_1) \bullet F_2(As_2) \cdots \bullet \cdots F_n(As_n) \bullet F_g(A_g) \quad (2)$$

$$F_1(As_1) \bullet e^{F_g(A_g)} \quad (3)$$

Where A is the state of arousal of the nervous system at any moment, the As designation refers to various specific forms of arousal and Ag refers to generalized CNS arousal. These simplified formulations are intended to show the type of problem we have to solve by obtaining the appropriate experimental data.

Second. What about the internal operations and mechanisms within arousal systems in the brain itself? We can think of three equations that conceptualize small changes in a collective arousal mechanism (M) as an increasing function (G) of upward, ascending arousal pathways (U) in the CNS and descending (D) arousal pathways, both of them operating in a set of mechanisms that have a particular initial condition. Relations between ascending and descending pathways could be additive (Eq. (4)), multiplicative (Eq. (5)) or exponential (Eq. (6)).

$$\delta M = G_U(U) + G_D(D) \qquad (4)$$

$$\delta M = G_U(U) \bullet G_D(D) \qquad (5)$$

$$\delta M = G_U(U) \bullet e^{G_D(D)} \qquad (6)$$

Third, Outputs. We should think of the applications of generalized arousal states of the CNS combined with specific motivationally important stimuli to influence specific states of arousal and subsequently specific, biologically important behaviors. These output functions are represented theoretically (Eqs. (7)–(9)).

$$\delta As_1 = H_M(M) + H_{S_1}(S_1) \qquad (7)$$

$$\delta As_1 = H_M(M) \bullet H_{S_1}(S_1) \qquad (8)$$

$$\delta As_1 = H_M(M) \bullet e^{H_{S_1}(S_1)} \qquad (9)$$

Where changes in a specific form of arousal (As_1, e.g. sexual arousal) are conceived as functions (H) of the state of the collective arousal mechanism (M) and sensory inputs (S_1) relevant to that specific form of arousal (e.g. sexually relevant stimuli).

Note the interesting potential for feedback from the equations for outputs onto terms of the inputs equations. Insofar as these represent possible positive feedbacks, they could explain rapid changes of arousal states.

We recognize that different elements contributing to these nine equations could combine in various ways as we have schematized above and also in other ways heretofore unimagined. If the changes (the deltas) represented in the equations above are small with respect to the initial values, then all the terms in the multiplicative and exponential forms of these equations become ever smaller, leading to the inference that the additive relations may be the most important. Our goal has been to start with the simplest possible approach. The main purpose in presenting these equations is to encourage systematic, quantitative thinking about the mathematical structures of arousal mechanisms.

Brief summary of mechanisms underlying generalized arousal

In some respects, we know a lot about mechanisms for generalized arousal of the mammalian brain, and in other respects, we are just beginning. A comprehensive account is available (Pfaff, 2005). In descending order of sophistication of our knowledge, we briefly review some of the highlights of arousal system neuroanatomy, followed by examples from electrophysiological studies. Then the nascent field of arousal system functional genomics will be introduced.

Neuroanatomy. Regarding ascending pathways, five major neurochemical systems are classically recognized as contributing to the arousal of the forebrain: those signaled by norepinephrine, dopamine, serotonin, acetylcholine, and histamine. Less widely recognized are the crucial roles played by large reticular neurons in the pons and medulla, likely signaled by the release of the excitatory transmitter, glutamate. Also important are descending pathways, the most obvious of which are the axons descending from the paraventricular nucleus of the hypothalamus which affect autonomic, endocrine, cortical and behavioral aspects of arousal. Other prominent examples are the axons descending from the preoptic area which affect autonomic control mechanisms. Two 'arousal neurochemicals' par excellence are the neurotransmitter histamine and the neuropeptide orexin/hypocretin. These participate in both ascending and descending systems. Much of the neuroanatomical data can be encompassed by BBURP theory (Pfaff, 2005). These systems are *bilateral* — one side can make up for the loss of the other side. They are also *bidirectional* — ascending and descending systems both participate importantly and differently. These primitive systems are *universal* — we are emphasizing those neurochemical mechanisms which can be found among all mammals and probably among all vertebrates. Finally, activation of these generalized arousal systems always cause *response potentiation* — in some cases approach responses and in other cases, avoidance.

Of special importance to the regulation of CNS arousal are the reticular neurons along the ventral and medial borders of the medullary and pontine reticular formation. These are crucial to the life of the organism in that they respond to pain (Suzuki et al., 2004), to genital sensations (Hubscher and Johnson, 1996), to CO_2 levels in the blood (de-Oliveira et al., 1996), to changes in body temperature, and (partly as a result) help to regulate respiration (Mulkey et al., 2004) and cardiovascular functions. Some of the actions of these reticular formation neurons involve projections forward to the paraventricular nucleus of the hypothalamus (Ciriello and Caverson, 1984), which then projects back (Pyner and Coote, 1999; Yang and Coote, 1999; Yang et al., 2001). Other medullary reticular neurons project to locus coeruleus neurons, with implications for cardiac (Guyenet and Young, 1987; Guyenet et al., 1996) and respiratory functions (Huangfu et al., 1992). Arousal-related orexin neurons in the hypothalamus also help to stimulate breathing through these medullary regions (Young et al., 2005). Important transmitters in this region are likely to include glutamate (Weston et al., 2004) and GABA (Schreihofer and Guyenet, 2003). While initial studies have shown estrogen influences on baroreflexes (Pamidimukkala et al., 2005), the linkages of molecular neuroendocrine mechanisms to these medullary neurons largely remain to be discovered.

Electrophysiology. To participate in generalized arousal pathways, neurons should be responsive to stimuli in a variety of sensory modalities, and not be limited to one sub-modality or have a narrow sensory field. Excellent examples come in the recordings of Peggy Mason, from neurons in the ventromedial medullary reticular formation (Leung and Mason, 1998, 1999), and from the so-called 'omnipause neurons' of the pons and midbrain (Evinger et al., 1982; Pare and Guitton, 1998; Phillips et al., 1999). A more controversial example comes through the recordings of midbrain dopamine neurons by Jon Horvitz (Horvitz et al., 1997). While some other researchers interpreted the firing rates of these neurons as specifically signaling reward expectation, the full range of data, exemplified by Horvitz's work, show that reward is simply one way of making a stimulus salient. These neurons can respond to a variety of salient stimuli and probably are most closely linked to the initiation of directional movements with respect to those stimuli.

Functional genomics. As of early 2005, more than 120 genes could be argued to participate directly in the regulation of CNS arousal (Pfaff, 2005). Why so many? Consider all of the neurotransmitters and neuropeptides involved. Then, consider that each of them implies genes involved in their synthesis, their receptors (for serotonin, e.g., 14 genes), their transporters and their breakdown by catabolic enzymes. We have used gene knockout technology (Garey et al., 2003), antisense oligo technology (Mong et al., 2003a) and molecular pharmacology (Easton et al., 2004) to show contributions from a nuclear receptor gene, an enzyme and a neurotransmitter, respectively. The very multiplicity of these genomic controls raises the question explored in the next section.

From generalized CNS arousal to specific forms of arousal

In no way can generalized arousal by itself produce organized, biologically adaptive courtship and sex behaviors. At least four levels of gene–gene interactions are involved. First, physiological regulation of reproductive processes for their own sake involves large numbers of genes. Second, consider seasonal and reproductive rhythms. Third, there are environmental constraints upon sex behaviors, including fear, pain, etc. Fourth, certain neuropeptides bear on affiliative behaviors that, though explicitly sexual, do bring animals and humans into a social context that fosters eventual reproduction. These interactions are schematized in Fig. 4.

The logical hierarchies of these interactions and their quantitative natures are unknown. The latter is very important if you consider the mechanism underlying behavior to form an algorithm with discoverable mathematical and physical properties. However, we know these interactions exist, and are beginning to discover their mechanisms. For example, at least three neurotransmitters signal generalized CNS arousal and affect electrical activity in the ventromedial hypothalamic neurons essential for lordosis behavior. They are histamine, norepinephrine and the opioid peptide enkephalin. Mechanisms through which they can affect hypo-

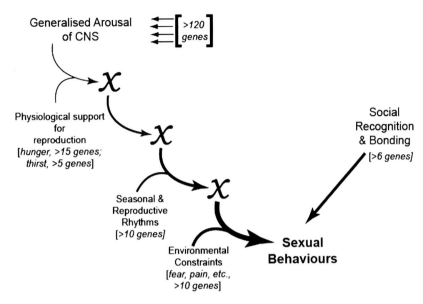

Fig. 4. At least four levels of gene–gene interactions are involved in producing organized, biologically adaptive courtship and sex behaviors from generalized arousal: genes involved in the physiological regulation of reproductive processes, seasonal and reproductive rhythms, environmental constraints upon sex behaviors (including fear, pain, etc.), and social recognition and bonding. Adapted from Pfaff et al., (2005).

thalamic neurons are reviewed in the next three sections.

Molecular biology of histamine receptors in CNS

An arousal-related neurotransmitter par excellence is histamine, produced by neurons in the tuberomammillary nucleus of the hypothalamus. It signals through three receptor sub-types.

The histamine H_1 receptor

Distribution in CNS. Long before histamine receptor subtypes were identified, antihistamines were developed for clinical use in allergy treatment. One side effect of antihistamines is central sedation. We now know that central histamine H_1 receptors are responsible for the sedative effects of antihistamines. H_1 receptors have a widespread distribution in the CNS, as shown by autoradiography studies using selective H_1 receptor antagonists (Palacios et al., 1981; Bouthenet et al., 1988). Particularly high levels of H_1 receptors are detected in brain areas involved in arousal, such as cerebral cortex (especially higher in layer IV), thalamus, tegmentum area and basal forebrain as well as locus coeruleus and raphe nuclei. High densities of H_1 receptors are also found in the limbic system. In the hippocampus, H_1 receptors display a laminated pattern of distribution and are the most abundant in the dentate gyrus and subiculum. In the amygdala, high densities are found in the medial nuclei. In the hypothalamus, H_1 labeling is highly heterogeneous with high densities in the preoptic area, ventromedial and most posterior nuclei, including tuberomammillary nuclei where histaminergic fibers are generated from. Nucleus accumbens and cerebellum also show high densities of H_1 receptors.

Molecular properties. The H_1 receptor was cloned in 1991 (Yamashita et al., 1991), representing a 491 amino acid protein with a molecular weight of 56 kDa. Similar to other G-protein-coupled receptors (GPCR), the H_1 receptor protein has seven transmembrane domains with a very large third intracellular loop and relatively short intracellular C terminus. The third cytoplasmic loop of the H_1

receptor is thought to interact with G protein and has many serine and threonine residues that may serve as sites for phosphorylation by protein kinases. Those potential sites of phosphorylation may play an important role in regulating signal transduction through the receptor molecule (Yamashita et al., 1991). The selective H_1 receptor agonists include 2-pyridlyethylamine and 2-thiazolylethylamine. A wide array of potent and selective antagonists is also available. Compounds such as mepyramine and triprolidine are highly potent H_1 antagonists and have great importance in pharmacological investigations.

Signaling pathways (also see Fig. 5). The primary signal transduction event induced by H_1 receptor activation in the brain is the activation of phospholipase C (PLC) via a pertussis toxin (PTx)-insensitive $G_{q/11}$ protein (Leurs et al., 1994; Leopoldt et al., 1997). When $G_{q/11}$ is activated by histamine binding to the H_1 receptor, it leads to the stimulation of PLC, which in turn

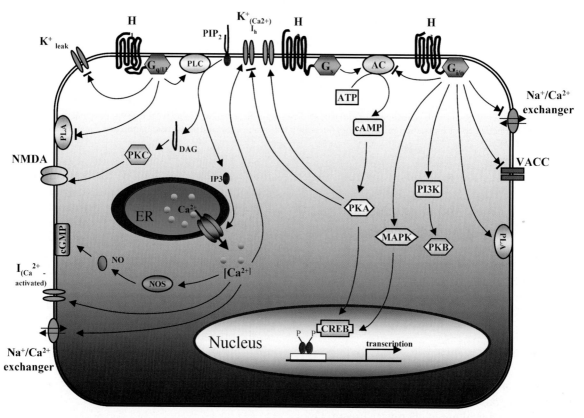

Fig. 5. Signaling pathways activated by histamine receptors. H_1 receptors couple to $G_{q/11}$ proteins to stimulate PLC, leading to the production of IP_3 and DAG from PIP_2. IP_3 further releases Ca^{2+} from internal stores, which leads to several events. DAG stimulates PKC activity. H_2 receptors couple to G_s proteins leading to the production of cAMP. cAMP further activates PKA. H_3 receptors couple to $G_{i/o}$ proteins to inhibit the production of cAMP from ATP, therefore depressing cAMP-activated signaling pathways (See text for details). *Abbreviations*: AA, arachidonic acid; AC, adenylyl cyclase; CREB, cyclic-AMP-response element-binding protein; DAG, diacylglycerol; GC, guanylyl cyclase; $G_{i/o}$, G protein i/o type; $G_{q/11}$, G protein q/11 type; G_s, G protein s type; H_1, histamine receptor 1; H_2, histamine receptor 2; H_3, histamine receptor 3; HVACC, high voltage-activated Ca^{2+} channel; $I_{K(Ca)}$, small conductance Ca^{2+}-dependent K^+ current; I_h hyperpolarization-activated cation current; $I_{K,leak}$, leak K^+ current; IP_3, inostol-1,4,5-triphosphate; MAPK, mitogen-activated protein kinase; NMDA, *N*-methyl-D-aspartate; NO, nitric oxide; NOS, nitric oxide synthase; PI3 K, 1, 4, 5-phosphoinositol kinase; PIP_2, phosphatidylinositol-4,5-bisphosphate; PKA, protein kinase A; PKC, protein kinase C; PLA, phospholipase A; PLC, phospholipase C.

hydrolyses phosphatidyl-4, 5-biphosphate (PIP$_2$) to DAG and IP$_3$. DAG potentiates PKC activity, while IP$_3$ binds to its own receptors located on the endoplasmic reticulum, allowing the release of stored Ca^{2+} into the cytoplasm. Among other signaling pathways activated by H$_1$ receptor activation, many of them appear to be secondary to changes in intracellular Ca^{2+} concentration or activation of PKC. Increased intracellular Ca^{2+} could further stimulate: (1) Ca^{2+}-activated cation channels or electrogenic Na$^+$/Ca^{2+} exchange, as in supraoptic vasopressin neurons, resulting in H$_1$ receptor-mediated dose-dependent depolarization (Smith and Armstrong, 1996); (2) the opening of Ca^{2+}-dependent K$^+$-channels (Fioretti et al., 2004), which causes afterhyperpolarization and accommodation resulting in hyperpolarization (Weiger et al., 1997); (3) stimulate nitric oxide synthase activity via a Ca^{2+}/calmodulin-dependent pathway and subsequent activation of soluble guanylyl cyclase in vitro (Richelson, 1978; Leurs et al., 1991). Phosphorylation of NMDA receptors by PKC reduces the sensitivity of voltage-dependent Mg^{2+} block leading to larger NMDA-mediated currents in cortical neurons (Chen and Huang, 1992; Payne and Neuman, 1997). Arachidonic acid release and the synthesis of arachidonic acid metabolites such as thromboxane A$_2$ can also be enhanced by H$_1$ receptor stimulation via phospholipase A$_2$ (PLA$_2$) (Leurs et al., 1994; van der Zee et al., 1995). Also notable is that histamine, via H$_1$ receptors linked to G protein, directly blocks a background leak K$^+$ current, which leads to depolarization and/or an increase in firing frequency. This effect has been found in many brain areas such as cortex (Reiner and Kamondi, 1994), thalamus (McCormick and Williamson, 1991), hypothalamus (Li and Hatton, 1996), and striatum (Munakata and Akaike, 1994) and might be the mechanism for antihistamine-induced sedation in the human brain (Reiner and Kamondi, 1994).

Histamine H$_2$ receptor

Distribution in CNS. Autoradiographic mapping of the H$_2$ receptor and in situ hybridization of its mRNA shows its heterogeneous distribution and laminated pattern (Ruat et al., 1990; Traiffort et al., 1992; Vizuete et al., 1997). Cerebral cortex, hippocampus and basal ganglia (which show low H$_1$ receptor density) are among the brain areas with high densities. H$_2$ receptors are present in most areas of the cerebral cortex, with the highest density in the superficial layers (I–III), which contain low H$_1$ receptors. In the hippocampus, H$_2$ receptors are densely labeled in granule cells of the dentate gyrus. In the thalamus, hypothalamus and cerebellum, the low density of H$_2$ receptors contrasts with the high density of H$_1$ receptors (Traiffort et al., 1992). These distributions suggest that H$_1$ and H$_2$ receptors may act synergistically in areas where they are colocalized, such as the cerebral cortex, while working independently in other areas, such as the thalamus.

Molecular properties. The H$_2$ receptor was initially identified based on the different pharmacological profile of histamine activation. Cloned in 1991 (Gantz et al., 1991), H$_2$ receptors also belong to the GPCR family and has seven transmembrane domains. Sequence homology of transmembrane domains between H$_1$ and H$_2$ (40.7%) is not higher than that between H$_1$ and m$_1$-muscarinic receptors (44.3%) (Shapiro et al., 1988; Yamashita et al., 1991). The most notable difference between the structure of H$_2$ and H$_1$ receptors is the much shorter third intracellular loop and the longer C-terminal of H$_2$ receptors. The pharmacological specificity of H$_2$ receptors reside on only a few key amino acids (Gantz et al., 1992). Dimaprit is a selective H$_2$ receptor agonist. It is almost as potent as histamine at H$_2$ receptors.

Signaling pathway. H$_2$ receptors couple with G$_s$ protein. Activation of G$_s$ protein leads to stimulation of adenylyl cyclase and increased production of cAMP and cAMP-dependent protein kinase A (PKA) (Baudry et al., 1975; Hegstrand et al., 1976). PKA can translocate to the nucleus, activate the transcription factor CREB, and then induce gene expression.

H$_2$ receptor activation leads to neuronal excitation, which is due to PKA's effect on the membrane, such as phosphorylation of target proteins on ion channels to inhibit conductance. For example, phosphorylation of Ca^{2+}-dependent K^+-channels leads to inhibition of its conductance, which is responsible for the accommodation of firing and afterhyperpolarization (Haas and Konnerth, 1983; Haug and Storm, 2000). In hippocampus CA1 pyramidal cells (Haas and Greene, 1986) and granule cells (Greene and Haas, 1990), histamine, by activating H_2 receptors and cAMP pathway, blocks Ca^{2+}-activated K^+ channels and therefore blocks both afterhyperpolarization and accommodation. This block action leads to a profound potentiation of excitatory signals. Activation of hyperpolarization-activated cation channels (I_h) via PKA in the thalamus (McCormick and Williamson, 1991) suggests another mechanism of H_2 receptor-mediated depolarization. This current (I_h) facilitates H_1 receptors in histamine-induced switch in firing mode in thalamic neurons.

Histamine H_3 receptor

Distribution in CNS: The inhibition of histamine on its own release revealed a class of receptor (H_3) pharmacologically distinct from H_1 and H_2 receptors. The H_3 receptor was described as a presynaptic autoreceptor located on tuberomammillary neuron cell bodies and axons regulating the release and synthesis of histamine (Arrang et al., 1983; Itoh et al., 1991). H_3 receptors have widespread distribution in the CNS (Pollard et al., 1993). High density of H_3 receptors is presented in the cerebral cortex (all areas and layers), thalamus and basal forebrain such as the nucleus accumbens, striatum, olfactory tubercles, and globus pallidus. Abundant H_3 receptors are also found in the CA3 and dentate gyrus of the hippocampus and amygdala. Moderate densities are found in the hypothalamus with slightly more abundance in the anterior and medial part. Its widespread and abundant neuronal expression in the brain highlights the significance of histamine as a general neurotransmitter modulator.

Molecular properties. The molecular architecture of the H_3 receptor was not revealed until it was cloned in 1999 (Lovenberg et al., 1999). Overall similarity between the H_3 and the H_1 and H_2 receptors is only 22 and 21%, respectively (Lovenberg et al., 1999; Leurs et al., 2005). H_3 receptors shares higher homology (~60%) with H_4 receptors which are primarily expressed in leukocytes and bone marrow (and therefore are not discussed here). Like H_1 and H_2 receptors, H_3 receptors also belong to the large superfamily of GPCR. Unlike H_1 and H_2 receptors, analysis of H_3 receptor gene showed the presence of several introns. The various H_3 receptor isoforms are derived from alternative splicing of the H_3 receptor gene.

Signaling pathway: Activation of H_3 receptor-mediated $G_{i/o}$ proteins leads to activation and inhibition of different signaling pathways. Activation of $G_{i/o}$ proteins decreases cAMP levels and reduces downstream gene transcription (Lovenberg et al., 1999). $G_{i/o}$ protein activation also results in activation of other second messenger pathways (Leurs et al., 2005): (1) mitogen-activated protein kinase (MAPK) pathway; (2) 1,4,5-phosphoinositol kinase (PI3K) pathway; (3) activation of PLA_2; (4) inhibition of the Na^+/H^+ exchanger and; (5) lowering intracellular Ca^{2+} levels by modulating voltage-gated ion channels. Of special interest is the mechanism of histamine's modulation of histamine release and the release of other neurotransmitters. H_3 receptors are exclusively presynaptic, located on histaminergic and other neurons, where they provide negative feedback to control histamine synthesis and release (Arrang et al., 1983), as well as other transmitters such as glutamine (Brown and Haas, 1999; Molina-Hernandez et al., 2001) and norepinephrine (Silver et al., 2002). Instead of activating inward-rectifying K^+ conductance as other $G_{i/o}$ protein-coupled autoreceptors do, inhibition of depolarization-induced Ca^{2+} entry through high voltage-activated Ca^{2+} channels appears to account for the effect of H_3 receptor inhibition on neurotransmitter release (Brown et al., 2001; Leurs et al., 2005).

α_{1B}-Noradrenergic receptor signaling

Noradrenergic signaling to the hypothalamus is estrogen-dependent (Kow and Pfaff, 1985, 1987;

Kow et al., 1992, 1995) and is important for lordosis behavior. The adrenergic receptor family includes nine receptor subtypes, which fall into three major groups, $\alpha 1$, $\alpha 2$, and β (Michelotti et al., 2000). Especially significant, according to the results of Etgen and her colleagues (Etgen, 2002) are alpha-1B adrenergic receptors (α_{1B}-AR).

α_{1B}-ARs, members of the GPCR superfamily of membrane proteins, are located on the cell surface and mediate the actions of the catecholamines norepinephrine and epinephrine (Piascik and Perez, 2001). Although α_{1B}-ARs are abundant in the brain, much of our knowledge on their function and signaling is based on studies of the sympathetic nervous system, particularly on cardiovascular regulatory activities.

It is well known that α_{1B}-AR activation results in increased intracellular Ca^{2+} via receptor coupling to the PTx-insensitive G protein, $G_{q/11}$ (Minneman, 1988; Kamp and Hell, 2000; Michelotti et al., 2000; Piascik and Perez, 2001; Koshimizu et al., 2002). α_{1B}-AR/$G_{q/11}$ coupling leads to activation of the enzyme PLC, generating IP_3 which mobilizes intracellular Ca^{2+}, and DAG which activates PKC. α_{1B}-ARs may also couple to other G proteins, such as G_h (Kang et al., 2002), G_s (Horie et al., 1995; Ruan et al., 1998), and G_i/G_o (Perez et al., 1993). There is compelling evidence that α_{1B}-ARs can activate other effectors such as MAPK pathway (Della Rocca et al., 1997; Piascik and Perez, 2001), cAMP (Petitti and Etgen, 1991; Horie et al., 1995), phospholipase D (PLD) (Ruan et al., 1998) and PLA_2 (Perez et al., 1993; Kreda et al., 2001). See Fig. 6 for a schematic.

$G_{q/11}$-coupled primary signaling pathway

α_{1B}-ARs can couple to a variety of G proteins and second messenger systems, but it is generally agreed that the primary pathway involves activation of the α-subunit of $G_{q/11}$, stimulating PLC in the plasma membrane (Kamp and Hell, 2000; Michelotti et al., 2000; Piascik and Perez, 2001; Koshimizu et al., 2002). Activated PLC hydrolyses membrane-bound PIP_2, generating second messengers IP_3 and DAG. IP_3 binds to specific receptors on the endoplasmic reticulum to mobilize intracellular stores of Ca^{2+}, while DAG (and intracellular Ca^{2+}) activates PKC, which then phosphorylates a variety of cellular substrates.

In cardiac myocytes, activated PKC is proposed to bind to anchoring proteins referred to as RACKs (receptor for activated C kinases) located near the high-voltage-activated L-type Ca^{2+} channel, which is then phosphorylated (Kamp and Hell, 2000). There is evidence that PKC can both stimulate and inhibit L-type Ca^{2+} channels depending on the cell type (Graham et al., 1996; Kamp and Hell, 2000). α_{1B}-ARs are also linked to Ca^{2+} influx through voltage-independent channels in DDT1 MF-2 and BC3H-1 cells (Han et al., 1992). Thus, activation of α_{1B}-ARs can increase Ca^{2+} by mobilizing intracellular stores of Ca^{2+}, as well as via voltage-dependent and -independent Ca^{2+} channels. However, Ca^{2+} influx is not necessary for α_{1B}-AR-mediated PLC activation, since it is resistant to chelation of extracellular Ca^{2+} with EGTA and to the L-type Ca^{2+} channel blocker nifedipine (Perez et al., 1993).

In immortalized hypothalamic neurons, an increase in Ca^{2+} can stimulate cytosolic PLA_2 and release arachidonic acid (AA) (Kreda et al., 2001). Activation of PLA_2 occurs downstream from PLC stimulation, triggered initially by an increase in cytoplasmic Ca^{2+} and further augmented by the influx of extracellular Ca^{2+} (Kreda et al., 2001). Increased Ca^{2+} has also been suggested to activate nitric oxide synthase, producing nitric oxide, which in turn stimulates cyclic guanosine monophosphate production by activating soluble guanylyl cyclase (Chu and Etgen, 1999).

Other G-protein coupled pathways

In addition to coupling to PTx-insensitive G proteins of the $G_{q/11}$ family, α_{1B}-ARs can also couple to PTx-sensitive G proteins (G_i or G_o), activating PLA_2 and increasing AA release in COS-1 and CHO cells (Perez et al., 1993). In Rat-1 fibroblasts, however, the increase in AA is mediated by the activation of PLD (Ruan et al., 1998).

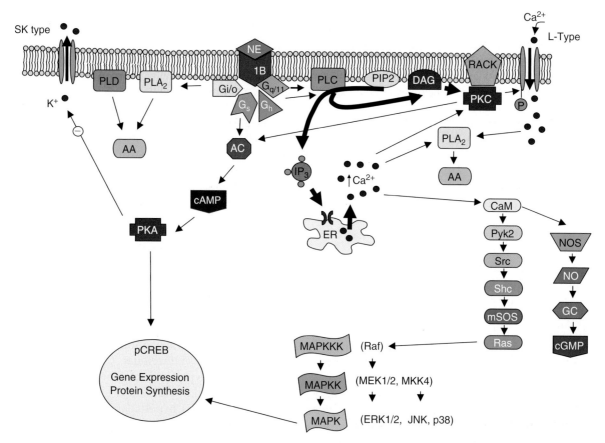

Fig. 6. Signaling pathways activated by alpha-1B adrenergic receptors. α_{1B}-ARs couple primarily to G_q proteins, stimulating PLC. PLC cleaves PIP_2, generating IP_3 and DAG. IP_3 mobilizes intracellular Ca^{2+} and DAG activates PKC. Mobilization of cytoplasmic Ca^{2+} stimulates PLA_2, releasing AA and activates NOS, producing NO, GC, and cGMP. PKC has been proposed to bind to RACKs near L-type Ca^{2+} channels, which it then phosphorylates. There is evidence that α_{1B}-ARs may also couple to G_h proteins, leading to production of IP_3 and DAG; to G_s proteins, leading to stimulation of PKA, inhibiting SK channel activity; and to G_i/G_o proteins, increasing AA through PLA_2 or PLD. Although the precise signaling mechanism is not known, α_{1B}-ARs can stimulate the MAPK pathway. α_{1B}-ARs can also activate ERKs, p38s, or JNKs depending upon the particular cell line or tissue it is expressed in. *Abbreviations*: AA, arachidonic acid; AC, adenylyl cyclase; AR, adrenergic receptors; CaM, calmodulin; cAMP, cyclic 3′, 5′-adenosine monophosphate; cGMP, cyclic guanosine monophosphate; DAG, diacylglycerol; ER, endoplasmic reticulum; ERK, extracellular signal-regulated kinases; GC, guanylyl cyclase; IP_3, inositol 1,4,5-trisphosphate; JNK, c-Jun N-terminal kinases; MAPK, mitogen-activated protein kinase; MAPKK, MAPK kinase; MAPKKK, MAPKK kinase; mSOS, murine sons of sevenless-1 protein; NE, norepinephrine; NO, nitric oxide; NOS, nitric oxide synthase; pCREB, cAMP-responsive element binding protein; PIP_2, phosphatidylinositol 4,5-biphosphate; PKA, PKC, protein kinase A, C; PLA_2, PLC, PLD, phospholipase A_2, C, D; Pyk2, proline-rich tyrosine kinase 2; RACK, receptor for activated C kinases; SK, small conductance Ca^{2+}-activated K^+ channel; Shc, SH2-containing protein. From Pfaff et al., 2006.

More recent research suggests that α_{1B}-ARs can also activate another class of PTx-insensitive G proteins, termed G_h, also known as transglutaminase type II (TGII) (Kang et al., 2002). α_{1B}-AR/TGII coupling leads to stimulation of PLC-δ1, producing second messengers IP_3 and DAG. Stimulation of PLC-δ1 by the coupling of α_{1B}-AR with TGII releases intracellular stores of Ca^{2+} and maintains the elevation of Ca^{2+} (Kang et al., 2002).

α_{1B}-ARs can also couple to G_s, elevating intracellular cAMP and PKA (Horie et al., 1995; Ruan et al., 1998). However, the increase in cAMP is probably via a PKC-dependent potentiation of adenylyl cyclase activity (Petitti and Etgen, 1991; Perez et al., 1993). cAMP-dependent PKA phosphorylates the small conductance Ca^{2+}-activated K^+ (SK) channel, inhibiting its activity in hippocampal neurons (Pedarzani and Storm, 1996). α_{1B}-Adrenergic inhibition of SK current has been reported in hypothalamic preoptic neurons, resulting in reduced spike frequency adaptation and increased firing (Wagner et al., 2001). Thus, NA can decrease K^+ conductance through SK channels, leading to reduced afterhyperpolarization following action potentials, thus promoting excitability.

Mitogen-activated protein kinase pathway

Although the precise signaling cascade is unknown, α_{1B}-ARs can activate all three major subfamilies of MAPK: the extracellular signal-regulated kinases (ERKs), c-Jun N-terminal kinases (JNKs), and the p38 kinases, depending on the cell type (Piascik and Perez, 2001).

α_{1B}-ARs mediate ERK1 and ERK2 activation via a Ras-dependent mechanism in HEK-293 cells, and Ras-independent mechanism in COS-7 cells (Hawes et al., 1995; Della Rocca et al., 1997). The α_{1B}-AR-mediated signal is dependent on stimulation of PLC-β, leading to PKC activation, increased intracellular Ca^{2+} and activation of calmodulin (CaM) and proline-rich tyrosine kinase 2 (Pyk2) (Della Rocca et al., 1997; Koshimizu et al., 2003). Pyk2 stimulates Src, resulting in phosphorylation of the Shc adaptor protein, recruitment of the Grb2-Sos complex to the membrane, and activation of Ras (Della Rocca et al., 1997). Ras then activates Raf kinases, which activate MEK1 and MEK2. MEK1 and MEK2 phosphorylate ERK1 and ERK2 (Della Rocca et al., 1997). ERK then translocates to the cell nucleus, where it phosphorylates and activates other kinases and nuclear transcription factors, such as pCREB, and plays a role in gene expression and protein synthesis (Koshimizu et al., 2003).

Activation of α_{1B}-AR/$G\alpha_q$ can activate the JNK pathway, which involves PLC-β, Src family tyrosine kinases, Rho family small GTPases and MKK4 (Koshimizu et al., 2002). JNK phosphorylates transcription factors c-Jun and activating transcription factor 2 (ATF2) and is involved in a variety of biological functions, such as cell movement, development and cell cycle arrest (Koshimizu et al., 2003). In PC12 cells, α_{1B}-ARs activate ERKs and p38 kinases, but not JNKs (Zhong and Minneman, 1999), suggesting that activation of specific MAPK pathways may depend on the cell line or tissue it is expressed in (Piascik and Perez, 2001).

μ and δ opioid receptor signaling

The levels of mRNA coding for enkephalin are elevated in VMH neurons by estrogen treatment of ovariectomized female rats (Romano et al., 1988), and, in turn, opioid signaling in the hypothalamus is important for the regulation of lordosis behavior (Nicot et al., 1997).

Opioid receptors belong to the GPCR superfamily. They can be activated by endogenously produced opioid peptides as well as exogenously administered opioid drugs. This nature of opioids results in a wide variety of effects, which include analgesia, respiratory depression, euphoria, feeding, hormone release, inhibition of gastrointestinal transit, and anxiety. Opioid research became one of the fundamental questions in addiction biology and physiology, as it is important to explain why exogenous opioid drugs such as, morphine and heroin, induce tolerance, dependence, and addiction.

To date there are two major opioid receptor groups characterized: the classical opioid receptors consisting of μ, κ, and δ receptors (MOR, KOR, and DOR, respectively) (Lord et al., 1977), and a second branch consisting of opioid receptor-like (ORL-1) receptors (also named NOP for its endogenous ligand, nociceptin/orphanin FQ peptide) (Bunzow et al., 1994; Mollereau et al., 1994; Wang et al., 1994a; Meunier et al., 1995; Reinscheid et al., 1995). There are several other receptors thought to be opioid receptors as well, including σ (for SKF10047) receptors (Martin et al., 1976), β-endorphin-sensitive ε receptors (Wuster et al.,

1979), ζ receptors (Zagan et al., 1991), and high-affinity binding sites called λ sites (Grevel et al., 1985). However, as they are poorly studied, their place among the opioid receptors is yet to be established. In addition, there is evidence for subtypes of MORs, KORs and DORs (Traynor and Elliott, 1993; Wolleman et al., 1993; Pasternak, 2004). Opioid receptor genomic and cDNA clones are available from different species, such as guinea pig, zebra fish, human, rhesus monkey, mouse, rat, etc. (Waldhoer et al., 2004). In this review we will briefly explore MOR and DOR.

MOR distribution. MOR genes were identified and pharmacologically characterized in the 1970s on the basis of their high affinity for morphine (mu for morphine) (Martin et al., 1976). They are highly expressed in several brain regions: thalamus, hypothalamus, caudate putamen, neocortex, nucleus accumbens, amygdala, interpeduncular complex, and inferior and superior colliculi (Mansour et al., 1987), periaqueductal gray and raphe nuclei (Hawkins et al., 1988), as well as in the superficial layers of the dorsal horn of the spinal cord (Besse et al., 1990). MOR distribution varies by species. For example, in the hypothalamus, there is a higher density of MOR in mouse than in rat, while the opposite is true in the hippocampus (Kitchen et al., 1997). Their pattern of distribution may differ by species, but their physiological function remains the same: pain and analgesia, respiratory and cardiovascular functions, intestinal transit, feeding, mood, thermoregulation, hormone secretion and immune functions (Dhawan et al., 1996).

DOR distribution. The DOR gene was first identified in mouse vas deference (delta for deference) where it had a high affinity for endogenous opioid enkephalin (Lord et al., 1977). This receptor is highly expressed in the olfactory bulb, neocortex, caudate putamen, nucleus accumbens, and amygdala (Mansour et al., 1987) as well as in the thalamus and hypothalamus (Kitchen et al., 1997). It is also expressed in the dorsal horn of the spinal cord. Its expression can vary by species, as is the case in MOR. For example, in caudal regions of the interpeduncular nucleus and pontine nuclei, there is a higher density of DORs expressed in rat than in mouse (Kitchen et al., 1997). Compared to MORs, DOR physiology is less known, but has been shown to be involved in mediating analgesic effects and may be involved in gastrointestinal motility, mood and cardiovascular regulation (Reinscheid et al., 1995).

Pharmacology. The overall effect of opioid receptor activation is to inhibit neural transmission. Both MORs and DORs can mediate the inhibition of cAMP formation consistent with the employment of a G-protein effector system (Childers, 1991). Their activation normally increases K^+ conductance to produce neuronal hyperpolarization and the inhibition of action potential generation. Their highly selective ligands facilitate the characterization of the effects they mediate. Table 1 shows the most selective antagonists and agonists for these two receptor types.

Pharmacological studies suggest the presence of multiple MORs, referred to as μ1 and μ2 (Pasternak and Wood, 1986; Pasternak, 2004). μ1 receptors are characterized by their high affinity to naloxonazine and the ligands, morphine and DADLE, and are involved in supraspinal analgesia, prolactin release, decrease in acetylcholine turnover, and induction of catalepsy (Pasternak, 2005). μ2 receptors are characterized by a lower affinity for morphine and DADLE, and are involved in respiratory depression, decreased dopamine turnover, and delayed gastrointestinal tract transit. The cloned DOR has a high affinity for DADLE and the selective ligand DPDPE, but not for the MOR selective ligand DAMGO. Some behavioral and binding studies suggest existence of multiple DORs (Portoghese et al., 1992). DPDPE, DADLE, DALCE, and BNTX bind to δ1-sites with high affinity, while DSLET, [D-Ala^2]deltorphin II, 5′-NTII, and NTB bind to δ2-sites with high affinity (Minami and Satoh, 1995).

Signal transduction. Besides PTX-sensitive heterotrimeric G_i/G_o proteins, opioid receptors couple with PTx-insensitive G_s and G_z proteins too (Crain and Shen, 1990; Garzon et al., 1998; Hendry et al., 2000).

Table 1. Selective antagonists and agonists for MOR and DOR types

Receptor	Endogenous peptide	Peptide agonists	Peptide antagonists
MOR	Endomorphin-1	DAMGO	CTOP
	Endomorphin-2	PL-17	Octreotide
			(SMS201,995)
	β-endorphin	Dermorphin	
	β-neoendorphin		
	Dermorphin		
DOR	Leu^5-Enkephalin	DADLE	ICI 174,864
	Met^5-Enkephalin	DPDPE	TIPP
	Met^5-Enkephalin-Arg^6-Phe^7	DSLET	
	Met^5-Enkephalin Arg^{6}Gly^{7}Leu^8		
	Deltorphin Deltorphin I Deltorphin II		

The GTP-bound α-subunit and the combined βγ-subunits can initiate distal steps in the signaling pathway by inhibiting adenylyl cyclases and voltage-gated Ca^{2+} channels. This may cause the release of neurotransmitters and modulation of the function of several protein kinase families. The same mechanism stimulates G-protein-activated K^+ conductance and phospholipase (Standifer and Pasternak, 1997; Quock et al., 1999). The biphasic nature of DPDPE was observed in DRG neurons and neuroblastoma × DRG neuron hybrid F11 cells for K^+ conductance: at concentrations of <1 nM, conductance was inhibited (Fan et al., 1991), but at higher concentrations, conductance was increased (Fan and Crain, 1995).

Like some other GPCRs, MORs, and DORs can express basal levels of constitutive activity and can activate G proteins in the absence of agonists. For example, MORs exhibit basal signaling in SH-SY5Y cells and transfected HEK293 cells (Wang et al., 1994b; Burford et al., 2000) and display more elevated constitutive activity following chronic exposure to morphine (Wang et al., 1994b). This basal signaling could involve calmodulin. In nonagonist exposed state of the receptor, calmodulin binds to MORs and competes for G-protein coupling, thereby inhibiting constitutive activity (Waldhoer et al., 2004). This is also true for DORs when expressed endogenously in neuroblastoma cells (Costa et al., 1990).

Inhibition of cAMP generation was the initial observed second messenger effect of opioid receptors, suggesting that cAMP inhibition may account for the analgesic effects of opioid drugs (Collier and Roy, 1974). However, later studies showed that injections of dibutyryl-cAMP blocked both MOR- and DOR-mediated spinal analgesia (Wang et al., 1993).

Activation of either endogenous or cloned opioid receptors leads to the activation of PKC (Kramer and Simon, 1999). A brief DPDPE exposure (5 min) fails to stimulate PKA activity; however, extended incubation with DPDPE (24 h) causes a significant increase in PKA activity (Lou and Pei, 1997). Opioid receptors can also activate MAPK (Fukuda et al., 1996; Kramer and Simon, 2000).

Physiological mechanisms: desensitization, endocytosis, and downregulation

Receptor desensitization. Involves receptor phosphorylation by G-protein coupled receptor kinases (GRKs) and subsequent β-arrestin recruitment (Ferguson, 2001) (see Fig. 7 for a schematic). They directly contribute to rapid receptor desensitization by facilitating the uncoupling of the receptor from its G protein. However, this process can also take place independently from GPK- and β-arrestin mechanisms (Liu and Anand, 2001). Besides GRKs, many kinases such as PKA, PKC, and calcium/calmodulin-dependent protein kinase II, can desensitize receptors as well (Krupnick and Benovic, 1998; Ferguson, 2001). Opioid receptors can also be tyrosine phosphorylated (Bailey and Connor, 2005).

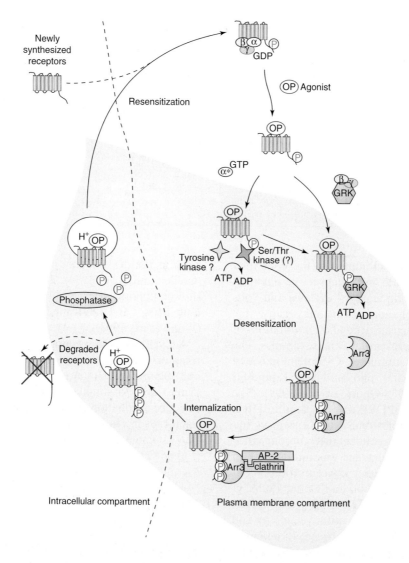

Fig. 7. Pathways for μ-opioid receptor activation, uncoupling, and internalization. The pink shaded area represents states of the receptor that are likely to be uncoupled from signaling. Although it is assumed that once MOR is phosphorylated and bound by arrestin it is functionally desensitized, alternative mechanisms of desensitization might also occur. Opioid receptors are basally phosphorylated, but the kinases responsible for this are unknown. A role for serine/threonine protein kinases such as protein kinase A, calmodulin-dependent protein kinase and protein kinase C in receptor desensitization and/or the recruitment of GRK-dependent receptor trafficking have been suggested, but they have yet to be defined in detail. Similarly, specific tyrosine kinases that phosphorylate the MOR have not been identified. Approximately 80% of MORs that undergo endocytosis are recycled; the remaining 20% are degraded in lysosomes. *Abbreviations*: AP-2, adaptor protein 2; Arr3, arrestin-3; P, phosphate. Reproduced with permission from Bailey and Connor (2005).

Endocytosis. After desensitization by GRKs and β-arrestins, opioid receptors are then rapidly endocytosed into an intracellular compartment. This process takes place with even brief agonist exposure and independently of signal transduction (Remmers et al., 1998; Zaki et al., 2001). The speed of this process is especially important for modulating signaling of endogenous ligands (neurotransmitters), which are released in a pulsatile manner. It has been proposed that endocytosis also mediates receptor 'resensitization' — meaning that receptors have been delivered to an endosome-associated phosphatase and then dephosphorylated receptors returned to the plasma membrane via rapid recycling way (Koch et al., 1998). Chronic exposure of opioid receptors to agonists, can also mediate receptor desensitization. This process involves PKA/PKC-mediated phosphorylation. Interestingly, PKA/PKC-phosphorylated receptors are not necessarily rapidly endocytosed, therefore receptors that have been desensitized by GRK-independent phosphorylation require different mechanism other than endocytosis to resensitize (Waldhoer et al., 2004).

Downregulation. Following endocytosis, receptors that do not resensitize (and therefore do not restore their functions) are further downregulated (Tsao and von Zastrow, 2000). Similar to endocytosis, not all the receptors are downregulated. For example, MORs are recycled following their endocytosis, but DORs are transported deeper into the endocytic pathway and degraded by lysosomes (Finn and Whistler, 2001; Whistler et al., 2002). It is important also to note that DORs, which interact with G-protein coupled receptor-associated sorting protein (GASP), are downregulated, whereas MORs, which do not interact with GASP, recycle and therefore are resensitized.

Endogenous antiopioid systems

An existence of antiopioid systems that compensate the opioid system and therefore maintain the homeostasis has been proposed. Among these systems are the Tyr-Pro-Leu-Gly-NH$_2$ peptide family, neuropeptide FF and orphanin FQ/nociceptin (Harrison et al., 1998), cholecystokinin (Noble and Roques, 1999), NMDA (Trujillo, 2000), and excitatory opioid receptors coupled through G$_s$ (Crain and Shen, 1990). There are no detailed molecular mechanisms of how antiopioid system upregulation works; however, there are some hypotheses. For example, the NMDA receptor antagonist, MK801, may inhibit the development of morphine tolerance and dependence (Trujillo and Akil, 1991). The NR2A and NR1 subunits of the NMDA receptor are upregulated following chronic morphine treatment, implicating NMDA receptor upregulation in the behavioral manifestation of tolerance (Marek et al., 1991; Inoue et al., 2003).

Summary and outlook

The strategic advantages inherent in analyzing steroid hormone-dependent behaviors coupled with the simplicity of lordosis behavior itself enabled the discovery of the complete neural circuit producing that behavior. The purpose of studying such a simple response was to see if the mechanisms for *any* mammalian behavior could be spelled out in detail. Because ERs are ligand-activated transcription factors, it was possible to find several genes whose transcript levels are elevated by estrogens and whose gene products foster lordosis behavior. Thus, in terms of the functional genomics of estrogen action in the brain, those molecular induction phenomena represent part of the mechanisms by which the hormone-dependent behavior is produced. The transcripts thus affected are not all similar to each other, but instead can theoretically be conceived in functional modules that make sense for the female's reproductive biology. The 'GAPPS' model presented last year emphasized the ability of sex hormones to make nerve cells *grow*, to set up their own *amplification*, to initiate *preparative* behavioral steps, to have *permissive* actions on the rest of the lordosis circuit, and to *synchronize* sex behavior with ovulation.

Having determined mechanisms for a hormone-dependent behavior, it was timely to enlarge the scope of this area of work to consider sexual arousal and the generalized arousal states of the CNS. These are also hormone-dependent. In doing so we could approach mechanisms for functions that underlie *all* behaviors. A theoretical summary of the neuroanatomy of generalized arousal of the mammalian CNS emphasizes its *bilaterality*, its *bidirectionality*, its *universality* and its roles in *response potentiation* be they approach or avoidance responses (BBURP theory). Increasingly, the electrophysiological and genomic underpinnings of CNS arousal are understood, and will provide an outlet for studies of hormone action, especially sex hormones, stress hormones, and thyroid hormones. Such studies are urgent to do, because generalized arousal is of overwhelming importance to neurological and psychiatric medicine and to certain issues of public health.

Looking forward, we must explore the large numbers of transcripts that yield interesting data during microarray studies of estrogenic effects in the CNS as well as sex differences in the hypothalamus and preoptic area. Another area of work explores the impact of generalized arousal on specific arousal functions and on specific hormone-dependent behaviors. It should be possible to spell out in physical detail exactly how neurochemicals such as histamine, norepinephrine and opioid peptides influence sex behavior-controlling neurons in the forebrain as well as to find out whether other arousal-related signals impact these hypothalamic and preoptic neurons.

Abbreviations

AA	arachidonic acid
AR	adrenergic receptors
ATF2	activating transcription factor 2
DAG	diacylglycerol
DOR	delta opioid receptor
ER	estrogen receptor
ERK	extracellular signal-regulated kinases
GASP	G-protein-coupled receptor associated sorting protein
GRK	G-protein-coupled receptor kinases
GnRH	gonadotropin-releasing hormone
GPCR	G-protein-coupled receptors
GS	glutamine synthase
IP_3	inositol 1,4,5-trisphosphate
JNK	c-Jun N-terminal kinases
KOR	kappa opioid receptor
LH	luteinizing hormone
MAPK	mitogen-activated protein kinase
MOR	mu opioid receptor
NA/NE	noradrenaline/norepinephrine
PGDS	prostaglandin D synthase
PIP_2	phosphatidylinositol 4,5-biphosphate
PKA	protein kinase A
PKC	protein kinase C
PLA_2	phospholipase A_2
PLC	phospholipase C
PLD	phospholipase D
PR	progesterone receptor
PTx	pertussis toxin
Pyk2	proline-rich tyrosine kinase 2
RACK	receptor for activated C kinases
SK	small conductance Ca^{2+}-activated K^+ channel
TGII	transglutaminase type II
VMH	ventromedial hypothalamic

References

Acosta-Martinez, M. and Etgen, A.M. (2002) Activation of mu-opioid receptors inhibits lordosis behavior in estrogen and progesterone-primed female rats. Horm. Behav., 41: 88–100.

Arbogast, L.A. and Voogt, J.L. (1993) Progesterone reverses the estradiol-induced decrease in tyrosine hydroxylase mRNA levels in the arcuate nucleus. Neuroendocrinology, 58: 501–510.

Arbogast, L.A. and Voogt, J.L. (1994) Progesterone suppresses tyrosine hydroxylase messenger ribonucleic acid levels in the arcuate nucleus on proestrus. Endocrinology, 135: 343–350.

Arrang, J.M., Garbarg, M. and Schwartz, J.C. (1983) Auto-inhibition of brain histamine release mediated by a novel class (H3) of histamine receptor. Nature, 302: 832–837.

Bailey, C.P. and Connor, M. (2005) Opioids: cellular mechanisms of tolerance and physical dependence. Curr. Opin. Pharmacol., 5: 60–68.

Ballabio, A. and Camerino, G. (1992) The gene for X-linked Kallmann syndrome: a human neuronal migration defect. Curr. Opin. Gen. Dev., 2: 417–421.

Baudry, M., Martres, M.P. and Schwartz, J.C. (1975) H1 and H2 receptors in the histamine-induced accumulation of cyclic AMP in guinea pig brain slices. Nature, 253: 362–364.

Besse, D., Lombard, M.C., Zajac, J.M., Roques, B.P. and Besson, J.M. (1990) Pre- and postsynaptic distribution of mu, delta and kappa opioid receptors in the superficial layers of the cervical dorsal horn of the rat spinal cord. Brain Res., 521: 15–22.

Beuckmann, C.T., Lazarus, M., Gerashchenko, D., Mizoguchi, A., Nomura, S., Mohri, I., Uesugi, A., Kaneko, T., Mizuno, N., Hayaishi, O. and Urade, Y. (2000) Cellular localization of lipocalin-type prostaglandin D synthase (beta-trace) in the

central nervous system of the rat. J. Comp. Neurol., 428: 62–78.
Bodnar, R., Commons, K. and Pfaff, D. (2002) Central Neural States Relating Sex and Pain. Johns Hopkins University Press, Baltimore.
Bouthenet, M.L., Ruat, M., Sales, N., Garbarg, M. and Schwartz, J.C. (1988) A detailed mapping of histamine H1-receptors in guinea-pig central nervous system established by autoradiography with [125I]iodobolpyramine. Neuroscience, 26: 553–600.
Brann, D.W., Chorich, L.P. and Mahesh, V.B. (1993) Effects of progesterone on galanin mRNA levels in the hypothalamus and the pituitary: correlation with the gonadotrophin surge. Neuroendocrinology, 58: 531–538.
Brown, R.E. and Haas, H.L. (1999) On the mechanism of histaminergic inhibition of glutamate release in the rat dentate gyrus. J. Physiol., 515(Pt 3): 777–786.
Brown, R.E., Stevens, D.R. and Haas, H.L. (2001) The physiology of brain histamine. Prog. Neurobiol., 63: 637–672.
Bunzow, J.R., Saez, C., Mortrud, M., Bouvier, C., Williams, J.T., Low, M. and Grandy, D.K. (1994) Molecular cloning and tissue distribution of a putative member of the rat opioid receptor gene family that is not a mu, delta or kappa opioid receptor type. FEBS Lett., 347: 284–288.
Burford, N.T., Wang, D. and Sadee, W. (2000) G-protein coupling of mu-opioid receptors (OP3): elevated basal signalling activity. Biochem. J., 348: 531–537.
Canfield, J.G. (2003) Temporal constraints on visually directed C-start responses: behavioral and physiological correlates. Brain Behav. Evol., 61: 148–158.
Carrer, H. and Aoki, A. (1982) Ultrastructural changes in the hypothalamic ventromedial nucleus of ovariectomized rats after estrogen treatment. Brain Res., 240: 221–233.
Chen, L. and Huang, L.Y. (1992) Protein kinase C reduces Mg^{2+} block of NMDA-receptor channels as a mechanism of modulation. Nature, 356: 521–523.
Childers, S.R. (1991) Opioid receptor-coupled second messenger systems. Life Sci., 48: 1991–2003.
Cho, B.N., Seong, J.Y., Cho, H. and Kim, K. (1994) Progesterone stimulates GnRH gene expression in the hypothalamus of ovariectomized, estrogen treated adult rats. Brain Res., 652: 177–180.
Choleris, E., Gustafsson, J.-A., Korach, K.S., Muglia, L.J., Pfaff, D.W. and Ogawa, S. (2003) An estrogen-dependent four-gene micronet regulating social recognition: a study with oxytocin and estrogen receptor-alpha and -beta knockout mice. Proc. Natl. Acad. Sci. USA, 100: 6192–6197.
Chu, H.P. and Etgen, A.M. (1999) Ovarian hormone dependence of alpha1-adrenoceptor activation of the nitric oxide-cGMP pathway: relevance for hormonal facilitation of lordosis behavior. J. Neurosci., 19: 7191–7197.
Ciriello, J. and Caverson, M.M. (1984) Ventrolateral medullary neurons relay cardiovascular inputs to the paraventricular nucleus. Am. J. Physiol. Regul. Integr. Comp. Physiol., 246: 968–978.
Cohen, R., Chung, S. and Pfaff, D. (1984) Alteration by estrogen of the nucleoli in nerve cells of the rat hypothalamus. Cell Tissue Res., 235: 485–489.
Cohen, R. and Pfaff, D. (1981) Ultrastructure of neurons in the ventromedial nucleus of the hypothalamus in ovariectomized rats with or without estrogen treatment. Cell Tissue Res., 217: 451–470.
Collier, H.O. and Roy, A.C. (1974) Morphine-like drugs inhibit the stimulation of E prostaglandins of cyclic AMP formation by rat brain homogenate. Nature, 248: 24–27.
Costa, T., Lang, J., Gless, C. and Herz, A. (1990) Spontaneous association between opioid receptors and GTP-binding regulatory proteins in native membranes: specific regulation by antagonists and sodium ions. Mol. Pharmacol., 37: 383–394.
Crain, S.M. and Shen, K.F. (1990) Opioids can evoke direct receptor-mediated excitatory effects on sensory neurons. Trends Pharmacol. Sci., 11: 77–81.
Della Rocca, G.J., van Biesen, T., Daaka, Y., Luttrell, D.K., Luttrell, L.M. and Lefkowitz, R.J. (1997) Ras-dependent mitogen-activated protein kinase activation by G protein-coupled receptors. Convergence of Gi- and Gq-mediated pathways on calcium/calmodulin, Pyk2, and Src kinase. J. Biol. Chem., 272: 19125–19132.
Dellovade, T., Hardelin, J.P., Pfaff, D.W. and Schwanzel-Fukada, M. (2003) Anosmin-I immunoreactivity during embryogenesis in a primitive eutherian mammal. Dev. Brain Res., 140: 157–167.
de-Oliveira, C.V., Assumpcao, J.A., Confessor, Y.Q., Guertzenstein, P. and Cravo, S. (1996) Evidence for neuronal release of isotopically labelled glycine from the rat ventral medullary surface in vivo. Braz. J. Med. Biol. Res., 29: 527–532.
Dhawan, B.N., Cesselin, F., Raghubir, R., Reisine, T., Bradley, P.B., Portoghese, P.S. and Hamon, M. (1996) International Union of Pharmacology. XII. Classification of opioid receptors. Pharmacol. Rev., 48: 567–592.
Easton, A., Norton, J., Goodwillie, A. and Pfaff, D.W. (2004) Sex differences in mouse behavior following pyrilamine treatment: role of histamine 1 receptors in arousal. Pharmacol. Biochem. Behav., 79: 563–572.
Etgen, A.M. (2002) Estrogen regulation of neurotransmitter and growth factor signaling in the brain. In: Pfaff, D.W., Arnold, A., Etgen, A.M., Fahrbach, S. and Rubin, R. (Eds.), Hormones, Brain and Behavior. Academic Press, San Diego, CA, pp. 381–440.
Evinger, C., Kaneko, C.R. and Fuchs, A.F. (1982) Activity of omnipause neurons in alert cats during saccadic eye movements and visual stimuli. J. Neurophysiol., 47: 827–844.
Faber, D.S., Korn, H. and Lin, J.W. (1991) Role of medullary networks and postsynaptic membrane properties in regulating Mauthner cell responsiveness to sensory excitation. Brain Behav. Evol., 37: 286–297.
Fan, S.F. and Crain, S.M. (1995) Dual regulation by mu, delta and kappa opioid receptor agonists of K+ conductance of DRG neurons and neuroblastoma X DRG neuron hybrid F11 cells. Brain Res., 696: 97–105.

Fan, S.F., Shen, K.F. and Crain, S.M. (1991) Opioids at low concentration decrease openings of K+ channels in sensory ganglion neurons. Brain Res., 558: 166–170.

Ferguson, J., Aldag, J.M., Insel, T.R. and Young, L.J. (2001) Oxytocin in the medial amygdala is essential for social recognition in the mouse. J. Neurosci., 21: 8278–8285.

Ferguson, J., Young, L. and Insel, T. (2002) The neuroendocrine basis of social recognition. Front. Neuroendocrinol., 2: 200–224.

Ferguson, S.S. (2001) Evolving concepts in G protein-coupled receptor endocytosis: the role in receptor desensitization and signaling. Pharmacol. Rev., 53: 1–24.

Finn, A.K. and Whistler, J.L. (2001) Endocytosis of the mu opioid receptor reduces tolerance and a cellular hallmark of opiate withdrawal. Neuron, 32: 829–839.

Fioretti, B., Catacuzzeno, L., Tata, A.M. and Franciolini, F. (2004) Histamine activates a background, arachidonic acid-sensitive K channel in embryonic chick dorsal root ganglion neurons. Neuroscience, 125: 119–127.

Flanagan-Cato, L. (2000) Estrogen-induced remodeling of hypothalamic neural circuitry. Front. Neuroendocrinol., 4: 309–329.

Flanagan-Cato, L.M., Calizo, L.H. and Daniels, D. (2001) The synaptic organization of VMH neurons that mediate the effects of estrogen on sexual behavior. Horm. Behav., 40: 178–182.

Frankfurt, M. (1994) Gonadal steroids and neuronal plasticity: studies in the adult rat hypothalamus. Ann. NY Acad. Sci., 743: 45–60.

Frohlich, J., Morgan, Ogawa, S., Burton, L. and Pfaff, D. (2001) Statistical analysis of measures of arousal in ovariectomized female mice. Horm. Behav., 39: 39–47.

Fukuda, K., Kato, S., Morikawa, H., Shoda, T. and Mori, K. (1996) Functional coupling of the delta-, mu-, and kappa-opioid receptors to mitogen-activated protein kinase and arachidonate release in Chinese hamster ovary cells. J. Neurochem., 67: 1309–1316.

Gantz, I., DelValle, J., Wang, L.D., Tashiro, T., Munzert, G., Guo, Y.J., Konda, Y. and Yamada, T. (1992) Molecular basis for the interaction of histamine with the histamine H2 receptor. J. Biol. Chem., 267: 20840–20843.

Gantz, I., Munzert, G., Tashiro, T., Schaffer, M., Wang, L., DelValle, J. and Yamada, T. (1991) Molecular cloning of the human histamine H2 receptor. Biochem. Biophys. Res. Commun., 178: 1386–1392.

Garey, J., Goodwillie, A., Frohlich, J., Morgan, M., Gustafsson, J.-A., Smithies, O., Korach, K.S., Ogawa, S. and Pfaff, D.W. (2003) Genetic contributions to generalized arousal of brain and behavior. Proc. Natl. Acad. Sci. USA, 100: 11019–11022.

Garzon, J., Castro, M. and Sanchez-Blazquez, P. (1998) Influence of Gz and Gi2 transducer proteins in the affinity of opioid agonists to mu receptors. Eur. J. Neurosci., 10: 2557–2564.

Graham, R.M., Perez, D.M., Hwa, J. and Piascik, M.T. (1996) alpha1-adrenergic receptor subtypes. Molecular structure, function, and signaling. Circ. Res., 78: 737–749.

Greene, R.W. and Haas, H.L. (1990) Effects of histamine on dentate granule cells in vitro. Neuroscience, 34: 299–303.

Grevel, J., Yu, V. and Sadee, W. (1985) Characterization of a labile naloxone binding site (lambda site) in rat brain. J. Neurochem., 44: 1647–1656.

Gu, G., Varoqueaux, F. and Simerly, R.B. (1999) Hormonal regulation of glutamate receptor gene expression in the anteroventral periventricular nucleus of the hypothalamus. J. Neurosci., 19: 3213–3222.

Guyenet, P.G., Koshiya, N., Huangfu, D., Baraban, S.C., Stornetta, R.L. and Li, Y.-W. (1996) Role of medulla oblongata in generation of sympathetic and vagal outflows. Prog. Brain Res., 107: 127–144.

Guyenet, P.G. and Young, B.S. (1987) Projections of nucleus paragigantocellularis lateralis to locus coeruleus and other structures in rat. Brain Res., 406: 171–184.

Haas, H.L. and Greene, R.W. (1986) Effects of histamine on hippocampal pyramidal cells of the rat in vitro. Exp. Brain Res., 62: 123–130.

Haas, H.L. and Konnerth, A. (1983) Histamine and noradrenaline decrease calcium-activated potassium conductance in hippocampal pyramidal cells. Nature, 302: 432–434.

Han, C., Esbenshade, T.A. and Minneman, K.P. (1992) Subtypes of alpha 1-adrenoceptors in DDT1 MF-2 and BC3H-1 clonal cell lines. Eur. J. Pharmacol., 226: 141–148.

Harrison, L.M., Kastin, A.J. and Zadina, J.E. (1998) Opiate tolerance and dependence: receptors, G-proteins, and antiopiates. Peptides, 19: 1603–1630.

Haug, T. and Storm, J.F. (2000) Protein kinase A mediates the modulation of the slow $Ca(2+)$-dependent $K(+)$ current, I(sAHP), by the neuropeptides CRF, VIP, and CGRP in hippocampal pyramidal neurons. J. Neurophysiol., 83: 2071–2079.

Hawes, B.E., van Biesen, T., Koch, W.J., Luttrell, L.M. and Lefkowitz, R.J. (1995) Distinct pathways of Gi-and Gq-mediated mitogen-activated protein kinase activation. J. Biol. Chem., 270: 17148–17153.

Hawkins, K.N., Knapp, R.J., Gehlert, D.R., Lui, G.K., Yamamura, M.S., Roeske, L.C., Hruby, V.J. and Yamamura, H.I. (1988) Quantitative autoradiography of [3H]CTOP binding to mu opioid receptors in rat brain. Life Sci., 42: 2541–2551.

Hegstrand, L.R., Kanof, P.D. and Greengard, P. (1976) Histamine-sensitive adenylate cyclase in mammalian brain. Nature, 260: 163–165.

Hendry, I.A., Kelleher, K.L., Bartlett, S.E., Leck, K.J., Reynolds, A.J., Heydon, K., Mellick, A., Megirian, D. and Matthaei, K.I. (2000) Hypertolerance to morphine in G(z alpha)-deficient mice. Brain Res., 870: 10–19.

Horie, K., Itoh, H. and Tsujimoto, G. (1995) Hamster alpha 1B-adrenergic receptor directly activates Gs in the transfected Chinese hamster ovary cells. Mol. Pharmacol., 48: 392–400.

Horvitz, J.C., Stewart, T. and Jacobs, B.L. (1997) Burst activity of ventral tegmental dopamine neurons is elicited by sensory stimuli in the awake cat. Brain Res., 759: 251–258.

Huangfu, D., Verberne, A.J.M. and Guynet, P.G. (1992) Rostral ventrolateral medullary neurons projecting to locus coeruleus have cardiorespiratory inputs. Brain Res., 598: 67–75.

Hubscher, C.H. and Johnson, R.D. (1996) Responses of medullary reticular formation neurons to input from the male genitalia. J. Neurophysiol., 76: 2474–2482.

Inoue, M., Mishina, M. and Ueda, H. (2003) Locus-specific rescue of GluRepsilon1 NMDA receptors in mutant mice identifies the brain regions important for morphine tolerance and dependence. J. Neurosci., 23: 6529–6536.

Insel, T. and Young, L. (2000) Neuropeptides and the evolution of social behavior. Curr. Opin. Neurobiol., 10: 784–789.

Itoh, Y., Oishi, R., Nishibori, M. and Saeki, K. (1991) Characterization of histamine release from the rat hypothalamus as measured by in vivo microdialysis. J. Neurochem., 56: 769–774.

Jensen, E.V. and Jacobson, H.I. (1962) Basic guides to the mechanism of estrogen action. Recent Prog. Horm. Res., 18: 387–414.

Jones, K., C.A., H., Chikaraishi, D.M. and Pfaff, D.W. (1990) Steroid hormone regulation of ribosomal RNA in rat hypothalamus: early detection using in situ hybridization and precursor-product ribosomal DNA probes. J. Neurosci., 10: 1513–1521.

Kamp, T.J. and Hell, J.W. (2000) Regulation of cardiac L-type calcium channels by protein kinase A and protein kinase C. Circ. Res., 87: 1095–1102.

Kang, S.K., Kim, D.K., Damron, D.S., Baek, K.J. and Im, M. (2002) Modulation of intracellular Ca(2+) via alpha(1B)-adrenoceptor signaling molecules, G alpha(h) (transglutaminase II) and phospholipase C-delta 1. Biochem. Biophys. Res. Commun., 293: 383–390.

Kia, H.K., Yen, G., Krebs, C.J. and Pfaff, D.W. (2002) Co-localization of estrogen receptor alpha and NMDA-2D mRNAs in amygdaloid and hypothalamic nuclei of the mouse brain. Bran Res. Mol. Brain Res., 104: 47–54.

Kitchen, I., Slowe, S.J., Matthes, H.W. and Kieffer, B. (1997) Quantitative autoradiographic mapping of mu-, delta- and kappa-opioid receptors in knockout mice lacking the mu-opioid receptor gene. Brain Res., 778: 73–88.

Koch, T., Schulz, S., Schroder, H., Wolf, R., Raulf, E. and Hollt, V. (1998) Carboxyl-terminal splicing of the rat mu opioid receptor modulates agonist-mediated internalization and receptor resensitization. J. Biol. Chem., 273: 13652–13657.

Koshimizu, T., Tanoue, A., Hirasawa, A., Yamauchi, J. and Tsujimoto, G. (2003) Recent advances in alpha1-adrenoceptor pharmacology. Pharmacol. Ther., 98: 235–244.

Koshimizu, T., Yamauchi, J., Hirasawa, A., Tanoue, A. and Tsujimoto, G. (2002) Recent progress in alpha1-adrenoceptor pharmacology. Biol. Pharm. Bull., 25: 401–408.

Kow, L., Weesner, G. and Pfaff, D. (1992) Alpha 1-adrenergic agonists act on the ventromedial hypothalamus to cause neuronal excitation and lordosis facilitation: electrophysiological and behavioral evidence. Brain Res., 588: 237–245.

Kow, L.-M. and Pfaff, D.W. (1985) Estrogen effects on neuronal responsiveness to electrical and neurotransmitter stimulation: an in vitro study on the ventromedial nucleus of the hypothalamus. Brain Res., 347: 1–10.

Kow, L.-M. and Pfaff, D.W. (1987) Responses of ventromedial hypothalamic neurons in vitro to norepinephrine: dependence on dose and receptor type. Brain Res., 413: 220–228.

Kow, L.M., Tsai, Y.F., Weiland, N.G., McEwen, B.S. and Pfaff, D.W. (1995) In vitro electro-pharmacological and autoradiographic analyses of muscarinic receptor subtypes in rat hypothalamic ventromedial nucleus: implications for cholinergic regulation of lordosis. Brain Res., 694: 29–39.

Kramer, H.K. and Simon, E.J. (1999) Role of protein kinase C (PKC) in agonist-induced mu-opioid receptor down-regulation: II. Activation and involvement of the alpha, epsilon, and zeta isoforms of PKC. J. Neurochem., 72: 594–604.

Kramer, H.K. and Simon, E.J. (2000) mu and delta-opioid receptor agonists induce mitogen-activated protein kinase (MAPK) activation in the absence of receptor internalization. Neuropharmacology, 39: 1707–1719.

Kreda, S.M., Sumner, M., Fillo, S., Ribeiro, C.M., Luo, G.X., Xie, W., Daniel, K.W., Shears, S., Collins, S. and Wetsel, W.C. (2001) alpha(1)-adrenergic receptors mediate LH-releasing hormone secretion through phopholipases C and A(2) in immortalized hypothalamic neurons. Endocrinology, 142: 4839–4851.

Krezel, W., Dupont, S., Krust, A., Chambon, P. and Chapman, P.F. (2001) Increased anxiety and synaptic plasticity in estrogen receptor-deficient mice. Proc. Natl. Acad. Sci. USA, 98: 12278–12282.

Krupnick, J.G. and Benovic, J.L. (1998) The role of receptor kinases and arrestins in G protein-coupled receptor regulation. Ann. Rev. Pharmacol. Toxicol., 38: 289–319.

Kuiper, C.G., Enmark, E., Pelto-Huikko, M., Nilsson, S. and Gustafsson, J.A. (1996) Cloning of a novel receptor expressed in rat prostate and ovary. Proc. Natl. Acad. Sci. USA, 93: 5925–5930.

Kume, K., Kume, S., Park, S.K., Hirsh, J. and Jackson, F.R. (2005) Dopamine is a regulator of arousal in the fruit fly. J. Neurosci., 25: 7377–7384.

Lee, R.K. and Eaton, R.C. (1991) Culospinal neurons of the adult zebrafish, Brachydanio rerio. J. Comp. Neurol., 304: 34–52.

Lee, R.K., Eaton, R.C. and Zottoli, S.J. (1993) Segmental arrangement of reticulospinal neurons in the goldfish hindbrain. J. Comp. Neurol., 329: 539–556.

Legouis, R., Hardelin, J.P., Levilliers, J., Claverie, J.M., Compain, S., Wunderle, V., Millasseau, P., LePaslier, D., Cohen, D., Caterina, D., Bougueleret, L., Delemarre-Van de Waal, H., Lutfalla, G., Weissenbach, J. and Petit, C. (1991) The candidate gene for the X-linked Kallmann syndrome encodes a protein related to adhesion molecules. Cell, 67: 423–435.

Leopoldt, D., Harteneck, C. and Nurnberg, B. (1997) G proteins endogenously expressed in Sf 9 cells: interactions with mammalian histamine receptors. Naunyn Schmiedebergs Arch. Pharmacol., 356: 216–224.

Leung, C.G. and Mason, P.A. (1998) A physiological survey of medullary raphe and magnocellular reticular neurons in the anesthetized rat. J. Neurophysiol., 80: 1630–1646.

Leung, C.G. and Mason, P. (1999) Physiological properties of raphe magnus neurons during sleep and waking. J. Neurophysiol., 81: 584–595.

Leurs, R., Bakker, R.A., Timmerman, H. and de Esch, I.J. (2005) The histamine H3 receptor: from gene cloning to H3 receptor drugs. Nat. Rev. Drug Discov., 4: 107–120.

Leurs, R., Brozius, M.M., Jansen, W., Bast, A. and Timmerman, H. (1991) Histamine H1-receptor-mediated cyclic GMP production in guinea-pig lung tissue is an L-arginine-dependent process. Biochem. Pharmacol., 42: 271–277.

Leurs, R., Traiffort, E., Arrang, J.M., Tardivel-Lacombe, J., Ruat, M. and Schwartz, J.C. (1994) Guinea pig histamine H1 receptor. II. Stable expression in Chinese hamster ovary cells reveals the interaction with three major signal transduction pathways. J. Neurochem., 62: 519–527.

Li, Z. and Hatton, G.I. (1996) Histamine-induced prolonged depolarization in rat supraoptic neurons: G-protein-mediated, $Ca(2+)$-independent suppression of $K+$ leakage conductance. Neuroscience, 70: 145–158.

Liu, J.G. and Anand, K.J. (2001) Protein kinases modulate the cellular adaptations associated with opioid tolerance and dependence. Brain Res. Brain Res. Rev., 38: 1–19.

Lord, J.A., Waterfield, A.A., Hughes, J. and Kosterlitz, H.W. (1977) Endogenous opioid peptides: multiple agonists and receptors. Nature, 267: 495–499.

Lou, L.G. and Pei, G. (1997) Modulation of protein kinase C and cAMP-dependent protein kinase by delta-opioid. Biochem. Biophys. Res. Commun., 236: 626–629.

Lovenberg, T.W., Roland, B.L., Wilson, S.J., Jiang, X., Pyati, J., Huvar, A., Jackson, M.R. and Erlander, M.G. (1999) Cloning and functional expression of the human histamine H3 receptor. Mol. Pharmacol., 55: 1101–1107.

Lydon, J.P., DeMayo, F.J., Fun, C.R., Mani, S.K., Hughes, A.R., Montgomery, C.A.J., Shyamala, G., Conneely, O.M. and O'Malley, B.W. (1995) Mice lacking progesterone receptor exhibit pleiotropic reproductive abnormalities. Genes Dev., 9: 2266–2278.

MacColl, G., Quinton, R. and Bouloux, P.M.G. (2002) GnRH neuronal development: insights into the mechanisms of hypogonadotropic hypogonadism. Trends Endocrinol. Met., 13: 112–118.

Mani, S.K., Blaustein, J.D., Allen, J.M., Law, S.W., O'Malley, B.W. and Clark, J.H. (1994) Inhibition of rat sexual behavior by antisense oligonucleotides to the progesterone receptor. Endocrinology, 135: 1409–1414.

Mansour, A., Khachaturian, H., Lewis, M.E., Akil, H. and Watson, S.J. (1987) Autoradiographic differentiation of mu, delta, and kappa opioid receptors in the rat forebrain and midbrain. J. Neurosci., 7: 2445–2464.

Marek, P., Ben-Eliyahu, S., Gold, M. and Liebeskind, J.C. (1991) Excitatory amino acid antagonists (kynurenic acid and MK-801) attenuate the development of morphine tolerance in the rat. Brain Res., 547: 77–81.

Martin, W.R., Eades, C.G., Thompson, J.A., Huppler, R.E. and Gilbert, P.E. (1976) The effects of morphine- and nalorphine- like drugs in the nondependent and morphine-dependent chronic spinal dog. J. Pharmacol. Exp. Ther., 197: 517–532.

McCarthy, M., Chung, S., Ogawa, S. and Pfaff, D.W. (1991) Behavioral effects of oxytocin: is there a unifying principle? In: Jard, J.R. (Ed.) Vasopressin, Vol. 208. Colloque INSERM/John Libbey Eurotext Ltd, New York, pp. 195–212.

McCarthy, M.M., McDonald, C.H., Brooks, P.J. and Goldman, D. (1996) An anxiolytic action of oxytocin is enhanced by estrogen in the mouse. Physiol. Behav., 60: 1209–1215.

McCormick, D.A. and Williamson, A. (1991) Modulation of neuronal firing mode in cat and guinea pig LGNd by histamine: possible cellular mechanisms of histaminergic control of arousal. J. Neurosci., 11: 3188–3199.

Menard, C.S. and Dohanich, G.P. (1990) Physostigmine facilitation of lordosis in naturally cycling female rats. Pharmacol. Biochem. Behav., 36: 853–858.

Meunier, J.C., Mollereau, C., Toll, L., Suaudeau, C., Moisand, C., Alvinerie, P., Butour, J.L., Guillemot, J.C., Ferrara, P. and Monsarrat, B. (1995) Isolation and structure of the endogenous agonist of opioid receptor-like ORL1 receptor. Nature, 377: 532–535.

Michelotti, G.A., Price, D.T. and Schwinn, D.A. (2000) Alpha1-adrenergic receptor regulation: basic science and clinical implications. Pharmacol. Ther., 88: 281–309.

Minami, M. and Satoh, M. (1995) Molecular biology of the opioid receptors: structures, functions and distributions. Neurosci. Res., 23: 121–145.

Minneman, K.P. (1988) Alpha1-adrenergic receptor subtypes, inositol phosphates, and sources of cell Ca^{2+}. Pharmacol. Rev., 40: 87–119.

Molina-Hernandez, A., Nunez, A., Sierra, J.J. and Arias-Montano, J.A. (2001) Histamine H3 receptor activation inhibits glutamate release from rat striatal synaptosomes. Neuropharmacology, 41: 928–934.

Mollereau, C., Paramentier, M., Mailleux, P., Butour, J.L., Moisand, C., Chalon, P., Caput, D., Vassart, G. and Meunier, J.C. (1994) ORL1, a novel member of the opioid receptor family cloning, functional expression and localization. FEBS Lett., 341: 33–38.

Mong, J.A., Devidze, N., Frail, D.E., O'Connor, L.T., Samuel, M., Choleris, E., Ogawa, S. and Pfaff, D.W. (2003a) Estradiol differentially regulates lipocalin-type prostaglandin D synthase transcript levels in the rodent brain: evidence from high-density oligonucleotide arrays and in situ hybridization. Proc. Natl. Acad. Sci. USA, 100: 318–323.

Mong, J.A., Devidze, N., Goodwillie, A. and Pfaff, D.W. (2003b) Reduction of lipocalin-type prostaglandin D synthase in the preoptic area of female mice mimics estradiol effects on arousal and sex behavior. Proc. Natl. Acad. Sci. USA, 100: 15206–15211.

Mong, J.A. and Pfaff, D.W. (2004) Hormonal symphony: steroid orchestration of gene modules of sociosexual behaviors. Mol. Psychiatr., 9: 550–556.

Moss, R. and McCann, S. (1973) Induction of mating behavior in rats by luteinizing hormone-releasing factor. Science, 181: 177–179.

Mulkey, D.K., Stornetta, R.L., Weston, M.C., Simmons, J.R., Parker, A., Bayliss, D.A. and Guynet, P.G. (2004) Respiratory control by ventral surface chemoreceptor neurons in rats. Nat. Neurosci., 7: 1360–1369.

Munakata, M. and Akaike, N. (1994) Regulation of K^+ conductance by histamine H1 and H2 receptors in neurones dissociated from rat neostriatum. J. Physiol., 480(Pt 2): 233–245.

Nicot, A., Ogawa, S., Berman, Y., Carr, K.D. and Pfaff, D.W. (1997) Effects of an intrahypothalamic injection of antisense oligonucleotides for preproenkephalin mRNA in female rats: evidence for opioid involvement in lordosis reflex. Brain Res., 777: 60–68.

Nilsson, S. and Gustafsson, J.A. (2002) Biological role of estrogen and estrogen receptors. Crit. Rev. Biochem. Mol. Biol., 37: 1–28.

Noble, F. and Roques, B.P. (1999) CCK-B receptor: chemistry, molecular biology, biochemistry and pharmacology. Prog. Neurobiol., 58: 348–379.

Nomura, M., Durback, L., Chan, J., Gustafsson, J.-A., Smithies, O., Korach, K.S., Pfaff, D.W. and Ogawa, S. (2002a) Genotype/age interactions on aggressive behavior in gonadally intact estrogen receptor knockout (betaERKO) male mice. Horm. Behav., 41: 288–296.

Nomura, M., McKenna, E., Korach, K.S., Pfaff, D.W. and Ogawa, S. (2002b) Estrogen receptor-beta regulates transcript levels for oxytocin and arginine vasopressin in the hypothalamic paraventricular nucleus of male mice. Brain. Res. Mol. Brain. Res., 109: 84–94.

Ogawa, S., Olazabal, U.E., Parhar, I.S. and Pfaff, D.W. (1994) Effects of intrahypothalamic administration of antisense DNA for progesterone receptor mRNA on reproductive behavior and progesterone receptor immunoreactivity in female rat. J. Neurosci., 14: 1766–1774.

Palacios, J.M., Wamsley, J.K. and Kuhar, M.J. (1981) The distribution of histamine H1-receptors in the rat brain: an autoradiographic study. Neuroscience, 6: 15–37.

Pamidimukkala, J., Xue, B., Newton, L.G., Lubahn, D.B. and Hay, M. (2005) Estrogen receptor-alpha mediates estrogen facilitation of baroreflex heart rate responses in conscious mice. Am. J. Physiol. Heart Circ. Physiol., 288: H1063–H1070.

Pare, M. and Guitton, D. (1998) Brain stem omnipause neurons and the control of combined eye-head gaze saccades in the alert cat. J. Neurophysiol., 79: 3060–3076.

Pasternak, G.W. (2004) Multiple opiate receptors: deja vu all over again. Neuropharmacology, 47: 312–323.

Pasternak, G.W. (2005) Molecular biology of opioid analgesia. J. Pain Symptom Manage., 29: 2–9.

Pasternak, G.W. and Wood, P.J. (1986) Multiple mu opiate receptors. Life Sci., 38: 1889–1898.

Payne, G.W. and Neuman, R.S. (1997) Effects of hypomagnesia on histamine H1 receptor-mediated facilitation of NMDA responses. Br. J. Pharmacol., 121: 199–204.

Pedarzani, P. and Storm, J.F. (1996) Interaction between alpha- and beta-adrenergic receptor agonists modulating the slow $Ca(2+)$-activated $K+$ current IAHP in hippocampal neurons. Eur. J. Neurosci., 8: 2098–2110.

Perez, D.M., DeYoung, M.B. and Graham, R.M. (1993) Coupling of expressed alpha 1B- and alpha 1D-adrenergic receptors to multiple signaling pathways is both G protein and cell type specific. Mol. Pharmacol., 44: 784–795.

Petersen, S.L., Keller, M.L., Carder, S.A. and McCrone, S. (1993) Differential effects of estrogen and progesterone on levels of POMC mRNA levels in the arcuate nucleus: relationship to the timing of LH surge release. J. Neuroendocrinol., 5: 643–648.

Petersen, S.L. and LaFlamme, K.D. (1997) Progesterone increases levels of mu-opioid receptor mRNA in the preoptic area and arcuate nucleus of ovariectomized, estradiol-treated female rats. Mol. Brain Res., 52: 32–37.

Petersen, S.L., McCrone, S., Keller, M. and Shores, S. (1995) Effects of estrogen and progesterone on luteinizing hormone-released hormone messenger ribonucleic acid levels: consideration of temporal and neuroanatomical variables. Endocrinology, 136: 3604–3610.

Petitti, N. and Etgen, A.M. (1991) Protein kinase C and phospholipase C mediate alpha1- and beta-adrenoceptor intercommunication in rat hypothalamic slices. J. Neurochem., 56: 628–635.

Petitti, N., Karkanias, G.B. and Etgen, A.M. (1992) Estradiol selectively regulates alpha 1B-noradrenergic receptors in the hypothalamus and preoptic area. J. Neurosci., 12: 3869–3876.

Pfaff, D. (1973) Luteinizing hormone releasing factor (LRF) potentiates lordosis behavior in hypophysectomized ovariectomized female rats. Science, 182: 1148–1149.

Pfaff, D., Choleris, E. and Ogawa, S. (2005) Genes for sex hormone receptors controlling mouse aggression. In: Bock, G. and Goode, J. (Eds.), Molecular Mechanisms Influencing Aggressive Behaviours, No. 268. Wiley, Hoboken, NJ, pp. 78–95.

Pfaff, D.W. (1980) Estrogens and Brain Function: Neural Analysis of a Hormone-Controlled Mammalian Reproductive Behavior. Springer, New York.

Pfaff, D. (1999) Drive: Neurobiological and Molecular Mechanisms of Sexual Motivation. The MIT Press, Cambridge.

Pfaff, D.W. (2005) Brain Arousal and Information Theory: Neural, Hormonal, and Genetic Analysis. Harvard University Press, Cambridge.

Pfaff, D.W., Sakuma, Y., Kow, L.-M., Lee, A.W. and Easton, A. (2006) Hormonal, neural, and genomic mechanisms for female reproductive behaviors, motivation, and arousal. In: Neill, J.D. (Ed.), Knobil and Neill's Physiology of Reproduction. 3rd ed., Vol. 1. Elsevier, Boston, MA, pp. 1825–1920.

Pfaff, D.W., Schwartz-Giblin, S., McCarthy, M.M. and Kow, L.-M. (1994) Cellular and molecular mechanisms of female reproductive behaviors. In: Knobil, E. and Neill, J. (Eds.), The Physiology of Reproduction. Raven, New York, pp. 107–220.

Pfaus, J. and Gorzalka, B. (1987a) Opioid and sexual behavior. Neurosci. Biobehav. Rev., 11: 1–34.

Pfaus, J. and Gorzalka, B.B. (1987b) Selective activation of opioid receptors differentially affects lordosis behavior in female rats. Peptides, 8: 309–317.

Pfaus, J. and Pfaff, D. (1992) Mu-, delta- and kappa-opioid receptor agonists selectively modulate sexual behaviors in the female rat: differential dependence on progesterone. Horm. Behav., 26: 457–473.

Phillips, J.O., Ling, L. and Fuchs, A.F. (1999) Action of the brain stem saccade generator during horizontal gaze shifts. I. Discharge patterns of the omnidirectional pause neurons. J. Neurophysiol., 81: 1284–1295.

Piascik, M.T. and Perez, D.M. (2001) Alpha1-adrenergic receptors: new insights and directions. J. Pharmacol. Exp. Ther., 298: 403–410.

Pollard, H., Moreau, J., Arrang, J.M. and Schwartz, J.C. (1993) A detailed autoradiographic mapping of histamine H3 receptors in rat brain areas. Neuroscience, 52: 169–189.

Pollio, G., Xue, P., Zanisi, M., Nicolin, A. and Maggi, A. (1993) Antisense oligonucleotide blocks progesterone-induced lordosis behavior in ovariectomized rats. Brain Res. Mol. Brain Res., 19: 135–139.

Portoghese, P.S., Sultana, M., Nelson, W.L., Klein, P. and Takkemori, A.E. (1992) Delta opioid antagonist activity and binding studies of regioisomeric isothiocyanate derivatives of naltrindole: evidence for delta receptor subtypes. J. Med. Chem., 35: 4086–4091.

Priest, C.A., Eckersell, C.B. and Micevych, P.E. (1995) Estrogen regulates preproenkephalin-A mRNA levels in the rat ventromedial nucleus: temporal and cellular aspects. Mol. Brain Res., 28: 251–262.

Pyner, S. and Coote, J.H. (1999) Identification of an efferent projection from the paraventricular nucleus of the hypothalamus terminating close to spinally projecting rostral ventrolateral medullary neurons. Neuroscience, 88: 949–957.

Quinones-Jenab, V., Jenab, S., Ogawa, S., Inturrisi, C. and Pfaff, D.W. (1997) Estrogen regulation of mu-opioid receptor mRNA in the forebrain of female rats. Mol. Brain Res., 47: 134–138.

Quock, R.M., Burkey, T.H., Varga, E., Hosohata, Y., Cowell, S.M., Slate, C.A., Ehlert, F.J., Roeske, W.R. and Yamamura, H.I. (1999) The delta-opioid receptor: molecular pharmacology, signal transduction, and the determination of drug efficacy. Pharmacol. Rev., 51: 503–532.

Reiner, P.B. and Kamondi, A. (1994) Mechanisms of antihistamine-induced sedation in the human brain: H1 receptor activation reduces a background leakage potassium current. Neuroscience, 59: 579–588.

Reinscheid, R.K., Nothacker, H.P., Bourson, A., Ardati, A., Henningsen, R.A., Bunzow, J.R., Grandy, D.K., Langen, H., Monsma, F.J.J. and Civelli, O. (1995) Orphanin FQ: a neuropeptide that activates an opioidlike G protein-coupled receptor. Science, 270: 792–794.

Remmers, A.E., Clark, M.J., Liu, X.Y. and Medzihradsky, F. (1998) Delta opioid receptor down-regulation is independent of functional G protein yet is dependent on agonist efficacy. J. Pharmacol. Exp. Ther., 287: 625–632.

Richelson, E. (1978) Histamine H1 receptor-mediated guanosine 3′, 5′-monophosphate formation by cultured mouse neuroblastoma cells. Science, 201: 69–71.

Romano, G.J., Harlan, R.E., Shives, B.D., Howells, R.D. and Pfaff, D.W. (1988) Estrogen increases proenkephalin messenger ribonucleic acid levels in the ventromedial hypothalamus of the rat. Mol. Endocrinol., 2: 1320–1328.

Rossmanith, W.G., Marks, D.L., Clifton, D.K. and Steiner, R.A. (1996) Induction of galanin mRNA in GnRH neurons by estradiol and its facilitation by progesterone. J. Neuroendocrinol., 8: 185–191.

Ruan, Y., Kan, H., Parmentier, J., Fatima, S., Allen, L.F. and Malik, K.U. (1998) Alpha-1A adrenergic receptor stimulation with phenylephrine promotes arachidonic acid release by activation of phospholipase D in Rat-1 fibroblasts: inhibition by protein kinase A. J. Pharm. Exp. Ther., 284: 576–585.

Ruat, M., Traiffort, E., Bouthenet, M.L., Schwartz, J.C., Hirschfeld, J., Buschauer, A. and Schunack, W. (1990) Reversible and irreversible labeling and autoradiographic localization of the cerebral histamine H2 receptor using [125I]iodinated probes. Proc. Natl. Acad. Sci. USA, 87: 1658–1662.

Schreihofer, A.M. and Guyenet, P. (2003) Baro-activated neurons with pulse-modulated activity in the rat caudal ventrolateral medulla express GAD67 mRNA. J. Neurophysiol., 89: 1265–1277.

Schwanzel-Fukada, M., Bick, D. and Pfaff, D.W. (1989) Luteinizing hormone-releasing hormone (LHRH)-expressing cells do not migrate normally in an inherited hypogonadal (Kallmann) syndrome. Mol. Brain Res., 6: 311–326.

Schwanzel-Fukada, M. and Pfaff, D.W. (1989) Origin of luteinizing hormone-releasing hormone neurons. Nature, 338: 161–164.

Shapiro, R.A., Scherer, N.M., Habecker, B.A., Subers, E.M. and Nathanson, N.M. (1988) Isolation, sequence, and functional expression of the mouse M1 muscarinic acetylcholine receptor gene. J. Biol. Chem., 263: 18397–18403.

Shughrue, P.J., Lane, M.V. and Merchanthaler, I. (1997) Comparative distribution of estrogen receptor-alpha and -beta RNA in the rat central nervous system. J. Comp. Neurol., 388: 507–525.

Silver, R.B., Poonwasi, K.S., Seyedi, N., Wilson, S.J., Lovenberg, T.W. and Levi, R. (2002) Decreased intracellular calcium mediates the histamine H3-receptor-induced attenuation of norepinephrine exocytosis from cardiac sympathetic nerve endings. Proc. Natl. Acad. Sci. USA, 99: 501–506.

Smith, B.N. and Armstrong, W.E. (1996) The ionic dependence of the histamine-induced depolarization of vasopressin neurones in the rat supraoptic nucleus. J. Physiol., 495(Pt 2): 465–478.

Standifer, K.M. and Pasternak, G.W. (1997) G proteins and opioid receptor-mediated signalling. Cell Sigal., 9: 237–248.

Suzuki, R., Rygh, L.J. and Dickenson, A.H. (2004) Bad news from the brain: descending 5-HT pathways that control spinal pain processing. Trends Pharmacol. Sci., 25: 613–617.

Thomas, A., Shughrue, P.J., Merchanthaler, I. and Amico, J.A. (1999) The effects of progesterone on oxytocin mRNA levels in paraventricular nucleus of the female rat can be altered by the administration of Diazepam or RU486. J. Neuroendocrinol., 11: 137–144.

Traiffort, E., Pollard, H., Moreau, J., Ruat, M., Schwartz, J.C., Martinez-Mir, M.I. and Palacios, J.M. (1992) Pharmacological characterization and autoradiographic localization of histamine H2 receptors in human brain identified with [125I]iodoaminopotentidine. J. Neurochem., 59: 290–299.

Traynor, J.R. and Elliott, J. (1993) delta-opioid receptor subtypes and cross-talk with mu-receptors. Trends Pharmacol. Sci., 14: 84–86.

Trujillo, K.A. (2000) Are NMDA receptors involved in opiate-induced neural and behavioral plasticity? A review of preclinical studies. Psychopharmacol. (Berl.), 151: 121–141.

Trujillo, K.A. and Akil, H. (1991) Inhibition of morphine tolerance and dependence by the NMDA receptor antagonist MK-801. Science, 251: 85–87.

Tsao, P. and von Zastrow, M. (2000) Downregulation of G protein-coupled receptors. Curr. Opin. Neurobiol., 10: 365–369.

Ueno, R., Ishikawa, Y., Nakayama, T. and Hayaishi, O. (1982) Prostaglandin D2 induces sleep when microinjected into the preoptic area of conscious rats. Biochem. Biophys. Res. Commun., 576–582.

Unda, R., Brann, D.W. and Mahesh, V.B. (1995) Progesterone suppression of glutamic acid decarboxylase (GAD67) mRNA levels in the preoptic area: correlation to luteinizing hormone surge. Neuroendocrinology, 62: 562–570.

van der Zee, L., Nelemans, A. and den Hertog, A. (1995) Arachidonic acid is functioning as a second messenger in activating the Ca^{2+} entry process on H1-histaminoceptor stimulation in DDT1 MF-2 cells. Biochem. J., 305(Pt 3): 859–864.

van Swinderen, B. and Greenspan, R.J. (2003) Salience modulates 20–30 Hz brain activity in Drosophila. Nat. Neurosci., 6: 570–586.

Vasudevan, N., Zhu, Y.S., Daniel, S., Koibuchi, N., Chinn, W.W. and Pfaff, D.W. (2001) Crosstalk between oestrogen receptors and thyroid hormone receptor isoforms results in differential regulation of the preproenkephalin gene. J. Neuroendocrinol., 13: 779–790.

Vizuete, M.L., Traiffort, E., Bouthenet, M.L., Ruat, M., Souil, E., Tardivel-Lacombe, J. and Schwartz, J.C. (1997) Detailed mapping of the histamine H2 receptor and its gene transcripts in guinea-pig brain. Neuroscience, 80: 321–343.

Wagner, E.J., Ronnekleiv, O.K. and Kelly, M.J. (2001) The noradrenergic inhibition of an apamin-sensitive, small-conductance Ca^{2+}-activated K^+ channel in hypothalamic gamma-aminobutyric acid neurons: pharmacology, estrogen sensitivity, and relevance to the control of the reproductive axis. J. Pharm. Exp. Ther., 299: 21–30.

Wahlestedt, C., Salmi, P., Good, L., Kela, J., Johnsson, T., Hokfelt, T., Broberger, C., Porreca, F., Lai, J., Ren, K., Ossipov, M., Koshkin, A., Jakobsen, N., Skouv, J., Oerum, H., Jacobson, M.H. and Wengel, J. (2000) Potent and nontoxic antisense oligonucleotides containing locked nucleic acids. Proc. Natl. Acad. Sci. USA, 97: 5633–5638.

Waldhoer, M., Bartlett, S.E. and Whistler, J.L. (2004) Opioid receptors. Ann. Rev. Biochem., 73: 953–990.

Walter, P., Green, S., Greene, G., Krust, A., Bornert, J.-M., Jeltsch, J.-M., Staub, A., Jensen, E.V., Scrace, G., Waterfield, M. and Chambon, P. (1985) Cloning of the human estrogen receptor cDNA. Proc. Natl. Acad. Sci. USA, 82: 7889–7893.

Wang, J.B., Johnson, P.B., Imai, Y., Persico, A.M., Ozenberger, B.A., Epler, C.M. and Uhl, G.R. (1994a) cDNA cloning of an orphan opiate receptor gene family member and its splice variant. FEBS Lett., 348: 75–79.

Wang, J.F., Shun, X.J., Yang, H.F., Ren, M.F. and Han, J.S. (1993) Suppression by [D-Pen2, D-Pen5]enkephalin on cyclic AMP dependent protein kinase-induced, but not protein kinase C-induced increment of intracellular free calcium in NG108-15 cells. Life Sci., 52: 1519–1525.

Wang, Z., Bilsky, BE.J., Porreca, F. and Sadee, W. (1994b) Constitutive mu opioid receptor activation as a regulatory mechanism underlying narcotic tolerance and dependence. Life Sci., 54: L339–L350.

Watanabe, T., Inoue, S., Hiroi, H., Orimo, A. and Muramatsu, M. (1999) NMDA receptor type 2D gene as target for estrogen receptor in the brain. Brain Res. Mol. Brain Res., 63: 375–379.

Weiger, T., Stevens, D.R., Wunder, L. and Haas, H.L. (1997) Histamine H1 receptors in C6 glial cells are coupled to calcium-dependent potassium channels via release of calcium from internal stores. Naunyn Schmiedebergs Arch. Pharmacol., 355: 559–565.

Weston, M.C., Stornetta, R.L. and Guyenet, P.G. (2004) Glutamatergic neuronal projections from the marginal layer of the rostral ventral medulla to the respiratory centers in rats. J. Comp. Neurol., 473: 73–85.

Whistler, J.L., Enquist, J., Marley, A., Fong, J., Gladher, F., Tsuruda, P., Murray, S.R. and von Zastrow, M. (2002) Modulation of postendocytic sorting of G protein-coupled receptors. Science, 297: 615–620.

Wolleman, M., Benyhe, S. and Simon, J. (1993) The kappa-opioid receptor: evidence for the different subtype. Life Sci., 52: 599–611.

Woolley, C.S. and Cohen, R.S. (2002) Sex steroids and neuronal growth in adulthood. In: Pfaff, D.W., Arnold, A., Etgen, A., Fahrbach, S. and Rubin, R. (Eds.), Hormones, Brain and Behavior. Academic Press, San Diego, CA, pp. 7171–7777.

Wuster, M., Schulz, R. and Herz, A. (1979) Specificity of opioids towards the mu-, delta- and epsilon-opiate receptors. Neurosci. Lett., 15: 193–198.

Xu, M., Urban, J.H., Hill, J.W. and Levine, J.E. (2000) Regulation of hypothalamic neuropeptide Y1 receptor gene expression during the estrous cycle: role of progesterone receptors. Endocrinology, 141: 3319–3327.

Yamashita, M., Fukui, H., Sugama, K., Horio, Y., Ito, S., Mizuguchi, H. and Wada, H. (1991) Expression cloning of a cDNA encoding the bovine histamine H1 receptor. Proc. Natl. Acad. Sci. USA, 88: 11515–11519.

Yang, Z., Bertram, D. and Coote, J.H. (2001) The role of glutamate and vasopressin in the excitation of RVL neurones by paraventricular neurones. Brain Res., 908: 99–103.

Yang, Z. and Coote, J.H. (1999) The influence of the paraventricular nucleus on baroreceptor dependent caudal ventrolateral medullary neurones of the rat. Eur. J. Physiol., 438: 47–52.

Young, J.K., Wu, M., Manaye, K.F., Ke, P., Allard, J.S., Mack, S.O. and Haxhiu, M.A. (2005) Orexin stimulates breathing via medullary and spinal pathways. J. Appl. Physiol., 98: 1387–1395.

Zagan, I.S., Gibo, D.M. and McLaughlin, P.J. (1991) Zeta (zeta), a growth-related opioid receptor in developing rat cerebellum: identification and characterization. Brain Res., 551: 28–35.

Zaki, P.A., Keith Jr., D.E., Thomas, J.B., Carroll, F.I. and Evans, C.J. (2001) Agonist-, antagonist-, and inverse agonist-regulated trafficking of the delta-opioid receptor correlates with, but does not require, G protein activation. J. Pharmacol. Exp. Ther., 298: 1015–1020.

Zhong, H. and Minneman, K.P. (1999) Differential activation of mitogen-activated protein kinase pathways in PC12 cells by closely related alpha1-adrenergic receptor subtypes. J. Neurochem., 72: 2388–2396.

Zhu, Y.-S., Cai, L.Q., You, X., Duan, Y., Imperato-McGinley, J., Chin, W.W. and Pfaff, D.W. (2001) Molecular analysis of estrogen induction of preproenkephalin gene expression and its modulation by thyroid hormones. Mol. Brain Res., 91: 23–33.

Zottoli, S.J. and Faber, D.S. (2000) The Mauthner cell: what has it taught us? Neuroscientist, 6: 25–38.

SECTION III

Future Directions

CHAPTER 13

Implications for the practice of psychiatry

Elisabeth B. Binder[1,2,*] and Charles B. Nemeroff[1]

[1]*Department of Psychiatry and Behavioral Sciences, Emory University School of Medicine, 101 Woodruff Circle, Suite 4000, Atlanta, GA 30322, USA*
[2]*Department of Human Genetics, Emory University School of Medicine, Atlanta, GA 30322, USA*

Introduction

Compared to some other medical specialties, psychiatric medicine is still hampered by the lack of objective, molecular or biological criteria for diagnosis or treatment evaluation. Although the use of biomarkers is a routine practice for certain other medical disciplines, psychiatrists still have to solely rely on relatively subjective-symptom severity rating scales and on broad diagnostic criteria for therapeutic decisions. The fact that the pathophysiology of the major psychiatric disorders and the mechanism of action of the effective treatments remain largely obscure, render it even more difficult to apply molecular diagnostic strategies that have been successful in other areas of medicine to psychiatry. However, it is important to recognize that much progress has been made in this area, including several attempts to identify biological markers in psychiatric disorders. Disease-related changes in the concentration of a variety of neurotransmitters and their metabolites in the cerebrospinal fluid (CSF) of psychiatric patients have been sought to identify potential biological markers. For example CSF concentrations of the neuropeptide neurotensin (NT) have repeatedly been reported to be reduced in subgroups of schizophrenic patients, and after effective antipsychotic treatment, CSF NT concentrations return to normal levels (Binder et al., 2001 for review). Others have used neuroendocrine challenge tests as measures of CNS dysfunction such as the dexamathasone suppression or combined dexamethasone suppression/CRH stimulation test as putative biological markers. Hyperactivity of the hypothalamus–pituitary–adrenal (HPA) axis in depression and its normalization with clinical remission is one of the most-replicated finding in biological psychiatry (Nemeroff, 1996; Holsboer, 2000). In the last several years, functional brain-imaging studies focusing on limbic–cortical network models have emerged as a promising approach to identify biological markers that could improve diagnostic accuracy and guide treatment selection for individual patients (Mayberg, 2003).

Despite 30–40 years of intense research on a series of promising markers, none have thus far been validated as diagnostic tools or predictors of treatment response. In addition, most of the past approaches were hypothesis-driven, relying on our limited knowledge of the pathophysiology of psychiatric disorders. With the sequence of the human genome being publicly available since February 2001 (Lander et al., 2001; Venter et al., 2001), an array of novel research tools has become available that may yield unbiased, hypothesis-free insight into the pathophysiological underpinnings of certain psychiatric disorders. These novel tools combine knowledge of the sequence of the human genome with miniaturized assays amenable for high-throughput processing for a parallel analysis of the whole genome. Using these, one can investigate the whole genome at the level of the DNA

*Corresponding author. E-mail: ebinder@genetics.emory.com

(genomics), all expressed mRNA (expressomics or more commonly expression array or microarray analysis) and all proteins (proteomics) in a single experiment. These three approaches have to deal with increasing levels of complexity, because the approximately 25,000 predicted human genes are expected to give rise to at least 10 times as many protein isoforms. Using these unbiased whole-genome-based approaches, novel pathways and molecules involved in the pathogenesis of psychiatric disorders may be identified. This knowledge may then help to increase diagnostic specificity by creating more homogeneous diagnostic subtypes based on biological similarities, to predict treatment response and to develop, new, more effective and individualized psychiatric treatments.

Proteomics

Although DNA is in general stable throughout life and the same in all cells, mRNA expression is dependent on the cell type and its level of activity. In addition, mRNA can be modified after transcription through alternate splicing to create isoforms containing different sets of the coding sequence all derived from the same gene. Proteins are subjected to even more modifications. Precursor proteins can be cleaved to a set of different peptides in a cell type-specific manner. Most proteins are modified by glycosylation or phosphorylation, steps required for their optimal activity. Proteins also often become functional only when associating with other proteins to form protein complexes; their own function can often be determined by the composition of the complex they are part of. Proteomics, thus involves the analysis of the complete pattern of the expressed proteins and their post-translational modifications in a cell or tissue. Comparison of the proteome of diseased and healthy tissue may represent a very powerful method to unravel the pathogenesis of disease, to identify therapeutic targets and to develop diagnostic tests. Compared to expressomics and genomics, proteomics measures the entities most directly related to cell function. Nonetheless, this method is the most technically difficult one (see other chapters for more detail).

Difficulties in psychiatric proteomics research

One of the main limitations of proteomics is its dependence on sufficient quantities of starting material. Unlike DNA or RNA, which utilize amplification techniques such as PCR that can almost indefinitely multiply starting material, no such method exist for proteins. Due to the relatively limited amount of protein in tissue and other body samples, the methods employed for protein identification and analysis of post-translational modifications need to be very sensitive. In recent years, the ability to measure ultra-low concentrations of protein has been achieved, largely due to advances in the mass spectrometry (MS) analysis of peptides and proteins (see other chapters for more detail).

As noted in the Introduction, the proteome is dynamic and dependent on the specific environment. To obtain meaningful results for psychiatric research, one has to carefully consider the target tissue. Except for rare biopsy material, brain tissue from living psychiatric patients is inaccessible, so that researchers have to resort to alternatives for proteomics research, such as animal brain tissue, postmortem human brain tissue, CSF or peripheral cells, such as peripheral blood monocytes (PBMCs). In the following section, we will briefly discuss the advantages and disadvantages of each of these approaches. Several of these issues not only relate to proteomics research but also apply to gene-expression studies.

Animal brain tissue
Using brain tissue from animal models of psychiatric disorders has the advantage that brain tissue can be collected from the region of interest in different experimental groups for which all conditions except the investigated one can be kept similar. This also includes the genome by the use of inbred strains. Optimal animal models for any psychiatric disorder, however, have not been developed as of yet, undoubtedly at least partly due to the intrinsically human nature of these complex behavioral phenotypes. One approach has been to focus on certain behavioral or endocrinological dimensions of these disorders that can be modeled in animals, but do not necessarily represent the

complexity of the disorder in humans. In addition, neurons within a given brain region exhibit a very heterogeneous expression of neurotransmitters, receptors and connections to other brain regions, likely leading to differential alterations of protein expression and modifications in neighboring neurons after exposure to the same stimulus. To obtain meaningful data, subtypes of neurons may have to be studied. This represents a great challenge for proteomics because this technique remains very dependent on sufficient sample quantities. Approaches such as laser-capture microdissection (LCM) have been used for proteomic approaches (Mouledous et al., 2003a, b). However, the required minimum number of 10^5–10^6 cells for a single experiment may require up to 40 h of dissection (Kim et al., 2004). In addition, the detection of less-abundant proteins usually requires microgram quantities of starting material (Freeman and Hemby, 2004) rendering LCM a less than optimal solution to increase cell-specificity for proteomics studies.

Postmortem human brain tissue
Postmortem human brain tissue has the advantage that we can indeed examine the proteome of patients afflicted with the psychiatric disorder of interest. Even though this is the most-direct approach to investigate the samples of interest, we must deal with a number of potential confounds, most notably clinical heterogeneity related to diagnostic subtypes and disease severity. In the absence of biological markers for diagnostic subtypes and disease progression in psychiatric disorders, matching patients and controls for investigating a system as dynamic as the proteome, is definitely a challenge. In addition to clinical heterogeneity, differences due to postmortem interval, brain pH and agonal state have been shown to independently influence protein degradation (Fountoulakis et al., 2001; Franzen et al., 2003; Hynd et al., 2003). Most important is the potential confound associated with the effects of psychopharmacological treatment and its effect on protein expression. Finally, similar to the issues discussed above in animal tissue, such as differences in proteomes among cell subtypes within a specific brain region have to be considered and would again necessitate cell-specific collection methods.

Cerebrospinal fluid
Human CSF is in direct contact with brain extracellular fluid. Besides the secretion of CSF by the choroid plexus, CSF is also derived from the ependymal lining of the ventricular system, glial membranes and blood vessels in the arachnoid. Biochemical alterations within the CNS are believed to be often reflected in CSF. The advantage of using CSF is that it is more readily available in psychiatry patients and that serial sampling of the same patient at various stages of the disorder is possible. In addition, CSF only contains soluble proteins. The investigation of membrane-bound proteins' proteomics has been hampered by their poor solubilization (Santoni et al., 2000). Recent technological developments may partially obviate this problem (Churchward et al., 2005). One of the major concerns of working with CSF is that there is a large range of protein concentrations and that this limits the ability to identify low-abundance proteins. These proteins are, however, of considerable importance because the high-abundance proteins within the CSF are likely serum and not brain proteins. Maccarrone et al. (2004) reported on the technical difficulties related to mining the CSF proteome. By using immunodepletion of the CSF they dramatically reduced the concentrations of the most-abundant CSF proteins, i.e. serum albumin, transferrin, haptoglobin, IgG, IgA and antitrypsin, leading to a much-improved visualization of low-abundance proteins on 2-D gels. By using shotgun MS on the immunodepleted sample, the authors were able to identify about 200 distinct proteins. Even though many of these proteins were apparently still derived from serum, the authors were able to also identify proteins expressed in brain such as amyloid protein, neural cell adhesion molecule and neuronal pentraxin receptor isoform 2. Yuan and Desiderio (2005) also reported that pre-fractionating CSF using a reversed- and solid-phase extraction cartridge increases the ability to detect low-abundance, and thus possibly disease-specific CSF proteins. An additional confounding

factor is the potential contamination of CSF by blood during sample collection, which not only distorts the pattern of identified proteins but also increases protein degradation (You et al., 2005).

Because the main challenge of using CSF proteomics in psychiatric research is that the most-abundant CSF proteins are serum proteins and these are the most reliably identified proteins, the relevance of proteomics in identifying brain-specific disease or progression markers in psychiatric disorders remains unclear at least until major technical improvements increase the sensitivity of these approaches to detect low-abundance proteins.

Use of peripheral cells: example peripheral blood monocytes

While PBMCs or lymphocytes are not likely to be the cell type causally involved in psychiatric disorders, they may well serve as markers for disease state. Indeed there is considerable communication between the immune system and the CNS. Many cytokine receptors have been located within the CNS and IL2 mRNA and T-cell receptors have been specifically detected in neurons (Funke et al., 1987; Shimojo et al., 1993). Lymphocytes, in contrast also express several neurotransmitter and hormone receptors, including dopamine, cholinergic and serotonergic receptors and gluco- and mineralocorticoid receptors and their chaperones (Gladkevich et al., 2004). Lymphocytes are directly influenced by glucocorticoids as well as cathecholamines and these two systems are perturbed in several psychiatric disorders. A number of studies have reported abnormalities in the immune system of psychiatric patients (see Gladkevich et al., 2004 for review). There appears to be a subgroup of schizophrenic patients that show a shift from T-helper 1 like cellular to T-helper 2-like humoral immune reactivity, accompanied by elevated IL6 levels. This pattern may be characteristic of schizophrenic patients with predominantly negative symptoms and poor therapeutic outcome (Schwarz et al., 2001). One of the most common immune system findings in patients with major depression is activation of T cells and increased production of several pro-inflammatory cytokines (Gladkevich et al., 2004). Studying lymphocytes in psychiatric disorders may thus yield information on disease-specific immune changes, and changes in lymphocytic immune function may serve as markers of disease progression. In addition, some receptor systems may show similar abnormalities in the lymphocyte and the brain. CNS glucocorticoid receptor (GR) resistance and its resolution with antidepressant treatment is one of the most consistent biological findings in major depression (Pariante and Miller, 2001). Steroid resistance has also been reported for the activation of T cells and monocytes in major depression and bipolar disorder (Pariante and Miller, 2001), suggesting comparable GR impairment in immune and CNS cells. In addition, genetic polymorphisms may similarly affect the function of molecules that are expressed in both lymphocytes and brain. Binder et al. (2004) reported that polymorphisms in FKBP5, a GR-regulating co-chaperone of hsp90, are associated with increased lymphocytic levels of FKBP5 protein. Increased FKBP5 expression has been shown to alter GR function (Scammell et al., 2001). The authors could show that the increase of FKBP5 protein in PBMC was paralleled by an altered response of the HPA axis and an accelerated response to antidepressant treatment, suggesting that the functional effects of these polymorphisms were not limited to immune cells but also affected CNS function (Binder et al., 2004). These data suggest that proteomics using peripheral lymphocytes may detect CNS-relevant changes in protein expression due to genetic polymorphisms.

Proteomics studies in psychiatry

Few studies have thus far measured proteomics patterns in psychiatric disorders and most of these have focused on schizophrenia.

Animal models

Few groups have used proteomics in animal models of psychiatric disorders so far. Paulson et al. in series of studies, used acute and subchronic treatment with the NMDA receptor antagonist MK801 in rats (as an animal model for psychosis) to study treatment-related changes in the proteome of the frontal cortex and thalamus of these animals (Paulson et al., 2003, 2004a, b). Using 2-D polyacrylamide gel electrophoresis (PAGE) and MS, the researchers found a series of proteins to be altered in expression by this treatment with both universal and region-specific changes observed (Paulson et al., 2003, 2004b). Length of the treatment influenced the type of altered protein (Paulson et al., 2004a). Of the proteins altered by MK801 treatment, hsp60, hsp72, albumin, DRP-2, aldolase c and malate dehydrogenase, have also been previously reported to be altered in schizophrenia (Schwarz et al., 1999; Johnston-Wilson et al., 2000; Middleton et al., 2002; Yao et al., 2000). Khawaja et al. (2004) investigated changes in the proteome of rat hippocampus following a 2-week treatment with either venlafaxine or fluoxetine. This group also used a combination of 2-D PAGE and MS. They identified 33 protein spots that were modulated by both drugs. Most of these proteins were related to neurogenesis (insulin-like growth factor 1 (IGFl), glial maturation factor beta for example) or synaptic plasticity (e.g. Ras-related protein 4a and 1b and hsp 10). These changes in protein expression were paralleled by a drug-induced increase in hippocampal neurogenesis. Finally Kromer et al., (2005) examined the difference in the proteome of mice selectively and bidirectionally bred for high- vs. low-anxiety behaviors (HAB-M vs. LAB-M), again using 2D-PAGE and MS for protein identification. They identified glyoxalase I, an enzyme involved in the detoxification of methyl-glyoxal as being expressed at consistently higher levels by LAB-M in several brain regions as well as in erythrocytes. This increase in protein expression was accompanied by an increase in mRNA expression. Interestingly, the glyoxalase I locus shows possible linkage or association with unipolar depression (Tanna et al., 1989).

CSF studies
Two groups have used proteomics methods to investigate CSF in psychiatric disorders so far and both have focused on schizophrenia. In an early study, Johnson et al. (1992a) performed 2-D gel electrophoresis on the CSF patients with schizophrenia or Alzheimer's disease and identified an isoform of alpha-2 haptoglobin and two other unidentified polypeptides to be increased in both disorders. They also reported a 21% increase in the number of detected CSF protein spots following haloperidol treatment that normalized after discontinuation of the treatment (Johnson et al., 1992b). Although these early studies employed 2-D gel electrophoresis, they were limited in the identification of the altered proteins because MS analysis with database-sequence comparison was not yet available. Jiang et al. (2003) used 2-D gel electrophoresis and MS to identify protein differences in the CSF of schizophrenic patients. Of the 54 proteins that exhibited altered expression levels in schizophrenic patients compared to controls, most were plasma proteins and only apolipoprotein A-IV levels were significantly decreased in the schizophrenic patients. This protein is mainly synthesized by the intestine and secreted into plasma, its role in the CNS is unknown. None of the studies conducted thus far have used methods to deplete serum proteins from CSF, so that the relevance of these earlier studies to brain-specific CSF protein remains questionable.

Postmortem brain studies
Again only two groups have currently used proteomic approaches to investigate postmortem brain tissues of psychiatric patients. Edgar et al. (1999) studied the hippocampus of seven schizophrenic patients, seven Alzheimer's disease patients and seven controls using 2-D gel electrophoresis of homogenized tissues. An altered expression of 16 protein spots was observed in the schizophrenia samples. One of these was identified to be the diazepam-binding inhibitor protein using N-terminal sequencing. This protein was also found to be decreased in the hippocampus of the Alzheimer's disease patients (Edgar et al., 1999). In a follow-up analysis, three more proteins altered in the hippocampus of the schizophrenic patients were identified, including manganese superoxide dismutase, T-complex protein I and collapsin-response mediator protein 2 (Edgar et al., 2000). Johnston-Wilson et al. (2000)

investigated the frontal cortex of 89 brains from the Stanley Foundation containing equal numbers of schizophrenic, bipolar, unipolar depressed and control subjects using 2-D gel electrophoresis and electrospray MS for protein identification. Several protein spots showed decreased intensity in all patient groups. Most of these turned out to be isoforms of the glial fibrillary acidic protein. In addition, dihydropyrimidinase-related protein 2 was decreased and fructose biphosphate aldolase C and aspartate amino transferase increased in all three disorders. Ubiquinone cytochrome c reductase core protein I and carbonic anhydrase 1 were specifically decreased in unipolar depressed patients.

The interpretation of these results is difficult. Overall most altered proteins seem to be common to several psychiatric and even neurological disorders, thus questioning their specificity for the pathophysiology of a certain disorder. In addition, neither of the studies has used cell-type specific sampling procedures. Thus a mix of neuronal, glial, immune and blood vessel cells was analyzed, obscuring smaller but possibly relevant differences in specific cell types. Finally, the question of diagnostic heterogeneity has not been addressed so that findings can be impacted by differences in disease severity or subtype.

mRNA expression arrays (expressomics)

Another possible approach is to identify candidate genes from genome-wide analyses of changes in mRNA expression using the so called "microarrays" (Lockhart and Winzeler, 2000). Methodological details of this technique are reviewed elsewhere in this volume.

Although recent methodological developments have significantly reduced the technical problems of microarrays, their application in psychiatric research remains challenging. Similar to the concerns cited for proteomics, patient and control subject selection as well as their matching are a major issue. In addition, the selection of brain regions relevant to specific psychiatric disorders is crucial in both human and animal studies. Pierce and Small (2004) suggest the use of brain imaging approaches to select brain regions for microarray experiments. Although more and more brain structures have been implicated in the pathophysiology of psychiatric disorders, the available data points to dysregulations affecting brain circuit involving several brain regions as oppose to single brain regions. Changes of expression in one area may only be disease-relevant when accompanied by changes in other structures in the implicated circuit. This circuit-based approach, however, poses novel problems for the already complicated data-analysis in expression microarray studies. In addition, smaller but relevant changes in a subpopulation of cells may be diluted, and thus not recognized if analyzing a whole region. In contrast to proteomics, the analysis of expression changes in single cell types is more feasible for mRNA expression. A combination of LCM and mRNA amplification techniques (see Ginsberg, 2004 for review) allows the comparison of expression changes in single cells. This strategy has already been successfully applied in schizophrenia research. Hemby et al. (2002) compared gene expression in stellate cells of layer II of the entorhinal cortex (ERC) of eight schizophrenic patients and nine matched controls. The authors identified changes in the expression of G-protein-coupled receptors, glutamate receptors and synaptic-related genes. Some of the genes regulated in stellate neurons of layer II, were also regulated when comparing expression pattern from whole-brain regions of schizophrenic patients to controls (Mirnics et al., 2000; Middleton et al., 2002).

As for proteomics research, different types of tissues can be used for expression analysis in psychiatric research. Below we summarize expression array studies performed in PBMCs, animal and human brain tissue

Expression array studies in psychiatric research

Expression analysis of PBMCs

As discussed above, lymphocytes are unlikely the cells causally involved in psychiatric disorders. Nonetheless, expression changes may serve as indicators of diagnostic subgroups, or as predictors of disease course and treatment response, especially when temporally examined. Recent studies have investigated disease-related expression changes in lymphocytes in psychiatric disorders using expression arrays. Vawter et al. (2004a) identified nine genes differentially regulated between schizophrenic and control subjects. Two of these were verified by real-time PCR analysis; the neuropeptide Y receptor 1 and guanine nucleotide-binding regulatory protein Go-alpha. Zvara et al (2005) reported upregulation of the RNA of the dopamine type 2 receptor and of an inwardly rectifying potassium channel (Kir 2.3) in PBMCs of schizophrenic patients. Cui et al. (2005) observed an upregulation of the adenomatous polyposis coli (APC) gene in the PBMCs of schizophrenic patients and this effect was independent of antipsychotic treatment. Interestingly the APC gene maps to chromosome 5q21–22, a locus that has shown linkage to schizophrenia (Lewis et al., 2003). The authors also genotyped polymorphisms within this gene in a Chinese sample of schizophrenic patients and controls, and showed an association of three exonic SNPs with the disease. Tsuang et al. (2005) reported that eight genes with differentially regulated mRNAs in schizophrenic patients, bipolar patients and controls discriminated among these three groups with 95–97% accuracy. Finally, Middleton et al. (2005a) used the PBMCs of discordant sibpairs from families with schizophrenia or bipolar disorder with linkage to chromosomes 5q and 6q, respectively. They identified a set of differentially regulated genes in these cells that showed a high degree of predictive power for classifying subjects according to diagnosis. Data from these studies suggest that expression analysis in PBMCs may be useful for diagnostic subtyping, as well as for candidate gene identification in psychiatric disorders.

Animal models
A range of animal models of psychiatric disorders have been used for expression array analysis. These include inherent strain differences in a relevant behavior (i.e. pre-pulse inhibition of the acoustic startle reflex (Grottick et al., 2005)), selective bidirectional breeding for a specific phenotype (Kromer et al., 2005) and most commonly, pharmacological challenges. For the latter, drugs-mimicking aspects of psychiatric disorders such as PCP (Kaiser et al., 2004), cocaine (Yuferov et al., 2003; Hayase et al., 2004) and methamphetamine (Ogden et al., 2004) have been used, as well as therapeutic agents, such as antidepressants (Yamada et al., 2000, 2001; Landgrebe et al., 2002; Drigues et al., 2003; Palotas et al., 2004) or mood stabilizers (Ogden et al., 2004). Although several expression studies in animal models of psychiatric disorders, including schizophrenia, depression or mania have been published, it is difficult to compare across studies due to the lack of similar protocols, with divergence in examined brain regions, treatment type and length and animals used. Because none of these animal models truly reflect the complexities of the human disorder, there likely are a high number of false-positive results, with genes showing promising regulation patterns that are likely not related to the pathophysiology of the disorder. In order to reduce the large number of false-positive associations, data from animal microarray studies must be supplemented by other sets of genomic and functional data in order to focus on the smaller number of potentially relevant genes.

Postmortem brain tissue
A series of studies have investigated disease-related expression changes in postmortem brain tissue of psychiatric patients using expression arrays and these are summarized in Table 1. Most studies have been performed on postmortem brain tissue from schizophrenic patients, and several pathways were found to be affected in this disorder. These include presynaptic-related genes (Mirnics et al., 2001a, 2000), energy metabolism (Iwamoto et al., 2004b; Middleton et al., 2002; Vawter et al., 2004b), ubiquitin degradation (Vawter et al., 2001; Middleton et al., 2002), protease activity (Vawter et al., 2004b), oligodendrocyte-specific markers (Hakak et al., 2001; Tkachev et al., 2003; Aston et al., 2004; Sugai et al., 2004;), Apolipoprotein L (Mimmack et al., 2002), heat-shock proteins (HSPA 12A and B (Pongrac et al., 2004)) and 14-3-3 genes (Vawter et al., 2001; Middleton et al., 2005b). Fewer studies have investigated other

Table 1. Summary of gene expression studies in postmortem human brain

Study	Disorder	Source of samples	Brain region	Array type	Perturbed pathway
Mirnics et al. (2000)	Schizophrenia	University of Pittsburgh's Center for the Neuroscience of Mental Disorders Brain Bank	Prefrontal cortex area 9	Custom cDNA array > 7800 cDNAs	Presynaptic secretory function + RGS4
Vawter et al. (2001)	Schizophrenia	Stanley Foundation	Prefrontal cortex, cerebellum, middle temporal gyrus	Custom cDNA array 15,000 cDNAs	Synaptic signaling and proteolytic function
Hakak et al. (2001)	Schizophrenia	Pilgrim Psychiatric Center (Long Island, NY, USA) and controls from nursing homes	Dorsolateral prefrontal cortex (area 9)	Affymetrix (HuGeneFL chip) > 6000 transcripts	Myelination-related genes (synaptic plasticity, neuronal development, neurotransmission and signal transduction)
Middleton et al. (2002)	Schizophrenia	University of Pittsburgh's Center for the Neuroscience of Mental Disorders Brain Bank	Dorsolateral prefrontal cortex (area 9)	UniGEM V cDNA microarray (Incyte Genomics Inc.) > 7800 transcripts	Ornithine and polyamine metabolism, the mitochondrial malate shuttle system, the transcarboxylic acid cycle, aspartate, alanine and ubiquitin metabolism
Mimmack et al. (2002)	Schizophrenia	Stanley Foundation and New Zealand/Japanese samples	Prefrontal cortex	Custom cDNA array 300 transcripts	Apolipoprotein L1
Hemby et al. (2002)	Schizophrenia	Mental Health Clinical Research Center on Schizophrenia at the University of Pennsylvania, Philadelphia	Entorhinal cortex layer II stellate neurons	Custom array > 15,000 transcripts	G-protein subunit i (alpha)1, glutamate receptor3, NMDA receptor 1, synaptophysin, SNAP 23 and 25
Aston et al. (2004)	Schizophrenia	Stanley Foundation	Temporal cortex (area 21)	Affymetrix HgU95A microarrays > 12,000 transcripts	Neurodevelopment, circadian pacemaker, chromatin function and signaling
Sugai et al. (2004)	Schizophrenia	Japanese sample	Dorsoprefrontal cortex (area 47)	Human Arrays (Clontech) > 1300 genes	Oligodendrocyte- and astrocyte-related genes
Bezchlibnyk et al. (2001)	Bipolar disorder	Stanley Foundation	Frontal cortex	Atlas Human 1.2 array (Clontech) 1200 genes	Transforming growth factor 1, caspase 8 precursor, transducer of erbB2
Tkachev et al. (2003)	Schizophrenia and bipolar disorder	Stanley Foundation	Dorsolateral prefrontal cortex (area 9)	Affymetrix HU133A chip, 22,283 transcript	Schizophrenia and bipolar: oligodendrocyte-

Table 1 (continued)

Study	Disorder	Source of samples	Brain region	Array type	Perturbed pathway
					and myelin-related genes
Konradi et al. (2004)	Schizophrenia and bipolar disorder	Harvard Brain Tissue Resource Center	Hippocampus	Affymetrix HgU95A microarrays >12,000 genes	Bipolar disorder: mitochondrial energy metabolism and downstream deficits of ATP-dependent processes
Iwamoto (2004b)	Bipolar disorder, schizophrenia, major depression	Stanley Foundation	Prefrontal cortex (area 10)	Affymetrix HgU95A microarrays >12,000 genes	Bipolar disorder: receptors, channels or transporters, stress-response proteins or molecular chaperones
Iwamoto et al. (2005)	Schizophrenia and bipolar disorder	Stanley Foundation	Prefrontal cortex (area 46)	Affymetrix HU133A chip, 22,283 transcripts	Schizophrenia and bipolar disorder: mitochondrial genes
Evans et al. (2004)	Major depression	Brain Donor Program at the UC Irvine	Dorsolateral prefrontal cortex (area 9)	Affymetrix HgU95A microarrays >12,000 genes	Fibroblast growth factor system
Aston et al. (2005)	Major depression	Stanley Foundation	Temporal cortex (area 21)	Affymetrix HgU95A microarrays >12,000 genes	Oligodendrocyte function
Lewohl et al. (2000)	Alcohol dependence	University of Queensland?	Superior frontal cortex	cDNA and oligoarrays >4000 genes	Myelin-related and cell-cycle genes
Mayfield et al. (2002)	Alcohol dependence	Waggoner Center for Alcohol and Addiction Research	Frontal and motor cortex	UniGEM V2.0 microarrays (Incyte) 10,000 transcripts	Myelin-related genes and genes involved in protein trafficking
Iwamoto (2004a)	Alcohol dependence	Stanley Foundation	Prefrontal cortex (area 10)	Affymetrix HgU95A microarrays >12,000 genes	Myelin-related genes and molecular chaperones
Lehrmann et al. (2003)	Cocaine abuse	Brain repository of the Clinical Brain Disorders Branch, NIMH	Prefrontal cortex (area 46)	Custom cDNA array 1152 transcripts	Guanine nucleotide-binding protein beta polypeptide 1, synaptic transmission
Albertson et al. (2004)	Cocaine abuse	Wayne State University's Human Investigation Committee	Nucleus accumbens	Affymetrix HU133A and B arrays, >39,000 transcripts	Myelin-related genes

major psychiatric disorders (see Table 1 for summary). For bipolar disorder, decreases in mitochondria-related genes, oligodendrocyte-specific markers but also in receptors and transporters, ion channels, stress proteins and chaperones have been identified (Bezchlibnyk et al., 2001; Tkachev et al., 2003; Iwamoto et al., 2004b; Konradi et al., 2004; Iwamoto et al., 2005). Two genes, LIM and HSPFI, have been reported to be altered in expression in brain tissue as well as lymphoblastoid cell lines of bipolar patients (Iwamoto et al., 2004b). Three studies have investigated alcohol abuse (Lewohl et al., 2000; Mayfield et al., 2002; Iwamoto et al., 2004a) and two studies cocaine abuse (Lehrmann et al., 2003; Albertson et al., 2004). Two studies scrutinized major depression (Evans et al., 2004; Aston et al., 2005). To address the issue of disease-specificity, Iwamoto et al. (2004b) compared expression pattern in the prefrontal cortex (PFC) of patients with bipolar disorder to patients with unipolar depression and schizophrenia, as well as healthy controls. Interestingly, most of the altered gene expressions in each disease were not shared by one another, suggesting the molecular distinctiveness of these mental disorders.

Validation of candidates from expression array studies

Following initial steps in the validation of a promising candidate from expression array studies, such as quantitative PCR or measures of protein levels, it is important to identify possible mechanisms for the altered expression pattern in patients. Functional genetic variants in regulatory regions or the 3′ untranslated region (UTR) of the gene could for example affect transcript abundance. A series of researchers have used single-nucleotide polymorphism (SNP) association studies to confirm candidate genes from expression studies (see section on Use of convergent evidence). So far, none of them have, however, been able to identify the "causal variant". While associated SNPs may represent the causal variant, they are more likely in linkage disequilibrium (LD) with this variant. The extent of LD depends on the number of meiotic recombination events that occurred between the markers, with LD being higher, the lower the number of recombination events. The identification of the causal variant is necessary to test functional relevance in vitro or in transgenic animal models. Depending on the extent of LD within a gene, this will be more or less challenging. Here, association studies in older populations with limited LD, such as African populations, may be helpful as the number of SNPs in LD with the causal variant will be smaller, reducing the set of SNPs that need to be tested for functional relevance. Genetic variants altering transcription efficiency or mRNA stability are, however, not the only possible explanation for disease-related expression changes. Recently Weaver et al. (2004) showed that perturbation in maternal behavior altered the methylation pattern in the promoter of the GR gene in hippocampal cells leading to changes in GR expression and ultimately in an altered stress response in the offsprings. These effects are likely due to methylation-induced altered histone acetylation and transcription factor (NGFI-A) binding to the GR promoter.

Environment-induced changes in methylation pattern are not likely related to any genetic variants and methods detecting changes in DNA-methylation status should be considered as an important validation tool for candidates genes in psychiatric research as gene–environment interactions emerge as a major etiological factor in the pathophysiology of these disorders (Caspi et al., 2002, 2003, 2005). Finally, RNA interference (RNAi) is an additional mechanism to regulate transcript abundance. RNAi is a post-transcriptional method of gene silencing, originally discovered in *Caenorhabditis elegans,* in which short double-stranded RNA (microRNA) mediates the destruction of mRNAs in a sequence-specific fashion (Fire et al., 1998). Several groups have demonstrated the importance of this pathway for transcriptional regulation in mammalian cells (Mattick and Makunin, 2005 for review). Surprisingly, RNAi has not yet been investigated in psychiatric disorders. The finding that FMRP, the gene causally related to the development of fragile X syndrome, which can be accompanied by psychiatric symptoms, regulates the mRNA translation of its target by facilitating RNAi (Jin et al., 2004), suggests that RNAi may be an important pathway to consider in the pathophysiology of psychiatric disorders. Even though our knowledge about RNAi and the abundance of microRNA in the human genome is still limited, researchers have composed a custom microarray platform for analysis of microRNA gene expression and observed the temporal regulation of a large class of microRNAs during embryonic development (Thomson et al., 2004). Such tools will allow for the comprehensive investigation of microRNAs in psychiatric disorders.

Whole genome SNP association studies

The knowledge of the sequence of the human genome, that has been publicly available to all researchers since February 2001 (Lander et al., 2001; Venter et al., 2001), has dramatically changed the possibilities for human genetic association studies. To investigate the whole genome, frequent and evenly distributed genetic markers are needed. In the last few years, SNPs have become the most-promoted genetic markers for complex or common phenotypes (Risch and Merikangas, 1996; Kwok and Chen 1998; Collins et al., 1999;). These polymorphisms consist of a single-base exchange and occur about every 100 bp. SNPs can either be functionally relevant themselves or serve as markers for other nearby mutations with which they are in LD (Brookes, 1999). The NIH maintained public SNP database, dbSNP (www.ncbi.nlm.nih.gov/SNP) now contains over 5 million validated SNPs. A main advantage of using SNPs as genetic markers is that they can be genotyped using high-throughput methods, allowing the rapid and affordable investigation of many SNPs, which is indispensable for a genome-wide approach (Kwok, 2000; Syvanen, 2001). The number of SNPs necessary to realize a genome screen covering all genes is still under debate, but current estimates range from 100,000 to over 500,000 SNPs. Platforms allowing the parallel genotyping of this magnitude of SNPs are already available from several companies. Affymetrix (www.affymetrix.com) offers a genotyping chip containing over 100,000 SNP assays, covering most of the genome with at least 1 SNP for every 100 kb. A panel of 500,000 SNP assays is currently under development. One of the problems in selecting a SNP panel covering the whole genome is that randomly selected markers (such as the ones for the Affymetrix GeneChip® Mapping Sets) may not cover all genes in a sufficient density to ensure that most of the possible variations are included. In regions of high LD, a few markers may be sufficient to detect disease associations, whereas regions with low LD will need a denser spacing of markers. Illumina (www.illumina.com) is currently developing a SNP panel that makes use of the knowledge of varying LD pattern throughout the genome derived from the HapMap project (www.hapmap.org). The HapMap project provides genome-wide information on LD pattern in four different ethnic panels. The Illumina SNP panel will include about 250,000 markers, with SNP densities adjusted to the extent of the local LD. Using LD information allows the comprehensive analysis of the genome without having to genotype redundant variants, thus reducing genotyping effort without loosing valuable information. Whole genome SNP association studies have the advantage over classical linkage studies that they can be performed with independent cases and controls (as opposed to families) that are easier to recruit in the study of complex disorders. In addition, the increased resolution allows the identification of single-candidate genes and not candidate loci, spanning several centiMorgans (cM) (1 cM represents on average 1–2 million bp in humans) that necessitate extensive additional fine mapping. Truly genome-wide association studies for psychiatric disorders using SNP markers have not yet been published. It has to be mentioned though that classical, hypothesis-driven association studies as well as classical linkage analyses have already yielded promising candidate genes for psychiatric disorders. Particularly relevant to mood disorders is a functional insertion/deletion polymorphism (often referred to as 5HT transporter gene-linked polymorphic region (*5-HTTLPR*)) located in a repeat region of the promoter of the serotonin transporter gene (gene name SLC6A4) (Heils et al., 1996). This polymorphism has repeatedly been associated with a series of phenotypes relevant to mood disorders, including altered brain activation pattern, susceptibility to early life stress and response to antidepressant treatments (Caspi et al., 2003; Hariri and Weinberger, 2003; Serretti, 2004). For schizophrenia, classical linkage analyses with subsequent fine-mapping have identified several putative novel susceptibility genes including neuregulin, dysbindin and DISC1 (Harrison and Weinberger, 2005) that have given novel insight into the pathophysiology of this disorder.

Use of convergent evidence

Overall, all whole-genome-based approaches are hampered by the large number of false-positive results. So far each expression array study has yielded a large number of regulated genes, of which only a few, if any, will actually be true positives. Whole genome SNP analyses have an expected high number of false-positive associations due to the high degree of multiple testing.

The use of convergent evidence from a series of genomic and functional approaches may be a promising alternative to identify potential true positives. In recent years, an increasing number of investigators have come to not only rely on one type of whole-genome approach, but on combining several of these approaches in order to identify the most-promising candidates. A common approach to limit the number of candidate genes from a large set of genes from proteomics and expression analysis as well genetic linkage studies has been to rely on previous data of a gene's potential pathophysiological involvement in the disorder of interest. With more novel candidate genes being confirmed for psychiatric disorders (e.g. Harrison and Weinberger, 2005), it becomes increasingly clear that relying solely on hypothesis-driven selection strategies may miss the most important genes. Strategies combining several hypothesis-free approaches may be more promising (see Fig. 1). John Kelsoe and his colleagues used an approach that combined microarray analysis of animal models of mania and linkage analysis in families with bipolar disorder to identify G-protein-coupled

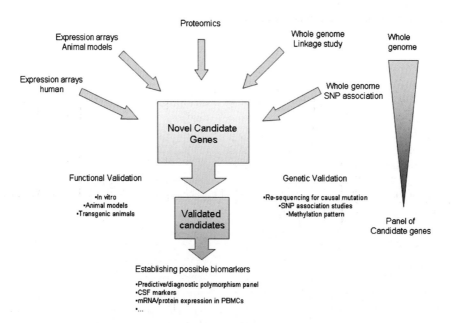

Fig. 1. Illustration of a convergent genomic approach to identify novel candidate genes for psychiatric disorders. Convergent evidence from several whole genome-wide approaches yield a set of candidate genes that is further refined through functional and genetic validation. Genes that have passed this selection process may serve for the development of novel diagnostic and predictive biomarkers in psychiatric disorders.

receptor kinase 3 (GRK3) as a promising candidate gene. This gene is involved in the homologous desensitization of G-protein-coupled receptors. The group had initially identified a linkage peak for bipolar disorder on chromosome 22q (Lachman et al., 1997; Kelsoe et al., 2001). That linked region did, however, span 32 cM, making it a challenging task to identify the causal gene by fine mapping strategies. The group then used microarray analysis of different brain regions in methamphetamine-treated rats and identified several genes that were regulated by this treatment that also mapped to previous linkage peaks with bipolar disorder (Niculescu et al., 2000). One of these was GRK3, which maps to 22q. In addition to being regulated in the animal model, protein levels for this gene were also found to be decreased in a subset of patient lymphoblastoid cell lines and the magnitude of the decrease correlated with disease severity. By using re-sequencing and SNP genotyping strategies, the group confirmed the association of 5′UTR and promoter variants of this gene with bipolar disorder (Barrett et al., 2003). Mirnics and associates used a combination of human postmortem brain microarrays and linkage, as well as association studies, to establish RSG4 as a major candidate gene for schizophrenia. In their initial study comparing ERC tissue from schizophrenic patients to matched controls, they identified that genes related to pre-synaptic secretory function were consistently altered in schizophrenics (Mirnics et al., 2000). The most-affected gene, however, varied from individual to individual. In a follow-up study, the group identified the GTP-ase activating protein RSG4 to be decreased in all but one schizophrenic individual (Mirnics et al., 2001b). This gene actually maps to chromosome 1q21–22, a locus that is supported by a large meta-analysis of schizophrenia linkage studies (Lewis et al., 2003). The possible relevance of RGS4 in the pathogenesis of schizophrenia was further confirmed by functional studies as well as in genetic association studies. RGS4 appears to be regulated by dopaminergic transmission (Bishop et al., 2002; Geurts et al., 2002; Geurts et al., 2003; Taymans et al., 2003; Taymans et al., 2004). In addition, RGS4 was shown to alter serotonergic transmission in vitro as well as in vivo via the 5HT1a autoreceptors (Beyer et al., 2004; Ghavami et al., 2004) and to be important for neuronal differentiation (Grillet et al., 2003). Nonetheless, mice lacking the rgs4 gene have normal neural development and are viable and fertile. Behavioral testing on mutant adults only revealed subtle sensorimotor deficits but no deficit in pre-pulse inhibition, one of the behavioral phenotypes long believed to be related to schizophrenia (Grillet et al., 2005). A series of SNP studies have confirmed an association of RGS4 with schizophrenia in different populations (Chowdari et al., 2002; Chen et al., 2004; Morris et al., 2004; Williams et al., 2004; Cordeiro et al., 2005). Furthermore, RGS4 variants were correlated with dorsolateral PFC volume in schizophrenic patients but not in controls (Prasad et al., 2005).

Recently Ogden et al. (2004) presented a model for ranking genes regulated in expression array studies from animal models of psychiatric disorders. To identify candidate genes for bipolar disorder, mice were treated with the stimulant methamphetamine, the mood stabilizer valproate or both and treatment-related mRNA expression changes were measured in several brain regions. Genes showing significant expression changes were then ranked by the following criteria: changed by methamphetamine and/or valproate, no change with co-treatment, change observed in multiple brain regions, disease relevant biological role, gene maps to known linkage peak for bipolar disorder or schizophrenia and gene expression has been found to be altered in the psychiatric disorder in postmortem tissue. Using these criteria the authors identified seven genes that fulfilled at least five of the six criteria. Topping the list of candidate genes, were DARPP-32 (dopamine- and cAMP-regulated phosphoprotein of 32 kDa) located at chromosome 17q12, PENK (preproenkephalin) located at 8qI2.1, and TACI (tachykinin I, substance P) located at 7q21.3. This more limited set of candidate genes now deserves further scrutiny and validation.

Future directions

Even though some promising candidates for psychiatric disorders have been identified using proteomics and functional genomics approaches, none have yet emerged as validated diagnostic or disease progression markers. Additional studies using convergent approaches are needed to identify more relevant candidate genes for psychiatric disorders. The most progress has been made in schizophrenia research, where a series of promising candidate genes are now available (Harrison and Weinberger, 2005). Other psychiatric disorders lag behind, with few if any strong candidates available at this time. Once a sufficient number of candidates are identified, sets of polymorphisms in several genes may be used to aid diagnostic subtyping and possibly prediction of disease course and treatment response. Expression arrays and proteomics studies in peripheral cells or CSF may also aid diagnosis and treatment-response monitoring. These approaches have the advantage that serial sampling is possible, so that changes in mRNA or protein expression can be monitored during treatment. These may then serve as early indicators of treatment response or resistance, allowing a more timely adjustment of the treatment regimen. Even though some genetic markers have already been associated with treatment response to psychotropic drugs (Malhotra et al., 2004), none has been subjected to an adequately powered prospective study. Besides their potential to serve as diagnostic or treatment course markers, novel candidates arising from proteomics and functional genetics studies may also lead the way to novel therapeutic strategies by uncovering pathogenic mechanisms. The hope is that in the future, molecular genetics tests will be utilized to identify the underlying biological cause of psychiatric disorders and will aid in selection of treatment that specifically interferes with the identified pathological pathway.

References

Albertson, D.N., Pruetz, B., Schmidt, C.J., Kuhn, D.M., Kapatos, G. and Bannon, M.J. (2004) Gene expression profile of the nucleus accumbens of human cocaine abusers: evidence for dysregulation of myelin. J. Neurochem., 88: 1211–1219.

Aston, C., Jiang, L. and Sokolov, B.P. (2004) Microarray analysis of postmortem temporal cortex from patients with schizophrenia. J. Neurosci. Res., 77: 858–866.

Aston, C., Jiang, L. and Sokolov, B.P. (2005) Transcriptional profiling reveals evidence for signaling and oligodendroglial abnormalities in the temporal cortex from patients with major depressive disorder. Mol. Psychiatry, 10: 309–322.

Barrett, T.B., Hauger, R.L., Kennedy, J.L., Sadovnick, A.D., Remick, R.A., Keck, P.E., McElroy, S.L., Alexander, M., Shaw, S.H. and Kelsoe, J.R. (2003) Evidence that a single nucleotide polymorphism in the promoter of the G protein receptor kinase 3 gene is associated with bipolar disorder. Mol. Psychiatry, 8: 546–557.

Beyer, C.E., Ghavami, A., Lin, Q., Sung, A., Rhodes, K.J., Dawson, L.A., Schechter, L.E. and Young, K.H. (2004) Regulators of G-protein signaling 4: modulation of 5-HT(1A)-mediated neurotransmitter release in vivo. Brain Res., 1022: 214–220.

Bezchlibnyk, Y.B., Wang, J.F., McQueen, G.M. and Young, L.T. (2001) Gene expression differences in bipolar disorder revealed by cDNA array analysis of post-mortem frontal cortex. J. Neurochem., 79: 826–834.

Binder, E.B., Kinkead, B., Owens, M.J. and Nemeroff, C.B. (2001) The role of neurotensin in the pathophysiology of schizophrenia and the mechanism of action of antipsychotic drugs. Biol Psychiatry, 50: 856–872.

Binder, E.B., Salyakina, D., Lichtner, P., Wochnik, G.M., Ising, M., Putz, B., Papiol, S., Seaman, S., Lucae, S., Kohli, M.A., Nickel, T., Kunzel, H.E., Fuchs, B., Majer, M., Pfennig, A., Kern, N., Brunner, J., Modell, S., Baghai, T., Deiml, T., Zill, P., Bondy, B., Rupprecht, R., Messer, T., Kohnlein, O., Dabitz, H., Bruckl, T., Muller, N., Pfister, H., Lieb, R., Mueller, J.C., Lohmussaar, E., Strom, T.M., Bettecken, T., Meitinger, T., Uhr, M., Rein, T., Holsboer, F. and Muller-Myhsok, B. (2004) Polymorphisms in FKBP5 are associated with increased recurrence of depressive episodes and rapid response to antidepressant treatment. Nat. Genet., 36: 1319–1325 Nov 21, Epub 2004.

Bishop, G.B., Cullinan, W.E., Curran, E. and Gutstein, H.B. (2002) Abused drugs modulate RGS4 mRNA levels in rat brain: comparison between acute drug treatment and a drug challenge after chronic treatment. Neurobiol. Dis., 10: 334–343.

Brookes, A.J. (1999) The essence of SNPs. Gene, 234: 177–186.

Caspi, A., McClay, J., Moffitt, T.E., Mill, J., Martin, J., Craig, I.W., Taylor, A. and Poulton, R. (2002) Role of genotype in the cycle of violence in maltreated children. Science, 297: 851–854.

Caspi, A., Moffitt, T.E., Cannon, M., McClay, J., Murray, R., Harrington, H., Taylor, A., Arseneault, L., Williams, B., Braithwaite, A., Poulton, R. and Craig, I.W. (2005) Moderation of the effect of adolescent-onset cannabis use on adult psychosis by a functional polymorphism in the catechol-O-methyltransferase gene: longitudinal evidence of a gene X environment interaction. Biol. Psychiatry, 57: 1117–1127.

Caspi, A., Sugden, K., Moffitt, T.E., Taylor, A., Craig, I.W., Harrington, H., McClay, J., Mill, J., Martin, J., Braithwaite, A. and Poulton, R. (2003) Influence-of life stress on depression: moderation by a polymorphism in the 5-HTT gene. Science, 301: 386–389.

Chen, X., Dunham, C., Kendler, S., Wang, X., O'Neill, F.A., Walsh, D. and Kendler, K.S. (2004) Regulator of G-protein signaling 4 (RGS4) gene is associated with schizophrenia in Irish high density families. Am. J. Med. Genet. B Neuropsychiatry Genet., 129: 23–236.

Chowdari, K.V., Mirnics, K., Semwal, P., Wood, J., Lawrence, E., Bhatia, T., Deshpande, S.N., B, K.T., Ferrell, R.E., Middleton, F.A., Devlin, B., Levitt, P., Lewis, D.A. and Nimgaonkar, V.L. (2002) Association and linkage analyses of RGS4 polymorphisms in schizophrenia. Hum. Mol. Genet., 11: 1373–1380.

Churchward, M.A., Butt, R.H., Lang, J.C., Hsu, K.K. and Coorssen, J.R. (2005) Enhanced detergent extraction for analysis of membrane proteomes by two-dimensional gel electrophoresis. Proteome Sci., 3: 5.

Collins, A., Lonjou, C. and Morton, N.E. (1999) Genetic epidemiology of single-nucleotide polymorphisms. Proc. Natl. Acad. Sci. USA, 96: 15173–15177.

Cordeiro, Q., Talkowski, M.E., Chowdari, K.V., Wood, J., Nimgaonkar, V. and Vallada, H. (2005) Association and linkage analysis of RGS4 polymorphisms with schizophrenia and bipolar disorder in Brazil. Genes Brain Behav., 4: 45–50.

Cui, D.H., Jiang, K.D., Jiang, S.D., Xu, Y.F. and Yao, H. (2005) The tumor suppressor adenomatous polyposis coli gene is associated with susceptibility to schizophrenia. Mol. Psychiatry, 10: 669–677.

Drigues, N., Poltyrev, T., Bejar, C., Weinstock, M. and Youdim, M.B. (2003) cDNA gene expression profile of rat hippocampus after chronic treatment with antidepressant drugs. J. Neural. Transm., 110: 1413–1436 Jan 1, Epub 2003.

Edgar, P.F., Douglas, J.E., Cooper, G.J., Dean, B., Kydd, R. and Faull, R.L. (2000) Comparative proteome analysis of the hippocampus implicates chromosome 6q in schizophrenia. Mol. Psychiatry, 5: 85–90.

Edgar, P.F., Schonberger, S.J., Dean, B., Faull, R.L., Kydd, R. and Cooper, G.J. (1999) A comparative proteome analysis of hippocampal tissue from schizophrenic and Alzheimer's disease individuals. Mol. Psychiatry, 4: 173–178.

Evans, S.J., Choudary, P.V., Neal, C.R., Li, J.Z., Vawter, M.P., Tomita, H., Lopez, J.F., Thompson, R.C., Meng, F., Stead, J.D., Walsh, D.M., Myers, R.M., Bunney, W.E., Watson, S.J., Jones, E.G. and Akil, H. (2004) Dysregulation of the fibroblast growth factor system in major depression. Proc. Natl. Acad. Sci. USA, 101: 15506–15511 Oct 13, Epub 2004.

Fire, A., Xu, S., Montgomery, M.K., Kostas, S.A., Driver, S.E. and Mello, C.C. (1998) Potent and specific genetic interference by double-stranded RNA in *Caenorhabditis elegans*. Nature, 391: 806–811.

Fountoulakis, M., Hardmeier, R., Hoger, H. and Lubec, G. (2001) Postmortem changes in the level of brain proteins. Exp. Neurol., 167: 86–94.

Franzen, B., Yang, Y., Sunnemark, D., Wickman, M., Ottervald, J., Oppermann, M. and Sandberg, K. (2003) Dihydropyrimidinase related protein-2 as a biomarker for temperature and time dependent post mortem changes in the mouse brain proteome. Proteomics, 3: 1920–1929.

Freeman, W.M. and Hemby, S.E. (2004) Proteomics for protein expression profiling in neuroscience. Neurochem. Res., 29: 1065–1081.

Funke, I., Hahn, A., Rieber, E.P., Weiss, E. and Riethmuller, G. (1987) The cellular receptor (CD4) of the human immunodeficiency virus is expressed on neurons and glial cells in human brain. J. Exp. Med., 165: 1230–1235.

Geurts, M., Hermans, E. and Maloteaux, J.M. (2002) Opposite modulation of regulators of G protein signalling-2 RGS2 and RGS4expression by dopamine receptors in the rat striatum. Neurosci. Lett., 333: 146–150.

Geurts, M., Maloteaux, J.M. and Hermans, E. (2003) Altered expression of regulators of G-protein signaling (RGS) mRNAs in the striatum of rats undergoing dopamine depletion. Biochem. Pharmacol., 66: 1163–1170.

Ginsberg, S.D., Elarova, I., Ruben, M., et al. (2004) Single-cell gene expression analysis: implications for neurodegenerative and neuropsychiatric disorders. Neurochem. Res., 29: 1053–1064.

Ghavami, A., Hunt, R.A., Olsen, M.A., Zhang, J., Smith, D.L., Kalgaonkar, S., Rahman, Z. and Young, K.H. (2004) Differential effects of regulator of G protein signaling (RGS) proteins on serotonin 5-HTIA, 5-HT2A, and dopamine D2 receptor-mediated signaling and adenylyl cyclase activity. Cell Signal, 16: 711–721.

Gladkevich, A., Kauffman, H.F. and Kor, J. (2004) Lymphocytes as a neural probe: potential for studying psychiatric disorders. Prog. Neuropsychopharmacol. Biol. Psychiatry, 28: 559–576.

Grillet, N., Dubreuil, V., Dufour, H.D. and Brunet, J.F. (2003) Dynamic expression of RGS4 in the developing nervous system and regulation by the neural type-specific transcription factor Phox2b. J. Neurosci., 23: 10613–10621.

Grillet, N., Pattyn, A., Contet, C., Kieffer, B.L., Goridis, C. and Brunet, J.F. (2005) Generation and characterization of Rgs4 mutant mice. Mol. Cell Biol., 25: 4221–4228.

Grottick, A.J., Bagnol, D., Phillips, S., McDonald, J., Behan, D.P., Chalmers, D.T. and Hakak, Y. (2005) Neurotransmission- and cellular stress-related gene expression associated with prepulse inhibition in mice. Brain Res. Mol. Brain Res., 14: 14.

Hakak, Y., Walker, J.R., Li, C., Wong, W.H., Davis, K.L., Buxbaum, J.D., Haroutunian, V. and Fienberg, A.A. (2001) Genomewide expression analysis reveals dysregulation of myelination-related genes in chronic schizophrenia. Proc. Natl. Acad. Sci. USA, 98: 4746–4751.

Hariri, A.R. and Weinberger, D.R. (2003) Functional neuroimaging of genetic variation in serotonergic neurotransmission. Genes Brain. Behav., 2(6): 341–349.

Harrison, P.J. and Weinberger, D.R. (2005) Schizophrenia genes, gene expression, and neuropathology: on the matter of their convergence. Mol. Psychiatry, 10: 40–68.

Hayase, T., Yamamoto, Y., Yamamoto, K., Muso, E. and Shiota, K. (2004) Microarray profile analysis of toxic cocaine-induced alterations in the expression of mouse brain gene sequences: a possible 'protective' effect of buprenorphine. J. Appl. Toxicol., l24: 15–20.

Heils, A., Teufel, A., Petri, S., Stober, G., Riederer, P., Bengel, D. and Lesch, K.P. (1996) Allelic variation of human serotonin transporter gene expression. J. Neurochem., 66: 2621–2624.

Hemby, S.E., Ginsberg, S.D., Brunk, B., Arnold, S.E., Trojanowski, J.Q. and Eberwine, J.H. (2002) Gene expression profile for schizophrenia: discrete neuron transcription patterns in the entorhinal cortex. Arch. Gen Psychiatry, 59: 631–640.

Holsboer, F. (2000) The corticosteroid receptor hypothesis of depression. Neuropsychopharmacology, 23: 477–501.

Hynd, M.R., Lewohl, J.M., Scott, H.L. and Dodd, P.R. (2003) Biochemical and molecular studies using human autopsy brain tissue. J. Neurochem., 85: 543–562.

Iwamoto, K., Bundo, M. and Kato, T. (2005) Altered expression of mitochondria-related genes in postmortem brains of patients with bipolar disorder or schizophrenia, as revealed by large-scale DNA microarray analysis. Hum. Mol. Genet., 14: 241–253 Nov 24, Epub 2004.

Iwamoto, K., Bundo, M., Yamamoto, M., Ozawa, H., Saito, T. and Kato, T. (2004a) Decreased expression of NEFH and PCP4/PEP19 in the prefrontal cortex of alcoholics. Neurosci. Res., 49: 379–385.

Iwamoto, K., Kakiuchi, C., Bundo, M., Ikeda, K. and Kato, T. (2004b) Molecular characterization of bipolar disorder by comparing gene expression profiles of postmortem brains of major mental disorders. Mol. Psychiatry, 9: 406–416.

Jiang, L., Lindpaintner, K., Li, H.F., Gu, N.F., Langen, H., He, L. and Fountoulakis, M. (2003) Proteomic analysis of the cerebrospinal fluid of patients with schizophrenia. Amino Acids, 25: 49–57.

Jin, P., Alisch, R.S. and Warren, S.T. (2004) RNA and microRNAs in fragile X mental retardation. Nat. Cell Biol., 6: 1048–1053.

Johnson, G., Brane, D., Block, W., van Kammen, D.P., Gurklis, J., Peters, J.L., Wyatt, R.J., Kirch, D.G., Ghanbari, H.A. and Merril, C.R. (1992a) Cerebrospinal fluid protein variations in common to Alzheimer's disease and schizophrenia. Appl. Theor. Electrophor., 3: 47–53.

Johnson, G., Brane, D., van Kammen, D.P., Gurklis, J., Peters, J.L., Perel, J.M., Ghanbari, H.A. and Merril, C.R. (1992b) Haloperidol-induced CSF protein variations in schizophrenic patients: as studied by two-dimensional electrophoresis. Appl. Theor. Electrophor., 3: 21–26.

Johnston-Wilson, N.L., Sims, C.D., Hofmann, J.P., Anderson, L., Shore, A.D., Torrey, E.F. and Yolken, R.H. (2000) Disease-specific alterations in frontal cortex brain proteins in schizophrenia, bipolar disorder, and major depressive disorder. The Stanley Neuropathology Consortium. Mol. Psychiatry, 5: 142–149.

Kaiser, S., Foltz, L.A., George, C.A., Kirkwood, S.C., Bemis, K.G., Lin, X., Gelbert, L.M. and Nisenbaum, L.K. (2004) Phencyclidine-induced changes in rat cortical gene expression identified by microarray analysis: implications for schizophrenia. Neurobiol. Dis., 16: 220–235.

Kelsoe, J.R., Spence, M.A., Loetscher, E., Foguet, M., Sadovnick, A.D., Remick, R.A., Flodman, P., Khristich, J., Mroczkowski-Parker, Z., Brown, J.L., Masser, D., Ungerleider, S., Rapaport, M.H., Wishart, W.L. and Luebbert, H. (2001) A genome survey indicates a possible susceptibility locus for bipolar disorder on chromosome 22. Proc. Natl. Acad. Sci. USA, 98: 585–590 Jan 9, Epub 2001.

Khawaja, X., Xu, J., Liang, 11. and Barrett, J.E. (2004) Proteomic analysis of protein changes developing in rat hippocampus after chronic antidepressant treatment: implications for depressive' disorders and future therapies. J. Neurosci. Res., 75: 451–460.

Kim, S.I., Voshol, H., van Oostrum, J., Hastings, T.G., Cascio, M. and Glucksman, M.J. (2004) Neuroproteomics: expression profiling of the brain's proteomes in health and disease. Neurochem. Res., 29: 1317–1331.

Konradi, C., Eaton, M., MacDonald, M.L., Walsh, J., Benes, F.M. and Heckers, S. (2004) Molecular evidence for mitochondrial dysfunction in bipolar disorder. Arch. Gen. Psychiatry, 61: 300–308.

Kromer, S.A., Kessler, M.S., Milfay, D., Birg, I.N., Bunck, M., Czibere, L., Panhuysen, M., Putz, B., Deussing, J.M., Holsboer, F., Landgraf, R. and Turck, C.W. (2005) Identification of glyoxalase-I as a protein marker in a mouse model of extremes in trait anxiety. J. Neurosci., 25: 4375–4384.

Kwok, P.Y. (2000) High-throughput genotyping assay approaches. Pharmacogenomics, 1: 95–100.

Kwok, P.Y. and Chen, X. (1998) Detection of single nucleotide variations. Genet. Eng. (N Y), 20: 125–134.

Lachman, H.M., Kelsoe, J.R., Remick, R.A., Sadovnick, A.D., Rapaport, M.H., Lin, M., Pazur, B.A., Roe, A.M., Saito, T. and Papolos, D.F. (1997) Linkage studies suggest a possible locus for bipolar disorder near the velo-cardiofacial syndrome region on chromosome 22. Am. J. Med. Genet., 74: 121–128.

Lander, E.S., Linton, L.M., Birren, B., Nusbaum, C., Zody, M.C., Baldwin, J., Devon, K., Dewar, K., Doyle, M., Fitz-Hugh, W., Funke, R., Gage, D., Harris, K., Heaford, A., Howland, J., Kann, L., Lehoczky, J., LeVine, R., McEwan, P., McKernan, K., Meldrim, J., Mesirov, J.P., Miranda, C., Morris, W., Naylor, J., Raymond, C., Rosetti, M., Santos, R., Sheridan, A., Sougnez, C., Stange-Thomann, N., Stojanovic, N., Subramanian, A., Wyman, D., Rogers, J., Sulston, J., Ainscough, R., Beck, S., Bentley, D., Burton, J., Clee, C., Carter, N., Coulson, A., Deadman, R., Deloukas, P., Dunham, A., Dunham, I., Durbin, R., French, L., Grafham, D., Gregory, S., Hubbard, T., Humphray, S., Hunt, A., Jones, M., Lloyd, C., McMurray, A., Matthews, L., Mercer, S., Milne, S., Mullikin, J.C., Mungall, A., Plumb, R., Ross, M., Shownkeen, R., Sims, S., Waterston, R.H., Wilson, R.K., Hillier, L.W., McPherson, J.D., Marra, M.A., Mardis, E.R., Fulton, L.A., Chinwalla, A.T., Pepin, K.H., Gish, W.R., Chissoe, S.L., Wendl, M.C., Delehaunty, K.D., Miner, T.L., Delehaunty, A., Kramer, J.B., Cook, L.L., Fulton, R.S., Johnson, D.L., Minx, P.J., Clifton, S.W., Hawkins, T., Branscomb, E., Predki, P., Richardson, P., Wenning, S., Slezak, T., Doggett, N., Cheng, J.F., Olsen, A., Luca, S.S., Elkin, C., Uberbacher, E., Frazier, M., et al. (2001) Initial sequencing and analysis of the human genome. Nature, 409: 860–921.

Landgrebe, J., Welzl, G., Metz, T., van Gaalen, M.M., Ropers, H., Wurst, W. and Holsboer, F. (2002) Molecular characterisation of antidepressant effects in the mouse brain using gene expression profiling. J. Psychiatr. Res., 36: 119–129.

Lehrmann, E., Oyler, J., Vawter, M.P., Hyde, T.M., Kolachana, B., Kleinman, J.E., Huestis, M.A., Becker, K.G. and Freed, W.J. (2003) Transcriptional profiling in the human prefrontal cortex: evidence for two activational states associated with cocaine abuse. Pharmacogenom. J., 3: 27–40.

Lewis, C.M., Levinson, D.F., Wise, L.H., DeLisi, L.E., Straub, R.E., Hovatta, I., Williams, N.M., Schwab, S.G., Pulver, A.E., Faraone, S.V., Brzustowicz, L.M., Kaufmann, C.A., Garver, D.L., Gurling, H.M., Lindholm, E., Coon, H., Moises, H.W., Byerley, W., Shaw, S.H., Mesen, A., Sherrington, R., O'Neill, F.A., Walsh, D., Kendler, K.S., Ekelund, J., Paunio, T., Lonnqvist, J., Peltonen, L., O'Donovan, M.C., Owen, M.J., Wildenauer, D.B., Maier, W., Nestadt, G., Blouin, J.L., Antonarakis, S.E., Mowry, B.J., Silverman, J.M., Crowe, R.R., Cloninger, C.R., Tsuang, M.T., Malaspina, D., Harkavy-Friedman, J.M., Svrakic, D.M., Bassett, A.S., Holcomb, J., Kalsi, G., McQuillin, A., Brynjolfson, J., Sigmundsson, T., Petursson, H., Jazin, E., Zoega, T. and Helgason, T. (2003) Genome scan metaanalysis of schizophrenia and bipolar disorder, part II: schizophrenia. Am. J. Hum. Genet., 73: 34–48 Jun 11, Epub 2003.

Lewohl, J.M., Wang, L., Miles, M.F., Zhang, L., Dodd, P.R. and Harris, R.A. (2000) Gene expression in human alcoholism: microarray analysis of frontal cortex. Alcohol Clin. Exp. Res., 24: 1873–1882.

Lockhart, D.J. and Winzeler, E.A. (2000) Genomics, gene expression and DNA arrays. Nature, 405: 827–836.

Maccarrone, G., Milfay, D., Birg, I., Rosenhagen, M., Holsboer, F., Grimm, R., Bailey, J., Zolotarjova, N. and Turck, C.W. (2004) Mining the human cerebrospinal fluid proteome by immunodepletion and shotgun mass spectrometry. Electrophoresis, 25: 2402–2412.

Malhotra, A.K., Murphy, G.M. and Kennedy, J.L. (2004) Pharmacogenetics of psychotropic drug response. Am. J. Psychiatry, 161: 780–796.

Mattick, J.S. and Makunin, I.V. (2005). Small regulatory RNAs in mammals. Hum. Mol. Genet., 14: RI21–RI32.

Mayberg, H.S. (2003) Modulating dysfunctional limbic–cortical circuits in depression: towards development of brain-based algorithms for diagnosis and optimised treatment. Br. Med. Bull., 65: 193–207.

Mayfield, R.D., Lewohl, J.M., Dodd, P.R., Herlihy, A., Liu, J. and Harris, R.A. (2002) Patterns of gene expression are altered in the frontal and motor cortices of human alcoholics. J. Neurochem., 81: 802–813.

Middleton, F.A., Mirnics, K., Pierri, I.N., Lewis, D.A. and Levitt, P. (2002) Gene expression profiling reveals alterations of specific metabolic pathways in schizophrenia. J. Neurosci., 22: 2718–2729.

Middleton, F.A., Pato, C.N., Gentile, K.L., McGann, L., Brown, A.M., Trauzzi, M., Diab, H., Morley, C.P., Medeiros, H., Macedo, A., Azevedo, M.H. and Pato, M.T. (2005a) Gene expression analysis of peripheral blood leukocytes from discordant sib-pairs with schizophrenia and bipolar disorder reveals points of convergence between genetic and functional genomic approaches. Am. J. Med. Genet. B Neuropsychiatry Genet., 136: 12–25.

Middleton, F.A., Peng, L., Lewis, D.A., Levitt, P. and Mirnics, K. (2005b) Altered expression of 14-3-3 genes in the prefrontal cortex of subjects with schizophrenia. Neuropsychopharmacology, 30: 974–983.

Mimmack, M.L., Ryan, M., Baba, H., Navarro-Ruiz, J., Iritani, S., Faull, R.L., McKenna, P.J., Jones, P.B., Arai, H., Starkey, M., Emson, P.C. and Bahn, S. (2002) Gene expression analysis in schizophrenia: reproducible up-regulation of several members of the apolipoprotein L family located in a high-susceptibility locus for schizophrenia on chromosome 22. Proc. Natl. Acad. Sci. USA, 99: 4680–4685.

Mirnics, K., Middleton, F.A., Lewis, D.A. and Levitt, P. (2001a) Analysis of complex brain disorders with gene expression microarrays: schizophrenia as a disease of the synapse. Trends Neurosci., 24: 479–486.

Mirnics, K., Middleton, F.A., Marquez, A., Lewis, D.A. and Levitt, P. (2000) Molecular characterization of schizophrenia viewed by microarray analysis of gene expression in prefrontal cortex. Neuron, 28: 53–67.

Mirnics, K., Middleton, F.A., Stanwood, G.D., Lewis, D.A. and Levitt, P. (2001b) Disease-specific changes in regulator of G-protein signaling 4 (RGS4) expression in schizophrenia. Mol. Psychiatry, 6: 293–301.

Morris, D.W., Rodgers, A., McGhee, K.A., Schwaiger, S., Scully, P., Quinn, J., Meagher, D., Waddington, J.L., Gill, M. and Corvin, A.P. (2004) Confirming RGS4 as a susceptibility gene for schizophrenia. Am. J. Med. Genet. B Neuropsychiatry Genet., 125: 50–53.

Mouledous, L., Hunt, S., Harcourt, R., Harry, J., Williams, K.L. and Gutstein, H.B. (2003a) Navigated laser capture microdissection as an alternative to direct histological staining for proteomic analysis of brain samples. Proteomics, 3: 610–615.

Mouledous, L., Hunt, S., Harcourt, R., Harry, J.L., Williams, K.L. and Gutstein, H.B. (2003b) Proteomic analysis of immunostained, laser-capture microdissected brain samples. Electrophoresis, 24: 296–302.

Nemeroff, C.B. (1996) The corticotropin-releasing factor (CRF) hypothesis of depression: new findings and new directions. Mol. Psychiatry, 1: 336–342.

Niculescu 3rd., A.B., Segal, D.S., Kuczenski, R., Barrett, T., Hauger, R.L. and Kelsoe, I.R. (2000) Identifying a series of candidate genes for mania and psychosis: a convergent functional genomics approach. Physiol. Genomics, 4: 83–91.

Ogden, C.A., Rich, M.E., Schork, N.J., Paulus, M.P., Geyer, M.A., Lohr, J.B., Kuczenski, R. and Niculescu, A.B. (2004) Candidate genes, pathways and mechanisms for bipolar (manic-depressive) and related disorders: an expanded convergent functional genomics approach. Mol. Psychiatry, 9: 1007–1029.

Palotas, M., Palotas, A., Puskas, L.G., Kitajka, K., Pakaski, M., Janka, Z., Molnar, J., Penke, B. and Kalman, J. (2004) Gene expression profile analysis of the rat cortex following treatment with imipramine and citalopram. Int. J. Neuropsychopharmacol., 7: 401–413 Jul 26, Epub 2004.

Pariante, C.M. and Miller, A.H. (2001) Glucocorticoid receptors in major depression: relevance to pathophysiology and treatment. Biol. Psychiatry, 49: 391–404.

Paulson, L., Martin, P., Ljung, E., Blennow, K. and Davidsson, P. (2004a) Effects on rat thalamic proteome by acute and subchronic MK-801-treatment. Eur. J. Pharmacol., 505: 103–109.

Paulson, L., Martin, P., Nilsson, C.L., Ljung, E., Westman-Brinkmalm, A., Blennow, K. and Davidsson, P. (2004b) Comparative proteome analysis of thalamus in MK-801-treated rats. Proteomics, 4: 819–825.

Paulson, L., Martin, P., Persson, A., Nilsson, C.L., Ljung, E., Westman-Brinkrnalm, A., Eriksson, P.S., Blennow, K. and Davidsson, P. (2003) Comparative genome- and proteome analysis of cerebral cortex from MK-801-treated rats. J. Neurosci. Res., 71: 526–533.

Pierce, A. and Small, S.A. (2004) Combining brain imaging with microarray: isolating molecules underlying the physiologic disorders of the brain. Neurochem. Res., 29: 1145–1152.

Pongrac, J.L., Middleton, F.A., Peng, L., Lewis, D.A., Levitt, P. and Mirnics, K. (2004) Heat shock protein 12A shows reduced expression in the prefrontal cortex of subjects with schizophrenia. Biol. Psychiatry, 56: 943–950.

Prasad, K.M., Chowdari, K.V., Nimgaonkar, V.L., Talkowski, M.E., Lewis, D.A. and Keshavan, M.S. (2005) Genetic polymorphisms of the RGS4 and dorsolateral prefrontal cortex morphometry among first episode schizophrenia patients. Mol. Psychiatry, 10: 213–219.

Risch, N. and Merikangas, K. (1996) The future of genetic studies of complex human diseases [see comments]. Science, 273: 1516–1517.

Santoni, V., Molloy, M. and Rabilloud, T. (2000) Membrane proteins and proteomics: un amour impossible? Electrophoresis, 21: 1054–1070.

Scammell, J.G., Denny, W.B., Valentine, D.L. and Smith, D.F. (2001) Overexpression of the FK506-binding immunophilin FKBP51 is the common cause of glucocorticoid resistance in three New World primates. Gen. Comp. Endocrinol., 124: 152–165.

Schwarz, M.J., Muller, N., Riedel, M. and Ackenheil, M. (2001) The Th2-hypothesis of schizophrenia: a strategy to identify a subgroup of schizophrenia caused by immune mechanisms. Med. Hypotheses, 56: 483–486.

Schwarz, M.J., Riedel, M., Gruber, R., Ackenheil, M. and Muller, N. (1999) Antibodies to heat shock proteins in schizophrenic patients: implications for the mechanism of the disease. Am. J. Psychiatry, 156: 1103–1104.

Serretti, A. (2004) Pharmacogenetics of antidepressants. Clin. Neuropsychiatry, 1: 79–90.

Shimojo, M., lmai, Y., Nakajima, K., Mizushima, S., Demura, A. and Kohsaka, S. (1993) Interleukin-2 enhances the viability of primary cultured rat neocortical neurons. Neurosci. Lett., 151: 170–173.

Sugai, T., Kawamura, M., lritani, S., Araki, K., Makifuchi, T., lmai, C., Nakamura, R., Kakita, A., Takahashi, H. and Nawa, H. (2004) Prefrontal abnormality of schizophrenia revealed by DNA microarray: impact on glial and neurotrophic gene expression. Ann. N Y Acad. Sci., 1025: 84–91.

Syvanen, A.C. (2001) Accessing genetic variation: genotyping single nucleotide polymorphisms. Nat. Rev. Genet., 2: 930–942.

Tanna, V.L., Wilson, A.F., Winokur, G. and Elston, R.C. (1989) Linkage analysis of pure depressive disease. J. Psychiatry Res., 23: 99–107.

Taymans, J.M., Kia, H.K., Claes, R., Cruz, C., Leysen, J. and Langlois, X. (2004) Dopamine receptor-mediated regulation of RGS2 and RGS4 mRNA differentially depends on ascending dopamine projections and time. Eur. J. Neurosci., 19: 2249–2260.

Taymans, J.M., Leysen, J.E. and Langlois, X. (2003) Striatal gene expression of RGS2 and RGS4 is specifically mediated by dopamine D1 and D2 receptors: clues for RGS2 andRGS4 functions. J. Neurochem., 84: 1118–1127.

Thomson, J.M., Parker, J., Perou, C.M. and Hammond, S.M. (2004) A custom microarray platform for analysis of microRNA gene expression. Nat. Methods, 1: 47–53 Sep 29, Epub 2004.

Tkachev, D., Mimmack, M.L., Ryan, M.M., Wayland, M., Freeman, T., Jones, P.B., Starkey, M., Webster, M.J., Yolken, R.H. and Bahn, S. (2003) Oligodendrocyte dysfunction in schizophrenia and bipolar disorder. Lancet, 362: 798–805.

Tsuang, M.T., Nossova, N., Yager, T., Tsuang, M.M., Guo, S.C., Shyu, K.G., Glatt, S.J. and Liew, C.C. (2005) Assessing the validity of blood-based gene expression profiles for the classification of schizophrenia and bipolar disorder: a preliminary report. Am. J. Med. Genet. B Neuropsychiatry Genet., 133: 1–5.

Vawter, M.P., Barrett, T., Cheadle, C., Sokolov, B.P., Wood 3rd.,, W.H., Donovan, D.M., Webster, M., Freed, W.J. and Becker, K.G. (2001) Application of cDNA microarrays to examine gene expression differences in schizophrenia. Brain Res. Bull., 55: 641–650.

Vawter, M.P., Ferran, E., Galke, B., Cooper, K., Bunney, W.E. and Byerley, W. (2004a) Microarray screening of lymphocyte gene expression differences in a multiplex schizophrenia pedigree. Schizophr. Res., 67: 41–52.

Vawter, M.P., Shannon Weickert, C., Ferran, E., Matsumoto, M., Overman, K., Hyde, T.M., Weinberger, D.R., Bunney, W.E. and Kleinman, J.E. (2004b) Gene expression of metabolic enzymes and a protease inhibitor in the prefrontal cortex are decreased in schizophrenia. Neurochem. Res., 29: 1245–1255.

Venter, J.C., Adams, M.D., Myers, E.W., Li, P.W., Mural, R.J., Sutton, G.G., Smith, H.O., Yandell, M., Evans, C.A., Holt, R.A., Gocayne, J.D., Amanatides, P., Ballew, R.M., Huson, D.H., Wortman, J.R., Zhang, Q., Kodira, C.D., Zheng, X.H., Chen, L., Skupski, M., Subramanian, G., Thomas, P.D., Zhang, J., Gabor Miklos, G.L., Nelson, C., Broder, S., Clark, A.G., Nadeau, J., McKusick, V.A., Zinder, N., Levine, A.J., Roberts, R.J., Simon, M., Slayman, C., Hunkapiller, M., Bolanos, R., Deicher, A., Dew, I.,

Fasulo, D., Flanigan, M., Florea, L., Halpern, A., Hannenhalli, S., Kravitz, S., Levy, S., Mobarry, C., Reinert, K., Remington, K., Abu-Threideh, J., Beasley, E., Biddick, K., Bonazzi, V., Brandon, R., Cargill, M., Chandramouliswaran, I., Charlab, R., Chaturvedi, K., Deng, Z., Di Francesco, V., Dunn, P., Eilbeck, K., Evangelista, C., Gabrielian, A.E., Gan, W., Ge, W., Gong, F., Gu, Z., Guan, P., Heiman, T.J., Higgins, M.E., Ji, R.R., Ke, Z., Ketchum, K.A., Lai, Z., Lei, Y., Li, Z., Li, J., Liang, Y., Lin, X., Lu, F., Merkulov, G.V., Milshina, N., Moore, H.M., Naik, A.K., Narayan, V.A., Neelam, B., Nusskern, D., Rusch, D.B., Salzberg, S., Shao, W., Shue, B., Sun, J., Wang, Z., Wang, A., Wang, X., Wang, J., Wei, M., Wides, R., Xiao, C., Yan, C., et al. (2001) The sequence of the human genome. Science, 291: 1304–1351.

Weaver, I.C., Cervoni, N., Champagne, F.A., D'Alessio, A.C., Sharma, S., Seckl, J.R., Dymov, S., Szyf, M. and Meaney, M.J. (2004) Epigenetic programming by maternal behavior. Nat. Neurosci., 7: 847–854 Jun 27, Epub 2004.

Williams, N.M., Preece, A., Spurlock, G., Norton, N., Williams, H.J., McCreadie, R.G., Buckland, P., Sharkey, V., Chowdari, K.V., Zammit, S., Nimgaonkar, V., Kirov, G., Owen, M.J. and O'Donovan, M.C. (2004) Support for RGS4 as a susceptibility gene for schizophrenia. Biol. Psychiatry, 55: 192–195.

Yamada, M., Yamazaki, S., Takahashi, K., Nara, K., Ozawa, H., Yamada, S., Kiuchi, Y., Oguchi, K., Kamijima, K., Higuchi, T. and Momose, K. (2001) Induction of cysteine string protein after chronic antidepressant treatment in rat frontal cortex. Neurosci. Lett., 301: 183–186.

Yamada, M., Yamazaki, S., Takahashi, K., Nishioka, G., Kudo, K., Ozawa, H., Yamada, S., Kiuchi, Y., Kamijima, K., Higuchi, T. and Momose, K. (2000) Identification of a novel gene with RING-H2 finger motif induced after chronic antidepressant treatment in rat brain. Biochem. Biophys. Res. Commun., 278: 150–157.

Yao, J.K., Leonard, S. and Reddy, R.D. (2000) Membrane phospholipid abnormalities in postmortem brains from schizophrenic patients. Schizophr. Res., 42: 7–17.

You, J.S., Gelfanova, V., Knierman, M.D., Witzmann, F.A., Wang, M. and Hale, J.E. (2005) The impact of blood contamination on the proteome of cerebrospinal fluid. Proteomics, 5: 290–296.

Yuan, X. and Desiderio, D.M. (2005) Proteomics analysis of prefractionated human lumbar cerebrospinal fluid. Proteomics, 5: 541–550.

Yuferov, V., Kroslak, T., Laforge, K.S., Zhou, Y., Ho, A. and Kreek, M.J. (2003) Differential gene expression in the rat caudate putamen after "binge" cocaine administration: advantage of triplicate microarray analysis. Synapse, 48: 157–169.

Zvara, A., Szekeres, G., Janka, Z., Kelemen, J.Z., Cimmer, C., Santha, M. and Puskas, L.G. (2005) Over-expression of dopamine D2 receptor and inwardly rectifying potassium channel genes in drug-naive schizophrenic peripheral blood lymphocytes as potential diagnostic markers. Dis. Markers, 21: 61–69.

CHAPTER 14

Human brain evolution

Hilliary Creely and Philipp Khaitovich*

Max-Planck Institute for Evolutionary Anthropology, Deutscher Platz, D-04103 Leipzig, Germany

Anatomical evolution

The most prominent feature distinguishing humans from all other organisms is our mind. From a biological perspective, study of the mind begins with a study of the brain. For our body size, the human brain is the largest among primate species, roughly three times the size that of our nearest living relative, the chimpanzee (Fig. 1). However, it is not likely that human-specific cognitive abilities can be explained by brain size alone. Several mammalian species, such as dolphins, whales and elephants, surpass humans in brain size but not in cognitive function. Furthermore, certain human congenital disorders lead to a drastic reduction in brain size without an equivalent decrease in cognitive abilities. Mutations in the *MCPH1* (*microcephalin*) and the *ASPM* (*abnormal spindle-like, microcephaly associated*) genes cause a disorder known as primary microcephaly, which is characterized by a severe decrease in human brain size, typically down to 430 g, with retention of overall brain structure (Kumar et al., 2002). Thus, brain size in primary microcephaly patients is comparable to the average adult chimpanzee brain size (410 g) (Herndon et al., 1999). Although patients with this disorder suffer from mental retardation that ranges from mild to severe, their mental abilities far exceed those of the chimpanzee. Nonetheless, the normal human brain is indisputably large. Its size is approximately five times than expected for a mammal of the same size and three times larger than expected for an ape of the same size (Woods et al., 2005). Having such a large brain is quite costly. At the metabolic level, the brain is an expensive organ that consumes 20–25% of total body energy at rest. At the organismal level, the enlarged cranium of a human fetus causes birthing difficulties resulting in high rate (0.5–1%) of labor-related maternal mortality, a rate much higher than observed in other mammals. As a consequence of this problem, the human female has a broader pelvis that decreases the efficiency of bipedal locomotion (reviewed in Gilbert et al., 2005). In light of such high fitness costs for the disproportionately large human brain, the increase in brain size must have conferred a great enough adaptive benefit to our ancestors to justify this cost. Thus, while brain size alone cannot account for the whole of human cognitive abilities, it appears to be one essential human brain feature.

Although decades of studies have brought us incrementally closer to deciphering the functional mechanisms underlying human-specific cognitive abilities, our knowledge remains very limited. Anatomical studies have provided some insight on functional differences between human and chimpanzee brain; however, these differences are not always obvious. On the anatomical level, all differences between humans and other primates known to date are quantitative, rather than qualitative. Compared to chimpanzees and other non-human primates, humans possess an enlarged cerebral cortex (Rilling and Insel, 1999). This enlargement is believed to reflect a longer period of neuronal formation during pre-natal development in humans

*Corresponding author. E-mail: khaitovich@eva.mpg.de

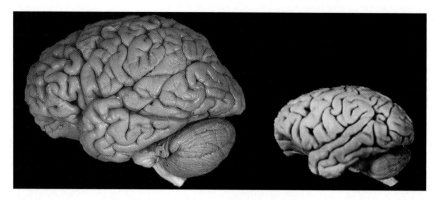

Fig. 1. Relative size comparison between adult human (left) and adult chimpanzee (right) brains.

(Kornack and Rakic, 1998). However, the precise molecular mechanisms determining the neuronal formation process in humans and other primates remain unknown. While brain size is one clearly definable feature, most functional differences distinguishing human brains from the brains of other primate species are much less obvious. For example, upper layers of the cerebral cortex are overrepresented in humans compared to other primates (Marin-Padilla, 1992). Similarly, special types of neurons called spindle cells are overrepresented in anterior cingulate cortex in the human brain compared to those of great apes (Nimchinsky et al., 1999). One of the most well-defined quantitative human-specific brain features is functional asymmetry of the cerebral cortex (Kandel et al., 2000). This asymmetry is most clearly reflected in fact that approximately 90% of humans are right handed. Other species, including chimpanzees, do have "handedness" on the individual level and some degree of right hand preference for certain tasks (Hopkins et al., 2005). However, they do not show the strong left–right asymmetry on the population level observed in humans. Distribution of other cognitive functions in humans is also asymmetric, most famously illustrated by language, which is preferentially localized in the left hemisphere (Broca, 1861; Wernicke, 1874). Indeed, functional specialization between human brain hemispheres may be a direct result of new cognitive abilities acquired during human evolution, with new functions, such as language, replacing an existing specialization in a given brain area (Corballis et al., 2000). This functional lateralization of the human brain is accompanied by anatomical asymmetry in the frontal and temporal lobes. Namely, the size of the cortical areas involved in language and speech production in humans, located in the frontal and temporal lobes, are typically larger in the left than in the right hemisphere (Geschwind and Levitsky, 1968; Amunts et al., 1999). These anatomical features were believed to be directly associated with language abilities and thus to be human-specific. However, recent studies indicate that chimpanzee brains show some asymmetries in the frontal and temporal lobes on the gross-anatomical level as well, although these asymmetries are less pronounced than those in human brains (Gannon et al., 1998; Hopkins et al., 1998; Cantalupo and Hopkins, 2001). Thus, the anatomical asymmetry observed in humans in the frontal and temporal lobes is another quantitative, rather than a qualitative, human-specific feature. With the exception of such quantitative differences, no anatomical features uniquely specific to the human brain have yet been identified.

In conclusion, although anatomical studies provided some important clues about human brain features that may be related to human-specific cognitive abilities, the molecular mechanisms underlying these cognitive abilities remain obscure. Another approach to understand these mechanisms is to identify molecular changes associated with the appearance of human-specific cognitive traits. Recent advances in large-scale analysis of gene structure and gene regulation in humans and non-human primates (predominantly chimpanzees) provide insight on how such approaches can help uncover changes important for human-specific cognitive abilities. These efforts, the new perspectives they create, and the difficulties that arise from them, will be the focus of the remainder of this chapter.

Protein sequence evolution

The molecular basis for the functional features that distinguish the human brain from that of our ancestors and other primate species lies in DNA sequence changes that happened on the human lineage after separation from the common human–chimpanzee ancestor 5–7 million years ago (Glazko and Nei, 2003). DNA sequence changes may influence phenotype by altering either amino acid sequence in proteins or changing gene expression regulation (Fig. 2). With the nearly completed human and chimpanzee genome sequences, all DNA sequence changes that determine phenotypic differences between species, including those related to cognitive functions, can, in principle, be identified. While changes in protein amino acid composition are relatively easy to identify based on genome sequence information, the vast majority of these changes are expected to be neutral with respect to phenotype. The notion that most amino acid substitutions that accumulate during evolution do not affect phenotype was first suggested by Motoo Kimura in 1968 and formally described in 1983 as the "Neutral Theory of Molecular Evolution" (Kimura, 1983). In practical terms, this means that only a small fraction of all DNA sequence changes observed between humans and chimpanzees are relevant to phenotypic differences between these two species. This makes identification of non-neutral changes a daunting task. As a consequence, only a very limited number of studies to date have been successful in bridging DNA sequence and phenotypic differences between species in general, and between humans and chimpanzees in particular. Nevertheless, there are several examples where protein changes have been linked to the evolution of human-specific cognitive features.

Arguably, the most prominent example of such changes discovered to date is given by the gene *FOXP2 (forkhead box P2)*. In humans, disruption of this gene causes a speech impairment, usually characterized by poor oralfacial movement control and accompanied by a disregulation of brain activity associated with speech (Lai et al., 2001; Vargha-Khadem et al., 2005). Although *FOXP2* is one of the most conserved genes in mammals, with only three amino acid substitutions occurring since the divergence of human and mouse species more

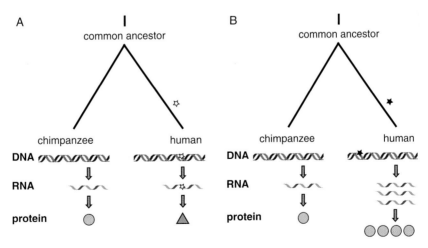

Fig. 2. Schematic representation of DNA sequence mutations on the human lineage (represented by a star) inducing structural (A) or regulatory (B) changes. Structural changes happen due to mutations within the protein-coding region and lead to an alteration in protein functional properties. Regulatory changes occur due to mutations that can take place almost anywhere in a genome and lead to an alteration in protein quantity.

than 80–120 million years ago, two of these substitutions have occurred on the human lineage within last 5–6 million years ago (Enard et al., 2002b). Additionally, an unusual pattern of nucleotide polymorphism around this gene is observed in extant humans and indicates that these mutations likely occurred within the last 200,000 years of human evolution, and then spread among the entire human population due to positive selection. These data strongly implicate *FOXP2* in the appearance of speech and language during recent human evolution, traits that likely provided a selective advantage to our ancestors. Interestingly, there are indications that *FOXP2* may be involved in evolution of vocal communication in other species as well. In mice, disruption of both copies of the *FOXP2* gene completely abolishes ultrasonic vocalization normally observed when pups are temporally removed from their mothers, while disruption of one copy of the gene significantly reduces these vocalizations (Shu et al., 2005). In song-learning birds such as zebra finches, the striatal nucleus Area X, necessary for vocal learning, expresses more *FOXP2* than the surrounding tissue during the period when vocal learning occurs (Haesler et al., 2004). Similarly, in adult canaries, transient increase in *FOXP2* expression in Area X is associated with instances of increased song variation. These examples strengthen the link between changes in FOXP2 and evolution of vocal communication, while highlighting the fact that even this trend cannot be characterized as uniquely human. Interestingly, while evolutionary adaptations in human *FOXP2* took place on the gene sequence level, in birds, these adaptations seem to have occurred on the level of gene expression regulation.

Other prominent examples of positive selection on the human lineage at the level of protein sequence are two genes involved in the regulation of brain size, *MCPH1* and *ASPM1*. Disruption of either of these genes in humans leads to primary microcephaly, a disorder described earlier in this chapter that is characterized by a dramatic reduction in brain size (Kumar et al., 2002). For both genes, an unusual increase in the number of amino acid substitutions is observed during ape evolution and particularly within the human lineage (Evans et al., 2004a, b). Moreover, adaptive differences between isoforms of these two proteins exist among extant humans (Evans et al., 2005; Mekel-Bobrov et al., 2005). Thus, although the major brain size increase on the human evolutionary lineage likely began as much as 2 million years ago (McHenry, 1994), more recent positive selection on *MCPH1* and *ASPM1* indicates that brain size regulation has probably played an important role in establishing modern brain functionality. It is particularly intriguing to think that the human brain may have been undergoing adaptive evolution in our very recent past. However, differences between distinct forms of *MCPH1* and *ASPM1* found in contemporary humans have not yet been directly linked to differences in cognitive abilities. Moreover, it is not exactly clear whether slight differences in cognition may confer any benefit to individuals in terms of reproductive success in contemporary society.

Other known examples of genes that show adaptive evolutionary changes at the protein sequence level include *MAOA* (*monoamine oxidase A*), *AHI1* (*Abelson helper integration site 1*), *GLUT2* (*glutamate dehydrogenase 2*) and *MYH16* (*myosin, heavy polypeptide 16*), genes potentially involved in the formation of human brain. Monoamine oxidase A (*MAOA*) has been implicated in different aspects of human behavior and cognition in a multitude of studies, though not without controversy (Andres et al., 2004; and references therein). Interestingly, patterns of polymorphism observed in this gene among contemporary humans indicate that it has undergone positive selection recently in human evolution (Gilad et al., 2002). This selection is human-specific and is not observed in apes (Andres et al., 2004). Structurally, the human form of *MAOA* contains only one non-synonymous amino acid substitution that may potentially influence protein dimerization properties. Although it is not clear whether selection acted on this structural change or on an as yet unknown regulatory change, the fact that *MAOA* was positively selected during human evolution certainly warrants further investigation of its potential role in the rise of human cognitive abilities.

Mutations disrupting the *AHI1* gene cause Joubert syndrome, a disorder characterized by abnormal axonal crossing in the brain that results in motor disturbances, such as severe clumsiness and abnormal breathing and eye movements. In addition, patients with Joubert syndrome suffer from cognitive impairments and autistic behaviors (Saraiva and Baraitser, 1992; Ozonoff et al., 1999). While *AHI1* evolved unusually rapidly in all apes at the amino acid sequence level, this trend is accelerated on the human lineage (Ferland et al., 2004). Although this acceleration cannot be taken as proof that changes in the *AHI1* gene were directly connected to the evolution of human brain functions, it provides an indication for the adaptive value of these changes during human evolution.

GLUT2 and its closely related homologue *GLUT1* encode two forms of an enzyme (glutamate dehydrogenase) essential for the recycling of glutamate, a major excitatory neurotransmitter (Plaitakis et al., 2003). While *GLUT1* encodes an evolutionarily old, housekeeping-form of an enzyme, *GLUT2* is an ape-specific gene that that is believed to have originated by retrotransposition of a processed *GLUT1* copy less than 23 million years ago (Burki and Kaessmann, 2004). *GLUT1* remained unchanged after this event, whereas *GLUT2* rapidly evolved resulting in an enzyme form that is adapted to functioning in brain. Thus, *GLUT2* may have contributed to the evolution of enhanced brain function by accelerating neurotransmitter turnover rates in humans and apes. This adaptation took place both on the structural level, through multiple substitutions in the amino acid sequence of the protein, and on the regulatory level, resulting in preferential expression of the modified enzyme version in neural tissue.

MYH16 is an isoform of a sarcomeric myosin, a main structural component of muscle tissue. *MYH16* has been inactivated by a fame shift mutation in humans and this inactivation is associated with marked reduction of chewing musculature. The emergence of this inactivating mutation was dated to 2.4 million years ago, a time point that coincides with the commencement of brain size expansion on the hominid lineage (Stedman et al., 2004). Thus, it has been speculated that loss of the massive chewing musculature attached to the scull may have facilitated brain expansion by allowing an unconstrained increase in the scull volume. However, this link is contested based on the observation that Neanderthals possessed massive chewing musculature and, at the same time, brains that were on average slightly greater in volume than the brains of modern humans (Pennisi, 2004). In addition, another study analyzed a larger data set and estimated the age of inactivating mutation at 5.3 million years ago (Perry et al., 2005). The new date puts a large time gap between loss of *MYH16* function and brain size expansion on the hominid limeade, making association between these two events less plausible.

The preceding examples highlight the fact that genes involved in the evolution of human cognition can potentially be identified. However, these individual gene studies do not provide any estimate of the overall number of proteins that had to undergo changes during human evolution in order for the human brain to reach its contemporary state. A recent study compared protein evolution rates in primates, including humans, and rodents in two manually selected gene sets: 214 proteins implicated in various aspects of human nervous system function and 95 housekeeping genes (Dorus et al., 2004). The set of nervous system proteins evolved 30% faster in primates than in rodents, while no significant difference was found for the housekeeping set. This finding would imply that evolution of the human nervous system required a large number of amino acid changes in many proteins. However, amino acid sequence comparisons between humans and chimpanzees for more than 13,000 proteins did not confirm this observation (Consortium, 2005). In contrast, this study found that genes involved in neuronal function or genes expressed in human brain tend to show a slower rate of change in amino acid composition compared to other genes (Consortium, 2005; Khaitovich et al., 2005). Despite this rate difference, genes expressed in brain tend to have more amino acid substitutions on the human lineage than on the chimpanzee lineage (Khaitovich et al., 2005). Thus, it appears that human brain evolution required substantially more changes in protein sequence than evolution of the chimpanzee brain. This finding may not be too surprising given that, since the separation from the most recent common ancestor, much more profound changes in brain size and capacity happened on the human than on the chimpanzee lineage. Further, it indicates that many more genes involved in the evolution of human cognition may be identified based on the patterns of protein structure changes between humans and other primates.

Gene expression evolution

As mentioned earlier in this chapter, the phenotypic differences we observe between species can be caused by changes in protein sequence as well as by changes in regulation of gene expression. Indeed, differences in gene expression can cause dramatic phenotypic changes. At all stages of development, an organism's genome remains unchanged. Thus, all of the organism's phenotypic changes throughout development, from egg to adult, are caused by changes in gene regulation. Similarly, within an organism, the great phenotypic and functional variation seen between different tissues is caused by differences in gene regulation. So, it is conceivable that in the most extreme case, all observed phenotypic differences between humans and chimpanzees could have been caused by differences in gene expression alone. The notion that the majority of phenotypic differences between humans and chimpanzees are caused by such gene regulation differences, and not by differences in protein sequence, was first introduced by Allen Wilson and Mary Claire King in 1975 (King and Wilson, 1975). Their pioneering work comparing both genome annealing kinetics and protein electrophoresis migration patterns in the two species revealed that the human and chimpanzee genomes differ, on average, in about 1% of their sequences and that a large proportion of proteins do not show any structural differences between species. Thirty years later, comparison between the completed human and chimpanzee genome sequences provided undisputable verification of these measurements. Indeed, human and chimpanzee genome sequences differ from one another by an average of 1.23% on the point mutation level and approximately 1/3 of the 13,454 proteins compared between the two species have no amino acid substitutions (Consortium, 2005). Nevertheless, an average sequence difference of 1.23% adds up to more than 30 million point mutations between humans and chimpanzees. In addition to these point mutations, genomic differences between the two species include a multitude of small and large sequence insertions, deletions, duplications and inversions. These rearrangements vary in scale from genome regions spanning only a few nucleotides to whole chromosome changes. Such large-scale rearrangement is most dramatically illustrated by the fusion of two ancestral chromosomes during human evolution producing modern human chromosome 2, thereby creating the chromosome number difference between humans and chimpanzees we observe today (Wienberg et al., 1994). All of these genomic differences can potentially contribute to differences in gene regulation between the two species. However, while differences in protein amino acid composition can be deduced from comparisons between the human and chimpanzee genomes, differences in gene regulation cannot, at present, be recognized based on genomic information alone. Several factors contribute to this limitation: first, regulatory sequence elements for a given gene can be located anywhere in a genome; second, multiple sequence elements and protein factors regulate expression of each gene; third, expression regulation differs between tissues; and fourth, expression regulation can change with time depending on input from within and from outside the organism.

Fortunately, the gene expression levels of all known genes can be measured directly using microarray technology and this information can be used to study evolution (Fig. 3). In recent years, several studies used microarrays to investigate expression differences between humans, chimpanzees and other primates in brain and other tissues. As a result of this work, the following conclusions emerge. First, few genes can be described as "brain-specific" in terms of their expression in humans (Su et al., 2002, 2004). The vast majority of genes are expressed in many tissues and in many different brain regions. Thus, differences in gene expression between tissues are mainly quantitative rather than qualitative. Second, approximately 5–10% of genes expressed in adult brain are expressed at significantly different levels in humans and chimpanzees (Enard et al., 2002a, b; Caceres

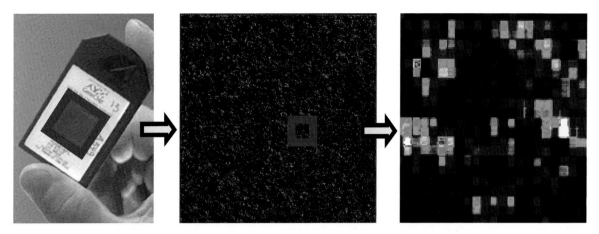

Fig. 3. Schematic representation of an Affymetrix® microarray, one of the most common types of microarrays currently used in large-scale gene expression analysis. While the size of the microarray field highlighted in red on the first panel is approximately 1 × 1 cm, the size of features corresponding to different microarray probes is approximately 10 × 10 μm.

et al., 2003; Enard et al., 2002a; Khaitovich et al., 2004a; Uddin et al., 2004). This may seem like a small percentage of genes but, given that more than half of all known human genes are expressed in brain, this adds up to approximately 1000 differently expressed genes. The proportion of genes differently expressed between humans and chimpanzees in brain is similar to that observed in other tissues like heart, kidney or liver (Khaitovich et al., 2005). However, the amplitude of expression differences observed in brain in significantly smaller than that observed in the other tissues studied to date, indicating that brain is one of the most conserved tissues in terms of gene regulation. Interestingly, genes expressed in brain show significantly fewer amino acid differences between humans and chimpanzees compared to genes expressed in the other tissues (Khaitovich et al., 2005). Thus, brain-expressed genes are under more constraint on both regulatory and structural levels than genes expressed in other tissues. Third, when expression differences observed between humans and chimpanzees are divided into two categories, one that occurred on the human lineage and the other one that occurred on the chimpanzee lineage, the differences appear to be distributed unequally. Namely, in brain, noticeably more regulatory changes have occurred on the human than on the chimpanzee lineage (Caceres et al., 2003; Enard et al., 2002a; Gu and Gu, 2003; Hsieh et al., 2003; Khaitovich et al., 2005). This asymmetry is observed in other tissues as well, but never to the extent seen in brain. Fourth, distinct brain regions involved in very different cognitive functions, such Broca's area, prefrontal, cingulate and primary visual regions of the cerebral cortex, and even non-cortical areas such as caudate nucleus and cerebellum, show almost the same expression differences between humans and chimpanzees (Fig. 4) (Khaitovich et al., 2004a). Thus, functionally distinct areas of the human brain appear to diverge equally far on the gene regulation level between humans and chimpanzees. Fifth, within brain, differences between various brain regions appear to be highly conserved in primate species. For example, all expression differences observed between various cortical regions in humans are also found in chimpanzees and even in rhesus macaque brains (Khaitovich et al., 2004a). Additionally, more ancient regions of the brain show more expression differences compared to younger brain regions (Fig. 5) (Khaitovich et al., 2004b). Hence, regulatory divergence of different brain areas does occur with time, but at a much slower pace than regulatory divergence between different species. This is not surprising given that regulatory changes within the brain must occur on the same genotypic background while between species genetic

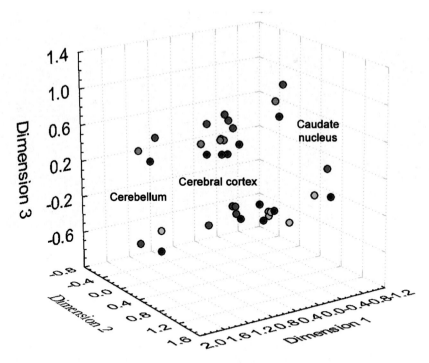

Fig. 4. Multidimensional scaling plot representing expression differences observed in brain within and between species. Each circle represents one sample. Chimpanzee samples are shown in red and human in blue. Different shades of color represent different individuals. Different brain regions form the same pattern in both species with most expression differences found between cerebellum, cortex and caudate nucleus. Despite large differences in expression between these regions within species, they all differ between species to the same extent.

divergence can cause more rapid accumulation of regulatory differences. However, genotypic changes between species are the result of random mutations that should be largely neutral or deleterious for the organism and only very rarely lead to adaptive changes. Thus, the question once asked for structural changes during the early days of genome studies now arises for studies of regulatory evolution: How many of the regulatory differences observed between species are relevant to differences in phenotype and how many differences are neutral?

Theory of gene expression evolution

In the 1950–1960s, at the dawn of DNA sequence studies, every mutation resulting in an amino acid substitution was thought to yield a functional change that would be either deleterious or advantageous for the organism. Within a population, it was believed that advantageous mutations would be quickly swept to fixation by Darwinian (positive) selection, while deleterious mutations would be weeded out by purifying (negative) selection (Darwin, 1859). This view persisted until the end of 1960s, when Motoo Kimura proposed the neutral theory of molecular evolution. As mentioned earlier in this chapter, this groundbreaking theory postulated that the vast majority of DNA sequence substitutions observed both within and between species have no effect on an organism's phenotype and thus are evolutionary neutral (Kimura, 1983). This theory stemmed from the observation that the number of amino acid substitutions between species far exceeds the number expected, assuming total functionality for each substitution (Kimura, 1968). Kimura's theory that most mutations are evolutionary neutral solves this contradiction and, at the same time, predicts that DNA mutations leading to amino acid substitutions should accumulate linearly with time and show little

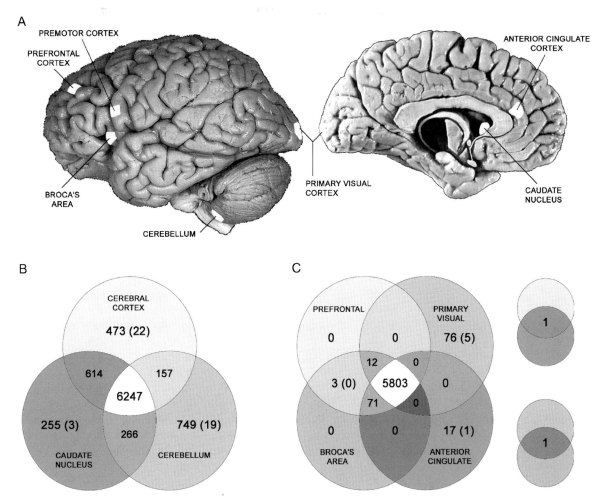

Fig. 5. Gene expression analysis of six brain regions in humans and chimpanzees. (A) Location of studied brain regions in the human brain. (B and C) Diagram showing numbers of significant expression differences observed between different brain regions in humans. Numbers in parenthesis indicate numbers of genes where differences between brain regions found in humans were not seen in chimpanzees. While different brain regions like cortex, cerebellum and caudate nucleus are substantially different from each other in their expression, almost no differences are seen between different cortical regions. Moreover, the vast majority of expression differences between brain regions found in humans are also observed in chimpanzees.

correlation with phenotypic changes. Decades of experimental studies unambiguously confirmed these predictions, making the neutral theory of molecular evolution a widely accepted null-hypothesis for every observed structural change. Importantly, this theoretical model of neutral evolution (also called the neutral model) allows for the development of tests designed to find significant deviation from the substitution patterns predicted by the model. Such unusual substitution patterns likely indicate sequence differences influenced by positive selection and thus suggest functional significance for the change. In the decades following Kimura's initial observation, several statistical tests to identify such patterns were introduced and have proven instrumental in finding genomic patterns bearing signatures of functional adaptations (Hudson et al., 1987; Tajima, 1989; McDonald and Kreitman, 1991; Fu and Li, 1993; Fay and Wu, 2003). In fact, all examples of genes involved in human brain evolution given thus far in this chapter were identified using these tests.

As described earlier, the introduction of microarray technology in the late 1990s allowed the simultaneous measurement of expression levels for thousands of genes. This important technical advance led to a multitude of studies measuring changes in gene expression due to physiological differences in test organisms, for example, comparisons between control and diseased individuals in cancer studies or between different cell cycle stages in yeast. In most of these cases, expression changes were observed between groups with the same genomic sequence, thus expression differences reflect regulatory variation due to phenotypic or environmental differences. Evolutionary studies, in contrast, usually involve comparisons between groups representing individuals from different species and thus differ not only in their phenotypes but also in their genomic sequences. However, in accordance with the neutral model, the vast majority of genomic sequence differences may not affect phenotype.

Should we then expect that all expression differences observed within and between species cause phenotypic differences? This question is comparable to the one asked in the late 1960s, namely: Should all amino acid differences seen within and between species cause phenotypic differences? In the latter case, as discussed earlier in this chapter, it is widely believed that the answer is no. However, in the case of gene expression, the answer may be less clear. Even for closely related species, such as humans and chimpanzees or different species of mice, at least 5% of all expressed genes show significantly different expression levels between species in every tissue studied thus far (Khaitovich et al., 2005). Adding further complexity to the question, unlike amino acid substitutions, there is no theory to date that allows estimation of the number of expected, functional expression differences. However, it is possible to test whether the predictions of the neutral theory formulated for amino acid changes can also be applied to gene expression evolution. More specifically, do expression changes accumulate linearly with time and show little correlation with phenotypic changes? Experiments in several naturally occurring populations of teleost fish of the genus *Fundulus* showed that, for most genes, expression variation within populations positively correlates with expression variation between populations (Oleksiak et al., 2002). This fits well with predictions of the neutral theory of evolution, which postulates that for most genes variation accumulates with time at the same rate within and between populations. However, a large proportion of expression variation between fish populations in this study depended on the environment in which these populations lived. Interestingly, this expression variation persisted even when all fish were raised in the same environment. This observation contradicts expectations from the neutral theory of evolution and indicates that a measurable proportion of expression changes can correlate with functional alterations. Despite this contradiction, most evolutionary gene expression studies show strong agreement with the neutral theory. Experiments in fruit fly showed that the expression differences between two developmental stages increased proportionally with the genetic distance between the strains and species studied (Rifkin et al., 2003). Similarly, different naturally occurring strains of yeast diverge in their gene expression proportionally to the genetic distance for the vast majority of genes, although a few genes showed a correlation with phenotypic adaptations (Fay et al., 2004). These findings support the notion that expression changes, like genetic changes, accumulate linearly with evolutionary time and that the majority of these changes are likely to be neutral with respect to phenotype. This idea was explicitly tested using expression data from several primate and mouse species. As predicted by the neutral theory, expression divergence between these species accumulates linearly with time in the tissues studied and expression variation within species correlates with expression divergence between species (Khaitovich et al., 2004). These findings provide strong support for the idea that most expression changes, like structural/sequence changes, comply with the neutral theory of evolution.

Here it must be noted that on both the gene expression and gene structure levels, many changes that occur in the population are not neutral but rather deleterious to the organism. Such changes will not become fixed in a population, but will be eliminated by negative selection. Indeed, studies in nematodes, fruit flies and primate tissues clearly demonstrate that negative selection has a strong effect on evolutionary gene expression changes (Denver et al., 2005; Khaitovich et al., 2005; Rifkin et al., 2005). Research in model organisms provides additional support for the observation that most expression changes that are not removed by negative selection do not lead to phenotypic changes, and are therefore evolutionary neutral. These studies indicate that up to 32% of yeast genes and up to 85% of nematode genes can be individually inactivated without any detectable phenotypic alterations (see, for instance, Brookfield, 1997; Kim, 2001). Most expression changes seen between populations or between species are much less drastic than complete gene inactivation and are thus even less likely to induce phenotypic changes. Additionally, expression divergence between species does not appear to reflect functional divergence. Different human brain regions are known to perform specialized cognitive tasks. While some regions, like Broca's area, are involved in human-specific functions like speech and language, others, like primary visual cortex, are thought to be involved in functions conserved among most primate species. Despite this difference in functional divergence, these distinct regions appear to have the same divergence between human and chimpanzees at the level of gene expression (Khaitovich et al., 2004a).

Taken together, these observations suggest that a large proportion of expression changes are neutral with respect to phenotype. Thus, use of the neutral model as a null hypothesis for gene expression evolution is warranted. As described for gene structure evolution, significant deviation in gene expression from patterns expected under the neutral model would indicate changes influenced by positive selection. However, the current absence of a solid theoretical basis for modeling expression evolution and the lack of reliable markers allowing estimation of the neutral change rate prevents assessment of the actual proportion of expression changes influenced by positive selection.

Adaptive human brain evolution

While the genetic distance between humans and our closest living relatives, chimpanzees, just slightly exceeds 1%, the phenotypic differences between the two species are far greater than between any other pair of primate species with similar genetic divergence. Based on current knowledge about the two species' common ancestor, most of these phenotypic changes, including increase in brain size, happened on the human lineage. This is especially interesting because genetic mutation, the molecular substrate of phenotypic change, occurs at approximately the same frequency in humans and chimpanzees. While the majority of genetic mutations have either no effect on phenotypic traits or are deleterious, a small proportion of mutations can confer an individual with a selective advantage. Such mutations spread rapidly within populations due to Darwinian (positive) selection. The apparent excess of phenotypic changes on the human lineage indicates that positive selection, expected to cause rapid fixation of advantageous phenotypes, played a substantial role during human evolution.

If positive selection was pervasive throughout human evolution, we might expect to see more functional changes occurring on the human lineage than on that of the chimpanzee. On the level of overall genomic divergence, however, no such effect is apparent. This is expected since, according to the neutral theory of molecular evolution, the overwhelming majority of genomic mutations separating the human and chimpanzee genomes have no effect on phenotype and accumulate in proportion to divergence time. However, when differences in protein sequence and expression regulation for genes expressed in human brain are considered, there are indeed more changes on the human lineage than on the chimpanzee lineage for genes expressed in this tissue, but not in any other tissues studied (Enard et al., 2002a; Caceres et al., 2003; Khaitovich et al., 2005). Thus, during human evolution, a detectable amount of genetic change affecting human brain function gave a selective advantage to the ancestors of modern humans. Further, the genetic features that conferred this advantage can potentially be identified.

Why is it important to identify positively selected changes? One of the central aims of modern biology is identification of gene structure and gene regulation changes responsible for phenotypic alteration. This statement is just as applicable to studies of the genetic basis for human diseases as it is to evolutionary studies. In disease research, detection of genes responsible for dysfunction may shed light on the molecular mechanisms of human disorders. Similarly, in evolutionary studies of human brain, finding genes related to human-specific brain functions would illuminate the molecular mechanisms involved with cognition. In the case of human brain, such information is particularly valuable since the molecular pathways underlying human cognition are largely unknown. How can one identify genes involved with such pathways? Studies of human psychiatric and neurological disorders affecting human-specific cognitive functions can lead to identification of genetic changes. However, genes identified in this way are not necessarily causal to the affected function and may also be involved in more basic neural functions not limited to humans. Here, evolutionary data may provide critical, additional information. If a gene found through analysis of a psychiatric or neurological disorder also bears a signature of recent-positive selection, this would strongly indicate its direct involvement in human-specific cognitive mechanisms. Several such cases of recent positive selection in genes associated with human cognition, including *FOXP2*, *MCPH1*, *ASPM1*, *MAOA*, *AHI1*, *GLUT2* and *MYH16* were discussed earlier in this chapter.

The only known example of positive selection at the level of gene regulation was shown in the altered regulation of *prodynorphin* (*PDYN*), a precursor molecule for a number of endogenous opioids and neuropeptides with roles in perception, behavior and memory (Rockman et al., 2005). This study showed that the promoter variant causing elevated expression of *PDYN* was positively selected during recent human evolution prior to the human spread from Africa. Furthermore, positive selection on different promoter variants continued in different human populations, potentially yielding regulatory differences adaptive to certain environmental or cultural conditions. The fact that only one example of positive selection on the gene regulation level has been discovered does not necessarily mean that regulatory changes were subjected to positive selection with less frequency during human evolution than structural changes. Rather, it may reflect the fact that positive selection on regulatory changes is more difficult to identify because regulatory sequences can be located almost anywhere in the genome and regulatory changes themselves are much less well defined than the structural ones. Given that in brain, but not in other tissues, there are more regulatory changes overall on the human lineage than on chimpanzee lineage, a substantial influence of positive selection on gene regulation likely occurred in our evolutionary history. On the other hand, the excess of regulatory changes observed on the human lineage in brain may indicate relaxation of negative selection, allowing genes expressed in human brain to vary more without any detrimental effect on phenotype. Given the increased complexity and capacity of the human brain, the latter explanation may not seem the most plausible but our limited knowledge of molecular mechanisms determining human brain functions prevents it from being ruled out. It is, however, possible to distinguish between these two alternative explanations using data on genome-wide patterns of human polymorphism. These patterns can illuminate positive selection events that took place in the last 200,000–500,000 years. If positive selection played substantial role in shaping regulatory changes that took place during recent human evolution, we would expect to see a positive correlation between such patterns and expression differences that can be currently observed between humans and other primates. Indeed, such a correlation can be seen in brain, but not in the other tissues studied (Khaitovich et al., 2006). This finding indicates that regulatory changes played a substantial role in shaping human-specific cognitive abilities. Identification of these changes remains a challenge, but this challenge is no longer insurmountable.

Conclusion

Studies of human brain function and dysfunction have provided a wealth of data, including the identification of many genes essential for the normal human condition. Though less extensive by comparison in terms of scope and amplitude, molecular evolution studies have provided evidence that changes in some of these genes gave a selective advantage to our ancestors, and are thus primary candidates in the search for genes responsible for the evolution of human-specific cognitive functions. Of course, many more genes involved in the evolution of human cognition, with changes affecting both gene structure and gene regulation, are yet to be identified. Since the rise of new cognitive abilities during human evolution should have given the individuals possessing these abilities a selective advantage, recently selected changes can be recognized in the nucleotide polymorphism patterns among contemporary humans. Identification of genes affected by positive selection, either on regulatory or structural levels, may help us discover the molecular mechanisms underlying human-specific brain functions. The advantage of this evolutionary approach is that it is not limited to genes already implicated in neurological function through disease studies, but rather allows the identification of completely new and unexpected functional changes. There is no doubt that dramatic expansion and improvement of genetic and regulatory data in humans and non-human primates, combined with advances in medical and neurobiological studies, will provide new insights into human cognition — perhaps even before this book is published.

Acknowledgments

We thank Svante Pääbo for guidance, and all members of our research group for their practical assistance and inspiring discussions. Special thanks to Wolfgang Enard, Ines Hellmann and Michael Lachmann for their insight and helpful suggestions. The original work described in this review was financially supported by the Max Planck Society and the Bundesministerium für Bildung und Forschung.

References

Amunts, K., Schleicher, A., Burgel, U., Mohlberg, H., Uylings, H.B., et al. (1999) Broca's region revisited: cytoarchitecture and intersubject variability. J. Comp. Neurol., 412: 319–341.

Andres, A.M., Soldevila, M., Navarro, A., Kidd, K.K., Oliva, B., et al. (2004) Positive selection in MAOA gene is human exclusive: determination of the putative amino acid change selected in the human lineage. Hum. Genet., 115: 377–386.

Broca, P.P. (1861) Perte de la parole, ramollissement chronique et destruction partielle du lobe antérieur gauche. [Sur le siège de la faculté du langage.]. Bull. Soc. Anthropol., 2: 235–238.

Brookfield, J.F. (1997) Genetic redundancy: screening for selection in yeast. Curr. Biol., 7: R366–R368.

Burki, F. and Kaessmann, H. (2004) Birth and adaptive evolution of a hominoid gene that supports high neurotransmitter flux. Nat. Genet., 36: 1061–1063.

Caceres, M., Lachuer, J., Zapala, M.A., Redmond, J.C., Kudo, L., et al. (2003) Elevated gene expression levels distinguish human from non-human primate brains. Proc. Natl. Acad. Sci. USA, 100: 13030–13035.

Cantalupo, C. and Hopkins, W.D. (2001) Asymmetric Broca's area in great apes. Nature, 414: 505.

Consortium, C.S.a.A. (2005) Initial sequence of the chimpanzee genome and comparison with the human genome. Nature, **437**: 69–87.

Corballis, P.M., Funnell, M.G. and Gazzaniga, M.S. (2000) An evolutionary perspective on hemispheric asymmetries. Brain Cogn., 43: 112–117.

Darwin, C. (1859). The Origin of Species by Means of Natural Selection; or, the Preservation of Favored Races in the Struggle for Life. Murray, London.

Denver, D.R., Morris, K., Streelman, J.T., Kim, S.K., Lynch, M., et al. (2005) The transcriptional consequences of mutation and natural selection in Caenorhabditis elegans. Nat. Genet., 37: 544–548.

Dorus, S., Vallender, E.J., Evans, P.D., Anderson, J.R., Gilbert, S.L., et al. (2004) Accelerated evolution of nervous system genes in the origin of Homo sapiens. Cell, 119: 1027–1040.

Enard, W., Khaitovich, P., Klose, J., Zollner, S., Heissig, F., et al. (2002a) Intra- and interspecific variation in primate gene expression patterns. Science, 296: 340–343.

Enard, W., Przeworski, M., Fisher, S.E., Lai, C.S., Wiebe, V., et al. (2002b) Molecular evolution of FOXP2, a gene involved in speech and language. Nature, 418: 869–872.

Evans, P.D., Anderson, J.R., Vallender, E.J., Choi, S.S. and Lahn, B.T. (2004a) Reconstructing the evolutionary history of microcephalin, a gene controlling human brain size. Hum. Mol. Genet., 13: 1139–1145.

Evans, P.D., Anderson, J.R., Vallender, E.J., Gilbert, S.L., Malcom, C.M., et al. (2004b) Adaptive evolution of ASPM, a major determinant of cerebral cortical size in humans. Hum. Mol. Genet., 13: 489–494.

Evans, P.D., Gilbert, S.L., Mekel-Bobrov, N., Vallender, E.J., Anderson, J.R., et al. (2005) Microcephalin, a gene regulating brain size, continues to evolve adaptively in humans. Science, 309: 1717–1720.

Fay, J.C. and Wu, C.I. (2003) Sequence divergence, functional constraint, and selection in protein evolution. Annu. Rev. Genomics Hum. Genet., 4: 213–235.

Fay, J.C., McCullough, H.L., Sniegowski, P.D., Eisen, M.B., et al. (2004) Population genetic variation in gene expression is associated with phenotypic variation in *Saccharomyces cerevisiae*. Genome Biol., 5: R26.

Ferland, R.J., Eyaid, W., Collura, R.V., Tully, L.D., Hill, R.S., et al. (2004) Abnormal cerebellar development and axonal decussation due to mutations in AHI1 in Joubert syndrome. Nat. Genet., 36: 1008–1013.

Fu, Y.X. and Li, W.H. (1993) Statistical tests of neutrality of mutations. Genetics, 133: 693–709.

Gannon, P.J., Holloway, R.L., Broadfield, D.C. and Braun, A.R. (1998) Asymmetry of chimpanzee planum temporale: humanlike pattern of Wernicke's brain language area homolog. Science, 279: 220–222.

Geschwind, N. and Levitsky, W. (1968) Human brain: left–right asymmetries in temporal speech region. Science, 161: 186–187.

Gilad, Y., Rosenberg, S., Przeworkski, M., Lancet, D. and Skorecki, K. (2002) Evidence for positive selection and population structure at the human MAO-A gene. Proc. Natl. Acad. Sci. USA, 99: 862–867.

Gilbert, S.L., Dobyns, W.B. and Lahn, B.T. (2005) Genetic links between brain development and brain evolution. Nat. Rev. Genet., 6: 581–590.

Glazko, G.V. and Nei, M. (2003) Estimation of divergence times for major lineages of primate species. Mol. Biol. Evol., 20: 424–434.

Gu, J. and Gu, X. (2003) Induced gene expression in human brain after the split from chimpanzee. Trends Genet., 19: 63–65.

Haesler, S., Wada, K., Nshdejan, A., Morrisey, E.E., Lints, T., et al. (2004) FoxP2 expression in avian vocal learners and non-learners. J. Neurosci., 24: 3164–3175.

Herndon, J.G., Tigges, J., Anderson, D.C., Klumpp, S.A. and McClure, H.M. (1999) Brain weight throughout the life span of the chimpanzee. J. Comp. Neurol., 409: 567–572.

Hopkins, W.D., Marino, L., Rilling, J.K. and MacGregor, L.A. (1998) Planum temporale asymmetries in great apes as revealed by magnetic resonance imaging (MRI). Neuroreport, 9: 2913–2918.

Hopkins, W.D., Russell, J.L., Cantalupo, C., Freeman, H. and Schapiro, S.J. (2005) Factors influencing the prevalence and handedness for throwing in captive chimpanzees (Pan troglodytes). J. Comp. Psychol., 119: 363–370.

Hsieh, W.P., Chu, T.M., Wolfinger, R.D. and Gibson, G. (2003) Mixed-model reanalysis of primate data suggests tissue and species biases in oligonucleotide-based gene expression profiles. Genetics, 165: 747–757.

Hudson, R.R., Kreitman, M. and Aguade, M. (1987) A test of neutral molecular evolution based on nucleotide data. Genetics, 116: 153–159.

Kandel, E.R., Schwartz, J.H. and Jessell, T.M. (2000) Principles of Neural Science. McGraw-Hill Health Professions Division, New York.

Khaitovich, P., Hellmann, I., Enard, W., Nowick, K., Leinweber, M., et al. (2005) Parallel patterns of evolution in the genomes and transcriptomes of humans and chimpanzees. Science, 309: 1850–1854.

Khaitovich, P., Muetzel, B., She, X., Lachmann, M., Hellmann, I., et al. (2004a) Regional patterns of gene expression in human and chimpanzee brains. Genome Res., 14: 1462–1473.

Khaitovich, P., Tang, K., Franz, H., Kelso, J., Hellmann, I., et al. (2006) Positive selection on gene expression in the human brain. Curr. Biol., 16: R356–R358.

Khaitovich, P., Weiss, G., Lachmann, M., Hellmann, I., Enard, W., et al. (2004b) A neutral model of transcriptome evolution. PLoS Biol., 2: E132.

Kim, S.K. (2001) *C. elegans*: mining the functional genomic landscape. Nat. Rev. Genet., 2: 681–689.

Kimura, M. (1968) Evolutionary rate at the molecular level. Nature, 217: 624–626.

Kimura, M. (1983) The neutral theory of molecular evolution. Cambridge University Press, Cambridge.

King, M.C. and Wilson, A.C. (1975) Evolution at two levels in humans and chimpanzees. Science, 188: 107–116.

Kornack, D.R. and Rakic, P. (1998) Changes in cell-cycle kinetics during the development and evolution of primate neocortex. Proc. Natl. Acad. Sci. USA, 95: 1242–1246.

Kumar, A., Markandaya, M. and Girimaji, S.C. (2002) Primary microcephaly: microcephalin and ASPM determine the size of the human brain. J. Biosci., 27: 629–632.

Lai, C.S., Fisher, S.E., Hurst, J.A., Vargha-Khadem, F. and Monaco, A.P. (2001) A forkhead-domain gene is mutated in a severe speech and language disorder. Nature, 413: 519–523.

Marin-Padilla, M. (1992) Ontogenesis of the pyramidal cell of the mammalian neocortex and developmental cytoarchitectonics: a unifying theory. J. Comp. Neurol., 321: 223–240.

McDonald, J.H. and Kreitman, M. (1991) Adaptive protein evolution at the Adh locus in Drosophila. Nature, 351: 652–654.

McHenry, H.M. (1994) Tempo and mode in human evolution. Proc. Natl. Acad. Sci. USA, 91: 6780–6786.

Mekel-Bobrov, N., Gilbert, S.L., Evans, P.D., Vallender, E.J., Anderson, J.R., et al. (2005) Ongoing adaptive evolution of ASPM, a brain size determinant in Homo sapiens. Science, 309: 1720–1722.

Nimchinsky, E.A., Gilissen, E., Allman, J.M., Perl, D.P., Erwin, J.M., et al. (1999) A neuronal morphologic type unique to humans and great apes. Proc. Natl. Acad. Sci. USA, 96: 5268–5273.

Oleksiak, M.F., Churchill, G.A. and Crawford, D.L. (2002) Variation in gene expression within and among natural populations. Nat. Genet., 32: 261–266.

Ozonoff, S., Williams, B.J., Gale, S. and Miller, J.N. (1999) Autism and autistic behavior in Joubert syndrome. J. Child Neurol., 14: 636–641.

Pennisi, E. (2004) Human evolution. The primate bite: brawn versus brain? Science, 303: 1957.

Perry, G.H., Verrelli, B.C. and Stone, A.C. (2005) Comparative analyses reveal a complex history of molecular evolution for human MYH16. Mol. Biol. Evol., 22: 379–382.

Plaitakis, A., Spanaki, C., Mastorodemos, V. and Zaganas, I. (2003) Study of structure-function relationships in human glutamate dehydrogenases reveals novel molecular mechanisms for the regulation of the nerve tissue-specific (GLUD2) isoenzyme. Neurochem. Int., 43: 401–410.

Rifkin, S.A., Houle, D., Kim, J. and White, K.P. (2005) A mutation accumulation assay reveals a broad capacity for rapid evolution of gene expression. Nature, 438: 220–223.

Rifkin, S.A., Kim, J. and White, K.P. (2003) Evolution of gene expression in the *Drosophila melanogaster* subgroup. Nat. Genet., 33: 138–144.

Rilling, J.K. and Insel, T.R. (1999) The primate neocortex in comparative perspective using magnetic resonance imaging. J. Hum. Evol., 37: 191–223.

Rockman, M.V., Hahn, M.W., Soranzo, N., Zimprich, F., Goldstein, D.B., et al. (2005) Ancient and recent positive selection transformed opioid cis-regulation in humans. PLoS Biol., 3: e387.

Saraiva, J.M. and Baraitser, M. (1992) Joubert syndrome: a review. Am. J. Med. Genet., 43: 726–731.

Shu, W., Cho, J.Y., Jiang, Y., Zhang, M., Weisz, D., et al. (2005) Altered ultrasonic vocalization in mice with a disruption in the Foxp2 gene. Proc. Natl. Acad. Sci. USA, 102: 9643–9648.

Stedman, H.H., Kozyak, B.W., Nelson, A., Thesier, D.M., Su, L.T., et al. (2004) Myosin gene mutation correlates with anatomical changes in the human lineage. Nature, 428: 415–418.

Su, A.I., Cooke, M.P., Ching, K.A., Hakak, Y., Walker, J.R., et al. (2002) Large-scale analysis of the human and mouse transcriptomes. Proc. Natl. Acad. Sci. USA, 99: 4465–4470.

Su, A.I., Wiltshire, T., Batalov, S., Lapp, H., Ching, K.A., et al. (2004) A gene atlas of the mouse and human protein-encoding transcriptomes. Proc. Natl. Acad. Sci. USA, 101: 6062–6067.

Tajima, F. (1989) Statistical method for testing the neutral mutation hypothesis by DNA polymorphism. Genetics, 123: 585–595.

Uddin, M., Wildman, D.E., Liu, G., Xu, W., Johnson, R.M., et al. (2004) Sister grouping of chimpanzees and humans as revealed by genome-wide phylogenetic analysis of brain gene expression profiles. Proc. Natl. Acad. Sci. USA, 101: 2957–2962.

Vargha-Khadem, F., Gadian, D.G., Copp, A. and Mishkin, M. (2005) FOXP2 and the neuroanatomy of speech and language. Nat. Rev. Neurosci, 6: 131–138.

Wernicke, K. (1874) Der Aphasische Symptomencomplex. Eine Psychologische Studie auf anatomischer Basis. M. Crohn und Weigert, Breslau.

Wienberg, J., Jauch, A., Ludecke, H.J., Senger, G., Horsthemke, B., et al. (1994) The origin of human chromosome 2 analyzed by comparative chromosome mapping with a DNA microlibrary. Chromosome Res., 2: 405–410.

Woods, C.G., Bond, J. and Enard, W. (2005) Autosomal recessive primary microcephaly (MCPH): a review of clinical, molecular, and evolutionary findings. Am. J. Hum. Genet., 76: 717–728.

Subject Index

2D-DIGE 55–56
  2D-DIGE workflow 57
2Dimensional gel electrophoresis 88–89, 92
  mass spectrometry 89
  protein microarrays 90
  tandem mass spectrometry 89–90
2D-PAGE 88
Acridine orange (AO) histofluorescence 199
Adenomatous polyposis coli (APC) gene 281
Affinity removal 45
Aged CA1 and CA3 hippocampal neurons, single cell profiling 213–214
Aging, gene profiling in 212–213
  normal aging within the frontal cortex, gene regulation during 214–216
Alzheimer's disease (AD) 198, 199
  frontal and temporal neocortex in, regional gene expression profiling 202–203
  galanin hyperinnervated CBF neurons in 205–206
  gene expression profiling in 206–207
  hippocampal NFTs in, single cell gene analysis 200–201
  hippocampal senile plaques in, single cell gene array analysis 199
  neurofibrillary tangles in, 199
  regional gene expression profiling in 201–202
  related brain regions, regional gene expression profiling 203
Amygdala (AMYG) 153
  on trisynaptic path integration in SZ 154
Amyotrophic lateral sclerosis (ALS) 198
Analgesia 247
Animal models of psychiatric disorders 281
Antisense RNA (aRNA) 28–31, 42
Arousal systems 250
  from generalized CNS arousal to specific forms of arousal 252–253
  generalized arousal, mechanisms underlying 251–252
ASPM1 gene 298
Atmospheric pressure ionization (API) 73

b-Ions 92
Bio-spin column (Bio-Rad) 45

Bipolars
  bipolar disorder (BPD) 130–137, 168
  regulation of apoptosis in the limbic lobe of 153–169
Blood–brain barrier (BBB) 44, 211
Brain-derived neurotrophic factor (BDNF) 176
Bzrp gene 237

Candidate gene list 120–121
Capillary isoelectric focusing (CIEF) 63
cDNA array analysis 200–201, 204
Central nervous system (CNS) 42, 253–256
  CNS arousal 250
  histamine H_1 receptor in 253–255, *see also separate entry*
  histamine H_2 receptor 255–256, *see also separate entry*
  histamine H_3 receptor 256, *see also separate entry*
Cerebrospinal fluid (CSF) 44–45, 275, 277–278
Cholinergic basal forebrain (CBF) neurons 198, 203–205
Chronic obstructive pulmonary disease (COPD) 9–10
Chronic temporal-lobe epilepsy humans and genetic epilepsy in rat 229–230
Clinical neurosciences 83–106, *see also under* Functional genomics
Cocaine abuse, genome and proteome profiles in 173–192
  cocaine addiction, neuroanatomy 175
  cocaine-overdose victims (COD) 181
  functional genomics 175–181, *see also separate entry*
  identified and matched proteins 182–187
  proteomics 181–191
Cocaine-overdose victims (COD) 181
  representative proteins exhibiting decreased abundance in 189
  representative proteins exhibiting increased abundance in 188
Controlled cortical impact (CCI) 229
Coomassie brilliant blue (CBB) 54
Creutzfeld–Jakob disease (CJD) 199, 212
Cyclic AMP response element-binding protein (CREB) 199

De novo analysis 91–92
DeCyder™ 58
Desalting techniques 45

Desensitization 261–263
Dialysis method 45
DNA sequence mutations 297
Dopamine transporter (DAT) 18
Dorsolateral prefrontal cortex (DLPFC) 141–148
  DLPFC in schizophrenia, dysfunction of 142
  in schizophrenia, transcriptome alterations in, causes 142–143
  in schizophrenia, transcriptome alterations in, consequences 142–143, 145–148
  in schizophrenia, transcriptome alterations in, types 142–143
Downregulation 261–263
Downstream genes to behavioral change

Endocytosis 261–263
Epileptic process 233
Epileptogenesis-related genes 223–238
  animal models and gene array analysis, description 226–230
  early and middle post-injury time windows 237
  early post-injury time window 236
  GO terms referring to biologic process 231–233, see also GO terms
  late post-injury time window 237
  methods 224–231
  middle post-injury time window 236–237
  search for common genes 230–231
  selection of papers for analysis and creation of gene lists 224–226
  similarly altered genes in different post-injury time windows 233–237
  status epilepticus 226–229
Expression array studies in psychiatric research 280–284
Expression proteomics 52–64
  analysis, non-gel-based approach 60
  functional proteomics 64
  iTRAQ^{TM} technique 61–62
  peptide labeling with $H_2^{16}O/H_2^{18}O$ 62
  quantitative analysis 59
  tandem ion exchange/reverse-phase chromatography 62
  top–down proteomics 63–64
  two-dimensional gel electrophoresis 52–58, see also under Gel electrophoresis
Extracellular fluid (ECF) 44

False discovery rate (FDR) 27, 113
Family-wise error rate (FWER) 113
Fast ion bombardment (FAB) 71
Female mating behaviors 243–249
Female reproductive behavior 244, see also Lordosis

Fischer Exact Test 163–164
Fluorescent stains 55
FOXP2 gene 298
Frontotemporal dementia with Parkinsonism (FTDP) 209
'Full' transcriptome-profiling study 21
Functional genomics 15–33, 175–181, see also under RNA
  analyzing massive datasets 26–27
  biological replicates or technical replicates in 21
  expression difference detection 19–20
  expression-level changes in the brain, magnitude 20
  expression-profiling analysis, verification 21–22
  expression-profiling datasets, potential for comparison 21–22
  false-discovery rate (FDR), estimation 27
  gene expression analysis 22–23
  gene expression profiling 17–19
  high-throughput genomic method 18
  human postmortem brain tissue, transcriptions in 20
  human post-mortem studies 180–181
  in situ hybridization 22–23
  Massive parallel signature sequencing (MPSS) 24–25
  microarray platforms 26
  microarray technology, considerations 31–32
  minimum starting material for 20–21
  negative data, interpretation of 22
  non-human primates 179–180
  Northern hybridization 22
  preliminary evaluation versus a fullblown microarray study 21
  qPCR assay 23
  regional and single cell assessment 27–28
  RNA amplification 28, see also RNA amplification strategies
  RNase protection assay 22
  rodent studies, non-contingent administration 175–177
  rodent studies, self-administration 177–179
  sample size and transcript representation 18–19
  sensitivity levels to detect the molecules of interest 19–20
  sequencing by hybridization (SBH) 25
  serial analysis of gene expression (SAGE) 23–24
  total analysis of gene expression (TOGA) 25
  transgenic mouse studies 178–179
Functional genomics and proteomics in the clinical neurosciences 83–106
  2Dimensional gel electrophoresis 88–89, see also separate entry
  CodeLink bioarrays 102
  data analysis 90–92

de novo analysis 91–92
experimental methods 85–90, *see also under* Microarrays
gene-ranking methods 105
global normalization methods 87
interpretation and validation 105–106
mass spectrometry analysis 91
microarray case study 102–105
normalization 87–88
outlier removal 86–87
proteomic technology 88–90
statistical analysis and pattern classification 92–102, *see also separate entry*

Galanin hyperinnervated CBF neurons in 205–206
'GAPPS' model 249
Gastric inhibitory peptide (GIP) 176
Gel electrophoresis 52–58
  2D-DIGE 55–56
  Coomassie brilliant blue (CBB) 54
  fluorescent stains 55
  gel electrophoresis, 2D, *see under* 2Dimensional gel electrophoresis
  isoelectric focusing (IEF) 53–54
  proteins, multi-dimensional separation 58–59
  quantitative image analysis 58
  sample preparation 53
  SDS-PAGE 54
  SDS-PAGE 54–58
  silver staining method 54
  traditional stains 54
Gene expression profiling (GEP) 153–169
  functional biopathways/clusters of genes 160–162
  in brain systems, analyzing 160
  in hippocampus of bipolar and schizophrenic subjects 164–169
  low stringency approach for analyzing 162–164
  quality controls 155–160
GenMapp pathway 164, 167
Genomic methodologies, *see under* Functional genomics
Genomic responses to hormones by functional modules (GAPPS) 249
Genomic technology 85–86
GLUT2 gene 299
Glutamic acid decarboxylase (GABA) 141, 145–148
Glycero-3-phosphate dehydrogenase (G3PDH) 155
Glycosylation 67–68
GO terms 223–233, 238
  analysis of representation 230
  chronic epilepsy 232–233
  day 1 after brain insult 232
  days 3–8 after insult 232

first 24 h after insult 231–232
over-represented at various stages of the epileptic process 233
referring to biologic process 231–233
referring to molecular function 233
search for common genes 230–231
Gonadotropin-releasing hormone (GnRH) 247
G-protein coupled pathways 257–259
G-protein coupled receptorassociated sorting protein (GASP) 263
G-protein-coupled receptors (GPCR) 143–144
Guanine diphosphate (GDP) 143–144
Guanine triphosphate (GTP) 143–144

Hierarchical clustering 93–94, 120
High pH method 50
Hippocampus (HIPP) 153
Histamine H_1 receptor in 253–255
  distribution in 253
  molecular properties 253–254
  signaling pathways 254
Histamine H_2 receptor 255–256
  distribution in CNS 255
  molecular properties 255
  signaling pathway 255
Histamine H_3 receptor 256
  molecular properties 256
  signaling pathway 256
Human brain evolution 295–307
  adaptive human brain evolution 305–306
  anatomical evolution 295–297
  gene expression evolution 300–302
  gene expression evolution, theory of 302–305
  protein sequence evolution 297–300
Human Genome Project 84
Human post-mortem studies 180–181
Human Proteome Organization (HUPO) 11
Hypothalamus–pituitary–adrenal (HPA) axis 275

ICAT™-labeled peptides 63
Immobilized metal affinity chromatography (IMAC) column 65
Immune function 233
Immunoblotting 65
In situ hybridization 22–23
Inner nuclear membrane (INM) 51
Ion homeostasis term 232
Isoelectric focusing (IEF) 53–54
  solution IEF 54
Isothermal amplification 29–30
Isotope-coded affinity tags (ICAT and iTRAQ) 59–62

k-nearest neighbors (kNN) 97

Laser capture microdissection (LCM) 178, 209, 277
Lectin-affinity technology 68
Linear discriminant analysis (LDA) 97–98
Linkage disequilibrium (LD) 284
Liquid-phase isoelectric focusing 46
LocusLink ID(s) 231
Lordosis 244
    from lordosis to sexual arousal to generalized CNS arousal 249–252

Major depressive disorder (MDD) 130
Major histocompatibility complex (MHC) 211
Mass analyzer 73
    ion trap mass analyzers 74
    quadrupole mass analyzer 74
    time-of-flight (TOF) mass analyzers 73–74
Massive parallel signature sequencing (MPSS) 18, 24–25
Matrix-assisted laser desorption-ionization (MALDI) mass spectra 71–73, 75–76
MCPH1 gene 298
Microarray studies, reproducibility of 109–124
    aim and scope 116
    approach 116–117
    choice of disease marker 114–116
    computational setup 117
    computing expression measures, methods for 117–118
    control of multiplicity 113
    correlation analysis 111–112
    correlation coefficients 118
    data analysis pipeline 110–114
    data quality, assessment of 114–116
    differentially expressed genes, detection of 111–113
    expression measures, computation of 111
    FDR control, procedures for 118
    functional interpretation 113–114
    impact of analysis protocol on functional interpretation 119–122
    implications for data mining 123–124
    outlier removal 114
    performance comparison 116–122
    pre-processing 110–111
    probe set definitions 110
    probe set definitions 117
    protocols, combinations of analysis methods 118–119
    quality control 110–111
    validation 122
    variable types 111
Microarrays
    cDNA microarrays 85–86
    microarray technology 90–91

    multi-channel microarray 87
    oligonucleotide microarrays 86
    protein microarrays 90
    quality control of microarrays 86–88
Mild cognitive impairment (MCI) 204
Mini Mental State Examination (MMSE) 202
Mitogen-activated protein kinase (MAPK) 215, 259
MMSE marker, of disease progression 114–116
Mood disorders, genomics of 129–137
    animal models, clues from 136–137
    bipolar disorder (BPD) 130
    chromosomal aberrations 131
    genetics of 130–131
    neurobiological and neuroanatomical substrates of 131–133
    pathophysiology of 133–136
    unipolar depression 130
mRNA expression arrays (expressomics) 280–285
Multiple sclerosis (MS) 211
    gene profiling in 211–212
Multi-protein complex (protein– protein interactions) 70–71
Muscarinic receptors 246
Myelin basic protein (MBP) 180
Myelin-associated oligodendrocyte basic protein (MOBP) 181
MYH16 gene 299

Nerve growth factor (NGF) 205
Neural and genomic mechanisms, for female mating behaviors 243–249
    amplification by progesterone 246–247
    anxiety reduction 248
    direct effects of estrogens 245–246
    gonadotropin-releasing hormone (GnRH) 247
    indirect effects of estrogens 246–248
    neuronal growth 246
    newer findings 248
    social recognition and aggression 248
Neurodegenerative disease, molecular biology 197–217, *see also under* Neuronal gene expression profiling
Neurofibrillary tangles (NFTs) 199
Neuronal gene expression profiling 197–217
    development 198
    regional analysis 198
Neuronal nitric oxide synthase (NOS1) 165
Neuronal plasticity 236
Neuropeptide neurotensin (NT) 275
Neuroscience, proteomics methods in 41–76, *see also under* Proteomics methods
NFT marker, of disease progression 114–116
Nitration 69–70

No cognitive impairment (NCI) 204
Non-human primates 179–180
α_{1B}-Noradrenergic receptor signaling 256–259
   Gq/11-coupled primary signaling pathway 257
Northern hybridization method 22
Nuclear pore complex (NPC) 51
Nucleolus 52
Nucleotide triphosphates (NTPs) 28
Nucleus accumbens (NAc) 173–192
μ and δ Opioid receptor signaling 259–263
   DOR distribution 260
   downregulation 263
   endogenous antiopioid systems 263
   MOR distribution 260
   pharmacology 260
   physiological mechanisms 261–263
   signal transduction 260–261

Outer nuclear membrane (ONM) 51

Parkinson's disease (PD) 198, 207
   Lewy body-containing SN_{pc} neurons in 209–210
   substantia nigra in, regional gene profiling 207–209
PDQuest™ 58
Peptide labeling with $H_2^{16}O/H_2^{18}O$ 62
Peptide mass fingerprinting (PMF) 75
Peripheral blood monocytes (PBMCs) 276, 278
   expression analysis 280–281
Phosphorylation 64–65
Polymerase chain reaction (PCR) 85–86
Positron emission tomography (PET) 131–132, 153
Postmortem human brain 277, 282
Postmortem intervals (PMIs) 3–7
   and RIN 8
Postmortem tissue 20
Post-translational modifications (PTM) 42–43
Prefrontal cortex (PFC) 18
Principal component analysis (PCA) 95
Progressive supranuclear palsy (PSP) 209
Pro-Q Diamond stain 65, 68
Protein arrays 76
Protein depletion 45
Protein identification 75
Protein kinase C (PKC) 215
Protein precipitation 45
Proteolipid protein 1 (PLP) 180
Proteome profiles in cocaine abuse 173–192, *see also under* Cocaine abuse
Proteomics methods, in Neuroscience 41–76
   affinity removal 45
   bio-spin column (Bio-Rad) 45
   cerebrospinal fluid (CSF) 44–45
   desalting techniques 45
   detergents 50
   dialysis method 45
   expression proteomics 52–64, *see also separate entry*
   glycosylation 67–68
   high pH method 50
   ion source 71
   liquid-phase isoelectric focusing 46
   mass spectrometry 71
   membranes 49
   multi-protein complex (protein– protein interactions) 70–71
   nitration 69–70
   nuclear envelope (NE) 50–51
   nuclei, mitochondria, cytoplasm and membrane 47–50
   nucleus 50–52
   organic acid 49–50
   organic solvents 50
   phosphoprotein analysis 64–65
   protein depletion 45
   protein precipitation 45
   solid-phase extraction 46
   subcellular fractionation 43–52
   synaptosomes and post-synaptic density 46–47
   ubiquitination 68–69
   ultrafiltration 45
   using ESI 73
   using mass analyzer 73, *see also separate entry*
Psychiatric disorders, animal models of 281
Psychiatry practice, implications 275–288
   animal brain tissue 276–277
   animal models 278–279
   convergent evidence 286–287
   CSF studies 279
   expression array studies in 280–284
   expression array studies, validation of candidates from 284–285
   mRNA expression arrays (expressomics) 280–285
   postmortem brain tissue 281–284
   postmortem human brain tissue 277
   proteomics 276–280
   whole genome SNP association studies 285–286
'Purine nucleotide binding' 233

q RT-PCR
   FRET-based quantitative q RT-PCR 157
qPCR assay 23

Receptor desensitization 261
recursive feature elimination (RFE) 100
Ribo-SPIA technology 29

RNA amplification strategies
　aRNA amplification 28–31
　isothermal amplification 29–30
　terminal continuation (TC) RNA amplification 30–31
RNA integrity 7–10
　and brain pH 10–11
　factors affecting 7–10
　RNA integrity number (RIN) 3–7
RNA
　degradation 20
　input sources of 16–17
　quantification 22
　RNA amplification 28, see also RNA amplification strategies
RNase protection assay 22
Rodent studies 175–179

Schizophrenia (SZ) 198, 210
　entorhinal cortex in, single cell gene profiling in 211
　frontal cortex in, regional gene expression profiling 210–211
　regulation of apoptosis in the limbic lobe of 153–169
　transcriptome alterations in 141–148, see also under Dorsolateral prefrontal cortex
Self-organizing map (SOM) 94–95
Sequencing by hybridization (SBH) 25
Serial analysis of gene expression (SAGE) 18, 23–24
Sex hormone-dependent neuroendocrine systems,
　functional genomics 243–264
　electrophysiology 252
　functional genomics 252
　mitogen-activated protein kinase pathway 259
　neural and genomic mechanisms for female mating behaviors 243–249
　neuroanatomy 251
　α_{1B}-noradrenergic receptor signaling 256–259
　μ and δ opioid receptor signaling 259–263, see also separate entry
Shotgun/multidimensional protein identification technology (MuDPIT) proteomics 49, 59, 62
Silver staining method 54
Single cell gene analysis 199–201
Single-nucleotide polymorphism (SNP) 284
S-nitrosylation 70
Solid-phase capture-and-release method 61
Solid-phase extraction 46
Specific fluorescence-based detection methods 65
Spin2c gene 237
Statistical analysis and pattern classification, in functional genomics and proteomics 92–102
　cross validation and error-estimation methods 99–100
　feature combinations and global search methods 101–102
　feature ranking with classifiers 100
　feature ranking with hypothesis testing 96–97
　hierarchical clustering 93–94
　k-nearest neighbors (kNN) 97
　linear discriminant analysis (LDA) 97–98
　performance evaluation 99–101
　principal component analysis 95
　receiver operating characteristic (ROC) curve 100–101
　self-organizing map (SOM) 94–95
　supervised methods 95–98
　support vector machines (SVM) 98
　unsupervised methods 92–95
　Wilcoxon rank-sum test 97
Status epilepticus (SE) 223, 226–229
Strong cation exchange (SCX) 61
Subcellular fractionation 43–52
　caveats for 43
Superoxide dismutase (SOD) 165
Surface-enhanced laser desorption ionization (SELDI) technique 64
Synapses 46
Synaptosomes 46–47

Tandem affinity purification (TAP) tag system 70
Tandem ion exchange/reverse-phase chromatography 62
Tandem mass spectrometry 74–75, 92
TaqMan assay 23
'Template matching' method 123
Temporal-lobe epilepsy (TLE) 230
Terminal continuation (TC) RNA amplification 30–31
Thymidine triphosphates (TTPs) 28
Tissue preparation and banking 3–11
　collection and harvesting tissue 4–6
　control subjects 4
　demographic database 6
　designating cohorts 6
　dissection and storage 5–6
　documenting 6–7
　ethical issues 4
　freezing/fixing 4
　handling tissue 4
　identifying subjects 3–4
　matching groups in a cohort 6–7
　neuropathological screening 5
　prospective/retrospective assessment 4
　protein integrity 10–11
　recruitment 3–4
　RNA integrity 7–10, see also individual entry
　tissue inventories 7
　toxicology 5

Total analysis of gene expression (TOGA) 25
Traditional stains 54
'Transcriptional activator' 233
Transcriptomics 15–16, 123
Transgenic mouse studies 178–179
Traumatic brain injury (TBI) 223, 229
Tyrosine nitration 69–70

Ubiquitination 68–69 importance 69
Ultra filtration 45

Uni-Gene ID(s) 231
Unipolar depression 130–137

Ventral tegmental area (VTA) 173–192
Ventromedial hypothalamic (VMH) 243–246

Wilcoxon rank-sum test 97

y-Ions 92